Current Topics in Microbiology 242 and Immunology

Editors

R.W. Compans, Atlanta/Georgia
M. Cooper, Birmingham/Alabama
J.M. Hogle, Boston/Massachusetts · Y. Ito, Kyoto
H. Koprowski, Philadelphia/Pennsylvania · F. Melchers, Basel
M. Oldstone, La Jolla/California · S. Olsnes, Oslo
M. Potter, Bethesda/Maryland · H. Saedler, Cologne
P.K. Vogt, La Jolla/California · H. Wagner, Munich

Springer
Berlin
Heidelberg
New York
Barcelona
Hong Kong
London
Milan
Paris
Singapore
Tokyo

The Hepatitis C Viruses

Edited by C.H. Hagedorn
and C.M. Rice

With 47 Figures and 15 Tables

Curt H. Hagedorn, M.D.
Division of Digestive Diseases and
Genetics – Winship Cancer Center
1639 Pierce Drive, Rm 2101 WMRB
Emory University School of Medicine
Atlanta, GA 30322
USA

Charles M. Rice, M.D.
Washington University School of Medicine
Department of Molecular Microbiology
600 South Euclid Avenue
St. Louis, MO 63110
USA

Cover Illustration: The map provides estimates of the worldwide prevalence of chronic hepatitis C *(see* DL Thomas; Hepatitis C Epidemiology*).* Contact lavanchyd@who.ch regarding the global collection of more precise data on hepatitis C epidemiology *(see Weekly Epidemiological Record 1997, 72, 341–348).*

Cover Design: design & production GmbH, Heidelberg

ISSN 0070-217X
ISBN 3-540-65358-9 Springer-Verlag Berlin Heidelberg New York

This work is subject to copyright. All rights are reserved, whether the whole or part of the material is concerned, specifically the rights of translation, reprinting, reuse of illustrations, recitation, broadcasting, reproduction on microfilm or in any other way, and storage in data banks. Duplication of this publication or parts thereof is permitted only under the provisions of the German Copyright Law of September 9, 1965, in its current version, and permission for use must always be obtained from Springer-Verlag. Violations are liable for prosecution under the German Copyright Law.

© Springer-Verlag Berlin Heidelberg 2000
Library of Congress Catalog Card Number 15-12910
Printed in Germany

The use of general descriptive names, registered names, trademarks, etc. in this publication does not imply, even in the absence of a specific statement, that such names are exempt from the relevant protective laws and regulations and therefore free for general use.

Product liability: The publishers cannot guarantee the accuracy of any information about dosage and application contained in this book. In every individual case the user must check such information by consulting other relevant literature.

Typesetting: Scientific Publishing Services (P) Ltd, Madras

Production Editor: Angélique Gcouta

SPIN: 10633172   27/3020 – 5 4 3 2 1 0 – Printed on acid-free paper

Preface

Understanding the world we live in can be a challenging task. It is currently estimated that at least 170 million people are chronically infected with the hepatitis C virus (HCV). Chronic infections lead to cirrhosis and liver failure or hepatocellular cancer in many instances. Although HCV is perhaps an ancient human virus, it was only recognized in biologic terms in the 1970s as a major cause of post-transfusion hepatitis. Molecular cloning of the viral genome in the late 1980s led to the development of assays for serodiagnosis of HCV infection and an explosion of research on this important pathogen. Nonetheless, many key questions remain to be answered. The goal of this volume is to provide helpful reviews in some of the major areas where progress has been made. Some of the challenges that we still face include gaining a better understanding of the biology of HCV and the molecular details of viral replication, as well as developing effective vaccines or new small molecule therapeutics for treating chronic hepatitis C. It is our hope that the contents of this volume will spark new and innovative approaches to HCV research by bringing together information and interested readers from different disciplines. We sincerely thank the many authors who took time from their research efforts to write chapters for this collection. Thanks are also given to Richard Compans for his suggestions and to Doris Walker for her invaluable help in assembling this volume.

July 1999

CURT H. HAGEDORN
Atlanta, Georgia
CHARLES M. RICE
St. Louis, Missouri

List of Contents

D.W. Bradley
Studies of Non-A, Non-B Hepatitis and Characterization
of the Hepatitis C Virus in Chimpanzees 1

D.L. Thomas
Hepatitis C Epidemiology 25

D. Theodore and M.W. Fried
Natural History and Disease Manifestations of Hepatitis
C Infection 43

K.E. Reed and C.M. Rice
Overview of Hepatitis C Virus Genome Structure,
Polyprotein Processing, and Protein Properties 55

R.C.A. Rijnbrand and S.M. Lemon
Internal Ribosome Entry Site-Mediated Translation
in Hepatitis C Virus Replication 85

M.M.C. Lai and C.F. Ware
Hepatitis C Virus Core Protein: Possible Roles
in Viral Pathogenesis 117

J. Dubuisson
Folding, Assembly and Subcellular Localization
of Hepatitis C Virus Glycoproteins 135

R. De Francesco and C. Steinkühler
Structure and Function of the Hepatitis C Virus
NS3-NS4A Serine Proteinase. 149

A.D. Kwong, J.L. Kim, and C. Lin
Structure and Function of Hepatitis C Virus
NS3 Helicase 171

M.J. Korth and M.G. Katze
Evading the Interferon Response: Hepatitis C Virus
and the Interferon-Induced Protein Kinase, PKR 197

C.H. Hagedorn, E.H. van Beers, and C. De Staercke
Hepatitis C Virus RNA-Dependent RNA Polymerase
(NS5B Polymerase). 225

N. Kato and K. Shimotohno
Systems to Culture Hepatitis C Virus 261

M.E. Major and S.M. Feinstone
Characterization of Hepatitis C Virus Infectious
Clones in Chimpanzees: Long-Term Studies 279

B. Rehermann and F.V. Chisari
Cell Mediated Immune Response to the Hepatitis C Virus . 299

M. Houghton
Strategies and Prospects for Vaccination Against 327
the Hepatitis C Viruses

J.N. Simons, S.M. Desai, and I.K. Mushahwar
The GB Viruses 341

Subject Index 377

List of Contributors

(Their addresses can be found at the beginning of their respective chapters.)

Bradley, D.W.　1
Chisari, F.V.　299
De Francesco, R.　149
De Staercke, C.　225
Desai, S.M.　341
Dubuisson, J.　135
Feinstone, S.M.　279
Fried, M.W.　43
Hagedorn, C.H.　225
Houghton, M.　327
Kato, N.　261
Katze, M.G.　197
Kim, J.L.　171
Korth, M.J.　197
Kwong, A.D.　171
Lai, M.M.C.　117

Lemon, S.M.　85
Lin, C.　171
Major, M.E.　279
Mushahwar, I.K.　341
Reed, K.E.　55
Rehermann, B.　299
Rice, C.M.　55
Rijnbrand, R.C.A.　85
Shimotohno, K.　261
Simons, J.N.　341
Steinkühler, C.　149
Theodore, D.　43
Thomas, D.L.　25
van Beers, E.H.　225
Ware, C.F.　117

Studies of Non-A, Non-B Hepatitis and Characterization of the Hepatitis C Virus in Chimpanzees

D.W. Bradley

1 Introduction	1
2 Development of Primate Model	2
3 Pathogenesis and Course of Disease in Chimpanzees	3
4 Evidence that PT-NANBH Was Caused by a Virus	5
5 Evidence for the Existence of Multiple NANBH Agents	6
6 Attempts to Detect NANBH (Virus)-Specific Antigens, Antibodies, and Enzymes	7
7 Infectivity and Virus-Like Particles: Peripheral Blood Lymphocytes, Cultured Chimpanzee Liver Cells, Liver Homogenates, and Blood Fractions	9
8 Comparative Morphology of Ultrastructural Alterations in Chimpanzee Hepatocytes	10
9 Viral Interference	12
10 Physicochemical Properties of the Major PT-NANBH Agent	13
11 Development of High-Titered Liver and Large-Volume Plasma Pools	16
12 Discussion	18
References	19

1 Introduction

This year marks the end of the second decade of research on post-transfusion non-A, non-B hepatitis (or PT-NANBH), now primarily known to be caused by hepatitis C virus (HCV; Choo et al. 1989). Approximately 10 years ago, the successful molecular cloning of HCV was publicly announced and ushered in a new era of epidemiological, medical, and laboratory research virtually unparalleled by any other area in virology. During the last 10 years, according to a January 1998 search of the NLM-Medline-PubMed database, more than 14,000 publications listing hepatitis C have been cited. This astounding growth was a result of the diligent and persistent efforts of at least two major laboratories and the accumulated knowledge generated by many other scientists in other laboratories who struggled to isolate,

2938 Kelly Court, Lawrenceville, GA 30044, USA

characterize, and clone the major agent (virus) responsible for PT-NANBH. One Australian scientist who previously worked in the field of PT-NANBH remarked (in 1990) that "over 80 laboratories at one time or another worked on non-A, non-B hepatitis...and that by the end of 1987 virtually only three or four laboratories were seriously pursuing research in this area." This perspective highlights the extremely frustrating nature of earlier research in the field of PT-NANBH and reveals the trend of a rapidly diminishing number of active laboratories over time during the mid-1980s.

PT-NANBH was first recognized by PRINCE et al. (1974) and ALTER et al. (1975) as a form of viral hepatitis that was distinct from disease caused by hepatitis A virus (HAV) or hepatitis B virus (HBV). Unlike HAV and HBV, however, the agent responsible for PT-NANBH proved to be elusive and defied the efforts of electron microscopists to visualize disease-associated virus-like particles and also remained undetectable by even the most sensitive serologic tests available at the time. Research in the late 1970s to early to mid-1980s revealed the true nature of the problem. The titer(s) of the infectious agent was so low in infectious plasma, serum, blood products, and liver tissue that many investigators began to accept the fact that they were not likely to isolate and characterize the presumptive viral agent.

This chapter summarizes much of what was learned about the major agent responsible for human PT-NANBH through the use of chimpanzees, the only biomedical model for this disease. More expansive reviews of the broad area of PT-NANBH by DIENSTAG (1983) and FAGAN and WILLIAMS (1984) provide a contemporaneous overview of this field.

2 Development of Primate Model

The chimpanzee is the only primate proven to be susceptible to infection with human NANBH viruses, primarily HCV (hepatitis E virus, HEV, is not considered within the scope of this review). Although chimpanzees had been used previously for experimental transmission of human HAV and HBV, it was not until 1978 that it was discovered that this species of primate was also susceptible to infection with one or more presumptive viral agents of NANBH. Prior to the latter studies, the use of untreated factor VIII concentrates (anti-hemophilic factor, FVIII) for the treatment of hemophilia A was known to carry with it the risk of transmitting either (or both) HBV or NANBH. In the latter instance, HRUBY and SCHAUF (1978) reported the transmission of short-incubation period NANBH in pediatric patients following infusion of FVIII concentrates, while CRASKE and SPOONER (1978) reported evidence for the existence of two distinct types of FVIII-associated non-B hepatitis in transfused hemophiliacs. ALTER et al. (1978) and TABOR et al. (1978) first reported the susceptibility of chimpanzees to human NANBH. HOLLINGER et al. (1978) also showed that human NANBH could be readily transmitted to chimpanzees. Laboratory studies initiated at the U.S. Centers for Disease Control in 1977, in response to an outbreak of NANBH among recipients of three different lots of commercially manufactured FVIII, led to the experimental transmission of disease (NANBH) to four colony-born chimpanzees in early 1978 using the above implicated lots of FVIII materials. BRADLEY et al. (1979) showed that intravenous infusion of 15–30ml of one or more lots of FVIII induced relatively short incubation period NANBH as evidenced by the rapid rise of alanine aminotransferase (ALT) activity following inoculation. Examination of acute-phase liver biopsy specimens revealed histologic evidence of disease that was consistent with a diagnosis of viral hepatitis. WYKE et al. (1979) showed that contaminated factor IX concentrates could also induce acute NANBH that was consistent with a diagnosis of viral hepatitis. Shortly thereafter, YOSHIZAWA et al. (1980) reported the transmission of NANBH to chimpanzees using suspect fibrinogen linked to the transmission of disease in humans. The above studies conclusively proved the extraordinary value of chimpanzees as a biomedical model for human NANBH. Numerous (largely unreported) attempts by our (Centers for Disease Control, CDC) laboratory and other laboratories to induce human-origin NANBH in other species, including tamarins, marmosets, and cynomolgus macaques, failed to demonstrate that any species of primate, other than the chimpanzee, was susceptible to infection with one or more proven human NANBH agents.

3 Pathogenesis and Course of Disease in Chimpanzees

Long-term follow-up studies of NANBH infected chimpanzees by BRADLEY et al. (1981, 1982) provided unequivocal evidence for the occurrence of persistent viremia and disease that was remarkably similar to that observed in human PT-NANBH, as described by KORETZ et al. (1976), GALBRAITH et al. (1979), RAKELA and REDEKER (1979), BERMAN et al. (1979), and ALTER et al. (1982). BRADLEY et al. (1981, 1982) also showed that approximately 70% of chimpanzees observed at CDC for 5–7 years after acute NANBH were found to display enzymatic, histologic, and electron microscopic evidence of chronic disease or silent infection (see below) (Fig. 1). One frequently observed pattern of disease development (pathogenesis) in chimpanzees was the initial rise and fall of ALT values accompanied by apparent histologic resolution of disease. However, longer-term studies of numerous chimpanzees revealed that while "resolution" of overt disease could be seen for up to 2 to 3 years (after inoculation) in some animals, others demonstrated slowly increasing and/or intermittent elevations of ALT activity indicative of persistent infection. Consistent with the above findings by BRADLEY et al. (1981, 1982), TABOR et al. (1980) and ALTER and DIENSTAG (1984) reported that persistent viremia could exist in the absence of elevated ALT activity in human blood donors. In fact, ALTER and DIENSTAG (1984) estimated that approximately 70% of donors capable of transmitting NANBH may not demonstrate abnormal ALT activity. These projections were corroborated by BRADLEY et al. (1982), who proved that several

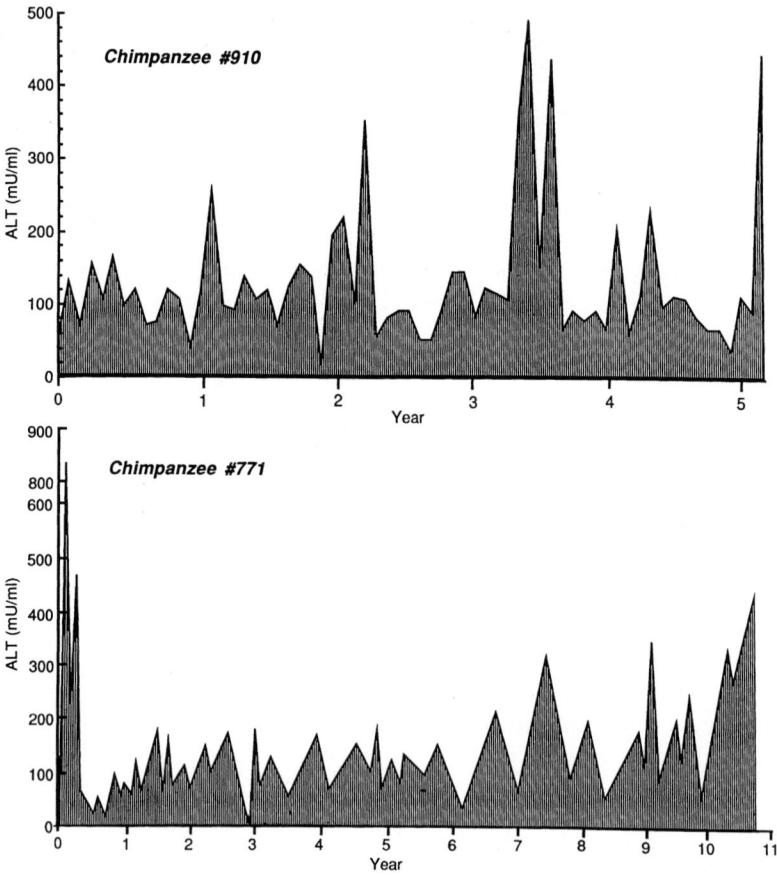

Fig. 1. Alanine aminotransferase (ALT) patterns in two chimpanzees persistently infected with the major agent of post-transfusion non-A, non-B hepatitis (PT-NANBH; HCV)

plasma units obtained from four animals with neither electron microscopic nor enzymatic evidence of persistent infection were infectious when intravenously inoculated into colony-born chimpanzees. Bradley et al. (unpublished findings) also demonstrated the transmission of biochemically and histologically "silent" disease to two chimpanzees using 1×10^{-7}g of NANBH-infected liver tissue. Electron microscopic (EM) evidence of infection was found in one of the above chimpanzees after careful examination of serial (weekly) liver biopsy specimens obtained over a period of months. The remaining animal showed no EM evidence of infection with the major agent of NANBH, but was shown to be refractory to infection by proven-infectious inocula (see Sect. 9, Viral Interference, below). The above findings suggested that the major etiologic agent of PT-NANBH could establish a subclinical course of infection in chimpanzees, and presumably in humans, in the absence of elevated ALT activity and histologic or EM evidence of infection.

Although these earlier studies could also be interpreted to mean that prior infection with the agent of NANBH induced neutralizing (protective) antibodies, the combined findings more strongly suggested that the agent of PT-NANBH normally causes persistent infection in which additional homologous challenge virus does not perturb or otherwise cause any alteration in the established disease state.

Spontaneous recrudescence of relatively severe disease was documented by BRADLEY et al. (1982) in persistently infected chimpanzees. In fact, one NANBH-infected chimpanzee developed both enzymatic and EM evidence of acute disease after a 2-year period of quiescence, a surprising and sobering finding that indicated that resolution of overt disease does not necessarily signal recovery from infection. The observation by BROTMAN et al. (1985) and BURK et al. (1984), that rechallenged chimpanzees exhibited renewed elevations of ALT activity due to reinfection in the absence of neutralizing antibodies, was at one time interpreted to mean that the co-occurrence of elevated enzymes was due to exacerbation of chronic, underlying disease. More recent studies, however, have shown that reinfection can occur after either homologous or heterologous virus challenge, as evidenced by the appearance of type-specific nucleotide sequences of HCV in serum or plasma. In spite of the latter findings, studies conducted by BRADLEY et al. at the CDC showed that NANBH-infected chimpanzees not subjected to rechallenge exhibited a variety of histologic, enzymatic, and EM patterns of disease and that silent infections, recurrent disease, and persistent infection with or without elevated ALT activity and histologic evidence of viral hepatitis could occur. All of the above findings, taken together, suggested that if there were any parallels between the course of disease observed in chimpanzees and humans, it was reasonable to assume that total resolution of NANBH may occur in only a small proportion of infected individuals.

4 Evidence that PT-NANBH Was Caused by a Virus

As noted and described above, there were several convincing lines of experimental and clinical evidence that suggested the etiologic agent of PT-NANBH was virus-like in nature. This evidence included the transmissibility of disease to both humans and chimpanzees by intravenous or percutaneous inoculation of contaminated donor blood, blood fractions, acute-phase liver homogenates, red blood cells, or "purified" chronic-phase plasma preparations, as previously reported by ALTER et al. (1978), TABOR et al. (1978), HAUGEN (1979), BRADLEY et al. (1979, 1980, 1981), BRADLEY and MAYNARD (1983), WYKE et al. (1979), YOSHIZAWA et al. (1980), and HOLLINGER et al. (1978). Infectivity of the major etiologic agent (also previously referred to as the "tubule forming agent," or TFA, due to the formation of peculiar hepatocyte cytoplasmic tubules in chimpanzees; see below) was also shown be destroyed by treatment with chloroform (BRADLEY and MAYNARD 1983; FEINSTONE et al. 1983), treatment with 1:1000 formalin (TABOR and GERETY 1980; YOSHIZAWA et al. 1982), inactivation with β-propiolactone and ultraviolet irradiation (PRINCE et al. 1985), or heating at 100°C for 60min (YOSHIZAWA et al. 1982) or 60°C for 10h (HOLLINGER et al. 1984). Other studies (some described below) demonstrated that the TFA would pass through an 80nm sharp cut-off polycarbonate filter (BRADLEY et al. 1985a) or partially through a 50nm filter (HE et al. 1987). It is also worth noting that persistent PT-NANBH infection in chimpanzees was shown by BRADLEY et al. (1983c), TSIQUAYE et al. (1983), and BROTMAN et al. (1983) to interfere with superinfection by two other hepatotropic viruses, namely HAV and HBV. The latter phenomenon strongly suggested that viral interference was the culprit and further supported the growing (or accepted) notion that the TFA was a virus. Furthermore, liver biopsy specimens obtained from both human patients and experimentally infected chimpanzees revealed a variety of characteristic light microscopic changes that were entirely consistent with a diagnosis of acute, chronic, or persistent viral hepatitis, as described by DIENES et al. (1982), BIANCHI and GUDAT (1983), and HOOFNAGLE and ALTER (1984). Finally, the hepatocyte ultrastructural alterations observed in PT-NANBH infected chimpanzees were reported by BRADLEY et al. (1985b) to be identical or similar to those found in cells infected by RNA, but not DNA viruses (see below).

5 Evidence for the Existence of Multiple NANBH Agents

Aside from the enterically transmitted form of NANBH, now known to be caused by hepatitis E virus (HEV; BRADLEY et al. 1988; BRADLEY 1990; REYES et al. 1990), earlier studies suggested that PT-NANBH might be caused by a variety of unrelated hepatotropic agents (viruses). This view was supported by: (1) the observation of multiple attacks of viral hepatitis in patients transfused with blood products, as reported by MOSLEY et al. (1977), HRUBY and SCHAUF (1978), CRASKE and SPOONER (1978), and NORKRANS et al. (1980); (2) the occurrence of both long and short incubation period disease in transfused patients, as reported by CRASKE et al. (1975), AACH et al. (1978), and GUYER et al. (1979); (3) the observation of unique or distinct patterns of ALT activity seen in infected patients, as reported by TATEDA et al. (1979); (4) the results of cross-challenge studies in chimpanzees that had been experimentally infected with a variety of infectious materials, including blood, blood fractions, or blood products, as reported by BRADLEY et al. (1980), HOLLINGER et al. (1980), TSIQUAYE and ZUCKERMAN (1979), YOSHIZAWA et al. (1981), and TABOR et al. (1980); (5) the appearance of characteristic hepatocyte ultrastructural alterations in some, but not all, NANBH-infected chimpanzees, as reported by BRADLEY et al. (1983), YOSHIZAWA et al. (1981), and SHIMIZU et al. (1979); and (6) the observation by BRADLEY et al. (1983) of two, sequential episodes of NANBH in chimpanzees after the inoculation of chloroform-resistant and chloroform-sensitive agents, respectively (see below).

In view of the reported high incidence of persistent PT-NANBH in many infected individuals and experimentally infected chimpanzees, it now appears that many of the earlier cross-challenge studies were most likely confounded by the probable effects of viral interference. Other, more recent discoveries of other bloodborne viruses, namely, hepatitis G virus (HGV), reported by LINNEN et al. (1996), and GBV-C (closely related to HGV), reported by SIMONS et al. (1995), were foreshadowed in a review by ALTER and BRADLEY (1995) that described a form of NANBH unrelated to what is now known as HCV. The latter review described evidence for and against the existence of another major agent of NANBH. While the "true" hepatotropic nature and importance of the above two viruses still remain in doubt, it is clear that other agents of PT-NANBH may still be lurking in our blood supply system. In contradistinction to the above (and much more recent) findings, the turbulent course of PT-NANBH observed in many patients and chimpanzees could also explain the occurrence of many so-called second bouts of NANBH that in reality were simply exacerbations of chronic, underlying disease.

6 Attempts to Detect NANBH (Virus)-Specific Antigens, Antibodies, and Enzymes

Numerous laboratories during the late 1970s and early 1980s worked in vain to develop methods to detect virus-specific antigens and antibodies. Other laboratories sought, and even reported finding, virus-specific enzyme activities such as particle-associated reverse transcriptase activity (see below). Although numerous reports of antigen-antibody systems associated with NANBH appeared in the early literature, confirmation of their specificity was always problematic, if not impossible. Studies performed at the CDC (Bradley et al., unpublished work) encompassed the use of presumed convalescent human and chimpanzee sera, acute-phase sera, and chronic-phase sera (as well as the individual IgG and IgM fractions derived from these materials). More than 30,000 individual tests embodying RIA, immunodiffusion, counterimmunoelectrophoresis, or immunofluorescent probe configurations were employed. A variety of plasma and serum concentrates as well as liver homogenates (or fractions thereof) were tested by the above methods and all results were uniformly negative. Scientists in other major laboratories around the world suffered a similar sense of paroxysmal indignity when they unsuccessfully applied the then state-of-the-art methodologies to solve one of the most perplexing problems of the day.

As noted earlier, studies at the CDC and elsewhere strongly suggested that acute PT-NANBH frequently, if not usually, progressed to persistent liver disease and/or viremia. BRADLEY and MAYNARD (1983) and BRADLEY (1984) surmised from the above findings that antibody (regardless of source and type) of sufficient avidity and potency (titer) for the development of sensitive serologic tests for virus-specific antigens (and/or antibodies) was simply not available or could not be readily identified. It was generally known at the time that RIA and ELISA procedures required the use of high-avidity and/or high-titered antibody for optimum sensitivity. It was also known at the time that antibodies that could readily function in far less sensitive procedures, such as immunoelectronmicroscopy, agar gel diffusion, fluorescent probe assays, and (somewhat later) western-blotting, were unsuitable for use in RIA and ELISA tests. BRADLEY (1984) speculated that even if antibodies of sufficient avidity and/or titers were available for use in the above procedures, viral antigens (or virus) would still not be detectable due to the low titers of circulating virus in known infectious plasmas, sera, liver homogenates, and blood fractions. BRADLEY concluded that the root of the enigma was not necessarily the source of antibody, but rather, that the titer of the PT-NANBH agent (virus) was below the level of detectability using the current methodologies. The latter hypothesis was supported by available titration data in chimpanzees that showed the majority of inocula had titers less than 1×10^3 chimpanzee infectious doses (CID) per ml. For example, the FVIII materials used by BRADLEY et al. (1979) in their first primate transmission studies were found to have a titer of less than 1×10^3 CID/ml (BRADLEY et al. 1983b). The NANBH "F" strain used by FEINSTONE et al. (1981) was a chronic-phase plasma and was shown to have a titer of less than 1×10^2 CID/ml. YOSHIZAWA et al. (1982) described a fibrinogen preparation with a titer greater than 1×10^2 but less than 1×10^4 CID/ml; similarly, Tabor and Gerety (personal communication) found that one of their inocula had a titer of approximately 1×10^2 CID/ml. Of great interest was the finding by Overby et al. (Abbott Laboratories, N. Chicago; unpublished studies of the late 1970s and early 1980s) that none of ten human acute-phase plasma units had a titer greater than 1×10^3 CID/ml. One acute-phase human plasma ("H"-strain agent) described by FEINSTONE et al. (1981), was found to have a titer of approximately 1×10^6 CID/ml; however, plasma taken from the same individual 1week later was found to be non-infectious when intravenously inoculated into a presumably naive chimpanzee (H.J. Alter, personal communication). The latter finding was puzzling, but indicated that titers of the NANBH agent (virus) could change significantly over a short period of time during the acute phase of disease. Taken together these findings supported the view by some investigators that the difficulties in developing a standard serologic assay for NANBH (virus)-specific antigen was not going to be accomplished until and unless a much more sensitive assay was developed.

Evanescent excitement regarding the possible identification of a retrovirus-like agent responsible for human PT-NANBH was generated by the nearly simultaneous reports of particle-associated reverse-transcriptase (RT) activity in human sera and in cultured NANBH-infected chimpanzee liver cells, as reported by PRINCE et al. (1984), SETO et al. (1984), and IWARSON et al. (1985). SETO et al. (1984) detected RT activity in four human NANBH serum specimens and in two plasma products, all of which had been shown to transmit disease to chimpanzees. RT activity was also detected in 12 of 12 human acute- and chronic-phase NANBH sera, but not in 47 of 49 sera from healthy donors. NANBH infectivity was also found at a buoyant density of $1.14 g/cm^3$ in a sucrose gradient that was coincident with a peak of RT activity. In spite of these earlier, potentially exciting findings, workers at the CDC and in other laboratories were unable to detect elevated RT activity in well-documented cases of human PT-NANBH using pedigreed panels of sera. Furthermore, pelleted acute-phase human plasma fractions that contained the equivalent of at least 1×10^6CID of the NANBH TFA were shown to be completely devoid of RT activity (A.J. Weiner, and M.A. Houghton 1984, personal communication). These findings, combined with the knowledge that the NANBH TFA was <80nm in diameter (see below), strongly suggested that the agent was not a retrovirus. Furthermore, use of hybridization probes specific for conserved DNA polymerase gene sequences of both hepadnaviruses (i.e., HBV) and retroviruses (replicative intermediates) including human T-cell lymphocytotropic virus type III (HTLV-III), MuLV, RSV, and cauliflower mosaic virus failed to detect homology between any of these viruses with total and polyadenylated RNAs extracted from several, proven-infectious chimpanzee livers (M.A. Houghton, A.J. Weiner, D.W. Bradley 1984, unpublished findings).

7 Infectivity and Virus-Like Particles: Peripheral Blood Lymphocytes, Cultured Chimpanzee Liver Cells, Liver Homogenates, and Blood Fractions

In addition to conventional studies with human and chimpanzee sera and plasmas, NANBH infectivity was sought in peripheral blood cell preparations from both humans and chimpanzees. Inoculation of a colony-born chimpanzee at the CDC with 2.2×10^7 peripheral blood lymphocytes (PBLs) obtained from a proven NANBH carrier did not induce elevated ALT activity up to 95 days after inoculation, nor was there any electron microscopic evidence of infection by the NANBH TFA in serial (weekly) liver biopsy specimens obtained during the study period (Bradley et al., unpublished data). Challenge of this animal with a NANBH chronic-phase liver homogenate resulted in short-incubation period disease accompanied by the appearance of characteristic hepatocyte ultrastructural alterations. In contrast to these early findings, HELLINGS et al. (1985) reported the transmission of NANBH to a chimpanzee using mononuclear leukocytes from a patient with chronic NANBH. This discrepancy in experimental findings was never resolved.

PRINCE et al. (1984) reported finding 85–90nm diameter virus-like particles in cultured chimpanzee liver cells infected with the NANBH TFA. DERMOTT (1985), however, suggested that the above viruses might be similar or identical to the (latent) simian foamy viruses previously described by HOOKS and GIBBS (1975). IWARSON et al. (1985) reported the appearance of tubules in cultured NANBH-infected chimpanzee liver cells that were identical in morphology to the tubules seen in NANBH-infected chimpanzee hepatoctyes. The liver cell tubules referred to above were noted by BRADLEY et al. (1985b) as being most similar to the test-tube and ring-shaped structures (TRF and CTS) previously identified by SHAMOTO et al. (1981) and SIDHU et al. (1983) in lymphocytes obtained from patients with adult T-cell leukemia and acquired immune deficiency syndrome (AIDS), respectively.

Delta virus, a "defective" agent that requires the presence of HBV for replication, was also shown to induce similar tubular structures in chimpanzee hepatocytes but not in human hepatocytes (see below for further discussion). At best, the above findings were interesting phenomena but confounded an area of viral research that was already plagued by numerous blind alleys and "red-herrings." The questionable presence of NANBH infectivity in chimpanzee PBLs, the lack of detectable RT activity in proven-infectious materials, and the puzzling (non-specific) nature of the tubular structures observed in NANBH-infected chimpanzee hepatocytes, however, all mitigated against the possibility that the major agent of human NANBH was a retrovirus.

Earlier findings by BRADLEY et al. (1979) and YOSHIZAWA et al. (1980) that 27nm diameter virus-like particles (VLPs) were associated with infection of chimpanzees with either FVIII concentrates or fibrinogen, respectively, were greeted with little enthusiasm by a highly skeptical scientific community. The inability of other laboratories to reproduce the reported findings did little to support the association of these VLPs with NANBH. In retrospect, however, the recovery from acute-phase chimpanzee liver tissues of extremely fragile, often ragged 27nm diameter VLPs coated with what appeared to be an IgM isotype antibody is consistent with the notion that these VLPs were NANBH TFA (HCV) capsids aggregated by anti-C22 (core) antibodies that are now known to be among the first to appear in circulation in response to infection.

8 Comparative Morphology of Ultrastructural Alterations in Chimpanzee Hepatocytes

The NANBH (virus)-induced ultrastructural alterations observed in chimpanzee hepatocytes were extensively reviewed by BRADLEY et al. (1985). However, for purposes of historical perspective, selected aspects of the above review and other published works are described below and demonstrate the intense interest many scientists had in this area of NANBH research during the late 1970s to mid-1980s. SHIMIZU et al. (1979) first described the finding of double-walled tubular structures in the hepatocyte cytoplasm of chimpanzees infected with the "H" strain of the NANBH TFA (HCV). Subsequently, numerous other workers, including BRADLEY et al. (1980), TSIQUAYE et al. (1980), PFEIFER et al. (1980), and YOSHIZAWA et al. (1982), reported finding identical structures in the hepatocyte cytoplasm of chimpanzees infected with a wide variety of NANBH inocula. In total, six different types of ultrastructural alterations were observed in chimpanzees infected with the (chloroform-sensitive; see below) TFA. All of the changes were confined to the cytoplasm of infected or affected liver cells and included: (1) dense reticular inclusion bodies; (2) convoluted membranes derived from proliferated smooth endoplasmic reticulum; (3) characteristic tubular structures (already referred to above); (4) bundles of tightly-packed granular microtubules; (5) crystalline arrays of 25nm "particles" in endothelial cells; and (6) highly structured crystals of proteinaceous material resembling paracrystalline arrays of tubulin found in reovirus-infected cells. It is of interest to note, as described earlier, that many of these ultrastructural alterations were also found by CANESE et al. (1984) in the hepatocyte cytoplasm of chimpanzees infected by delta virus.

Comparative morphologic studies by BRADLEY et al. (1985b) of the TFA-induced ultrastructural alterations revealed the occurrence of similar changes in cells infected by other, well characterized viruses. For example, cytoplasmic tubules observed by ROBERTS and HARRISON (1970) in plant cells infected with strawberry latent ringspot virus (SLRV, a picornavirus) were strikingly similar to the tubular structures found in TFA-infected chimpanzee hepatocytes. Both structures were derived from proliferated smooth endoplasmic reticulum (SER) and consisted of double-walled tubules surrounded by a membranous sheath. Tubules or cylindrical structures have also been visualized by HARRISON et al. (1982) and MONATH et al. (1983) in the cytoplasm of mosquito cells and visceral target organs of hamsters infected with St. Louis encephalitis virus (SLEV), a member of the Flaviviridae family. These cylinders were shown to be associated with virus replication and were also derived from proliferated SER. Tubules identical in morphology to those observed in NANBH TFA-infected chimpanzee hepatocytes were observed by SIDHU et al. (1983) in lymphocytes of patients with AIDS; similarly, SHAMOTO et al. (1981) reported identical tubules in the cytoplasm of lymphocytes from cases of adult T-cell leukemia. Paired, convoluted membranes enclosing an osmiophilic substance (Erc) were also observed by HAMPTON et al. (1973) in plant cells infected with pea seed-borne mosaic virus (PsbMV), a rod-shaped, non-enveloped RNA virus. ERc identical in morphology to that described above was also seen by HARRISON et al. (1982) in the lamina propria of hamster ileum infected with SLEV. Structures identical or very similar to the dense reticular inclusion bodies seen in infected chimpanzee hepatocyte cytoplasm were also observed in the cytoplasm of cell infected by a variety of RNA viruses, namely, reovirus (DALES 1973), influenza virus (COMPANS and CHOPIN 1973), coronavirus (mouse hepatitis virus; DAVID-FERREIRA and MANAKER 1965), poliovirus (DALES et al. 1965), SLEV (MONATH et al. 1983), and seadog (Tyuleniy) virus (SDV, a member of the Flaviviridae family; Zhdanov 1982, personal communication). Of great interest was the fact that the dense reticular inclusion bodies most closely resembled those found in cells infected with SDV or SLEV, both members of the Flaviviridae family. The above inclusion bodies appeared to contain masses of highly convoluted and densely-stained 25nm diameter filaments entangled in amorphic, proteinaceous materials that were absolutely identical in morphology to those found in SDV-infected cells and in the mid-gut epithelium of *Culex pipiens* (mosquitoes) infected with SLEV (MONATH et al. 1983). In the latter case, these structures were found to be spatially related to virus particle formation and were thought to represent viroplasmic foci (virus "factories"). Microtubular aggregates also have their counterpart in other virus infections. Nearly identical aggregates were previously found by VIDANO (1970) in the mid-gut tissue of an insect vector (*Laodelphax striatellus*) infected with maize rough dwarf virus (MRDV9), a small, RNA-containing virus. The paracrystalline structures composed of proteinaceous materials, probably tubulin (the protein precursor of microtubules), were also observed in chimpanzee hepatocytes, as noted above. These same structures were described by DALES (1973) in cells infected with reovirus.

The constellation of disease-associated ultrastructural changes observed in chimpanzee hepatocytes was confined to the cytoplasm, and, as noted, bore a striking resemblance to ultrastructural alterations found in other kinds of cells infected with plant, insect vector, or animal viruses. The results of this early study of the comparative morphology of PT-NANBH-associated changes strongly supported the notion that the major agent (virus) had an RNA genome since all of the above viruses shared the common property of an RNA (and not DNA) genome.

9 Viral Interference

Early studies at the CDC and elsewhere showed that coinfection with HAV and HBV could occur in both humans (HINDMAN et al. 1977) and experimentally infected chimpanzees (DRUCKER et al. 1979) without obvious viral interference. In later studies, two patients with chronic HBV infections and intercurrent episodes of hepatitis A (in contrast to coinfection with HBV and HAV) were found to have depressed or even negative markers of HBV replication, including hepatitis B "e" antigen (HBeAg), HBV-DNA, and DNA polymerase activity with concomitant appearance of leukocyte (α)-interferon during the acute-phase of disease, as reported by DAVIS et al. (1984). Nevertheless, both patients exhibited significantly elevated ALT activity and early seroconversion to anti-HAV, indicating that chronic HBV infection did not interfere with superinfection by HAV.

In sharp contrast to the above findings, BRADLEY et al. (1983), TSIQUAYE et al. (1983), BROTMAN et al. (1983), and LIAW et al. (1982) documented the phenomenon of viral interference in chimpanzees and humans simultaneously or sequentially infected with HBV and the NANBH TFA or HAV and the NANBH TFA (HCV). Studies at CDC by BRADLEY et al. (1983) revealed a profound effect of persistent NANBH infection on superinfection by HAV. Neither of two NANBH-infected chimpanzees (both with persistent viremia and persistently or intermittently elevated ALT activity) developed additional or increased elevations in ALT activity when challenged with proven infectious HAV. In addition, both challenged animals demonstrated a delayed anti-HAV antibody response (28 and 43 days after inoculation) when compared to two control chimpanzees (14 days after inoculation) who had received the identical inoculum. Furthermore, neither superinfected chimpanzee was shown to have detectable HAV antigen by an immunofluorescent antibody assay (FA) in acute-phase liver biopsy specimens; all daily stool specimens, with one possible exception in one animal, were also negative for HAV antigen when tested by a sensitive RIA. Bradley et al. (unpublished data) also found that chimpanzees with biochemically and electron microscopically resolved

PT-NANBH could also interfere with HAV infection. The latter finding suggested that even presumed low-level replication of the NANBH TFA (HCV) was sufficient to interfere with superinfection by HAV. With regard to hepatitis B, BRADLEY et al. (1983) and TSIQUAYE et al. (1983) also showed that acute PT-NANBH infection in HBsAg carrier chimpanzees depressed the replication of HBV as judged by the decrease in surface antigen titer and serum DNA polymerase activity during the acute-phase of NANBH. BROTMAN et al. (1983) and Dolana et al. (1983, unpublished studies) showed that coinfection of chimpanzees with HBV and the NANBH TFA also delayed, moderated, or obviated the appearance of serologic markers of HBV infection. These studies showed that the NANBH TFA was a "dominant" agent and further supported the growing (if not already established) consensus that PT-NANBH was, indeed, caused by a virus.

10 Physicochemical Properties of the Major PT-NANBH Agent

Although numerous and persistent attempts were made at CDC and in other laboratories to visualize the virus of PT-NANBH, it became readily apparent by 1980–1981 that the titer of virus in most sera, plasmas, and liver/liver preparations was far too low to permit detection by electron microscopy or immune electron microscopy(IEM). In fact, the titers of most proven-infectious materials were found to be less than $1 \times 10^3 \text{CID/ml}$ as summarized by BRADLEY (1984). Since direct electron microscopy detection of virus particles generally requires particle concentrations on the order of $1 \times 10^{8-10}$/ml, there was little realistic hope of visualizing the presumed viral agent of PT-NANBH. Furthermore, even the use of a more sensitive technique, IEM, still required virus particle concentrations of approximately $1 \times 10^{6-7}$/ml, far above the majority of titers reported for proven-infectious inocula. As a consequence, routine examination of known infectious plasmas and liver materials never revealed the presence of virus or VLPs that could be reproducibly detected. The absence of any information on the morphology or properties of the major agent of PT-NANBH prompted BRADLEY et al. at CDC to consider, as early as 1980–1981, alternative approaches to the characterization of the causative agent(s). As noted earlier, BRADLEY et al. (1983) defined the possible existence of two distinct agents of PT-NANBH by noting their differential sensitivity to treatment with chloroform, a lipid solvent. Since it was obvious from all previous reports that there was one major agent of PT-NANBH, as judged by the consistent appearance in chimpanzees of the ultrastructural alterations described above, a decision was made at CDC to further define the properties of the PT-NANBH TFA (HCV). Furthermore, FEINSTONE et al. (1983) also reported that the TFA (as well as HBV) could be inactivated by treatment with chloroform, a finding that was consistent with the notion that the agent was a lipid-containing or enveloped virus. The combined findings by BRADLEY et al. and FEINSTONE et al., that the TFA was chloroform-sensitive, mitigated against any notion that the agent (virus) was either a viroid or a scrapie- or prion-like agent, since the latter agents were reported by PRUSINER et al. (1984) and MERZ et al. (1983) to be proteinaceous in nature. Since only certain families of viruses, both DNA and RNA, contain viruses with essential lipid, particle-sizing studies were initiated at CDC in 1983 in order to eliminate several possible families of viruses that contained lipid and exceeded a pre-determined diameter. Controlled pore (polycarbonate membrane) filters were used at CDC to sequentially filter pelleted plasma preparations through a final pore size of 80nm (BRADLEY et al. 1985a). Chronic-phase chimpanzee plasma was diluted 1:6 in TENB, pH 8.0 buffer (TENB: 0.05M Tris, 0.001M EDTA, 0.1M NaCl) and centrifuged at $120,000 \times g$ for 5.0h at 20°C to pellet the TFA. The pelleting procedure assumed the NANBH TFA was a virus with a sedimentation coefficient of approximately 150S. The pellet was resuspended in TENB buffer and successively passed through 450-, 200-, and 80-nm filters. The final filtrate was used to inoculate a naive (never-before-used) chimpanzee. The latter animal developed elevated ALT activity 38 days after inoculation with a peak value at day 65; ultrastructural alterations indicative of infection with the TFA agent were also detected by EM in hepatocyte cytoplasm (Fig. 2). This disease profile was typical for many of the chimpanzees infected with the TFA at the CDC. In a later study, HE et al. (1987) also reported that the TFA agent would pass an 80-nm filter and partially pass a 50-nm (but not 30-nm) filter suggesting that the agent had a diameter of between 40 and 60nm, consistent with the earlier hypothesis by BRADLEY et al. (1985a), that the major etiologic agent of PT-NANBH was a small, enveloped virus. Additional studies conducted at CDC and elsewhere ruled out the possibility

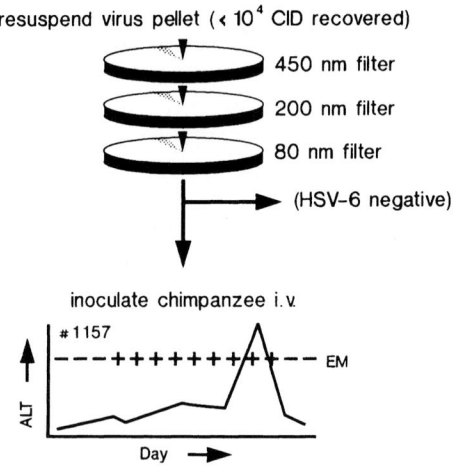

Fig. 2. Pelleting and microfiltration of the post-transfusion non-A, non-B hepatitis (PT-NANBH; HCV) agent

that the TFA was either a hepadnavirus or a delta-like virus (i.e., low or no significant nucleotide sequence homology, as revealed by carefully controlled hybridization assays). By a process of systematic virus-class "elimination" combined with the newly defined physicochemical properties of the TFA, BRADLEY further concluded that the agent was most like a small, enveloped RNA virus (in particular a flavi- or flavi-like virus) based in part on the results of an earlier comparative analysis of ultrastructural alterations induced in cells by well known flaviviruses. Figure 3 summarizes the process used at CDC to predict the most likely virus-candidate (i.e., family of virus) of the TFA (HCV) based on a combination of established physicochemical properties (not including its buoyant density in sucrose) and the findings of a comparative analysis of TFA-induced ultrastructural alterations observed in infected chimpanzee hepatocytes. The process was methodical and time-consuming; it also required the most expeditious, yet sparing, use of chimpanzees, a limited and highly expensive primate model for PT-NANBH research.

Other studies were initiated at CDC in 1987 to determine the buoyant density of the PT-NANBH TFA (HCV) in sucrose using standard isopycnic banding methods. A total of five chimpanzees were used in two different study phases (conducted over a period of approximately 3 years) to determine, by back-titration of gradient fraction pools, the distribution and amount of infectious virus (in CID/gradient fraction pool) throughout the sucrose gradient. BRADLEY et al. (1991) reported that one ml of a chronic-phase chimpanzee plasma (from animal #910) that contained 1×10^6 CID could be completely recovered in a gradient fraction pool that encompassed fractions with buoyant densities of $1.09-1.11 \text{g/cm}^3$ (Fig. 4). This buoyant density (combined with the previous findings described above and the concurrent knowledge that the TFA was itself HCV and that it shared genomic properties of both flavi- and pestiviruses) was consistent with the much earlier

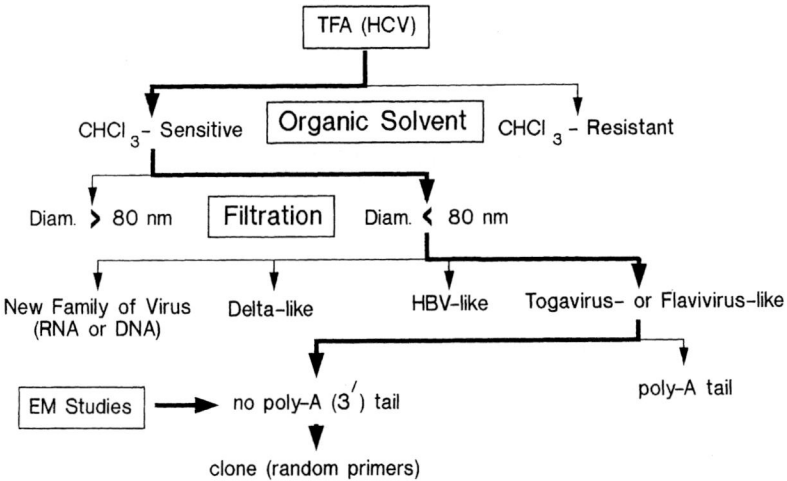

Fig. 3. Algorithm used at CDC to predict family of virus

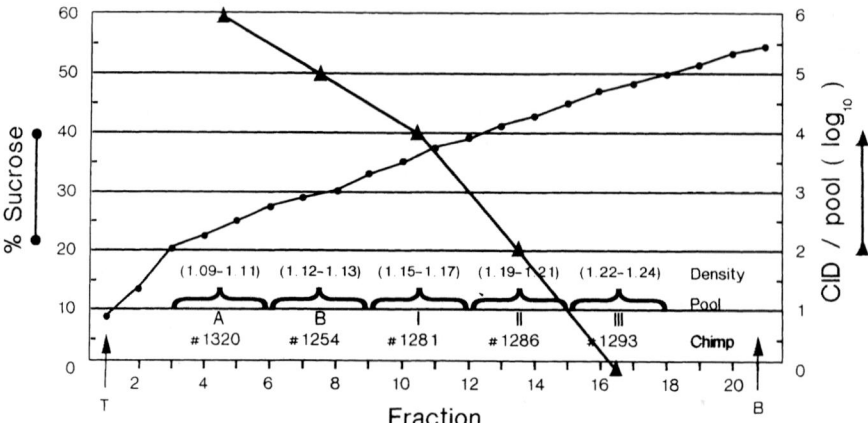

Fig. 4. Buoyant density determination of HCV (tubule forming agent, TFA); back-titration studies in chimpanzees

hypothesis that the TFA was a Togavirus (meaning either flavivirus or pestivirus). MIYAMOTO et al. (1992) confirmed the above findings when they found that HCV in human donor plasma also banded at a low density (i.e., $1.08 g/cm^3$). Both BRADLEY et al. (1991) and MIYAMOTO et al. (1992) acknowledged the fact that determination of this critical physicochemical property of HCV provided the means for the development of efficient purification procedures for HCV contained in large volumes of plasma (or tissue culture fluids/lysates).

11 Development of High-Titered Liver and Large-Volume Plasma Pools

The development of high-titered plasma pools and liver tissue was essential to successful molecular cloning of the TFA (HCV). Chimpanzee studies were initiated at the CDC in 1979 specifically for the purpose of generating high-titered materials that would enable: (1) visualization of the virus by electron microscopy or IEM, (2) purification of the virus by accepted or revised procedures according to virus properties, (3) development of polyspecific or monoclonal antibodies, (4) attempts to propagate the virus in tissue culture, and (5) molecular cloning of the viral genome. The latter purpose became a more specific focus of our studies once it became apparent in the early 1980s that few, if any, sources available at the time had the requisite titer and volume for successful molecular cloning of the viral genome. Several different approaches were used to achieve the above goal and included: (1) serial passage of the virus in chimpanzees to "adapt" the virus to the host (with consequent increase of virus in liver, as previously reported for HAV by

BRADLEY et al. 1984; HCV passage studies were initiated in the early 1980s); (2) immunosuppression of the host chimpanzee to increase the severity of disease (and, hopefully, titer of the TFA; study reported by BRADLEY et al. 1984); (3) titration of selected acute-phase and chronic-phase livers from chimpanzees; and (4) titration of highly selected pools of plasma from persistently (chronically) infected chimpanzees. A detailed summary of the chimpanzee studies that primarily involved the use of liver can be found in BRADLEY (1990).

Cloning studies (of the PT-NANBH TFA) were conducted in close collaboration with Chiron Corporation. From the beginning it was apparent that the relative proportion of nonviral nucleic acid found in even high-titered liver (approximately $1 \times 10^7 CID/g$ of tissue) would make cloning difficult and underscored the need to develop other sources of virus that would be less complex. Although other investigators found that acute-phase (and even one chronic-phase) plasma generally had TFA (HCV) titers equal to or less than $1 \times 10^3 CID/ml$, it was still apparent that, if achievable, a large-volume, high-titer plasma or plasma pool would be superior to liver as a source of virus for molecular cloning studies. The minimum acceptable infectivity titer of plasma (plasma pool) for cloning of the TFA genome was calculated to be $1 \times 10^5 CID/ml$, assuming that there were ten defective particles for every infectious particle (1CID). Previous studies of the course of disease in chimpanzees housed at the CDC showed that chronic-phase plasma might be a richer source of virus than acute-phase plasma, since disease in many chimpanzees appeared to worsen with time after initial inoculation. As a result, several chimpanzees were intensively followed for periods of time up to 11 years in order to prospectively collect large volumes of plasma for characterization of the virus and (later) molecular cloning of the viral genome. Two chimpanzees, namely Don (#771) and Rodney (#910), were most carefully studied (Fig. 1), including determination of their ALT activity at relatively frequent intervals of time, and examination of serial liver biopsy specimens from each animal for histologic and EM evidence of disease severity that would be indicative of increased viral replication in liver. Based on the hypothesis (by this author) that the TFA was cytopathic and that the highest titers of virus would be found in plasma during periods of exacerbated, chronic-phase disease, units with the highest levels of ALT activity (relative to other reposited and catalogued units of plasma) were pooled and titered in chimpanzees. It should be noted that collection of chronic-phase plasma units was initiated at the CDC in 1979. A large pool of plasma was originally generated from units of plasma collected from chimpanzee #771 over a 4-year period after the acute-phase of disease. The original criteria for the initiation of a given plasmapheresis were: (1) a periodic episode of elevated ALT activity, and (2) EM or histologic evidence of increased viral replication in liver. Units of plasma collected at the peaks of recrudescent disease (including those collected well beyond 4years after inoculation) were selected, pooled, aliquoted, catalogued, and stored at $-70°C$ until used. Inoculation of two chimpanzees with either of two different plasma pools showed that the titers of the TFA were $1 \times 10^4 CID/ml$ (first pool), and $1 \times 105 CID/ml$ (second pool). Units of plasma were also prospectively collected from chimpanzee #910 using the above-described process. One pool con-

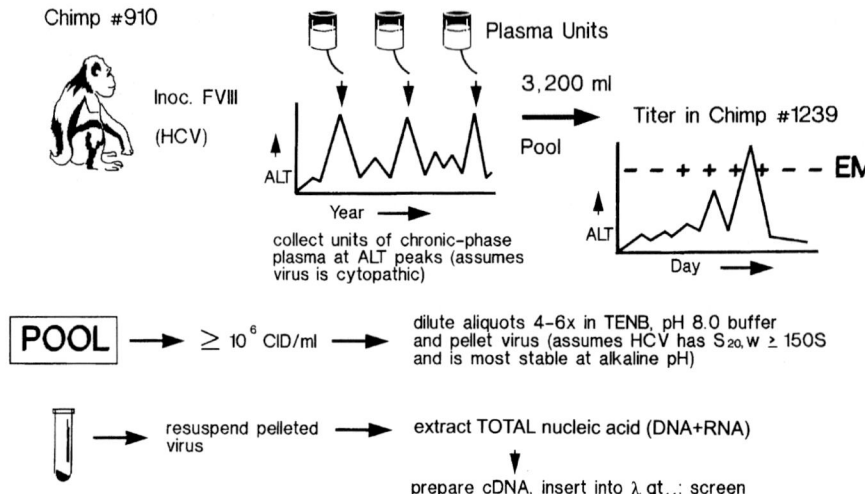

Fig. 5. Cloning of hepatitis C virus (HCV): production of large-volume, high-titer plasma pools

sisting of approximately 3,200ml was shown to have a titer of at least 1×10^6CID/ml; this pool contained all of the units of prospectively-collected plasma that were determined to have the highest levels of (chronic-phase) ALT activity (Fig. 5). A second pool of plasma consisting of units with a statistically lower (mean) ALT value was found to have a titer of 1×10^5CID/ml. The latter findings reaffirmed the hypothesis (and prediction) that large-volume, high-titered plasma pools could, in fact, be generated by a process that involved the collection of units of chronic-phase plasma from chimpanzees that had demonstrated the most severe disease, as indicated by relative values of ALT activity, EM evidence of worsening disease, and histologic evidence (in some instances) of more severe disease. The major criterion for the selection of any unit of plasma was its ALT value. It is of interest to note that liver tissue obtained from chimpanzee #910 during one period of exacerbated disease was shown to have a titer of 1×10^7CID/g.

12 Discussion

The isolation, characterization, and eventual molecular cloning of HCV have made it possible to: (1) develop blood donor or patient screening methods (for virus-specific antibodies and/or nucleic acid, i.e., RNA); (2) attempt propagation of HCV in tissue culture; (3) design and develop strategies for the production of recombinant vaccines; (4) devise rational approaches to the construction of inhibitors of HCV replication (such as synthetic organic molecules for virally encoded proteins, i.e., proteases, helicase, RNA-dependent RNA polymerase, or sequence-specific oligonucleotides that could bind to the HCV genome). Although no tissue culture system has yet been developed that produces desirable quantities of HCV in vitro, the basic properties of HCV, including its buoyant density in sucrose, are now known and should facilitate the development of purification methods for HCV derived from tissue culture extracts. The latter methods, in all likelihood, will depend on some form of a quantitative test for viral RNA.

Nearly 14,000 articles have been published on one or more aspects of hepatitis C or HCV during the past decade. This explosive growth of information has provided the foundation for further exciting findings and developments in viral hepatitis C and reveals the level of interest in a field that was once considered to be moribund.

References

Aach RD, Lander JJ, Miller WV, Kahn RA, Gitnick GL, Hollinger FB, Werch J, Szmuness W, Stevens CE, Kellner AK, Weiner JM, Mosley JW (1978) Transfusion-transmitted viruses: interim analysis of hepatitis among transfused and non-transfused patients. In: Vyas GN, Cohen SH (eds) Viral hepatitis, a contemporary assessment of etiology, epidemiology, pathogenesis, and prevention. Franklin Institute Press, Philadelphia, pp 383–396

Alter HJ, Bradley DW (1995) Non-A, non-B hepatitis unrelated to the hepatitis C virus (Non-ABC). Semin Liver Dis 15:110–120

Alter HJ, Holland PV, Morrow AG, Purcell RH, Feinstone SM, Moritsugu Y (1975) Clinical and serological analysis of transfusion-transmitted hepatitis. Lancet 2:838–841

Alter HJ, Purcell RH, Holland PV, Popper H (1978) Transmissible agent in non-A, non-B hepatitis. Lancet 1:459–463

Alter HJ, Purcell RH, Feinstone SM, Tegtmeier GE (1982) Non-A, non-B hepatitis: its relationship to cytomegalovirus, to chronic hepatitis, and to direct and indirect test methods. In: Szmuness W, Alter HJ, Maynard JE (eds) Viral hepatitis 1981 international symposium. Franklin Institute Press, Philadelphia, pp 279–294

Alter HJ, Dienstag JL (1984) The evolving spectrum of non-A, non-B hepatitis. In: Chisari FV (ed) Advances in hepatitis research. Masson, New York, pp 281–292

Berman M, Alter HJ, Ishak KG, Purcell RH, Jones EA (1979) The chronic sequelae of non-A, non-B hepatitis. Ann Intern Med 91:1–6

Bianchi L, Gudat FG (1983) Histo- and immunopathology of viral hepatitis. In: Deinhardt F, Deinhardt J (eds) Viral hepatitis: laboratory and clinical science. Marcel Dekker, New York, pp 335–382

Bradley DW, Cook EH, Maynard JE, McCaustland KA, Ebert JW, Dolana GH, Petzel RA, Kantor RJ, Heilbrunn A, Fields HA, Murphy BL (1979) Experimental infection of chimpanzees with antihemophilic (Factor VIII) materials: recovery of virus-like particles associated with non-A, non-B hepatitis. J Med Virol 3:253–269

Bradley DW, Mayanard JE, Cook EH, Ebert JW, Gravelle CR, Tsiquaye KN, Kessler H, Zuckerman AJ, Miller MF, Ling C-M, Overby LR (1980) Non-A/non-B hepatitis in experimentally infected chimpanzees: cross-challenge and electron microscopic studies. J Med Virol 6:185–201

Bradley DW, Maynard JE, Popper H, Ebert JW, Cook EH, Fields HA, Kemler BJ (1981) Persistent non-A, non-B hepatitis in experimentally infected chimpanzees. J Infect Dis 143:210–218

Bradley DW, Maynard JE, Krawczynski KZ, Popper H, Cook EH, Gravelle CR, Ebert JW (1982) Non-A, non-B hepatitis in chimpanzees infected with a Factor VIII agent: evidence of persistent hepatic disease. In: Szmuness W, Alter HJ, Maynard JE (eds) Viral hepatitis 1981 international symposium. Franklin Institute Press, New York, pp 319–329

Bradley DW, Krawczynski KZ, Cook EH, Gravelle CR, Ebert JW, Maynard JE (1982) Recrudescence of non-A, non-B hepatitis in persistently infected chimpanzees. In: Hopkins R, Fileds S (eds) Viral hepatitis, 2nd international workshop. Nuclear Enterprises, Edinburgh, pp 43–48

Bradley DW, Maynard JE (1983) Non-A, non-B hepatitis: research progress and current perspectives. Dev Biol Stand 54:463–473

Bradley DW, Maynard JE, Popper H, Cook, EH, Ebert JW, McCaustland KA, Schable CA, Fields HA (1983) Post-transfusion non-A, non-B hepatitis: physicochemical properties of two distinct agents. J Infect Dis 148:254–265

Bradley DW, Maynard JE, McCaustland KA, Murphy BL, Cook EH, Ebert JW (1983) Non-A, non-B hepatitis in chimpanzees: interference with acute hepatitis A virus and chronic hepatitis B virus infections. J Med Virol 11:207–213

Bradley DW, Schable CA, McCaustland KA, Cook EH, Murphy BL, Fields HA, Ebert JW, Wheeler CM, Maynard JE (1984) Hepatitis A virus: growth characteristics of in vivo and in vitro propagated wild and attenuated virus strains. J Med Virol 14:373–386

Bradley DW (1984) Transmission, etiology, and pathogenesis of viral hepatitis non-A, non-B in nonhuman primates. In: Chisari FV (ed) Advances in hepatitis research, chap 31. Masson, New York, pp 268–280

Bradley DW, McCaustland KA, Cook EH, Ebert JW, Maynard JE (1984) Non-A, non-B hepatitis is chimpanzees: effects of immunosuppression on course of disease and recovery of tubule-forming agent from infected liver. Viral hepatitis and liver disease, chap 39. Grune and Stratton, New York, pp 451–458

Bradley DW, McCaustland KA, Cook EH, Schable CA, Ebert JW, Maynard JE (1985a) Post-transfusion non-A, non-B hepatitis in chimpanzees: physicochemical evidence that the tubule forming agent is a small, enveloped virus. Gastroenterology 88:773–779

Bradley DW, Cook EH, Miller MF, Schaff Z, Gravelle CR, McCaustland KA, Maynard JE (1985b) Non-A, non-B hepatitis in experimentally infected chimpanzees: comparative morphology of virusinduced ultrastructural changes. In: Nishioka K, Blumber B, Ishida N, Koike K (eds) Hepatitis viruses and hepatocellular carcinoma: approaches through molecular biology and ecology. Academic, Tokyo, pp 225–250

Bradley DW, Andjaparidze A, Cook EH, McCaustland KA, Balayan M, Stetler H, Velazquez O, Robertson B, Humphrey C, Kane M, Weisfuse I (1988) Etiologic agent of enterically-transmitted non-A, non-B hepatitis. J Gen Virol 69:731–758

Bradley DW (1990) Hepatitis non-A, non-B viruses become identified as hepatitis C and E viruses. Prog Med Virol 37:101–135

Bradley DW, McCaustland KA, Krawczynski K, Spelbring J, Humphrey C, Cook EH (1991) Hepatitis C virus: buoyant density of the Factor VIII-derived isolate in sucrose. J Med Virol 34:206–208

Brotman B, Prince AM, Huima T, Richardson L, von den Ende MC, Pfeifer U (1983) Interference between non-A, non-B and hepatitis B virus infections in chimpanzees. J Med Virol 11:191–205

Brotman B, Prince AM, Huima T (1985) Non-A, non-B hepatitis: is there more than one single bloodborne strain? J Infect Dis 151:618–625

Brown P, Rohwer RG, Green EM, Gajdusek DC (1982) Effect of chemicals, heat, and histopathologic processing on high-infectivity hamster-adapted scrapie virus. J Infect Dis 145:683–687

Burk KH, Dreesman GR, Cabral GA, Peters RL (1984) Long-term sequelae of non-A, non-B hepatitis in chimpanzees. Hepatology 4:808–816

Canese MG, Rizzetto M, Novara R, London WT, Purcell RH (1984) Experimental infection of chimpanzees with the HBsAg associated delta virus: an ultrastructural study. J Med Virol 13:63–72

Choo Q-L, Kuo G, Weiner AJ, Overby LR, Bradley DW, Houghton M (1989) Isolation of a cDNA clone derived from a blood-borne non-A, non-B viral hepatitis genome. Science 244:359–362

Compans RW, Chopin PW (1973) Orthomyxoviruses and paramyxoviruses. In: Dalton AJ (ed) Ultrastructure of animal viruses and bacteriophages: an atlas. Academic, New York, pp 213–237

Craske J, Dilling N, Stern D (1975) An outbreak of hepatitis associated with intravenous injection of factor VIII concentrate. Lancet 2:221–223

Craske J, Spooner R (1978) Evidence for existence of two types of factor VIII-associated non-B transfusion hepatitis. Lancet 2:1051–1052

Dales S, Eggers HJ, Tamm I, Palade GE (1965) Electron microscopic study of the formation of poliovirus. Virology 26:379–389

Dales S (1973) The structure and replication of reoviruses. In: Dalton AJ (ed) Ultrastructure of animal viruses and bacteriophages: an atlas. Academic, New York, pp 155–171

David-Ferreira JR, Manaker RA (1965) An electron microscopic study of the development of a mouse hepatitis in tissue culture cells. J Cell Biol 24:57–78

Davis CL, Hoofnagle JE, Waggoner JB (1984) Acute type A hepatitis during chronic hepatitis B virus infection: association of depressed hepatitis B virus replication with appearance of endogenous alpha interferon. J Med Virol 14:141–147

Dermott E (1985) Non-A, non-B hepatitis virus Lancet 1:170

Dienes H, Popper H, Arnold W, Lobeck H (1982) Histologic observations in human hepatitis non-A, non-B. Hepatology 2:562–571

Dienstag JL (1983) Non-A, non-B hepatitis II: experimental transmission, putative virus agents and markers, and prevention. Gastroenterology 85:743–768

Drucker J, Tabor E, Gerety RJ, Jackson D, Barker LF (1979) Simultaneous acute infection with hepatitis A and hepatitis B viruses in a chimpanzee. J Infect Dis 139:338–342

Fagan EA, Williams R (1984) Non-A, non-B hepatitis. Semin Liver Dis 4:314–335

Feinstone SM, Alter HJ, Dienes HP, Shimizu Y, Popper H, Blackmore D, Sly D, London WT, Purcell RH (1981) Non-A, non-B hepatitis in chimpanzees and marmosets. J Infect Dis 144:588–598

Feinstone SM, Michalik KS, Kamimura T, Alter HJ, London WT, Purcell RH (1983) Inactivation of hepatitis B virus and non-A, non-B hepatitis by chloroform. Infect Immun 41:816–821

Galbraith RM, Dienstag JL, Purcell RH, Gower PH, Zuckerman AJ, Williams R (1979) Non-A, non-B hepatitis associated with chronic liver disease in a haemodialysis unit. Lancet 1(8123):951-953

Guyer B, Bradley DW, Bryan JA, Maynard JE (1979) Non-A, non-B hepatitis among participants in a plasmapheresis stimulation program. J Infect Dis 139:634–640

Hampton RO, Phillips S, Knesek JE, Mink GI (1973) Ultrastructural cytology of pea leaves and roots infected by pea seedborne mosaic virus. Arch Gesamte Virusforsch 42:242–253

Harrison AK, Murphy FA, Gardner JJ (1982) Visceral target organs in systemic St. Louis encephalitis virus infections in hamsters. Exp Mol Pathol 37:292–304

Haugen RK (1979) Hepatitis after the transfusion of frozen red cells. N Engl J Med 301:393–395

He L-F, Alling D, Popkin T, Shapiro M, Alter HJ, Purcell RH (1987) Determining the size of non-A, non-B hepatitis virus by filtration. J Infect Dis 156:636–640

Hellings JA, Van Der Veen-Du Prie J, Snelting-Van Densen R, Stute R (1985) Preliminary results of transmission experiments of non-A, non-B hepatitis by mononuclear leucocytes from a chronic patient. J Virol Methods 10:321–326

Hindman SH, Maynard JE, Bradley DW, Denes AE, Berquist KR (1977) Simultaneous infections with type A and B hepatitis viruses. Am J Epidemiol 105:135–139

Hollinger FB, Gitnick GL, Aach RD, Szmuness W, Mosley JW, Stevens CE, Peters RL, Weiner JM, Werch JB, Lander JJ (1978) Non-A, non-B hepatitis transmission in chimpanzees: a project of the transfusion transmitted viruses study group. Intervirology 10:60–68

Hollinger FB, Mosley JW, Szmuness W, Aach RD, Peters RL, Stevens C (1980) Transfusion-transmitted viruses study: experimental evidence for two non-A, non-B hepatitis agents. J Infect Dis 142:400–407

Hollinger FB, Dolana G, Thomas W, Gyorkey F (1984) Reduction in risk of hepatitis transmission by heat-treatment of a human factor VIII concentrate. J Infect Dis 150:250–262

Hoofnagle JH, Alter HJ (1984) Chronic viral hepatitis. In: Vyas GN, Dienstag JL, Hoofnagle JH (eds) Viral hepatitis and liver disease. Grune and Stratton, New York, pp 97–113

Hooks JJ, Gibbs CJ (1975) The foamy viruses. Bacteriol Rev 39:169–185

Hruby MA, Schauf V (1978) Transfusion-related short-incubation hepatitis in hemophilic patients. JAMA 240:1355–1357

Iwarson S, Schaff Z, Mitchell F, Gerety RJ (1985) Non-A, non-B hepatitis virus. Lancet 1:171

Koretz RL, Suffin SC, Gitnick GL (1976) Post-transfusion chronic liver disease. Gastroenterology 71:797–803

Liaw Y-F, Chu C-M, Chang-Chien C-S, Wu C-S (1982) Simultaneous acute infections with hepatitis non-A, non-B viruses. Dig Dis Sci 27:762–764

Linnen J, Wages J, Zhang_Keck Z-Y, Fry K, Krawczynski K, Alter H, Koonin E, Gallagher M, Alter M, Hadziyannis K, Karayiannis P, Fung K, Nakatsuji Y, Shih W-K, Young L, Piatak M, Hoover C, Fernandez J, Chen S, Zou J-C, Morris T, Hyams K, Ismay S, Lifson J, Hess G, Foung S, Thomas H, Bradley D, Margolis H, Kim J (1996) Molecular cloning and disease association of hepatitis G virus: a transfusion-transmissible agent. Science 271:505–508

Merz PA, Somerville RA, Wisniewski HM, Manuelidis L, Manuelidis EE (1983) Scrapie associated fibrils in Creutzfeldt-Jakob disease. Nature 306:474

Miyamoto H, Okamoto H, Sato K, Tanaka T, Mishiro S (1992) Extraordinarily low density hepatitis C virus estimated by sucrose density gradient centrifugation and the polymerase chain reaction. J Gen Virol 73:715–718

Monath TP, Cropp CB, Harrison AK (1983) Mode of entry of a neurotropic arbovirus into the central nervous system: reinvestigation of an old controversy. Lab Invest 48:399–410

Mosley JW, Redeker AG, Feinstone SM, Purcell RH (1977) Multiple hepatitis viruses in multiple attacks of acute viral hepatitis. N Engl J Med 296:75–78

Norkrans G, Frosner G, Hermodsson S, Iwarson S (1980) Multiple hepatitis attacks in drug addicts. JAMA 243:1056–1058

Pfeifer U, Thomssen R, Legler K, Bottcher Y, Gerlick W, Weinmann E, Klinge O (1980) Experimental non-A,non-B hepatitis: four types of cytoplasmic alteration in hepatocytes of infected chimpanzees. Virchows Arch B Cell Pathol 33:233–243

Prince AM, Brotman B, Grady GF, Kuhns WJ, Hazzi C, Levine RW, Millian SJ (1974) Long-incubation post-transfusion hepatitis without serological evidence of exposure to hepatitis-B virus. Lancet 2:241–246

Prince AM, Huima T, Williams BA, Bardina L, Brotman B (1984) Isolation of a virus from chimpanzee liver cell cultures inoculated with sera containing the agent of non-A, non-B hepatitis. Lancet 2:1071–1075

Prince AM, Stephan W, Dichtelmuller H et al. (1985) Inactivation of the Hutchinson strain of non-A, non-B hepatitis virus by combined use of beta-propiolactone and ultraviolet irradiation. J Med Virol 16:119–125

Prusiner SB (1982) Novel proteinaceous infectious particles cause scrapie. Science 216:136–144

Prusiner SB, Groth DF, Bolton DC, Kent SB, Hood LE (1984) Purification and structural studies of a major scrapie prion protein. Cell 38:127–134

Rakela JH, Redker AG (1979) Chronic liver disease after acute non-A, non-B viral hepatitis. Gastroenterology 77:1200–1202

Reyes GR, Purdy MA, Kim JP, Luk K-C, Young LM, Fry KE, Bradley DW (1990) Isolation of a cDNA from the virus responsible for enterically transmitted non-A, non-B hepatitis. Science 247:1335–1339

Roberts IM, Harrison BD (1970) Inclusion bodies and tubular structures in *Chenopodium amaranticolor* plants infected with strawberry latent ringspot virus. J Gen Virol 7:47–54

Seto B, Iwarson S, Coleman WG, Gerety RJ (1984) Detection of reverse transcriptase activity in association with the non-A, non-B hepatitis agent(s). Lancet 2:941–943

Shamoto M, Murakami S, Zenke T (1981) Adult T-cell leukemia in Japan: an ultrastructural study. Cancer 47:1804–1811

Shimizu YK, Feinstone SM, Purcell RH, Alter HJ, London WT (1979) Non-A, non-B hepatitis: ultrastructural evidence for two agents in experimentally infected chimpanzees. Science 205:197–200

Sidhu GS, Stahl RE, El-Sadr W, Zolla-Payner S (1983) Ultrastructural markers of AIDS. Lancet 1:990–991

Simons JN, Leary TP, Dawson GJ, Pilot-Matias TJ, Merhoff AS, Schlauder G, Desai SM, Mushawar IK (1995) Isolation of novel virus-like sequences associated with human hepatitis. Nat Med 1:564–569

Tabor E, Gerety RJ, Drucker JA, Seeff LB, Hoofnagle JH, Jackson DR, April M, Barker LF, Pineda-Tamondong G (1978) Transmission of non-A, non-B hepatitis from man to chimpanzee. Lancet 1:463–466

Tabor E, Seeff LB, Gerety RJ (1980) Chronic non-A, non-B hepatitis carrier state. N Engl J Med 303:139–143

Tabor E, Gerety RJ (1980) Inactivation of an agent of human non-A, non-B hepatitis by formalin. J Infect Dis 142:767–770

Tateda A, Kikuchi K, Numazaki Y, Shirachi R, Ishida N (1979) Non-A, non-B hepatitis in Japanese recipients of blood transfusions: clinical and serologic studies after the introduction of laboratory screening of donor blood to hepatitis B surface antigen. J Infect Dis 139:511–518

Tsiquaye KN, Zuckerman AJ (1979) New human hepatitis virus. Lancet 1:1135–1136

Tsiquaye KN, Byrd RG, Tovey G, Wyke RJ, Williams R, Zuckerman AJ (1980) Further evidence of cellular changes associated with non-A, non-B hepatitis. J Med Virol 5:63–71

Tsiquaye KN, Portmann B, Tovey G, Kessler H, Shanlian H, Xiao-zhen L, Zuckerman AJ, Craske J, Williams R (1983) Non-A, non-B hepatitis in persistent carriers of hepatitis B virus. J Med Virol 11:79–89

Vidano C (1970) Phases of maize rough dwarf virus multiplication in the vector Laodelphax striatellus. Virology 41:218–232

Wyke RJ, Tsiquaye KN, Thornton A, White Y, Portmann B, Das PK, Zuckerman AJ, Williams R (1979) Transmission of non-A, non-B hepatitis to chimpanzees by factor IX concentrates after fatal complications in patients with chronic liver disease. Lancet 1:520–524

Yoshizawa H, Akahane Y Itoh Y, Iwakiri S, Kitajima K, Morita M, Tanaka A, Nojiri T, Shimizu M, Miyakawa Y, Mayumi M (1980) Virus-like particles in a plasma fraction (fibrinogen) and in the circulation of apparently healthy blood donors capable of inducing non-A, non-B hepatitis in humans and chimpanzees. Gastroenterology 79:512–520

Yoshizawa H, Itoh Y, Iwakiri S, Kitajima K, Tanaka A, Nojiri T, Miyakawa Y, Mayumi M (1981) Demonstration of two different types of non-A, non-B hepatitis by reinjection and cross-challenge studies in chimpanzees. Gastroenterology 81:107–113

Yoshizawa H, Itoh Y, Iwakiri S, Kitajima K, Tanaka A, Tachibana T, Nakamura T, Miyakawa Y, Mayumi M (1982) Non-A, non-B (type 1) hepatitis agent capable of inducing tubular ultrastructures in the hepatocyte cytoplasm of chimpanzees: inactivation by formalin and heat. Gastroenterology 82:502–506

Hepatitis C Epidemiology

D.L. THOMAS

1	Introduction	25
1.1	Biologic Basis	26
1.2	Molecular Tools for Studying Hepatitis C Virus Transmission	26
2	Percutaneous Transmission	27
2.1	Transfusion of Blood Products	27
2.2	Organ Transplantation	28
2.3	Illicit Drug Use	28
2.4	Nosocomial Transmission	29
2.4.1	Patient to Patient	29
2.4.2	Patient to Health Care Worker	30
2.4.3	Health Care Worker to Patient	30
2.5	Miscellaneous Percutaneous Transmission	30
3	Sexual Transmission	31
4	Perinatal Transmission	31
5	Transmission Cofactors	32
6	Worldwide Epidemiology	33
6.1	Worldwide Occurrence	33
6.2	Highly Endemic Regions	35
6.3	The United States	35
7	Summary	36
References		36

1 Introduction

More than 170 million individuals in virtually every area of the world have hepatitis C virus (HCV) infection. HCV is most often transmitted by percutaneous exposure to blood. However, the predominant modes of transmission have changed over time and differ between and even within countries. Before economic development occurs, HCV may be transmitted through folk and traditional medical procedures and other percutaneous practices such as injection drug use, acupuncture, tattooing, and sharing razors. In economically developed countries, most

---

The Johns Hopkins Schools of Medicine and Hygiene and Public Health, 720 Rutland Avenue, Ross 1147, Baltimore, MD 21205, USA

new HCV infections are related to drug use, though blood transfused prior to HCV antibody screening has been an important source of infection. HCV may also be transmitted between sexual partners and from a mother to her infant, though this is uncommon.

1.1 Biologic Basis

Hepatitis C virus transmission requires that infectious virions contact susceptible cells that sustain replication. It is difficult to ascertain which body fluids contain infectious hepatitis C virions. Using sensitive techniques, HCV RNA can be detected in blood (including serum and plasma), saliva, tears, seminal fluid, ascitic fluid, and cerebrospinal fluid (CHEN et al. 1995; FIORE et al. 1995; LIOU et al. 1992; MENDEL et al. 1997; WANG et al. 1992). HCV RNA-containing blood is infectious when administered intravenously, for example, by transfusion or experimental inoculation of chimpanzees. In addition, one chimpanzee was infected by intravenous inoculation of saliva (ABE and INCHAUSPE 1991). However, it is unknown whether these other non-blood body fluids harbor infections virions, both because the experiments have not been performed and because accidental percutaneous exposures to non-blood body fluids are rare.

The second requirement for transmission is contact of infectious virions with a susceptible cell. HCV replication occurs in the hepatocyte and possibly elsewhere. However, it is not precisely known which cells are susceptible to HCV infection and how the virus enters the cell. As mentioned above, if HCV reaches the blood, infection commonly occurs, and transmission through the conjunctiva has been reported (SARTORI et al. 1993). Seminal fluid may contain HCV RNA, but sexual transmission is uncommon. Whether this discrepancy is due to a paucity of infectious virions in seminal fluid or insufficient numbers of susceptible cells in the genital mucosa is unknown.

1.2 Molecular Tools for Studying Hepatitis C Virus Transmission

The nucleotide sequence corresponding to the HCV envelope and some nonstructural proteins is highly variable, and at least six distinct HCV genotypes have been described (BUKH et al. 1995; SIMMONDS et al. 1993). The genetic heterogeneity of HCV strains is sufficiently high that detection of the same or nearly identical nucleotide sequences in two individuals is strong evidence for a common source of infection. For example, RNA sequences in the E1 gene of HCV infected infants have 98%–100% identity with their mothers, but less than 92% with one another (THOMAS et al. 1998). Similar comparisons have been used to demonstrate HCV transmission between sexual partners, within families, among patients, and from a health care worker to patients (ALLANDER et al. 1995; BRONOWICKI et al. 1997; ESTEBAN et al. 1996; THOMAS et al. 1995b). HCV genotype/subtype classification also may be used epidemiologically, but is less specific than nucleotide sequence analysis.

2 Percutaneous Transmission

2.1 Transfusion of Blood Products

When blood from HCV-antibody positive donors is transfused, more than 80% of recipients acquire HCV infection (ESTEBAN et al. 1991; VRIELINK et al. 1995). The high infectivity of direct intravenous administration of a large HCV inoculum also is evident in the high HCV prevalence found among multiply transfused thalassemics and hemophiliacs (Tables 1, 2) (BRETTLER et al. 1990; DE MONTALEMBERT et al. 1992; EYSTER et al. 1993; LAI et al. 1994).

Prior to screening blood donations for HCV antibodies and surrogate markers, approximately 17% of HCV infections in the United States were caused by transfusion (CDC 1991). With screening, the risk of transfusion-transmission of HCV has been reduced substantially (BLAJCHMAN et al. 1995; DONAHUE et al. 1992). Transfusion-transmission may still occur from donors with recent infection who have not yet developed antibodies and possibly from others who lose or never develop HCV antibody (WIDELL et al. 1996). However, this risk is estimated to be less than 1 in 100,000 (SCHREIBER et al. 1996). Accordingly, blood transfusion now causes less than 4% of HCV infections in the United States (CDC 1991).

HCV has been transmitted by intravenous administration of contaminated blood products including immunoglobulin (Ig) and clotting factors, as illustrated in several large outbreaks (BJORO et al. 1994; POWER et al. 1994; YAP et al. 1994). In the early 1970s, 417 recipients of HCV contaminated anti-D Ig were infected in Ireland (POWER et al. 1994). More recently, an outbreak of HCV-infection in the United States was linked with IV administration of an Ig preparation, Gammagard

Table 1. An outline of hepatitis C virus transmission routes

Percutaneous	
Transfusion of blood products	Red blood cell transfusion
	Intravenous immunoglobulin
Illicit drug use	Sharing needles and equipment
	?? Intranasal cocaine
Nosocomial	
Patient-to-patient	Hemodialysis
	Organ donation
	Other patient to patient (colonoscopy, administration of intravenous fluids, unknown)
Patients to health care worker	Needlestick
Health care worker to patient	Surgery
Miscellaneous	Tattoo
	Preventative health campaigns (vaccinations, injection therapy for schistosomiasis)
Sexual	Multiple sexual partners
	Commercial sex workers
Perinatal	Infants born to HCV-positive mothers

Table 2. Individuals at increased risk for hepatitis C virus infection

---
Illicit drug users
Recipients of blood transfusions prior to initiation of blood screening[a]
Individuals with multiple sexual partners
Individuals with HCV-infected sexual partners
Children of HCV-infected mothers
Residents of hyperendemic areas[b]
Individuals with known exposure (e.g., needlestick from an infected patient, hemodialysis patients, and organ transplant recipients)

---

[a] Refers to testing all blood donations for HCV antibody and discarding positive donations, which began in 1991 in the United States.
[b] Areas with a high prevalence of HCV infection include Egypt and some regions in Japan and Italy.

(BRESEE et al. 1996). Current Ig decontamination procedures and recombinant clotting factor use should substantially diminish the risk of further transmission by these products.

2.2 Organ Transplantation

Persons who receive organ transplants due to liver or kidney failure are at increased risk of HCV infection. One reason is that the donor organs themselves may contain HCV. In fact, organs from HCV-infected donors almost always transmit HCV infection to HCV negative recipients (PEREIRA et al. 1991, 1992; TERRAULT and WRIGHT 1995). In addition, reinfection with a second distinct viral strain has been demonstrated in HCV-infected patients receiving organs from HCV infected donors (KONIG et al. 1992).

2.3 Illicit Drug Use

Illicit use of drugs, especially by injection, accounts for the majority of HCV infections in many developed countries. Since 1992, at least two-thirds of new HCV infections in the United States have been due to drug use (ALTER 1997). Worldwide, 50%–95% of persons acknowledging drug use have HCV infection (Fig. 1) (BELL et al. 1990; BOLUMAR et al. 1996; GIRARDI et al. 1990; PATTI et al. 1993; THOMAS et al. 1995a; VAN AMEIJDEN et al. 1993). HCV infection generally occurs within months of initiating injection use of drugs. In one cohort, 80% of subjects acknowledging 2 or more years of injection use were HCV-infected, an incidence that was higher than human immunodeficiency virus (HIV) and even hepatitis B virus (HBV) infections (GARFEIN et al. 1996; THOMAS et al. 1995a). Injection drug users acquire HCV infection by sharing contaminated needles and drug-use equipment, sometimes among groups of persons (shooting galleries). New initiates into drug use are at highest risk for HCV infection if they are "mentored" by an older drug user (GARFEIN et al. 1997).

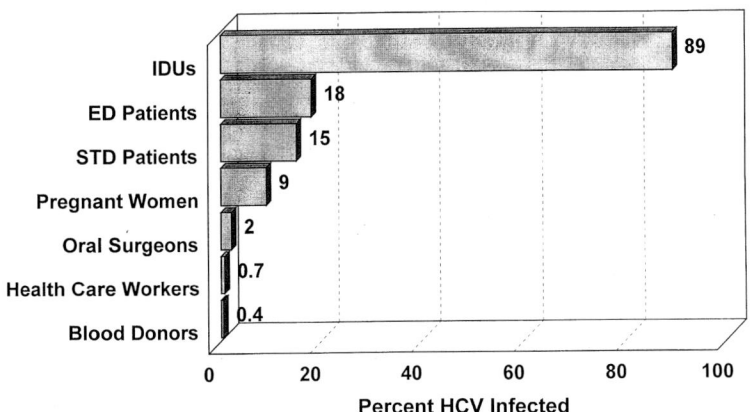

Fig. 1. The prevalence of hepatitis C virus (HCV) in Baltimore, Maryland. *IDUs*, injection drug users; *ED*, emergency department; *STD*, sexually transmitted diseases. HCV prevalence ascertained by detection of HCV antibody. (From KELEN et al. 1992; THOMAS et al. 1993, 1994, 1994, 1995a, 1996)

There is no evidence that unsafe sexual practices contribute substantially to HCV infections among drug users (THOMAS et al. 1995a; VILLANO et al. 1997). HCV infection that occurs in the context of drug use but without acknowledged injection use may be due to other blood exposures (such as sharing straws for intranasal ingestion of cocaine; CONRY-CANTILENA et al. 1996). However, unacknowledged injection use is difficult to exclude.

In some United States cities, there has been a reduction in drug-use-related HCV incidence (ALTER 1997). Although needle-exchange-programs have been associated with a reduction in HCV incidence, their use is probably too restricted to account for these nationwide trends (HAGAN et al. 1995).

2.4 Nosocomial Transmission

2.4.1 Patient to Patient

Patient to patient HCV transmission has been documented by molecular and traditional epidemiologic tools. In one example, two patients acquired HCV infection 8–10 weeks after a colonoscopic procedure, performed with the same colonoscope as was used hours earlier on an HCV-infected patient (BRONOWICKI et al. 1997). HCV isolates from all three patients had high nucleotide identity in a variable HCV genomic segment, essentially proving a common source of infection. Nosocomial HCV transmission also has been suggested by identification of clusters of untransfused patients with similar HCV nucleotide sequences. In one Swedish hematology ward, five clusters of identical or closely related viruses were found. All patients in each cluster had overlapping hospitalizations but not common sources of blood (ALLANDER et al. 1995). Similarly, there is evidence of patient to patient

HCV transmission in several dialysis centers (MUNRO et al. 1996; SAMPIETRO et al. 1995; SCHVARCZ et al. 1997; STUYVER et al. 1996).

2.4.2 Patient to Health Care Worker

Hepatitis C virus transmission occurs after 2%–8% of needle stick exposures to HCV-infected patients (KIYOSAWA et al. 1991; MITSUI et al. 1992; RIDZON et al. 1997). While hollow-bore needle stick exposures account for most transmission from patients to health care workers, HCV infection has also been reported from blood splashed on the conjunctiva, and a solid-bore needle stick (SARTORI et al. 1993). Nonetheless, the occurrence of HCV infection among dental and medical health care workers is commonly less than or similar to the general population, (CAMPELLO et al. 1992; GERBERDING 1994; KUO et al. 1993; POLISH et al. 1993; PURO et al. 1995; THOMAS et al. 1993, 1996).

2.4.3 Health Care Worker to Patient

Hepatitis C virus may be transmitted from health care providers to patients, though this is rare. In one instance, HCV infection was detected in six patients after cardiac surgery (ESTEBAN et al. 1996). Blood donors for these patients were HCV-negative. Five of six patients had a genetically similar, unusual HCV strain which was later found in the surgeon. No infection control breaches were identified. However, percutaneous injuries occurred occasionally when the surgeon tied wires to close the sternum.

In community-based studies, recent receipt or provision of health care is not commonly acknowledged by patients with new HCV infections (ALTER 1997). Thus, nosocomial HCV transmission must be rare in most areas. Breaks in infection control practices have been detected in some instances of nosocomial HCV transmission and are impossible to exclude in others. Strict adherence to these guidelines must be vigorously maintained, especially when mucosal barriers are frequently broken, such as in dialysis units.

2.5 Miscellaneous Percutaneous Transmission

Hepatitis C virus can be transmitted by other percutaneous exposures that occur too infrequently to be detected in many studies. Tattooing has been associated with HCV-infection (KO et al. 1992; SUN et al. 1996). Human bite and folk remedies such as acupuncture and scarification rituals have also been associated with HCV infection (DUSHEIKO et al. 1990) (see Sect. 6.1).

3 Sexual Transmission

Hepatitis C virus is infrequently transmitted by sexual intercourse. HCV RNA has been detected in semen and saliva (FIORE et al. 1995; LIOU et al. 1992; WANG et al. 1992). In addition, high rates of HCV infection have been found in persons with multiple sexual partners and commercial sex workers (NAKASHIMA et al. 1992; PETERSEN et al. 1992; THOMAS et al. 1994, 1995b; UTSUMI et al. 1995; VAN DOORNUM et al. 1991). Acute HCV infection also has been reported in instances where sexual, but not other exposures, are recognized (CAPELLI et al. 1997; HEALEY et al. 1995). However, sexual transmission is difficult to prove since exposures other than intercourse cannot be excluded.

For example, in studies of families of HCV-infected patients, sexual partners are generally the only contacts at increased infection risk, a risk which increases with the duration of the relationship (AKAHANE et al. 1994; CHAYAMA et al. 1995; KAO et al. 1992, 1996). High nucleotide identity is often found in the HCV strains of the sexual partners (CHAYAMA et al. 1995; KAO et al. 1992; PIAZZA et al. 1997; THOMAS et al. 1995b). While sexual transmission could explain these findings, other common exposures such as sharing razors or unacknowledged drug use cannot be ruled out. These cautions also apply to reports of high HCV prevalence rates among persons with multiple sexual partners and commercial sex workers (NAKASHIMA et al. 1992; OSMOND et al. 1993b; THOMAS et al. 1994, 1995b). Studies of long-term sexual partners of HCV-infected hemophiliacs and transfusion recipients generally show little or no HCV transmission, even if there was unprotected sexual intercourse (BRESTERS et al. 1993; BRETTLER et al. 1992; EVERHART et al. 1990; GORDON et al. 1992). HCV prevalence rates among homosexual men are generally lower than for other infections like HIV, HBV, and syphilis, for which sexual transmission is well established (BODSWORTH et al. 1996; DONAHUE et al. 1991; MELBYE et al. 1990; OSMOND et al. 1993a). HIV coinfection seems to increase HCV sexual transmission (EYSTER et al. 1991).

While the risk of transmission attributable to intercourse per se may never be precisely defined, individuals with HCV-positive sexual partners and especially those with multiple partners generally are at increased risk of infection. Individuals in long-term monogamous relationships should be informed of the low (and possibly negligible) risk of future transmission, and may elect not to use barrier precautions.

4 Perinatal Transmission

Hepatitis C virus is uncommonly transmitted from mother to infant. Estimates of the perinatal transmission frequency vary, but range from 0% to 8% in larger studies (LAM et al. 1993; NOVATI et al. 1992; OHTO et al. 1994b; REINUS et al. 1992;

Roudot Thoraval et al. 1993; Wejstal et al. 1992; Zanetti et al. 1995). The timing of transmission is not known. However, HCV RNA can be detected in non-breast-fed infant plasma within 1 month of delivery (Thomas et al. 1998). Because of passive transfer of maternal HCV antibody, infant HCV infection must be diagnosed through detection in infant serum of HCV RNA or HCV antibody after 15 months of age.

HIV coinfection has been associated with more frequent transmission of HCV from mother to infant in some studies (Weintrub et al. 1991; Zanetti et al. 1995). Higher maternal HCV viral load also has been associated with transmission of HCV from mother to infant (Lin et al. 1994; Matsubara et al. 1995; Moriya et al. 1995; Ohto et al. 1994b; Thomas et al. 1998). The effect of maternal HIV on perinatal HCV transmission may be through increasing the HCV viral load (Eyster et al. 1994; Sherman et al. 1993; Telfer et al. 1994; Thomas et al. 1996). Maternal HCV infection also has been associated with increased perinatal HIV infection (Hershow et al. 1997). There have been no well documented cases of HCV transmission from HCV antibody positive but RNA negative mothers to their infants. HCV RNA has been detected in breast milk (Ogasawara et al. 1993), and associations have been made between infant HCV infection and breast feeding (Ohto et al. 1994a,b; Wejstal et al. 1992). However, the risk of HCV transmission from breast feeding has not been substantiated (Lin et al. 1995) and most authorities do not recommend that breast feeding be discontinued because of HCV infection (CDC).

5 Transmission Cofactors (Table 3)

Individuals without detectable HCV RNA in their blood do not appear to be infectious. In a recent review of 2022 parenteral, sexual, and perinatal HCV exposures, HCV transmission only occurred from individuals with HCV RNA detected in blood (Dore et al. 1997). Moreover, nonparenteral HCV transmission is rare even when HCV RNA is detectable, provided the viral load is low (Thomas et al. 1998). HIV coinfection has been reported as a HCV transmission cofactor in some (Eyster et al. 1991; Weintrub et al. 1991; Zanetti et al. 1995), but not all, studies (Lam et al. 1993; Manzini et al. 1995). If an HCV-infected individual acquires HIV infection, the HCV viral load increases (Eyster et al. 1994; Sherman et al. 1993; Telfer et al. 1994; Thomas et al. 1996) and, in some studies, continues to increase as HIV infection progresses (Eyster et al. 1994; Thomas et al. 1996). Thus, the association of HIV coinfection with increased sexual and perinatal HCV transmission may relate to increased HCV viral load. In addition, studies not detecting an effect of HIV infection on HCV transmission may contain subjects with less advanced HIV infection. Anal-receptive intercourse and sexually transmitted diseases are cofactors for HIV transmission (Kingsley et al. 1990; Quinn et al. 1990). HCV infection is found in association with syphilis and some-

Table 3. Factors that may increase hepatitis C virus transmission

Factor	Observation	Comment	Reference
HCV viral load	1. HCV antibody positive but RNA negative persons are not known to transmit HCV infection. 2. Sexual partners and mothers with higher viral load appear to be more likely to transmit HCV.	Persons with false-positive antibody tests and probably those with self-limited infections are not infectious. Transmission is exceedingly rare after exposure to persons with low viral load, but not all exposures to persons with high viral load result in transmission.	DORE et al. 1997 LIN et al. 1994; MATSUBARA et al. 1995; MORIYA et al. 1995; OHTO et al. 1994b; THOMAS et al. 1998; THOMAS et al. 1995b
HIV coinfection	1. Maternal HIV infection increases detection of HCV infection in infant. 2. Infant HIV infection increases detection of HCV infection in infant. 3. Partners of HIV-HCV coinfected persons are more likely to have HCV than if partner has HCV alone.	May be related to HCV viral load. Mothers with more advanced HIV may be more likely to transmit both viruses. May be related to HCV viral load.	WEINTRUB et al. 1991; ZANETTI et al. 1995 THOMAS et al. 1998 EYSTER et al. 1991

times other sexually transmitted diseases (SHEV et al. 1995; THOMAS et al. 1994). However, these correlations appear more related to shared risk factors than enhanced transmission, and HCV infection is uncommon among gay men in most communities.

6 Worldwide Epidemiology

6.1 Worldwide Occurrence

It is estimated that there are more than 170 million persons infected with HCV worldwide (World Health Organization 1997) (Fig. 2). Precise estimates of HCV prevalence are not available in the general population of most countries. However, important general observations can be made.

Through August 1997, 130 countries had reported HCV prevalence rates to the World Health Organization or in the literature in at least one population. HCV infection was found in all but three countries, and it is difficult to imagine that infection would not be found there with further investigation. In developed nations, general population HCV prevalence rates are generally less than 3%, while among volunteer blood donors, they are less than 1%.

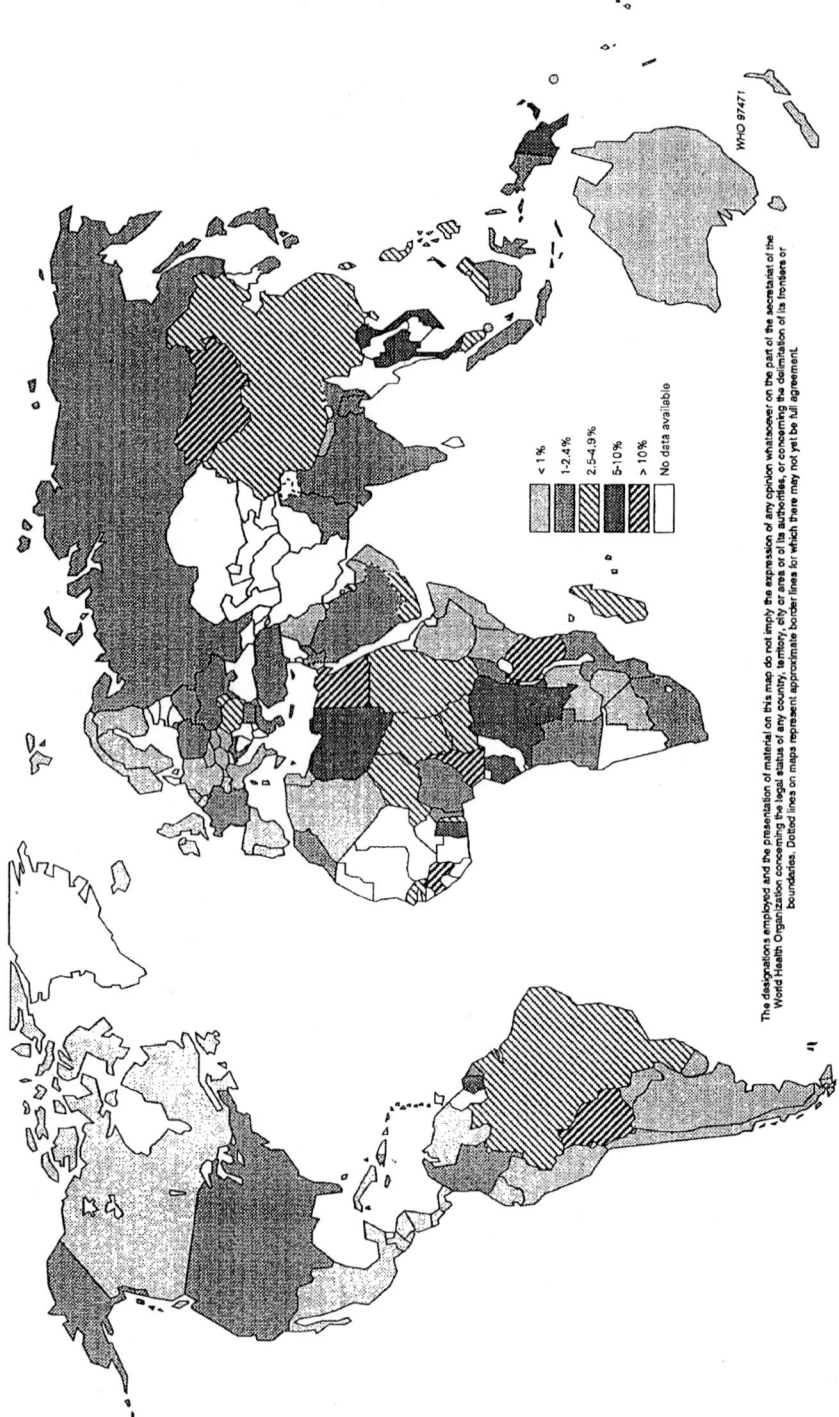

Fig. 2. The worldwide hepatitis C virus (HCV) prevalence. (From World Health Organization 1997)

6.2 Highly Endemic Regions

In a few nations and distinct regions within nations, HCV prevalence rates in the general population exceed 10%. In Egypt, HCV prevalence rates from 10% to 30% have been found (ABDEL-WAHAB et al. 1994; ARTHUR et al. 1997; DARWISH et al. 1993; EL-SAYED et al. 1996; HIBBS et al. 1993; KAMEL et al. 1992). In most highly endemic areas, HCV infection is prevalent among persons over 40 years of age, but uncommon in those less than 20 years of age (CHIARAMONTE et al. 1996; GUADAGNINO et al. 1997; NAKASHIMA et al. 1995; OSELLA et al. 1997). This cohort effect suggests a time-restricted exposure which in many instances appears to have been receipt of a medical procedure. While not yet confirmed, in Egypt it is suspected that a national campaign to treat schistosomiasis infections was responsible. In the 1970s, parenteral anti-schistosomiasis therapies were administered to entire villages, and needles were frequently reused. Similarly, in several areas in Italy and Japan, a high HCV prevalence among older persons is linked to receipt of medical care (CHIARAMONTE et al. 1996; GUADAGNINO et al. 1997; NOGUCHI et al. 1997; OSELLA et al. 1997; PRATI et al. 1997). In the isolated Arahiro region of Japan, 45% of individuals over 41 years of age had HCV infection (KIYOSAWA et al. 1994), whereas, in another area, the same-age prevalence was 2%. Folk remedies such as acupuncture and cutting of skin with non-sterilized knives were identified as likely transmission modes.

High rates of HCV infection also have been reported in urban areas of developed countries. In urban Baltimore, Maryland, HCV infection was found in 18% of patients attending an inner-city emergency department and 15% attending a nearby clinic for sexually transmitted diseases (KELEN et al. 1992; THOMAS et al. 1994). Injection drug use, not medical procedures, is chiefly responsible for transmission in this setting.

6.3 The United States

The epidemiology of HCV infection in the United States has been carefully studied by the Centers for Disease Control. In the 1980s, the yearly incidence of HCV infection was approximately 15/100,000, but since then has declined substantially (ALTER 1995). At least two-thirds of community-acquired HCV infections are related to injection drug use. Injection use of drugs in the 6 months prior to infection is acknowledged by approximately 38% of subjects (ALTER 1997; ALTER et al. 1990). However, non-injection drug use and other indicators of injection use are acknowledged by another 44%. Sexual or household exposure to HCV is detected in approximately 10% of individuals with acute HCV infection, while transfusions, occupational exposures, and other factors are infrequently (<4%) identified.

A precise estimate of the general population prevalence of HCV infection in the United States is available from the *Third National Health and Nutrition Examination*. Overall, there are an estimate 3.9 million individuals with HCV infection in the United States, or 1.8% of the general population (ALTER 1997). HCV prevalence is higher among minorities, persons of lower socioeconomic status, and those more than 40 years of age (ALTER 1997).

7 Summary

Hepatitis C virus infection occurs in all parts of the world. Infection is generally due to percutaneous exposures, though sexual and perinatal transmission may occur. While further study is needed to elucidate the biology of HCV transmission and develop vaccines for prevention, new HCV infections can be reduced by economic development and education regarding blood-borne infections.

References

Abdel-Wahab MF, Zakaria S, Kamel M, Abdel-Khaliq MK, Mabrouk MA, Salama H, Esmat G, Thomas DL, Strickland GT (1994) High seroprevalence of hepatitis C infection among risk groups in Egypt. Am J Trop Med Hyg 51:563–567
Abe K, Inchauspe G (1991) Transmission of hepatitis C by saliva. Lancet 337:248–248
Akahane Y, Kojima M, Sugai Y, Sakamoto M, Miyazaki Y, Tanaka T, Tsuda F, Mishiro S, Okamoto H, Miyakawa Y (1994) Hepatitis C virus infection in spouses of patients with type C chronic liver disease. Ann Intern Med 120:748–752
Allander T, Gruber A, Naghavi M, Beyene A, Söderström T, Björkholm M, Grillner L, Persson MAA (1995) Frequent patient-to-patient transmission of hepatitis C virus in a haematology ward. Lancet 345:603–607
Alter MJ (1995) Epidemiology of hepatitis C in the West. Semin Liver Dis 15:5–14
Alter MJ (1997) Epidemiology of hepatitis C. Hepatology 26:62S–65S
Alter MJ, Hadler SC, Judson FN, Mares A, Alexander WJ, Hu PY, Miller JK, Moyer LA, Fields HA, Bradley DW (1990) Risk factors for acute non-A, non-B hepatitis in the United States and association with hepatitis C virus infection. JAMA 264:2231–2235
Centers for Disease Control and Prevention (1998) Recommendations for prevention and control of hepatitis C virus infection and HCV-related chronic diseases. MMWR 47:1–39
Centers for Disease Control and Prevention (1991) Public Health Service interagency guidelines for screening blood, plasma, organs, tissue and semen for evidence of hepatitis B and C. MMWR 40:1–23
Arthur RR, Hassan NF, Abdallah MY, El-Sharkawy MS, Saad MD, Hackbart BG, Imam IZ (1997) Hepatitis C antibody prevalence in blood donors in different governorates in Egypt. Trans R Soc Trop Med Hyg 91:271–274
Bell J, Batey RG, Farrell GC, Crewe EB, Cunningham AL, Byth K (1990) Hepatitis C virus in intravenous drug users. Med J Aust 153:274–276
Bjoro K, Froland SS, Yun Z, Samdal HH, Haaland T (1994) Hepatitis C infection in patients with primary hypogammaglobulinemia after treatment with contaminated immune globulin. N Engl J Med 331:1607–1611
Blajchman MA, Bull SB, Feinman SV (1995) Post-transfusion hepatitis: Impact of non-A, non-B hepatitis surrogate tests. Lancet 345:21–25
Bodsworth NJ, Cunningham P, Kaldor J, Donovan B (1996) Hepatitis C virus infection in a large cohort of homosexually active men: independent associations with HIV-1 infection and injecting drug use but not sexual behaviour. Genitourin Med 72:118–122
Bolumar F, Hernandez-Aguado I, Ferrer L, Ruiz I, Aviñó M, Rebagliato M (1996) Prevalence of antibodies to hepatitis C in a population of intravenous drug users in Valencia, Spain, 1990–1992. Int J Epidemiol 25:204–209

Bresee JS, Mast EE, Coleman FJ, Baron MJ, Schonberger LB, Alter MJ, Jonas MM, Yu MYW, Renzi PM, Schneider LC (1996) Hepatitis C virus infection associated with administration of intravenous immune globulin – a cohort study. JAMA 276:1563–1567

Bresters D, Mauser-Bunschoten ED, Reesink HW, Roosendaal G, van der Poel CL, Chamuleau RAFM, Jansen PLM, Weegink CJ, Cuypers HTM, Lelie PN, Janden Berg HM (1993) Sexual transmission of hepatitis C. Lancet 342:210–211

Brettler DB, Alter HJ, Dienstag JL, Forsberg AD, Levine PH (1990) Prevalence of hepatitis C virus antibody in a cohort study of hemophilia patients. Blood 76:254–256

Brettler DB, Mannucci PM, Gringeri A, Rasko JE, Forsberg AD, Rumi MG et al. (1992) The low risk of hepatitis C virus transmission among sexual partners of hepatitis C infected hemophilic males: an international, multicenter study. Blood 80:540–543

Bronowicki JP, Venard V, Botté C, Monhoven N, Gastin I, Choné L, Hudziak H, Rhin B, Delanoë C, LeFaou A, Bigard MA, Gaucher P (1997) Patient-to-patient transmission of hepatitis C virus during colonoscopy. N Engl J Med 337:237–240

Bukh J, Miller RH, Purcell RH (1995) Genetic heterogeneity of hepatitis C virus: quasispecies and genotypes. Semin Liver Dis 15:41–63

Campello C, Majori S, Poli A, Pacini P, Nicolardi L, Pini F (1992) Prevalence of HCV antibodies in health-care workers from northern Italy. Infection 20:224–226

Capelli C, Prati D, Bosoni P, Zanuso F, Pappalettera M, Mozzi F, De Mattei C, Zanella A, Sirchia G (1997) Sexual transmission of hepatitis C virus to a repeat blood donor. Transfusion 37:436–440

Chayama K, Kobayashi M, Tsubota A, Koida I, Arase Y, Saitoh S, Ikeda K, Kumada H (1995) Molecular analysis of intraspousal transmission of hepatitis C virus. J Hepatol 22:431–439

Chen M, Yun Z-B, Sällberg M, Schvarcz R, Bergquist I, Berglund H-B, Sönnerborg A (1995) Detection of hepatitis C virus RNA in the cell fraction of saliva before and after oral surgery. J Med Virol 45:223–226

Chiaramonte M, Stroffolini T, Lorenzoni U, Minniti F, Conti S, Floreani A, Ntakirutimana E, Vian A, Ngatchu T, Naccarato R (1996) Risk factors in community-acquired chronic hepatitis C virus infection: a case-control study in Italy. J Hepatol 24:129–134

Conry-Cantilena C, Vanraden MT, Gibble J, Melpolder J, Shakil AO, Viladomiu L, Cheung L, Di Bisceglie A, Hoofnagle JH, Shih JW, Kaslow R, Ness PM, Alter HJ (1996) Routes of infection, viremia, and liver disease in blood donors found to have hepatitis C virus infection. N Engl J Med 334:1691–1696

Darwish MA, Raouf TA, Rushdy P, Constantine NT, Rao MR, Edelman R (1993) Risk factors associated with a high seroprevalence of hepatitis C virus infection in Egyptian blood donors. Am J Trop Med Hyg 49:440–447

de Montalembert M, Costagliola DG, Lefrere JJ, Cornu G, Lombardo T, Cosentino S, Perrimond H, Girot R (1992) Prevalence of markers for human immunodeficiency virus types 1 and 2, human T-lymphotropic virus type I, cytomegalovirus, and hepatitis B and C virus in multiply transfused thalassemia patients. The French Study Group On Thalassaemia. Transfusion 32:509–512

Donahue JG, Munoz A, Ness PM, Brown DE, Yawn DH, McAllister HA, Reitz BA, Nelson KE (1992) The declining risk of post-transfusion hepatitis C virus infection. N Engl J Med 327:369–373

Donahue JG, Nelson KE, Munoz A, Vlahov D, Rennie LL, Taylor EL, Saah AJ, Cohn S, Odaka NJ, Farzadegan H (1991) Antibody to hepatitis C virus among cardiac surgery patients, homosexual men, and intravenous drug users in Baltimore, Maryland. Am J Epidemiol 134:1206–1211

Dore GJ, Kaldor JM, McCaughan GW (1997) Systematic review of role of polymerase chain reaction in defining infectiousness among people infected with hepatitis C virus. Br Med J 315:333–337

Dusheiko GM, Smith M, Scheuer PJ (1990) Hepatitis C virus transmission by human bite. Lancet 336:503–504

El-Sayed NM, Gomatos PJ, Rodier GR, Wierzba TF, Darwish A, Khashaba S, Arthur RR (1996) Seroprevalence survey of Egyptian tourism workers for hepatitis B virus, hepatitis C virus, human immunodeficiency virus, and Treponema pallidum infections: association of hepatitis C virus infections with specific regions of Egypt. Am J Trop Med Hyg 55:179–184

Esteban JI, Gómez J, Martell M, Cabot B, Quer J, Camps J, González A, Otero T, Moya A, Esteban R, Guardia J (1996) Transmission of hepatitis C virus by a cardiac surgeon. N Engl J Med 334:555–560

Esteban JI, Lopez-Talavera JC, Genesca J, Madoz P, Viladomiu L, Muniz E, Martin-Vega C, Rosell M, Allende H (1991) High rate of infectivity and liver disease in blood donors with antibodies to hepatitis C virus. Ann Intern Med 115:443–449

Everhart JE, Di Bisceglie AM, Murray LM, Alter HJ, Melpolder JJ, Kuo G, Hoofnagle JH (1990) Risk for non-A, non-B (Type C) hepatitis through sexual or household contact with chronic carriers. Ann Intern Med 112:544–555

Eyster ME, Alter HJ, Aledort LM, Quan S, Hatzakis A, Goedert JJ (1991) Heterosexual co-transmission of hepatitis C virus (HCV) and human immunodeficiency virus (HIV). Ann Intern Med 115:764–768

Eyster ME, Diamondstone LS, Lien JM, Ehmann WC, Quan S, Goedert JJ (1993) Natural history of hepatitis C virus infection in multitransfused hemophiliacs: effect of coinfection with human immunodeficiency virus. The Multicenter Hemophilia Cohort Study. J Acquir Immune Defic Syndr 6:602–610

Eyster ME, Fried MW, Di Bisceglie AM, Goedert JJ (1994) Increasing hepatitis C virus RNA levels in hemophiliacs: relationship to human immunodeficiency virus infection and liver disease. Blood 84:1020–1023

Fiore RJ, Potenza D, Monno L, Appice A, DiStefano M, Giannelli A, LaGrasta L, Romanelli C, DiBari C, Pastore G (1995) Detection of HCV RNA in serum and seminal fluid from HIV-1 co-infected intravenous drug addicts. J Med Virol 46:364–367

Garfein RS, Doherty MC, Brown D, Thomas DL, Villano SA, Monterroso E, Vlahov D (1998) Hepatitis C virus infection among short-term injection drug users. J Acquir Immune Defic Syndr 18:S11–S19

Garfein RS, Vlahov D, Galai N, Doherty MC, Nelson KE (1996) Viral infections in short-term injection drug users: the prevalence of the hepatitis C, hepatitis B, human immunodeficiency, and human T-lymphotropic viruses. Am J Publ Health 86:655–661

Gerberding JL (1994) Incidence and prevalence of human immunodeficiency virus, hepatitis B virus, hepatitis C virus, and cytomegalovirus among health care personnel at risk for blood exposure: final report from a longitudinal study. J Infect Dis 170:1410–1417

Girardi E, Zaccarelli M, Tossini G, Puro V, Narciso P, Visco G (1990) Hepatitis C virus infection in intravenous drug users: prevalence and risk factors. Scand J Infect Dis 22:751–752

Gordon SC, Patel AH, Kulesza GW, Barnes RE, Silverman AL (1992) Lack of evidence for the heterosexual transmission of hepatitis C. Am J Gastroenterol 87:1849–1851

Guadagnino V, Stroffolini T, Rapicetta M, Costantino A, Kondili LA, Menniti-Ippolito F, Caroleo B, Costa C, Griffo G, Loiacono L, Pisani V, Foca A, Piazza M (1997) Prevalence, risk factors, and genotype distribution of hepatitis C virus infection in the general population: a community-based survey in southern Italy. Hepatology 26:1006–1011

Hagan H, Jarlais DCD, Friedman SR, Purchase D, Alter MJ (1995) Reduced risk of hepatitis B and hepatitis C among injection drug users in the Tacoma syringe exchange program. Am J Publ Health 85:1531–1537

Healey CJ, Smith DB, Walker JL, Holmes EC, Fleming KA, Chapman RWG, Simmonds P (1995) Acute hepatitis C infection after sexual exposure. Gut 36:148–150

Hershow RC, Riester KA, Lew J, Quinn TC, Mofenson L, Davenny K, Landesman S, Cotton D, Hanson C, Hillyer GV, Tang HB, Thomas DL (1997) Increased vertical transmission of human immunodeficiency virus from hepatitis C coinfected mothers. J Infect Dis 176:414–420

Hibbs RG, Corwin AL, Hassan NF, Kamel M, Darwish M, Edelman R, Constantine NT, Rao MR, Khalifa AS, Mokhtar S et al. (1993) The epidemiology of antibody to hepatitis C in Egypt. J Infect Dis 168:789–790

Kamel MA, Ghaffar YA, Wasef MA, Wright M, Clark LC, Miller FD (1992) High HCV prevalence in Egyptian blood donors. Lancet 340:427–427

Kao JH, Chen PJ, Yang PM, Lai MY, Sheu JC, Wang TH, Chen DS (1992) Intrafamilial transmission of hepatitis C virus: the important role of infections between spouses. J Infect Dis 166:900–903

Kao JH, Hwang YT, Chen PJ, Yang PM, Lai MY, Wang TH, Chen DS (1996) Transmission of hepatitis C virus between spouses: the important role of exposure duration. Am J Gastroenterol 91:2087–2090

Kelen GD, Green GB, Purcell RH, Chan DW, Qaqish BF, Sivertson KT, Quinn TC (1992) Hepatitis B and hepatitis C in emergency department patients. N Engl J Med 326:1399–1404

Kingsley LA, Rinaldo CR Jr, Lyter DW, Valdiserri RO, Belle SH, Ho M (1990) Sexual transmission efficiency of hepatitis B virus and human immunodeficiency virus among homosexual men. JAMA 264:230–234

Kiyosawa K, Sodeyama T, Tanaka E, Nakano Y, Furuta S, Nishioka K, Purcell RH, Alter HJ (1991) Hepatitis C in hospital employees with needlestick injuries. Ann Intern Med 115:367–369

Kiyosawa K, Tanaka E, Sodeyama T, Yoshizawa K, Yabu K, Furuta K, Imai H, Nakano Y, Usuda S, Uemra K, Furuta S, Watanabe Y, Watanabe J, Fukuda Y, Takayama T (1994) Transmission of hepatitis C in an isolated area in Japan: community-acquired infection. Gastroenterology 106:1596–1602

Ko YC, Ho MS, Chiang TA, Chang SJ, Chang PY (1992) Tattooing as a risk of hepatitis C virus infection. J Med Virol 38:288–291

Konig V, Bauditz J, Lobeck H, Lusebrink R, Neuhaus P, Blumhardt G, Bechstein WO, Neuhaus R, Steffen R, Hopf U (1992) Hepatitis C virus reinfection in allografts after orthotopic liver transplantation. Hepatology 16:1137–1143

Kuo MY, Hahn LJ, Hong CY, Kao JH, Chen DS (1993) Low prevalence of hepatitis C virus infection among dentists in Taiwan. J Med Virol 40:10–13

Lai ME, Mazzoleni AP, Argiolu F, De Virgilis S, Balestrieri A, Purcell, RH, Cao A, Farci P (1994) Hepatitis C virus in multiple episodes of acute hepatitis in polytransfused thalassaemic children. Lancet 343:388–390

Lam JPH, McOmish F, Burns SM, Yap PL, Mok JYQ, Simmonds P (1993) Infrequent vertical transmission of hepatitis C virus. J Infect Dis 167:572–576

Lin H-H, Kao J-H, Hsu H-Y, Ni Y-H, Chang M-H, Huang S-C, Hwang L-H, Chen P-J, Chen D-S (1995) Absence of infection in breast-fed infants born to hepatitis C virus-infected mothers. J Pediatr 126:589–591

Lin H-H, Kao J-H, Hsu H-Y, Ni Y-H, Yeh S-H, Hwang L-H, Chang M-H, Huang S-C, Chen P-J, Chen D-S (1994) Possible role of high-titer maternal viremia in perinatal transmission of hepatitis C virus. J Infect Dis 169:638–641

Liou TC, Chang TT, Young KC, Lin XZ, Lin CY, Wu HL (1992) Detection of HCV RNA in saliva, urine, seminal fluid, and ascites. J Med Virol 37:197–202

Manzini P, Saracco G, Cerchier A, Riva C, Musso A, Ricotti E, Palomba E, Scolfaro C, Verme G, Bonino F, Tovo PA (1995) Human immunodeficiency virus infection as risk factor for mother-to-child hepatitis C virus transmission; Persistence of anti-hepatitis C virus in children is associated with the mother's anti-hepatitis C virus immunoblotting pattern. Hepatology 21:328–332

Matsubara T, Sumazaki R, Takita H (1995) Mother-to-infant transmission of hepatitis C virus: a prospective study. Eur J Pediatr 154:973–978

Melbye M, Biggar RJ, Wantzin P, Krogsgaard K, Ebbesen P, Becker NG (1990) Sexual transmission of hepatitis C virus: cohort study (1981–9) among European homosexual men. BMJ 301:210–212

Mendel I, Muraine M, Riachi G, El Forzli F, Bertin C, Colin R, Brasseur G, Buffet-Janvresse C (1997) Detection and genotyping of the hepatitis C RNA in tear fluid from patients with chronic hepatitis C. J Med Virol 51:231–233

Mitsui T, Iwano K, Masuko K, Yamazaki C, Okamoto H, Tsuda F, Tanaka T, Mishiro S (1992) Hepatitis C virus infection in medical personnel after needlestick accident. Hepatology 16:1109–1114

Moriya T, Sasaki F, Mizui M, Ohno N, Mohri H, Mishiro S, Yoshizawa H (1995) Transmission of hepatitis C virus from mothers to infants: its frequency and risk factors revisited. Biomed Pharmacother 49:59–64

Munro J, Biggs JD, McCruden EAB (1996) Detection of a cluster of hepatitis C infections in a renal transplant unit by analysis of sequence variation of the NS5a gene. J Infect Dis 174:177–180

Nakashima K, Ikematsu H, Hayashi J, Kishihara Y, Mitsutake A, Kashiwagi S (1995) Intrafamilial transmission of hepatitis C virus among the population of an endemic area of Japan. JAMA 274:1459–1461

Nakashima K, Kashiwagi S, Hayashi J, Noguchi A, Hirata M, Kajiyama W et al (1992) Sexual transmission of hepatitis C virus among female prostitutes and patients with sexually transmitted diseases in Fukuoka, Kyushu, Japan. Am J Epidemiol 136:1132–1137

Noguchi S, Sata M, Suzuki H, Mizokami M, Tanikawa K (1997) Routes of transmission of hepatitis C virus in an endemic rural area of Japan – Molecular epidemiologic study of hepatitis C virus infection. Scand J Infect Dis 29:23–28

Novati R, Thiers V, Monforte AD, Maisonneuve P, Principi N, Conti M, Lazzarin A, Brechot C (1992) Mother-to-child transmission of hepatitis C virus detected by nested polymerase chain reaction. J Infect Dis 165:720–723

Ogasawara S, Kage M, Kosai K, Shimamatsu K, Kojiro M (1993) Hepatitis C virus RNA in saliva and breast milk of hepatitis C carrier mothers. Lancet 341:561–561

Ohto H, Okamoto H, Mishiro S (1994a) Vertical transmission of hepatitis C virus. Reply. N Engl J Med 331:400

Ohto H, Terazawa S, Nobuhiko S, Sasaki N, Hino K, Ishiwata C, Kako M, Ujiie N, Endo C, Matsui A, Okamoto H, Mishiro S (1994b) Transmission of hepatitis C virus from mothers to infants. N Engl J Infect Dis 330:744–750

Osella AR, Misciagna G, Leone A, Di Leo A, Fiore G (1997) Epidemiology of hepatitis C virus infection in an area of southern Italy. J Hepatol 27:30–35

Osmond DH, Charlebois E, Sheppard HW, Page K, Winkelstein W, Moss AR, Reingold A (1993a) Comparison of risk factors for hepatitis C and hepatitis B virus infection in homosexual men. J Infect Dis 167:66–71

Osmond DH, Padian NS, Sheppard HW, Glass S, Shiboski SC, Reingold A (1993b) Risk factors for hepatitis C virus seropositivity in heterosexual couples. JAMA 269:361–365

Patti AM, Santi AL, Pompa MG, Giustini C, Vescia N, Mastroeni I, Fara GM (1993) Viral hepatitis and drugs: a continuing problem. Int J Epidemiol 22:135–139

Pereira BJG, Milford EL, Kirkman RL, Levey AS (1991) Transmission of hepatitis C virus by organ transplantation. N Engl J Med 325:454–460

Pereira BJG, Milford EL, Kirkman RL, Quan S, Sayre KR, Johnson PJ, Wilber JC, Levey AS (1992) Prevelane of hepatitis C virus RNA in organ donors positive for hepatitis C antibody and in the recipients of their organs. N Engl J Med 327:910–915

Petersen EE, Clemens R, Bock HL, Friese K, Hess G (1992) Hepatitis B and C in heterosexual patients with various sexually transmitted diseases. Infection 20:128–131

Piazza M, Sagliocca L, Tosone G, Guadagnino V, Stazi MA, Orlando R, Borgia G, Rosa D, Abrignani S, Palumbo F, Manzin A, Clementi M (1997) Sexual transmission of the hepatitis C virus and efficacy of prophylaxis with intramuscular immune serum globulin – a randomized controlled trial. Arch Intern Med 157:1537–1544

Polish LB, Tong MJ, Co RL, Coleman PJ, Alter MJ (1993) Risk factors for hepatitis C virus infection among health care personnel in a community hospital. Am J Infect Control 21:196–200

Power JP, Lawlor E, Davidson F (1994) Hepatitis C viremia in recipients of Irish intravenous anti-D immunoglobulin. Lancet 344:1166–1167

Prati D, Capelli C, Silvani C, De Mattei C, Bosoni P, Pappalettera M, Mozzi F, Colombo M, Zanella A, Sirchia G (1997) The incidence and risk factors of community-acquired hepatitis C in a cohort of Italian blood donors. Hepatology 25:702–704

Puro V, Petrosillo N, Ippolito G, Aloisi MS, Boumis E, Ravá L, Arici C, Marchesi D, Bonaventura ME, Chiaretti B, Di Nardo V, Chiodera A, Cristini G, Corradi MP, Daglio M, Vlacos D, De Gennaro M, Angarano G, De Sanctis C, Ranchino M, Stuto A, Desperati M, Francavilla E, Cadrobbi P (1995) Occupational hepatitis C virus infection in Italian health care workers. Am J Publ Health 85:1272–1275

Quinn TC, Cannon RO, Glasser D, Groseclose S, Brathwaite WS, Fauci AS, Hook EWI (1990) The association of syphilis with risk of human immunodeficiency virus infection in patients attending sexually transmitted diseases clinics. Arch Intern Med 150:1297–1302

Reinus JF, Leikin EL, Alter HJ, Cheung L, Shindo M, Jett B, Piazza S, Shih W-K (1992) Failure to detect vertical transmission of hepatitis C virus. Ann Intern Med 117:881–886

Ridzon R, Gallagher K, Ciesielski C, Mast EE, Ginsberg MB, Robertson BJ, Luo CC, DeMaria A Jr (1997) Simultaneous transmission of human immunodeficiency virus and hepatitis C virus from a needle-stick injury. N Engl J Med 336:919–922

Roudot Thoraval F, Pawlotsky JM, Thiers V, Deforges L, Girollet PP, Guillot F, Huraux C, Aumont P, Brechot C, Dhumeaux D (1993) Lack of mother-to-infant transmission of hepatitis C virus in human immunodeficiency virus-seronegative women: a prospective study with hepatitis C virus RNA testing. Hepatology 17:772–777

Sampietro M, Badalamenti S, Salvadori S, Corbetta N, Graziani G, Como G, Fiorelli G, Ponticelli C (1995) High prevalence of a rare hepatitis C virus in patients treated in the same hemodialysis unit: evidence for nosocomial transmission of HCV. Kidney Int 47:911–917

Sartori M, La Terra G, Aglietta M, Manzin A, Navino C, Verzetti G (1993) Transmission of hepatitis C via blood splash into conjunctiva. Scand J Infect Dis 25:270–271

Schreiber GB, Busch MP, Kleinman SH, Korelitz JJ (1996) The risk of transfusion-transmitted viral infections. The Retrovirus Epidemiology Donor Study. N Engl J Med 334:1685–1690

Schvarcz R, Johansson B, Nyström B, Sönnerborg A (1997) Nosocomial transmission of hepatitis C virus. Infection 25:74–77

Sherman KE, O'Brien J, Gutierrez G, Harrison S, Urdea M, Neuwald P, Wilber J (1993) Quantitative evaluation of hepatitis C virus RNA in patients with concurrent human immunodeficiency virus infections. J Clin Microbiol 31:2679–2682

Shev S, Widell A, Bergström T, Hermodsson S, Lindholm A, Norkrans G (1995) Herpes simplex virus-2 may increase susceptibility of the sexual transmission of hepatitis C. Sexually Transmitted Dis 22:210–216

Simmonds P, Holmes EC, Cha T-A, Chan S-W, McOmish F, Irvine B, Beall E, Yap PL, Kolberg J, Urdea MS (1993) Classification of hepatitis C virus into six major genotypes and a series of subtypes by phylogenetic analysis of the NS-5 region. J Gen Virol 74:2391–2399

Stuyver L, Claeys H, Wyseur A, Van Arnhem W, De Beenhouwer H, Uytendaele S, Beckers J, Matthijs D, Leroux-Roels G, Maertens G, De Paepe M (1996) Hepatitis C virus in a hemodialysis unit: molecular evidence for nosocomial transmission. Kidney Int 49:889–895

Sun DX, Zhang FG, Geng YQ, Xi DS (1996) Hepatitis C transmission by cosmetic tattooing in women. Lancet 347:541–541

Telfer PT, Brown D, Devereux H, Lee CA, Dusheiko GM (1994) HCV RNA levels and HIV infection: evidence for a viral interaction in haemophilic patients. Br J Haematol 88:397–399

Terrault NA, Wright TL (1995) Hepatitis C virus in the setting of transplantation. Semin Liver Dis 15: 92–100

Thomas DL, Cannon RO, Shapiro C, Hook EWI, Alter MJ, Quinn TC (1994) Hepatitis C, hepatitis B, and human immunodeficiency virus infections among non-intravenous drug-using patients attending clinics for sexually transmitted diseases. J Infect Dis 169:990–995

Thomas DL, Factor S, Kelen G, Washington AS, Taylor E, Quinn TQ (1993) Hepatitis B and C in health care workers at the Johns Hopkins Hospital. Arch Intern Med 153:1705–1712

Thomas DL, Gruninger SE, Siew C, Joy ED, Quinn TC (1996) Occupational risk of hepatitis C infections among general dentists and oral surgeons in North America. Am J Med 100:41–45

Thomas DL, Nelson KE, Quinn TC (1994) The course of non-A, non-B hepatitis unrelated to transfusion (letter). Ann Intern Med 120:171–171

Thomas DL, Shih JW, Alter HJ, Vlahov D, Cohn S, Hoover DR, Cheung L, Nelson KE (1996) Effect of human immunodeficiency virus on hepatitis C virus infection among injecting drug users. J Infect Dis 174:690–695

Thomas DL, Villano SA, Reister K, Hershow R, Mofenson LM, Landesman SH, Hollinger FB, Davenny K, Riley L, Diaz C, Tang HB, Quinn TC (1998) Association of high viral load with perinatal transmission of hepatitis C virus: a study of mothers coinfected with human immunodeficiency virus. J Infect Dis (in press)

Thomas DL, Vlahov D, Solomon L, Cohn S, Taylor E, Garfein R, Nelson KE (1995a) Correlates of hepatitis C virus infections among injection drug users in Baltimore. Medicine 74:212–220

Thomas DL, Zenilman JZ, Alter HJ, Shih JW, Galai N, Quinn TC (1995b) Sexual transmission of hepatitis C virus among patients attending Baltimore sexually transmitted diseases clinics – an analysis of 309 sexual partnerships. J Infect Dis 171:768–775

Utsumi T, Hashimoto E, Okumura Y, Takayanagi M, Nishikawa H, Kigawa M, Kumakura N, Toyokawa H (1995) Heterosexual activity as a risk factor for the transmission of hepatitis C virus. J Med Virol 46:122–125

Van Ameijden EJ, van den Hoek JA, Mientjes GH, Coutinho RA (1993) A longitudinal study on the incidence and transmission patterns of HIV, HBV and HCV infection among drug users in Amsterdam. Eur J Epidemiol 9:255–262

van Doornum GJJ, Hooykaas C, Cuypers MT, van Der Lind MMD, Coutinho RS (1991) Prevalence of hepatitis C virus infections among heterosexuals with multiple partners. J Med Virol 35:22–27

Villano SA, Vlahov D, Nelson KE, Lyles CM, Cohn S, Thomas DL (1997) Incidence and risk factors for hepatitis C among injection drug users in Baltimore, Maryland. J Clin Micro 35:3274–3277

Vrielink H, van der Poel CL, Reesink HW, Zaaijer HL, Scholten E, Kremer LCM, Cuypers HTM, Lelie PN, Van Oers MHJ (1995) Look-back study of infectivity of anti-HCV ELISA-positive blood components. Lancet 345:95–96

Wang JT, Wang TH, Sheu JC, Lin JT, Chen DS (1992) Hepatitis C virus RNA in saliva of patients with posttransfusion hepatitis and low efficiency of transmission among spouses. J Med Virol 36:28–31

Weintrub PS, Veereman Wauters G, Cowan MJ, Thaler MM (1991) Hepatitis C virus infection in infants whose mothers took street drugs intravenously. J Pediatr 119:869–869

Wejstal R, Widell A, Mansson A-S, Hermodsson S, Norkrans G (1992) Mother-to-infant transmission of hepatitis C virus. Ann Intern Med 117:887–890

Widell A, Elmud H, Persson MH, Jonsson M (1996) Transmission of hepatitis C via both erythrocyte and platelet transfusions from a single donor in serological window-phase of hepatitis C. Vox Sanguinis 71:55–57

World Health Organization (1997) Hepatitis C: global prevalence. Weekly Epidemiol Rec 341–348

Yap PL, McOmish F, Webster ADB, Hammarstrom L, Smith CIE, Bjorkander J, Ochs HD, Fischer SH, Quinti I, Simmonds P (1994) Hepatitis C virus transmission by intravenous immunoglobulin. J Hepatol 21:455–460

Zanetti AR, Tanzi E, Paccagnini S, Principi N, Pizzocolo G, Caccamo ML, D'Amico E, Cambiè G, Vecchi L (1995) Lombardy Grp Vertic HCV Transmission. Mother-to-infant transmission of hepatitis C virus. Lancet 345:28–291

Natural History and Disease Manifestations of Hepatitis C Infection

D. THEODORE and M.W. FRIED

1 Introduction	43
2 Acute Infection	44
3 Chronic Infection	45
4 Natural History	47
5 Genotype and Disease Severity	48
6 Therapy of Hepatitis C	50
References	51

1 Introduction

Approximately 4 million persons in the United States are currently infected with the hepatitis C virus (HCV). The Centers for Disease Control and Prevention has estimated that during the 1980s, an average of 230,000 new infections occurred each year (Centers for Disease Control and Prevention 1998). Data from 1996 suggest that the annual incidence has declined to 36,000 new cases (ALTER 1997a). While changes in blood donation practices and screening of blood products with highly sensitive assays undoubtedly resulted in a significant reduction of new cases, the major change in incident cases comes from a reduction of cases in injecting-drug users (Centers for Disease Control and Prevention 1998). To better understand the rationale for treatment of this disease, an understanding of the natural history of HCV infection is imperative.

Research into the natural history of HCV infection has been hindered by several factors. First, most patients with the acute disease have few if any overt clinical signs or symptoms. Those patients who do present with obvious manifestations may represent a different subgroup. A second hurdle in defining the natural history of the disease results from an inability to identify the full spectrum of

---

Division of Digestive Diseases, University of North Carolina at Chapel Hill, Chapel Hill, NC 27514, USA

disease and thus our current knowledge undoubtedly suffers from spectrum bias. Perhaps the major encumbrance is the long latency period between infection and liver-related complications. Patients may not suffer any untoward effects for several decades, making longitudinal studies a daunting undertaking. Finally, the natural history of the disease has probably been changed in the group of patients who have received treatment.

2 Acute Infection

Hepatitis C virus is a single stranded, positive-sense RNA virus (LEMON and BROWN 1995). The transmission of this virus occurs by several routes, most often by direct percutaneous exposure through sharing needles among injecting drug users or as a recipient of contaminated blood. Sexual transmission of the virus appears to be very inefficient, as is maternal-infant transmission. As shown in Table 1 the prevalence of infection differs by risk group.

HCV accounts for about 20% of acute cases of hepatitis in the US (ALTER and MAST 1994), yet acute HCV infection is subclinical or mild in 65%–75% of patients. Roughly 20% of patients have jaundice and 10%–20% complain of malaise, anorexia, abdominal pain, or non-specific flu-like symptoms (Centers for Disease Control and Prevention 1998). Fulminant hepatic failure is reportedly rare. The incubation period averages about 50 days, but may be as short as 15 days and as long as 150 days (HSU et al. 1995). Evidence of HCV infection may be obtained by serologic tests such as commercial anti-HCV testing that encompasses antibodies to several structural and non-structural viral proteins: core, NS3, NS4 and NS5. The presence of such antibodies signifies infection and not immunity. The presence of

Table 1. Prevalence of HCV infection by risk group

Risk group	Prevalence of HCV (range)
Persons with large or repeated direct percutaneous exposure, e.g., hemophiliacs and injection drug users	60%–90%
Persons with smaller but repeated direct or inapparent percutaneous exposures, e.g., hemodialysis patients	10%–30%
Persons with inapparent parenteral or mucosal exposures	
Individuals with high risk sexual behaviors	5% (1%–18%)
Sexual contacts of persons with HCV	5% (0%–15%)
Household contacts of persons with HCV	4% (0%–11%)
Sporadic percutaneous exposures, e.g., health care workers	1%–2%
Blood donors	0.3%
Perinatal transmission	5% (0%–25%)
Perinatal transmission (mother HIV positive)	14% (5%–36%)

From Alter (1997b).

viremia and active infection can only be made by tests that detect viral nucleic acid, such as polymerase chain reaction. New tests for antibody to E2 look promising, and appear to correlate with serum RNA positivity. In an acute infection, the molecular techniques will be positive before the antibody tests.

In the acute phase HCV infection is characterized by a transient elevation in serum alanine aminotransferase (ALT) levels, followed shortly by symptoms should they occur. ALT levels ten times the upper limit of normal are observed in greater than 80% of affected individuals (HOOFNAGLE 1997). As with other hepatitis viruses illness generally lasts for 2–12 weeks.

3 Chronic Infection

Following acute infection it is estimated that only about 15%–20% of patients will clear the virus. Those who do clear the virus may lose serum antibody to HCV. In the US, the seroprevalence of anti-HCV is 1.8% as measured by NHANES III (1988–1994) which evaluated many different health measures from a representative population in the United States (Centers for Disease Control and Prevention 1998). HCV accounts for about 20% of acute cases of hepatitis in the US, but is by far the leading cause of chronic viral hepatitis. This is due to its unique ability to cause persistent infection in >75% of infected individuals. The prevalence of hepatitis C also varies around the world with one of the highest rates found in the Egyptian population (TIBBS 1997). The specific mechanisms underlying viral persistence are not known, but it appears that immunocompromised hosts are no more likely than healthy adults to develop persistent infection. A discussion of the immunopathogenesis of hepatitis C infection appears elsewhere in this volume.

Infection tends to be insidious and subclinical, most often discovered during the evaluation of abnormal liver function tests obtained on routine exams. Symptoms are commonly absent, minimal, intermittent or non-specific. The most common symptom reported is fatigue. Indeed, it may be difficult to distinguish between HCV-related fatigue and other more prevalent causes of fatigue. Some patients with more severe or advanced disease suffer from abdominal pain, anorexia, pruritus, weight loss, nausea and dark urine. Jaundice is not seen unless patients have advanced cirrhosis. Hoofnagle and colleagues published data on the frequency of symptoms suggestive of chronic hepatitis among a cohort of patients with chronic hepatitis C compared to a matched cohort of healthy blood donors (HOOFNAGLE 1997). Patients with chronic hepatitis C were without clinical evidence of cirrhosis. Hepatitis C was confirmed by the presence of antibody to HCV and detectable HCV RNA in serum. Symptoms were obtained by a self-administered questionnaire. Symptoms included in the survey were fatigue, pruritus, nausea, abdominal pain, anorexia and dark urine.

Overall, 70% of patients with chronic hepatitis C reported at least one of the aforementioned symptoms. However, this was not statistically different from the frequency of reported symptoms in healthy blood donors. Fatigue was reported most commonly in both cohorts – 62% of patients and 70% of blood donors. Nausea was present in 19% of patients and 16% of controls. Not unexpectedly, pruritus, abdominal pain and dark urine were more frequently reported among patients with chronic hepatitis C than among matched healthy blood donors.

The same study found no correlation between severity of symptoms and serum ALT levels. There was, however, a positive correlation between the peak serum ALT and histological scores. Unfortunately, a significant degree of overlap existed, making this correlation difficult to interpret in the individual patient. Serum ALT levels are abnormal or intermittently elevated in roughly two thirds of patients. The remaining one third have persistently normal serum ALT levels. This subgroup of patients seems to have a more benign disease course.

Epidemiological data also link HCV to a number of immune complex-mediated extrahepatic diseases. The best data exist for essential mixed cryoglobulinemia (EMC) and porphyria cutanea tarda (PCT). EMC is best thought of as a syndrome. Many patients with EMC complain of fatigue, weakness, arthralgias and myalgias. Frank arthritis may also occur. End organ damage results from deposition of circulating immune complexes in small- to medium-sized blood vessels. Involvement of the kidneys and nerves leads to glomerulonephritis and neuropathy, respectively. Several lines of evidence point to an association between hepatitis C infection and EMC. HCV antibodies and HCV RNA are detected in a majority of patients with EMC, compared to the low prevalence of such markers in patients with other rheumatologic diseases (AGNELLO et al. 1992; FERRI et al. 1991a,b,c). Some investigators have also been able to isolate HCV antibodies in the blood vessels in skin biopsy specimens from patients with vasculitis and EMC. Cryoglobulins are frequently found in patients with chronic hepatitis C, but infrequently observed in healthy persons or individuals with other disorders. Lastly, the skin lesions of EMC characteristically resolve or improve during treatment for chronic hepatitis C infection (DURAND et al. 1994; KHELLA et al. 1995; SCHIRREN et al. 1995). This resolution is associated with viral clearance. Unfortunately, relatively few patients have sustained virologic response to treatment, and relapse of hepatitis is associated with recurrence of cryoglobulinemia. Porphyria cutanea tarda also has been associated with HCV infection (CARPINTERO et al. 1997; CONRY-CANTILENA et al. 1995; CRIBIER et al. 1995; DABROWSKA et al. 1998; DECASTRO et al. 1993; ENGLISH et al. 1996; FARGION et al. 1992; FERRI et al. 1993; HERRERO et al. 1993; HUSSAIN et al. 1996; KOESTER et al. 1994; LACOUR et al. 1993; NAVAS et al. 1995; PIPERNO et al. 1992; SAMPIETRO et al. 1997; TSUKAZAKI et al. 1998). There is a high prevalence of HCV markers in patients with PCT. Like other forms of PCT, the mainstay of treatment for HCV-related PCT is phlebotomy to decrease iron stores. The relationship of PCT with chronic hepatitis C is particularly interesting since the recognition of the association between hepatic iron overload and response to interferon therapy (AKIYOSHI et al. 1997; DI MARCO et al. 1997; IKURA et al. 1996). The complex interactions of iron overload, disease manifestations, and response to therapy are under active investigation.

4 Natural History

Clinicians have long recognized that the clinical spectrum of chronic hepatitis C is very variable and the development of complications related to liver disease is unpredictable. The most frequent major complication of chronic hepatitis C infection is cirrhosis. While cirrhosis generally develops after decades of infection, reports clearly demonstrate a more aggressive course may occur wherein cirrhosis has developed within several years of infection. Whether all patients with chronic hepatitis C infection will eventually succumb to the complications of HCV is unknown but much debated. Prospective studies of transfusion-associated non-A, non-B (NANB) hepatitis have revealed that 20% of patients (range 8–24) develop cirrhosis after a mean follow-up of 8–14 years (DI BISCEGLIE et al. 1991; HOPF et al. 1990; KORETZ et al. 1993; MATTSSON et al. 1993; TREMOLADA et al. 1992).

Several studies have attempted to define the natural history of hepatitis C with conflicting results. In an early study of transfusion associated NANB hepatitis, there was a small but significant increase of liver related mortality in hepatitis patients over an 18 year follow-up from the time of transfusion (SEEFF et al. 1992). The majority of deaths in this study occurred in patients who were alcoholics. TONG et al. (1995) reviewed their experience with patients with transfusion associated hepatitis C at a tertiary referral center and found a high incidence of cirrhosis (51%) and death from liver failure or hepatocellular carcinoma (15%). These results were undoubtedly biased by the nature of the referral population and spectrum bias but still indicate that hepatitis C can be a progressive liver disease in some patients. This is further supported by the recognition that end-stage liver disease due to hepatitis C is the most common indication for liver transplantation in the United States.

Other studies have demonstrated that the clinical course of hepatitis C infection may not be as relentless. In studies of Irish women who were infected with hepatitis C from Rho-Gam, liver biopsies obtained two decades after infection identified fewer than 10% who progressed to cirrhosis and the majority had mild chronic hepatitis (CROWE et al. 1995). Similarly, patients with hepatitis C viremia may have persistently normal ALT activities during prolonged follow-up with minimal histologic changes (SHAKIL et al. 1994). A more benign course to HCV infection is also suggested in a study of well-compensated patients with cirrhosis already established whose 10 year survival probability was 79% (FATTOVICH et al. 1997). Thus, the outcome of HCV infection is not uniformly fatal nor predictable. The factors that contribute to the varied clinical course of this disease are largely unknown.

The hallmark of advanced liver disease and prelude to complications of portal hypertension and hepatocellular carcinoma is the development of fibrosis and cirrhosis. Again, the time frame over which hepatitis C disease progresses may be variable. Progression of liver fibrosis was evaluated in a cross-sectional study of European patients who had been enrolled in several large antiviral treatment protocols (POYNARD et al. 1997). Three populations of patients were identified who had median rates of development of fibrosis varying from 20 years or less (in 33% of patients) to those who were unlikely to develop cirrhosis even with 50 years of actuarial follow-up (31%). One third of the population studied were felt to have an average rate of cirrhosis developing 30 years after HCV infection. Three independent risk factors, older age at acquisition of hepatitis C, alcohol consumption greater than 50g per day, and male sex, were associated with more rapid progression of fibrosis. There was no association between HCV genotype and fibrosis. Although this study was very carefully designed, the duration of infection, a critical component needed to calculate fibrosis progression, could be established with some certainty in only 52% of the cohort based upon identification of risk factors and patient medical histories.

Additional information about factors associated with the natural history of hepatitis C can be derived from studies in patients with hemophilia, a population at highest risk for HCV infection. In a study from a single hemophilia treatment center in the United States, liver failure was found in 7% of HCV-infected patients and exclusively in those who were co-infected with human immunodeficiency virus (EYSTER et al. 1993). This study did not attempt to identify any other demographic variables possibly associated with progression of liver disease, such as alcohol use or concomitant medications. In addition, HCV RNA data were not provided for this cohort, although in a follow-up study that looked at a subset of these patients, HCV RNA levels were found to increase over a long period of follow-up with the greatest increases of HCV RNA following HIV seroconversion (EYSTER et al. 1994). The level of HCV RNA was indirectly correlated with levels of CD4 cells. Subsequent studies have also indicated that HIV co-infection increases levels of HCV RNA and the likelihood of hepatic decompensation (ROCKSTROH et al. 1996; SHERMAN et al. 1993). In a study of 138 hemophiliacs with hepatitis C in Great Britain, the presence of cirrhosis was evaluated on liver biopsies performed after correction of clotting abnormalities. Using multivariate analysis, development of cirrhosis was associated with positive HIV status, duration of HCV infection, and older age at acquisition of HCV (MAKRIS et al. 1996). It has been postulated that accelerated liver disease in co-infected patients is a result of failing cell-mediated immunity; this requires further investigation.

5 Genotype and Disease Severity

Due to a high rate of spontaneous mutations, HCV exists as a heterogeneous group of viruses that have been classified into six major genotypes and a number of subtypes based upon analysis of nucleotide sequences (LAU et al. 1996). The prototype HCV sequence first characterized in the United States was genotype 1a (OHNO et al. 1996). The relationship between genotype and disease severity has been reviewed previously (FRIED 1997).

Several methods are available to classify HCV genotypes. The most specific method relies on PCR amplification of a region of the HCV genome followed by sequencing of the PCR product. The nucleotide sequence of the sample is compared to various established sequences to determine the degree of homology from which genotype is assigned (OHNO et al. 1996). Genotyping can also be accomplished by performing PCR using genotype-specific primers and restriction fragment length polymorphism. A commercially available assay has simplified differentiation of HCV genotypes. In the line probe assay (INNO-LiPA, Innogenetics, Belgium), genotype-specific probes are embedded on a nitrocellulose strip. Amplified PCR products ($5'$ untranslated region) from patient sera hybridize to the embedded probe only when the sequence is complementary to the sequence for a given genotype. Studies that have compared these various assays have demonstrated that most methods are reliable for correctly assigning HCV genotype (LAU et al. 1996).

HCV genotypes appear to have a distinct international distribution. In a large study that evaluated 438 patients from several centers across the United States, genotype 1a and 1b accounted for 72% (split approximately evenly) of all HCV infections. Furthermore, the distribution of genotypes across the United States was similar at the different regional centers. In contrast, HCV infection in Europe and Japan is predominantly due to genotype 1b (NOUSBAUM et al. 1995; POL et al. 1995; DUSHEIKO et al. 1994). Interestingly, in Japan, genotype 1a is found almost exclusively in hemophiliacs who received factors imported from the United States (KINOSHITA et al. 1993). Genotype 4 was found most frequently in patients from the Middle East (DUSHEIKO et al. 1994). Since genotypic variation is due to mutation of HCV virus, over time, genotype distribution may also change. Among French hemodialysis patients, the frequency of genotype 1b decreased significantly after 1977, concurrent with the appearance of other genotypes (POL et al. 1995).

The recognition that hepatitis C was a heterogeneous virus resulted in numerous studies evaluating this virologic variable to explain differences in disease progression. It has been postulated that differences in nucleotide sequence could result in differential activity of HCV proteins that could alter the rate of HCV replication, endogenous interferon response to HCV infection, or pathogenicity of the virus (SIMMONDS 1995). As might be expected with such a complex question the results of these studies are not consistent. Some of the major findings are summarized in Table 2.

In two studies from France and Italy, genotype 1b was associated with a higher frequency of cirrhosis and abnormal serum aminotransferases than were other genotypes (NOUSBAUM et al. 1995; SILINI et al. 1995). In contrast, the largest survey performed in the United States showed similar levels of disease activity across HCV genotypes as determined by the line probe assay (LAU et al. 1996). A study from Germany that evaluated 97 patients also failed to show any relationship between histologic activity index and HCV genotype (ZEUZEM et al. 1996).

From the above discussion, it is evident that the clinical spectrum of chronic hepatitis C is variable and not entirely predictable. The studies that have been performed, thus far, have just begun to identify certain risk factors that may be associated with progressive HCV disease. It is unlikely that these risk factors are

Table 2. Summary of selected studies evaluating the relationship between HCV genotype and severity of chronic hepatitis C

Investigator (reference)	Location	Number of patients	Findings
Yamada et al. 1994	Japan	251	No correlation
Pozzato et al. 1994	Italy/Japan	111	Cirrhosis more frequent in genotype 1b
Lau et al. 1996	United States	121	Genotype 2 with increased HAI
Booth et al. 1995	United Kingdom	20	No correlation
Nousbaum et al. 1995	France/Italy	220	Cirrhosis more frequent in genotype 1b
Sillini et al. 1995	Italy	341	Cirrhosis more frequent in genotype 1b
Lau et al. 1996	United States	438	No correlation
Zeuzem et al. 1996	Germany	97	No correlation
Mihm et al. 1997	Germany	90	Fibrosis more frequent in genotype 1b

Adapted from Fried 1997. HAI, histologic activity index.

solely responsible for clinical outcomes. Complex interactions between virological variables and other host factors, including the immune response, immunogenetics, and host genomic variables that mediate expression of antiviral elements, must exist but are far more difficult to study.

6 Therapy of Hepatitis C

Therapeutic strategies for chronic hepatitis C are rapidly evolving. At the present time there are three formulations of interferon approved for treatment of chronic hepatitis C; interferon-α2b, interferon-α2a, and consensus interferon. When virologic endpoints are considered, the sustained response rates to all of these agents in patients who have never been treated with interferon previously are similar and uniformly disappointing, ranging between 10% and 20% in various studies (FRIED 1996).

The nucleoside analogue, ribavirin, recently has been approved for use in combination with interferon-α2b and appears to greatly improve the sustained response rates for patients with chronic hepatitis C. Recent studies have demonstrated the efficacy of interferon and ribavirin as initial therapy for patients with chronic hepatitis C. McHUTCHISON et al. (1998) treated over 900 patients with interferon and ribavirin or interferon alone for 24 or 48 weeks. The best sustained response rates were seen in patients treated with combination therapy for 48 weeks (38% vs 13% of those treated with interferon alone). Patients with genotype 1 were still relatively resistant to therapy with only 25% of those treated for 48 weeks exhibiting a sustained response compared to nearly 70% of those with other genotypes.

In a large, United States multicenter trial, patients who previously had a biochemical response to interferon monotherapy but subsequently relapsed were randomized to receive interferon (3MU TIW) plus placebo or interferon plus ribavirin (1000–1200mg/day) for 24 weeks. Patients treated with both drugs had a significantly higher virological sustained response rate (43% vs 4%) (DAVIS et al. 1998). Patients who have relapsed following cessation of interferon monotherapy may also have significant improvement in sustained response rates when treated with consensus interferon. In a randomized trial comparing 24- and 48-week regimens of consensus interferon at a dose of 15µg thrice weekly, approximately 50% of patients had a sustained response (HEATHCOTE et al. 1998).

Despite these improvements in treatment response, the majority of patients still fail to have a sustained response to any currently available therapy, reinforcing the need for additional treatment strategies and new agents. Ongoing studies are evaluating the efficacy of ribavirin in combination with high-dose induction regimens and with long-acting pegylated forms of interferon. In addition, a better understanding of the molecular virology of the hepatitis C virus has led to the promise of agents that inhibit various enzymes required for hepatitis C replication, such as the helicase and protease, which may hold the key to future, more effective therapies.

References

Agnello V, Chung RT, Kaplan LM (1992) A role for hepatitis C virus infection in type II cryoglobulinemia [see comments]. N Engl J Med 327:1490–1495
Akiyoshi F, Sata M, Uchimura Y, Suzuki H, Tanikawa K (1997) Hepatic iron stainings in chronic hepatitis C patients with low HCV RNA levels: a predictive marker of IFN therapy. Am J Gastroenterol 92:1463–1466
Alter MJ (1997a) Epidemiology of hepatitis C. Hepatology 26:62S–65S
Alter MJ (1997b) The epidemiology of acute and chronic hepatitis C. Clinics in Liver Disease 3:559–568
Alter MJ, Mast EE (1994) The epidemiology of viral hepatitis in the United States. Gastroenterol Clin North Am 23:437–455
Booth JCL, Foster GR, Kumar U, Galassini R, Goldin RD, et al. (1995) Chronic hepatitis C virus infections: predictive value of genotype and level of viraemia on disease progression and response to interferon ?. Gut 36:427–432
Carpintero P, DeCastro M, Garcia-Monzon C, Garcia-Buey L, Borque MJ, Garcia-Diez A, Moreno-Otero R (1997) Hepatitis C virus infection detected by viral RNA analysis in porphyria cutanea tarda. J Infect 34:61–64
Centers for Disease Control and Prevention (1998) Recommendations for prevention and control of hepatitis C virus (HCV) infection and HCV-related chronic disease. MMWR Morb Mortal Wkly Rep 47:1–39
Conry-Cantilena C, Vilamidou L, Melpolder JC, VanRaden M, Alter HJ (1995) Porphyria cutanea tarda in hepatitis C virus-infected blood donors. J Am Acad Dermatol 32:512–514
Cribier B, Petiau P, Keller F, Schmitt C, Vetter D, Heid E, Grosshans E (1995) Porphyria cutanea tarda and hepatitis C viral infection. A clinical and virologic study. Arch Dermatol 131:801–804
Crowe J, Doyle C, Fielding JG, Holloway H, Keegan M, Kelleher D, Kelly P, Leader M, Little M, McDonald G, McCarthy CF, McWeeney J, O'Keane C, Rajan E, Walsh LK, Weir DG, Whelton M (1995) Presentation of hepatitis C in a unique uniform cohort 17 years from inoculation. Gastroenterology 108:A1054

Dabrowska E, Jablonska-Kaszewska I, Falkiewicz B (1998) High prevalence of hepatitis C virus infection in patients with porphyria cutanea tarda in Poland [letter]. Clin Exp Dermatol 23:95–96

Davis GL, Esteban-Mur R, Rustgi V, Hoefs J, Gordon SC, et al. (1998). Interferon alfa-2b alone or in combination with ribavirin for the treatment of relapse of chronic hepatitis C. N Engl J Med 339:1493–1499

DeCastro M, Sanchez J, Herrera JF, Chaves A, Duran R, Garcia-Buey L, Garcia-Monzon C, Sequi J, Moreno-Otero R (1993) Hepatitis C virus antibodies and liver disease in patients with porphyria cutanea tarda [see comments]. Hepatology 17:551–557

Di Bisceglie AM, Goodman ZD, Ishak KG, Hoofnagle JH, Melpolder JJ, Alter HJ (1991) Long-term clinical and histopathological follow-up of chronic posttransfusion hepatitis. Hepatology 14:969–974

Di Marco V, Lo Iacono O, Almasio P, Ciaccio C, Capra M, et al. (1997) Long-term efficacy of alpha-interferon in beta-thalassemics with chronic hepatitis C. Blood 90:2207–2212

Durand JM, Cretel E, Kaplanski G, Lefevre P, Retornaz F, Soubeyrand J (1994) Long-term results of therapy with interferon alpha for cryoglobulinemia associated with hepatitis C virus infection. Clin Rheumatol 13:123–125

Dusheiko G, Schmilovitz-Weiss, Brown D, McOmish F, et al. (1994) Hepatitis C virus genotypes: an investigation of type-specific differences in geographic origin and disease. Hepatology 19:13–18

English JC, 3rd, Peake MF, Becker LE (1996) Hepatitis C and porphyria cutanea tarda. Cutis 57:404–408

Eyster ME, Diamondstone LS, Lien JM, Ehmann WC, Quan S, Goedert JJ (1993) Natural history of hepatitis C infection in multitransfused hemophiliacs: effect of coinfection with HIV. The Multicenter Hemophilia Cohort Study. JAIDS 6(6):602–610

Eyster ME, Fried MW, Di Bisceglie AM, Goedert JJ (1994) Increasing hepatitis C virus RNA levels in hemophiliacs: relationship to human immunodeficiency virus infection and liver disease. Multicenter Hemophilia Cohort Study. Blood 84:1020–1023

Fargion S, Piperno A, Cappellini MD, Sampietro M, Fracanzani AL, Romano R, Caldarelli R, Marcelli R, Vecchi L, Fiorelli G (1992) Hepatitis C virus and porphyria cutanea tarda: evidence of a strong association [see comments]. Hepatology 16:1322–1326

Fattovich G, Giustina G, Degos F, et al. (1997) Morbidity and mortality in compensated cirrhosis type C: a retrospective follow-up study of 384 patients [see comments]. Gastroenterology 112:463–472

Ferri C, Baicchi U, la Civita L, Greco F, Longombardo G, Mazzoni A, Careccia G, Bombardieri S, Pasero G, Zignego AL, et al. (1993) Hepatitis C virus-related autoimmunity in patients with porphyria cutanea tarda. Eur J Clin Invest 23:851–855

Ferri C, Greco F, Longombardo G, Palla P, Moretti A, Marzo E, Fosella PV, Pasero G, Bombardieri S (1991a) Antibodies against hepatitis C virus in mixed cryoglobulinemia patients. Infection 19:417–420

Ferri C, Greco F, Longombardo G, Palla P, Moretti A, Marzo E, Fosella PV, Pasero G, Bombardieri S (1991b) Antibodies to hepatitis C virus in patients with mixed cryoglobulinemia. Arthritis Rheum 34:1606–1610

Ferri C, Greco F, Longombardo G, Palla P, Moretti A, Marzo E, Mazzoni A, Pasero G, Bombardieri S, Highfield P, et al. (1991c) Association between hepatitis C virus and mixed cryoglobulinemia [see comment]. Clin Exp Rheumatol 9:621–624

Fried MW. Clinical Application of Hepatitis C virus Genotyping and Quantitation (1997) Clinics in liver disease 1(3):631–646

Fried MW (1996) Therapy for chronic viral hepatitis. Medical Clinics North America 80:957–972

Heathcote EJ, Keeffe EB, Lee SS, Feinman SV, Tong MJ, et al. (1998) Re-treatment of chronic hepatitis C with consensus interferon. Hepatology 27:1136–1143

Herrero C, Vicente A, Bruguera M, Ercilla MG, Barrera JM, Vidal J, Teres J, Mascaro JM, Rodes J (1993) Is hepatitis C virus infection a trigger of porphyria cutanea tarda? [see comments]. Lancet 341:788–789

Hoofnagle JH (1997) Hepatitis C: the clinical spectrum of disease. Hepatology 26:15S–20S

Hopf U, Moller B, Kuther D, Stemerowicz R, Lobeck H, Ludtke-Handjery A, Walter E, Blum HE, Roggendorf M, Deinhardt F (1990) Long-term follow-up of posttransfusion and sporadic chronic hepatitis non-A, non-B and frequency of circulating antibodies to hepatitis C virus (HCV). J Hepatol 10:69–76

Hsu HH, Feinstone SM, Hoofnagle JH (1995) Acute viral hepatitis. In: Mandell GL, Douglas RG, Bennett JE, Dolin R (eds) Mandell, Douglas and Bennett's principles and practice of infectious diseases. 4th ed. Churchill Livingstone, New York: 2 v

Hussain I, Hepburn NC, Jones A, O'Rourke K, Hayes PC (1996) The association of hepatitis C viral infection with porphyria cutanea tarda in the Lothian region of Scotland. Clin Exp Dermatol 21: 283–285

Ikura Y, Morimoto H, Johmura H, Fukui M, Sakurai M (1996) Relationship between hepatic iron deposits and response to interferon in chronic hepatitis C. Am J Gastroenterol 91:1367–1373

Khella SL, Frost S, Hermann GA, Leventhal L, Whyatt S, Sajid MA, Scherer SS (1995) Hepatitis C infection, cryoglobulinemia, and vasculitic neuropathy. Treatment with interferon alfa: case report and literature review. Neurology 45:407–411

Kinoshita T, Miyake K, Okamoto H, Mishiro S (1993) Imported hepatitis C virus genotypes in Japanese hemophiliacs. J Infectious Diseases 168:249–250 (letter)

Koester G, Feldman J, Bigler C (1994) Hepatitis C in patients with porphyria cutanea tarda. J Am Acad Dermatol 31:1054

Koretz RL, Abbey H, Coleman E, Gitnick G (1993) Non-A, non-B post-transfusion hepatitis. Looking back in the second decade [see comments]. Ann Intern Med 119:110–115

Lacour JP, Bodokh I, Castanet J, Bekri S, Ortonne JP (1993) Porphyria cutanea tarda and antibodies to hepatitis C virus. Br J Dermatol 128:121–123

Lau JYN, Davis GL, Prescott LE, Maertens G, et al. (1996) Distribution of hepatitis C virus genotypes determined by line probe assay in patients with chronic hepatitis C seen at tertiary referral centers in the United States. Ann Intern Med 124:868–876

Lemon SM, Brown EA (1995) Hepatitis C virus. In: Mandell GL, Douglas RG, Bennett JE, Dolin R (eds) Mandell, Douglas and Bennett's principles and practice of infectious diseases. 4th ed. Churchill Livingstone, New York: 2 v

Makris M, Preston FE, Rosendaal FR, Underwood JC, Rice KM, Triger DR (1996) The natural history of chronic hepatitis C in haemophiliacs [see comments]. Br J Haematol 94:746–752

Mattsson L, Sonnerborg A, Weiland O (1993) Outcome of acute symptomatic non-A, non-B hepatitis: a 13-year follow- up study of hepatitis C virus markers. Liver 13:274–278

McHutchison JG, Gordon SC, Schiff ER, Shiffman ML, Lee WM, et al. (1998). Interferon alf-2b alone or in combination with ribavirin as initial treatment for chronic hepatitis C. N Engl J Med 339:1485–1492

Mihm S, Fayyzai A, Hartmann H, Ramadori G (1997) Analysis of histopathological manifestations of chronic hepatitis C virus infection with respect to virus genotype. Hepatology 25:735–739

Navas S, Bosch O, Castillo I, Marriott E, Carreno V (1995) Porphyria cutanea tarda and hepatitis C and B viruses infection: a retrospective study. Hepatology 21:279–284

Nousbaum J-B, Stanislas P, Nalpas B, Landais P, et al. (1995) Hepatitis C virus type 1b (II) infection in France and Italy. Ann Intern Med 122:161–168

Ohno T, Lau JYN (1996) The "Gold-Standard," accuracy, and the current concepts: Hepatitis C virus genotype and viremia (editorial). Hepatology 24:1312–1315

Piperno A, D'Alba R, Roffi L, Pozzi M, Farina A, Vecchi L, Fiorelli G (1992) Hepatitis C virus infection in patients with idiopathic hemochromatosis (IH) and porphyria cutanea tarda (PCT). Arch Virol Suppl 4:215–216

Pol S, Thiers V, Nousbaum J-B, Legendre C, et al. (1995) The changing relative prevalence of hepatitis C virus genotypes: evidence in hemodialyzed patients and kidney recipients. Gastroenterology 108:581–583

Poynard T, et al. (1997) Natural history of liver fibrosis progression in patients with chronic hepatitis C. Lancet 349:825–832

Pozzato G, Kaneko S, Moretti M, Croce LS, et al. (1994) Different genotypes of hepatitis C virus are associated with different severity of chronic liver disease. J Med Virol 43:291–296

Rockstroh JK, et al. (1996) Immunosuppression may lead to progression of hepatitis C virus-associated liver disease in hemophiliacs coinfected with HIV. Am J Gastroenterol 91:2563–2568

Sampietro M, Fracanzani AL, Corbetta N, Amato M, Mattioli M, Molteni V, Fiorelli G, Fargion S (1997) High prevalence of hepatitis C virus type 1b in Italian patients with Porphyria cutanea tarda. Ital J Gastroenterol Hepatol 29:543–547

Schirren CA, Zachoval R, Schirren CG, Gerbes AL, Pape GR (1995) A role for chronic hepatitis C virus infection in a patient with cutaneous vasculitis, cryoglobulinemia, and chronic liver disease. Effective therapy with interferon-alpha. Dig Dis Sci 40:1221–1225

Seeff LB, Buskell-Bales Z, Wright EC, Durako SJ, Alter HJ, Iber FL, Hollinger FB, Gitnick G, Knodell RG, Perrillo RP, et al. (1992) Long-term mortality after transfusion-associated non-A, non-B hepatitis. The National Heart, Lung, and Blood Institute Study Group [see comments]. N Engl J Med 327:1906–1911

Shakil AO, et al. (1994) Liver histopathology in blood donors with hepatitis C and normal ALT levels. Gastroenterology 110:A981

Sherman K, et al. (1993) Quantitative evaluation of hepatitis C virus RNA in patients with concurrent HIV infection. J Clin Microbiol 21:2679–2682

Silini E, Bono F, Cividini A, Cerino A, et al. (1995) Differential distribution of hepatitis C virus genotypes in patients with and without liver function abnormalities. Hepatology 21:285–290

Simmonds P (1995) Variability of hepatitis C virus. Hepatology 21:570–583

Tibbs CJ (1997) Tropical aspects of viral hepatitis. Hepatitis C. Trans R Soc Trop Med Hyg 91:121–124

Tong MJ, el-Farra NS, Reikes AR, Co RL (1995) Clinical outcomes after transfusion associated hepatitis C. N Engl J Med 1995;332:1463–1466

Tremolada F, Casarin C, Alberti A, Drago C, Tagger A, Ribero ML, Realdi G (1992) Long-term followup of non-A, non-B (type C) post-transfusion hepatitis. J Hepatol 16:273–281

Tsukazaki N, Watanabe M, Irifune H (1998) Porphyria cutanea tarda and hepatitis C virus infection. Br J Dermatol 138:1015–1017

Yamada M, Kakumu S, Yoshioka K, Higashi Y, et al. (1994) Hepatitis C virus genotypes are not responsible for development of serious liver disease. Dig Dis Sci 39:234–239

Zeuzem S, Franke A, Lee J-H, Herrmann G, et al. (1996) Phylogenetic analysis of hepatitis C virus isolates and their correlation to viremia, liver function tests, and histology. Hepatology 24:1003–1009

Overview of Hepatitis C Virus Genome Structure, Polyprotein Processing, and Protein Properties

K.E. REED and C.M. RICE

1	Introduction	55
2	Classification	56
3	Genome Structure	57
3.1	The 5′ Nontranslated Region	57
3.2	The Open Reading Frame	58
3.3	The 3′ Nontranslated Region	58
4	Hepatitis C Virus Polyprotein Processing	59
4.1	Structural Region Processing	60
4.2	Nonstructural Region Processing	61
5	Features of Hepatitis C Virus-Encoded Proteins	61
5.1	The Capsid/Core (C) Protein	61
5.2	The E1 and E2 Glycoproteins	63
5.3	The NS2 Protein	64
5.4	The NS3 Protein	66
5.5	The NS4A Protein	69
5.6	The NS4B Protein	69
5.7	The NS5A Protein	69
5.8	The NS5B Protein	72
6	Conclusions	73
	References	74

1 Introduction

Hepatitis C was first recognized as a distinct form of liver disease in the mid-1970s with the advent of diagnostic tests for hepatitis A and B virus infection (ALTER et al. 1975; PRINCE et al. 1974). The etiologic agent of hepatitis C was proposed to be a small, enveloped virus based on demonstrations of its transmissibility to chimpanzees (ALTER et al. 1978; HOLLINGER et al. 1978; TABOR et al. 1978), small size (<80nm) (BRADLEY et al. 1985; HE et al. 1987), and sensitivity to chloroform (BRADLEY et al. 1983; FEINSTONE et al. 1983). The genome of hepatitis C virus

---

Department of Molecular Microbiology, Washington University School of Medicine, Campus Box 8230, 660 South Euclid Ave., St. Louis, MO 63110-1093, USA

(HCV) was first cloned in 1989 by screening a λgt11 cDNA expression library, derived from the plasma of a persistently infected chimpanzee, with hepatitis C patient serum (CHOO et al. 1989). Hybridization and nuclease digestion experiments indicated that the HCV genome consists of a single-stranded, positive-sense, RNA molecule (CHOO et al. 1989).

2 Classification

Analyses of the cloned sequences revealed that HCV is related to members of the family *Flaviviridae* (MILLER and PURCELL 1990). All of these viruses have small, enveloped virions and positive-sense RNA genomes that are translated as single, long polyproteins, with the structural proteins grouped together in the NH_2-terminal portion, followed by the nonstructural proteins. Their polyproteins are processed into individual viral proteins by a combination of host and viral proteases, including host signalase and a viral serine protease located in nonstructural protein 3 (NS3). Amino acid similarity among these viruses is limited to the serine protease and nucleoside triphosphatase (NTPase) domains of NS3 and the NS5/5B polymerase domain (MILLER and PURCELL 1990). Together, these similarities have since led to the classification of HCV in a separate genus (*Hepacivirus*) of the family *Flaviviridae*, which includes two other genera, *Flavivirus* and *Pestivirus* (FRANCKI et al. 1991).

HCV appears to be more closely related to pestiviruses than to flaviviruses, based on additional similarity in the nucleotide sequences (BUKH et al. 1992; HAN et al. 1991) and secondary structures (BROWN et al. 1992) of their 5′ non-translated regions (NTRs). Furthermore, both the HCV (TSUKIYAMA-KOHARA et al. 1992) and pestivirus (POOLE et al. 1995) 5′ NTRs appear to serve as internal ribosome entry sites (IRESs) for cap-independent translation (see below and elsewhere in this volume), unlike flavivirus 5′ NTRs, which are thought to bind to ribosomes via typical 5′ cap structures (CLEAVES and DUBIN 1979; WENGLER et al. 1978). However, HCV appears to be even more closely related to a group of recently cloned viruses GBV-A (MUERHOFF et al. 1995; SIMONS et al. 1995b), GBV-B (MUERHOFF et al. 1995; SIMONS et al. 1995b), and GBV-C/hepatitis G virus (LEARY et al. 1996; LINNEN et al. 1996; SIMONS et al. 1995a), known as the GB agents. The first members of this group (GBV-A and GBV-B) were cloned from tamarins that had received passaged serum originating from a surgeon (GB) with hepatitis of unknown etiology (SIMONS et al. 1995b). However, GBV-A appears to be a native tamarin virus (BUKH and APGAR 1997), and the relationship, if any, of GBV-B to the agent responsible for this surgeon's hepatitis remains unclear. Despite the similarities among the various members of the *Flaviviridae* and the GB agents, these viruses exhibit considerable diversity, even within a single genus. In this case of HCV, this diversity has led to further classification into at least six major genotypes and numerous subtypes (BUKH et al. 1995; SIMMONDS 1994).

3 Genome Structure

3.1 The 5' Nontranslated Region

The HCV genome RNA is approximately 9.6 kb in length and consists of a 5' NTR, a long open reading frame (ORF) encoding the viral polyprotein, and a 3' NTR (Fig. 1). The HCV 5' NTR is typically 341 nucleotides in length, although an additional sequence of eight nucleotides at the extreme 5' end has been reported (TROWBRIDGE and GOWANS 1998). The secondary structure of the 5' NTR appears to be highly conserved among HCV, GBV-B, and pestiviruses in the 220 nucleotides (nt) or so immediately upstream of the translation initiation codon. Similar features include a large stem-loop (III), a pseudoknot, and, in HCV and GBV-B, a smaller stem-loop (IV) which contains the translation initiation codon (HONDA et al. 1996a). However, the region corresponding to stem-loop IV of HCV and GBV-B appears to be largely single-stranded in pestivirus genomes (SIZOVA et al. 1998). Significant similarity has also been observed in the upstream sequence of HCV (nt 1–124) and GBV-B (nt 1–241) (HONDA et al. 1996a). Excluding two large insertions in the GBV-B sequence, the nucleotide identity between these upstream regions is about 60% (HONDA et al. 1996a). Furthermore, the genomes of both HCV and GBV-B are thought to contain a short stem-loop of 16–18 nt (I/Ia) near their extreme 5' ends (HONDA et al. 1996a). The IRES activity of HCV appears to require most of the 5' NTR, with the possible exception of stem-loop I (FUKUSHI et al. 1994; HONDA et al. 1996b; REYNOLDS et al. 1995; RIJNBRAND et al. 1995; TSUKI-YAMA-KOHARA et al. 1992; WANG et al. 1993). Sequences downstream of the initiation codon can also influence this activity (HONDA et al. 1996b; LU and WIMMER 1996; REYNOLDS et al. 1995) perhaps due to their effects on the stability of stem-loop IV (HONDA et al. 1996a) (see elsewhere in this volume). Ribosome binding is thought to occur at or immediately upstream of the translation initiation codon (REYNOLDS et al. 1996; RIJNBRAND et al. 1996) and, in contrast to the IRES of

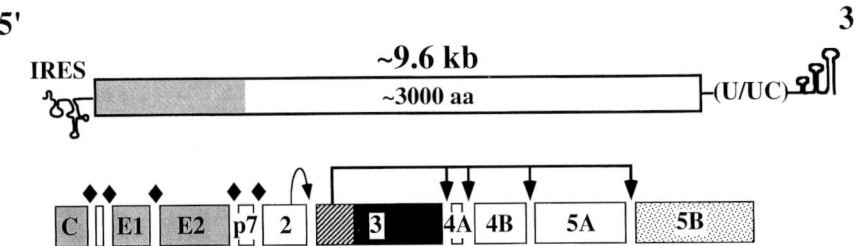

Fig. 1. Hepatitis C virus (HCV) genome organization. The HCV genome RNA is shown with its processed translation products *below*. The structural proteins are shaded in *gray*, and the nonstructural proteins are shown in *white*, except for the NS3 serine protease domain (*hatched*), the NS3 NTPase/helicase domain (*black*), and the NS5B polymerase domain (*speckled*) among the processed viral proteins. *Arched arrow*, autocatalytic cleavage site for the NS2-3 protease; *black diamonds* or the *connected arrows*, cleavage sites for host signalase or the viral NS3 serine protease, respectively

encephalomyocarditis virus (EMCV, a picornavirus), does not require any of the canonical eukaryotic initiation factors (eIFs) (PESTOVA et al. 1998), although it appears to be enhanced by specific interaction of regions of stem-loop III with eIF3 (SIZOVA et al. 1998). Interestingly, the HCV IRES can functionally substitute for that of poliovirus (LU and WIMMER 1996) and the pestivirus bovine viral diarrhea virus (BVDV) (FROLOV et al. 1998).

Other cellular proteins shown to bind to the 5′ NTR include polypyrimidine tract-binding protein (PTB) (ALI and SIDDIQUI 1995), heterogeneous nuclear protein L (HAHM et al. 1998), La (ALI and SIDDIQUI 1997), and unknown proteins of approximately 25 kDa (FUKUSHI et al. 1997) and 87 kDa (YEN et al. 1995). Interaction of these proteins with the 5′ NTR may be important for translation and/or replication of the HCV genome; however, analysis of their effects on replication awaits the development of an improved system for HCV replication in intact cells or one based on cellular extracts.

3.2 The Open Reading Frame

The polyprotein encoded by the HCV ORF ranges from 3008–3037 residues. At least ten discrete viral proteins are produced as a result of co- and post-translational proteolytic cleavage. In order from the NH2-terminus, they are the capsid or core protein C, the envelope glycoproteins E1 and E2, a small polypeptide of unknown function known as p7, and six nonstructural (NS) proteins: NS2, NS3, NS4A, NS4B, NS5A, and NS5B. The properties of these proteins are discussed in more detail below.

3.3 The 3′ Nontranslated Region

The first full-length HCV genome that was cloned contained a poly (A) tail at its 3′ terminus (HAN et al. 1991). However, this result was probably an artifact attributable to the use of oligo (dT) for isolation of poly (A) RNA and reverse transcription. Several other groups cloned HCV sequences that terminate with poly (U) by: (1) reverse transcription of HCV RNA with oligo (dA) (OKAMOTO et al. 1992), (2) tailing HCV RNA with poly (A) using *Escherichia coli* poly (A) polymerase, followed by reverse transcription with an oligo (dT)-containing primer (CHAYAMA et al. 1994; HAYASHI et al. 1993; KATO et al. 1991; OKAMOTO et al. 1991), or (3) tailing HCV cDNA with oligo (dA) using terminal nucleotidyl transferase (CHEN et al. 1992). The 3′ sequences of the resulting oligo (dA)-tailed cDNAs were then amplified by polymerase chain reaction (PCR) with an oligo (dT) primer or an adaptor primer to a non-homopolymeric region of the oligo (dT)-containing primer used for reverse transcription and an HCV-specific primer. Although reverse transcription of the HCV genome with an oligo (dA) primer is expected to generate clones that terminate with poly (U), the tailing method makes no assumption about the 3′ end sequence. However, these poly (A) or (dA) sequences are capable of annealing with the poly (U/T) tract and self-priming during reverse transcription or PCR, resulting in the elimination of sequences downstream of the poly (U) sequence. Probably for this reason, the complete 3' end of the HCV genome, which includes an additional 98 nt downstream of the U-rich sequence, was not cloned until alternative approaches were used, such as RNA linker ligation, followed by reverse transcription and PCR with adaptor and HCV-specific primers (KOLYK-HALOV et al. 1996; YAMADA et al. 1996) or tailing the HCV cDNA with dC instead of dA, followed by PCR with an oligo (dG)-containing adaptor and an internal HCV primer (TANAKA et al. 1995, 1996).

Thus, the full-length HCV 3' NTR consists of a poorly conserved sequence of approximately 40 nt and an internal poly (U)/polypyrimidine tract, followed by a highly conserved 98-nt sequence, the last 45 nt of which have been shown to form a highly stable stem-loop structure (BLIGHT and RICE 1997). The high percentage of nucleotide identity in this 98-nt region among various HCV isolates suggests that it may be important for the initiation of minus strand synthesis, a hypothesis that is supported by the inability of HCV RNAs lacking this region to replicate in chimpanzees (Kolykhalov et al., in preparation). However, low levels of replication in HuH-7 (Yoo et al. 1995) or HepG2 cells (DASH et al. 1997) have been reported for RNAs that terminate with poly (A) or within the poly (U) tract. Specific interaction of PTB with the HCV 3' NTR has been reported (GONTAREK et al. 1997; ITO and LAI 1997; TSUCHIHARA et al. 1997) and may have a role in RNA replication; however, investigation of this possibility awaits the development of efficient cell-free or cell-culture systems for HCV replication. PTB binding to the HCV 3' NTR has also been implicated in the mild stimulation of HCV or EMCV IRES activity observed as a result of fusion of the 3' terminal 98 nt of HCV to the 3' end of a luciferase reporter gene (ITO et al. 1998).

4 Hepatitis C Virus Polyprotein Processing

Processing of the HCV polyprotein is accomplished by at least three proteases, two viral and one cellular. The cellular enzyme, which is thought to be the signal peptidase residing in the lumen of the endoplasmic reticulum (ER), appears to be responsible for most of the cleavages in the structural region of the polyprotein, including the NH_2-terminus of E1 and the E1/E2, E2/p7, and p7/NS2 sites (see below). Further processing may occur at the COOH-terminus of the capsid; however, opinions differ as to the precise length of the mature capsid protein and the identity of the protease responsible for this additional cleavage (see below). The HCV NS2/3 (2/3) site appears to be cleaved autocatalytically by a zinc-stimulated viral protease activity located in the NS2-3 region (GRAKOUI et al. 1993b; HIJIKATA et al. 1993a), whereas the remaining cleavages in the NS region, the 3/4A, 4A/4B, 4B/5A, and 5A/5B sites, are catalyzed by a viral serine protease located in NS3 (BARTENSCHLAGER et al. 1993; ECKART et al. 1993; GRAKOUI et al. 1993a; HIJIKATA et al. 1993a; MANABE et al. 1994; TOMEI et al. 1993). NS4A is required as a cofactor for efficient processing at the 3/4A, 4A/4B, and 4B/5A sites and stimulates cleavage at the 5A/5B site (BARTENSCHLAGER et al. 1994; FAILLA et al. 1994; LIN et al. 1994b; TANJI et al. 1994a). Although the NS3 region is required for processing at the 2/3 site (GRAKOUI et al. 1993b; HIJIKATA et al. 1993a) as well as downstream sites, the protease responsible for these cleavages appear to be distinct, since mutations in the catalytic triad of the serine protease fail to disrupt 2/3 cleavage (BARTENSCHLAGER et al. 1993; ECKART et al. 1993; GRAKOUI et al. 1993a; HIJIKATA et al. 1993a; TOMEI et al. 1993). Furthermore, mutations at NS2 residues His-952 or Cys-993 of NS2 that inactivate the NS2-3 protease have little or no effect on processing at downstream sites (GRAKOUI et al. 1993b; HIJIKATA et al. 1993a).

4.1 Structural Region Processing

Signal peptidase cleavage at the NH_2-terminus of E1 and the E1/E2, E2/p7, and p7/NS2 sites was suggested by several lines of evidence, including: (1) the presence of hydrophobic regions immediately upstream of the C/E1, E1/E2, E2/p7 and p7/NS2 junctions that resemble signal sequences (HIJIKATA et al. 1991b; LIN et al. 1994a; MIZUSHIMA et al. 1994a,b; SELBY et al. 1994), (2) the requirement for membranes in cleavage at these sites (HIJIKATA et al. 1991b; Hüssy et al. 1996a; LIN et al. 1994a; MIZUSHIMA et al. 1994a; MIZUSHIMA et al. 1994b; SANTOLINI et al. 1994, 1995), (3) the disruption of processing at these sites by mutations known to inhibit signalase-dependent cleavages (MIZUSHIMA et al. 1994a,b), and (4) the dependence of cleavage at these sites on the signal recognition particle (SANTOLINI et al. 1994, 1995). Although cleavages at the NH_2-termini of E1 and E2 proceed rapidly to completion during or immediately after translation, cleavage is delayed at the E2/NS2 site and incomplete at the E2/p7 site, resulting in the production of fully processed E2 and uncleaved E2-p7 (LIN et al. 1994a; MIZUSHIMA et al. 1994a; SELBY et al. 1994). The significance of these two forms of E2 with respect to HCV virion assembly or other functions of the glycoproteins has yet to be determined. As previously mentioned, production of the mature capsid protein is thought to involve an additional cleavage at its COOH terminus. Leu-179 (HÜSSY et al. 1996a), Leu-182 (HÜSSY et al. 1996a), and Ser-173 (SANTOLINI et al. 1994) have been proposed as candidates for the P1 position of this cleavage site, but more evidence is needed to establish the true COOH terminus. Like the other cleavages in the structural region, this COOH-terminal trimming appears to be dependent on membranes (HÜSSY et al. 1996a; SANTOLINI et al. 1994) and the signal recognition particle (SANTOLINI et al. 1994), but it remains to be determined whether this cleavage is catalyzed by signal peptidase or another membrane-associated host protease.

4.2 Nonstructural Region Processing

As previously mentioned, two viral proteases, a zinc-stimulated NS2-3 protease and the NS3 serine protease are responsible for cleavages in the nonstructural region of the HCV polyprotein. The NS2-3 protease appears to be dedicated solely to cleavage at the 2/3 site, which occurs rapidly and apparently by an autocatalytic mechanism (GRAKOUI et al. 1993b; HIJIKATA et al. 1993a). Stimulation of this cleavage in vitro by $ZnCl_2$ and inhibition by EDTA initially led to the proposal that the NS2-3 protease is a metalloprotease (GRAKOUI et al. 1993b; HIJIKATA et al. 1993a). However, the absence of motifs typical of the active center of other known metalloproteases, the importance of His-952 and Cys-993 for 2/3 cleavage (GRAKOUI et al. 1993b; HIJIKATA et al. 1993a) and primary/secondary-structure modeling of the NS2 region (GORBALENYA and SNIJDER 1996) seem to be more consistent with the hypothesis that it is a cysteine protease. Resolution of this issue awaits further biochemical studies.

Serine protease-mediated cleavage at the 3/4A site also appears to be autocatalytic; however, processing at the 4A/4B, 4B/5A, and 5A/5B sites can occur in *trans* and perhaps also in *cis* (BARTENSCHLAGER et al. 1994; LIN et al. 1994b; TANJI et al. 1994b; TOMEI et al. 1993). Processing at the autocatalytically cleaved 2/3 and 3/4A sites, as well as the 5A/5B site, appears to be fairly rapid, as evidenced by the absence of detectable precursors after a brief pulse-labeling period. In contrast, the detection of various precursors from the NS4A-5A region in cultured cells expressing the HCV polyprotein suggests that cleavage at the 4A/4B and 4B/5A junctions is somewhat delayed relative to the other sites (BARTENSCHLAGER et al. 1994; LIN et al. 1994b; TANJI et al. 1994a). It should be noted, however, that the details of this processing scheme were determined by transient expression assays, in which the concentration of serine protease was likely to be much higher than in HCV-infected cells, raising the possibility that some of these details may differ in HCV-infected liver tissue.

5 Features of Hepatitis C Virus-Encoded Proteins

5.1 The Capsid/Core (C) Protein

The C protein is a highly conserved basic protein of $\sim$21 kDa that sometimes appears as a doublet (BURATTI et al. 1998; HÜSSY et al. 1996a; LIU et al. 1997; LO et al. 1995; SANTOLINI et al. 1994; YASUI et al. 1998). The slower migrating minor form is believed to be generated by cleavage only at the C/E1 signalase site; the faster migrating "mature" form by cleavage at a site near residue 173, as discussed above. A smaller form of the C protein with an apparent molecular mass of 16 kDa has also been reported (LO et al. 1994, 1995); however, this form has only been detected during expression of the HCV-1 isolate and has been largely attributed to an Arg-to-Lys mutation at position 9 (Lo et al. 1994).

In most studies, C has been found to be membrane-associated and localized in the vicinity of perinuclear and ER membranes (CHANG et al. 1994; HARADA et al. 1991; HIJIKATA et al. 1991b; KIM et al. 1994; MORADPOUR et al. 1996; RAVAGGI et al. 1994; SANTOLINI et al. 1994; SUZUKI et al. 1995). Nuclear localization of C, particularly truncated forms lacking COOH-terminal hydrophobic sequences, has also been reported (CHANG et al. 1994; LIU et al. 1997; Lo et al. 1995; RAVAGGI et al. 1994; SHIH et al. 1993; SUZUKI et al. 1995; YASUI et al. 1998). However, nuclear localization of the intact C protein is controversial and has not been observed in all studies (BARBA et al. 1997; BURATTI et al. 1998; MORADPOUR et al. 1996; MORADPOUR et al. 1998; SANTOLINI et al. 1994). Several functions for nuclear-localized C protein have been proposed, including the modulation of cellular gene transcription (RAY et al. 1995, 1997, 1998b), repression of transcription from the human immunodeficiency virus (HIV)-1 long-terminal repeat (LTR) sequences (SRINIVAS et al. 1996), and the suppression of hepatitis B virus (HBV) transcription and replication in HuH-7 cells (SHIH et al. 1993), the latter of which may be regulated by phosphorylation at Ser-99 and Ser-116 (SHIH et al. 1995). However, HBV replication does not appear to be suppressed in transgenic mice expressing the capsid protein (PASQUINELLI et al. 1997).

A number of cytoplasmic functions have also been reported for the capsid protein, including: (1) RNA binding (HWANG et al. 1995; SANTOLINI et al. 1994), (2) multimerization with other capsid protein molecules (MATSUMOTO et al. 1996; NOLANDT et al. 1997), and (3) interaction with E1 (Lo et al. 1996), all of which may be important for virion assembly; (4) association with 60S ribosomal subunits (SANTOLINI et al. 1994), which, in the case of alphaviruses, another group of small, positive-strand RNA viruses, is thought to be involved in the uncoating process; (5) association with lipid droplets and colocalization with apolipoprotein II (BARBA et al. 1997), which may be related to the observation that the capsid protein has been shown to induce steatosis in transgenic mice (MORIYA et al. 1997) or to the reported association of HCV virions with lipids in the bloodstream (THOMSSEN et al. 1992, 1993), and (6) interaction with the lymphotoxin-β receptor ($LT_\beta R$) (CHEN et al. 1997; MATSUMOTO et al. 1997) and tumor necrosis factor receptor (TNFR)-1 (ZHU et al. 1998), which are two TNFR family members involved in a number of host immune functions that may be modulated by HCV in order to promote its survival and propagation. The ability of the C protein to interact with TNRF family members may also be related to the observed correlations between C protein expression and cellular sensitivity to apoptosis, although conflicting effects have been reported (CHEN et al. 1997; RAY et al. 1996b, 1998a; RUGGIERI et al. 1997; ZHU et al. 1998), perhaps due to differences in the cell types or apoptotic stimuli used in the experiments. The C protein has also been reported to transform primary rat embryo fibroblasts (REF) in cooperation with Ras (RAY et al. 1996a) or an immortalized REF cell line called Rat-1 (CHANG et al. 1998) and recently, to induce hepatocellular carcinoma in transgenic mice (MORIYA et al. 1998), however, such tumors were not observed in previous studies of engineered mice with capsid transgenes (KAWAMURA et al. 1997; PASQUINELLI et al. 1997; WAKITA et al. 1998). Although the capsid protein has obviously been the subject of much research, more work is needed to determine the significance of these numerous observations with respect to specific steps of the viral life cycle and HCV-associated liver pathology.

5.2 The E1 and E2 Glycoproteins

Like many other viral envelope proteins, HCV E1 and E2 appear to undergo Asn-linked glycosylation at multiple sites (GRAKOUI et al. 1993c; HIJIKATA et al. 1991b; HSU et al. 1993; KOHARA et al. 1992; LANFORD et al. 1993; MATSUURA et al. 1992; NISHIHARA et al. 1993). The addition of these carbohydrate moieties slows the migration of E1 and E2 in SDS-polyacrylamide gels so that they appear to have molecular masses larger than those predicted based on their amino acid sequence (~30–35 kDa and 70 kDa). Both E1 and E2 have a COOH-terminal hydrophobic domain that appears to be inserted into the membrane of the ER, while the remainder is translocated into the lumen. This model of E1 and E2 as ER-localized type I integral membrane proteins was based on the following evidence: (1) COOH-terminal truncation of E1 and E2 results in their secretion, whereas the full-length proteins are localized to the cytoplasm (HÜSSY et al. 1996b; INUDOH et al. 1996; LESNIEWSKI et al. 1995; MATSUURA et al. 1994; MICHALAK et al. 1997; NISHIHARA et al. 1993; SELBY et al. 1994; SPAETE et al. 1992), (2) membrane association of E1 and E2 is disrupted by detergent but not high salt concentrations (RALSTON et al. 1993), (3) E1 and E2 expressed in mammalian cells or translated in vitro in the presence of microsomal membranes are resistant to proteinase K or trypsin digestion, indicating that they are protected within the lumen and/or membrane of the ER (HIJIKATA et al. 1991b; Lo et al. 1996), (4) the characteristics of E1 and E2 glycosylation are more consistent with ER than Golgi localization, e.g., their sensitivity to endoglycosidase H digestion (DUBUISSON et al. 1994; HIJIKATA et al. 1991a; LANFORD et al. 1993; RYU et al. 1995; SELBY et al. 1994) and lack of sialic acid (RALSTON et al. 1993; SPAETE et al. 1992) or fucose (DUBUISSON et al. 1994) moieties, and (5) E1-E2 complexes colocalize with protein disulfide isomerase, a resident ER protein (DELEERSNYDER et al. 1997). Extensive studies of E1 and E2 expressed transiently in mammalian cells have suggested that the folding of E2 and, in particular, E1 occurs slowly, resulting in the delayed formation of E1-E2 heterodimers (DELEERSNYDER et al. 1997; DUBUISSON et al. 1994; DUBUISSON and RICE 1996). Native E1-E2 complexes appear to be held together by noncovalent interactions (DELEERSNYDER et al. 1997; DUBUISSON et al. 1994; MATSUURA et al. 1994; RALSTON et al. 1993); however, heterogeneous disulfide-linked aggregates have also been observed (DUBUISSON et al. 1994; GRAKOUI et al. 1993c). Although some data have suggested that regions of E1 and E2 upstream of their COOH-terminal domains are sufficient for their interaction (HÜSSY et al. 1996b; LANFORD et al. 1993; MATSUURA et al. 1994; MICHALAK et al. 1997; YI et al. 1997), the formation of native E1-E2 complexes, as monitored by conformation-sensitive monoclonal antibodies, appears to require the COOH-terminal domain of E2 (COCQUEREL et al. 1998; MICHALAK et al. 1997), a finding that is probably related to the observation that full-length E2 is necessary for the proper folding of E1 (MICHALAK et al. 1997). By analyzing of a series of chimeric E2 proteins (using CD4 and a signal for GPI addition), recent work has shown that the COOH-terminal transmembrane domain of E2 functions as an ER retention signal (COCQUEREL et al. 1998). The presence of an E2 ER retention signal suggests that this signal must somehow be masked during assembly of HCV particles so that egress through the secretory pathway can occur.

Two other features of the E2 glycoprotein should be mentioned. First, the NH_2-terminal region of E2 exhibits a high degree of variability (called hypervariable region 1 or HVR1) (KUROSAKI et al. 1993; OGATA et al. 1991) that is apparently driven by antibody selection of immune escape variants (FARCI et al. 1994, 1996; KATO et al. 1993; SEKIYA et al. 1994; SHIMIZU et al. 1996; WEINER et al. 1992). Thus far, E2 HVR1 is the only defined target for neutralizing antibodies (HIJIKATA et al. 1991a; KATO et al. 1992; WEINER et al. 1991, 1992). In a recent study, the E2 ectodomain was shown to bind to an extracellular loop of human CD81, a tetraspanning cell surface membrane protein (PILERI et al. 1998). This interaction may play a key role in HCV binding and uptake into hepatocytes, but further studies will be needed to establish its importance.

As proposed for flaviviruses, HCV core particles may acquire their envelope with its accompanying glycoproteins by budding through the ER membrane, consistent with the observed ER localization of properly folded E1-E2 heterodimers. The virions thus assembled are then thought to travel through the host secretory pathway to the cell surface, where they are apparently released by exocytosis. However, studies on the structure, assembly and release of HCV particles are limited. Many groups have tried to produce higher levels of HCV virus-like particles (VLPs) with the aid of heterologous expression systems. Using the vaccinia virus-T7 transient system to express the full-length HCV coding region, particles of approximately 30 nm containing the C protein, assumed to be nonenveloped, core-like particles, and enveloped, 45-nm VLPs have been reported in HeLa G cells (MIZUNO et al. 1995). More recently, BAUMERT et al. (1998) reported the production of VLPs using the baculovirus system to express a portion of the 5' NTR and the structural region of a genotype 1b isolate in Sf9 cells. Enveloped VLPs were observed in intracellular vesicles late in infection (at 72–96 h) but were not secreted. These particles, which contained selectively encapsidated HCV RNA, could be released from baculovirus-infected cells only after mild detergent treatment and sonication.

5.3 The NS2 Protein

NS2 is a hydrophobic protein with an apparent molecular mass of 23 kDa. As mentioned previously, one of the major functions of NS2 appears to be cleavage at its own COOH terminus, in conjunction with NS3. Amino acids 827–1207

(GRAKOUI et al. 1993b) or 898–1325 (HIJIKATA et al. 1993a) are sufficient for 2/3 processing, but a polypeptide with the minimal NH_2-terminal and COOH-terminal boundaries of amino acids 898 and 1207 has yet to be tested. Cleavage at the 2/3 site in rabbit reticulocyte lysates is stimulated by the addition of microsomal membranes (GRAKOUI et al. 1993b; SANTOLINI et al. 1995), perhaps as a result of conformational changes in the NS2-3 region mediated by the membrane association of NS2, although the degree of this stimulation differs among HCV isolates. A study by SANTOLINI et al. (1995) has suggested that the signal recognition particle may be necessary for targeting NS2 to these membranes. These researchers further suggest that NS2 is inserted into the ER membrane with its COOH terminus in the lumen and its NH_2-terminal in the cytosol in a manner than is dependent on cleavage at the 2/3 site. However, this result is somewhat surprising given the location of a signal sequence at the p7/NS2 junction (GRAKOUI et al. 1993b; MIZUSHIMA et al. 1994b), and these experiments must be interpreted with caution, since this signal sequence was absent in many of the HCV polypeptides that were analyzed.

As mentioned in the section regarding HCV nonstructural protein processing, cleavage at the 2/3 site is stimulated by zinc (HIJIKATA et al. 1993a). However, at the present time it is unclear whether this zinc plays a catalytic or structural role in 2/3 processing. Interestingly, a tetrahedrally coordinated zinc-binding site has been identified in the region of NS3 required for NS2-3 protease activity, comprised of Cys-1123, Cys-1125, Cys-1171, and His-1175 (DEFRANCESCO et al. 1996; STEMPNIAK et al. 1997). Crystallographic (KIM et al. 1996; LOVE et al. 1996; YAN et al. 1998) and biochemical (DEFRANCESCO et al. 1996; STEMPNIAK et al. 1997) analyses have suggested that this zinc is essential for the structural integrity and, consequently, the activity of the NS3 serine protease (see elsewhere in this volume). Mutations at Cys-1123, Cys-1125, and Cys-1171 inhibit cleavage at the 2/3 site as well as downstream cleavages catalyzed by the serine protease, consistent with a role for this zinc in both types of processing (HIJIKATA et al. 1993a). Mutations at His-1175 have a more subtle effect (GRAKOUI et al. 1993b; HIJIKATA et al. 1993a; STEMPNIAK et al. 1997), possibly because this residue is not in direct contact with zinc (KIM et al. 1996; LOVE et al. 1996; YAN et al. 1998). However, additional experiments are needed to investigate the importance of this NS3-bound zinc or zinc atoms located elsewhere in the NS2-3 protease for processing at the 2/3 site. These and other such studies would be greatly facilitated by the development of a *trans* cleavage assay and purification scheme for this protease. However, the NS2-3 protease works inefficiently in *trans* (GRAKOUI et al. 1993b), and the rapid autocatalytic cleavage of this protease, together with the hydrophobic nature of NS2, are major obstacles to the purification of a soluble, intact protease. However, the discovery that cleavage at the 2/3 site can be post-translationally activated in rabbit reticulocyte lysates by the addition of detergents has permitted the analysis of some biochemical properties of the NS2-3 protease, such as its sensitivity to various protease inhibitors, in the absence of confounding effects on translation (PIERONI et al. 1997).

Cleavage at the 2/3 site is thought to occur by an autocatalytic mechanism, but cleavage of NS2-3 polypeptides with deletions or mutations that inactivate the NS2-3 protease has been observed upon coexpression with functional NS2 and/or NS3 protease subunits (GRAKOUI et al. 1993b; REED et al. 1995). However, this bimolecular cleavage is inefficient and seems to require the presence of a functional NS2 or NS3 region in *cis* (REED et al. 1995), suggesting that the precursor cannot achieve the proper conformation for cleavage unless it is capable of contributing at least one functional subunit to the formation of the protease responsible for that cleavage. This finding suggests that the 2/3 site must be located in the midst of a properly folded NS2-3 protease in order to be cleaved, but proper folding of the NS2-3 may be possible in the absence of a cleavable 2/3 site. This hypothesis is based on the observation that polypeptides containing an intact 2/3 site and defects in either NS2 or NS3 (but not both) which inactivate their NS2-3 protease activities are cleaved when they are coexpressed with a protease that cannot undergo autocatalytic cleavage as a result of deletion of the P1 and P1' residues of its 2/3 site (REED et al. 1995).

The importance of protein conformation in cleavage of the 2/3 site was also supported by site-directed mutagenesis studies. Many individual amino acid substitutions in the P5-P3' positions of the 2/3 site were well-tolerated by the NS2-3 protease, even at the P1 and P1' positions (HIROWATARI et al. 1993; REED et al. 1995), indicating that recognition of the 2/3 site is, for the most part, not based on single amino acid determinants. A few point mutations that severely inhibited or abolished 2/3 processing were observed (REED et al. 1995), but they were non-conservative changes with increased potential for disrupting the conformation of the region encompassing the 2/3 site. Together, these data suggest that conformation is more important than primary structure in recognition of the 2/3 site, as well as in the formation of an active NS2-3 protease.

5.4 The NS3 Protein

NS3 is a fairly hydrophilic protein of approximately 70 kDa with two functional domains: the aforementioned serine protease domain in the NH_2-terminal one-third of the protein, and an NTPase/helicase domain in the COOH-terminal two-thirds. The serine protease, as discussed above (see also elsewhere in this volume), is responsible for autocatalytic cleavage at the 3/4A site and *trans* (and perhaps *cis*) cleavage at the 4A/4B, 4B/5A, and 5A/5B sites. NS4A stimulates cleavage at the 5A/5B site and is essential for upstream, serine protease-dependent cleavages (BARTENSCHLAGER et al. 1994; FAILLA et al. 1994; LIN et al. 1994b; TANJI et al. 1995a). Activation of NS3 serine protease activity by NS4A requires the formation of a stable complex between these two proteins (BARTENSCHLAGER et al. 1995b; FAILLA et al. 1995; LIN et al. 1995; SATOH et al. 1995). The crystal structure of the NS3 serine protease domain has been solved in the presence (KIM et al. 1996; YAN et al. 1998) or absence (LOVE et al. 1996) of an NS4A peptide with cofactor activity, revealing a structure that is generally similar to that of trypsin, with two large domains made up largely of six-stranded β barrels separated by a cleft that contains the active site and substrate binding pocket. However, the HCV NS3 serine protease has some additional unique features. For instance, NS4A forms an integral part of this structure and interacts with the extreme NH_2-terminal residues of NS3 to form two additional, antiparallel β-strands (KIM et al. 1996; YAN et al. 1998). The major specificity determinant in the substrate-binding pocket appears to be Phe-1180, which is predicted to make a favorable interaction with the conserved Cys residues present in the P1 position of the 4A/4B, 4B/5A, and 5A/5B sites (KIM et al. 1996; LOVE et al. 1996; PIZZI et al. 1994; YAN et al. 1998). The remainder of the substrate-binding pocket has few distinguishing features. These findings were supported by a number of mutagenesis studies in which substitutions at the P1 position of these cleavage sites strongly inhibited or abolished cleavage, while substitutions at other positions were fairly well tolerated (BARTENSCHLAGER et al. 1995a; KOLYKHALOV et al. 1994; LEINBACH et al. 1994). However, the autocatalytically cleaved 3/4A site contains Thr at the P1 position, and, as in the case of the 2/3 site, specificity seems to be determined more by protein conformation than the side chains of individual amino acids flanking the cleavage site.

The NTPase/helicase activities of NS3 are thought to participate in the viral RNA replication process by unwinding regions of extensive secondary structure in the template or double-stranded RNAs resulting from the synthesis of the complementary strand. Although these activities have yet to be observed in the context of viral replication, NTP hydrolysis and helicase activity on synthetic substrates have been demonstrated for the COOH-terminal two-thirds of NS3 (HEILEK and PETERSON 1997; JIN and PETERSON 1995; KIM et al. 1995; PREUGSCHAT et al. 1996; SUZICH et al. 1993; TAI et al. 1996) or full-length NS3 (GALLINARI et al. 1998) expressed in *E. coli* and full-length NS3-NS4A complexes expressed in mammalian cells (HONG et al. 1996; MORGENSTERN et al. 1997) or in insect cells from recombinant baculoviruses (HONG et al. 1996; KYONO et al. 1998). Binding of NS3 to completely or partially single-stranded DNA or RNA has also been demonstrated, resulting in an increase in its ATPase activity. The minimal length of nucleic acid required for this binding appears to be between 7 and 20 nt. Poly (U), poly (dU), and poly (A) seem to have the greatest effect on the NS3 ATPase activity, consistent with their relatively high affinity for NS3 (GWACK et al. 1996, 1997; HONG et al. 1996; JIN and PETERSON 1995; Kanai et al. 1995; MORGENSTERN et al. 1997; PREUGSCHAT et al. 1996; SUZICH et al. 1993; TAI et al. 1996). The preference for poly (U) may be related to the presence of a poly (U) tract in the HCV 3′ NTR; however, specific binding between NS3 and the HCV 3′ NTR has yet to be demonstrated. The NTPase/helicase domain has been mapped to the region between amino acids 1209 and 1608 of the HCV polyprotein, but the regions necessary for RNA binding appear to be more limited, since ≥70% of the RNA binding activity is retained in mutants lacking an additional 16 or 85 amino acids, respectively, at the NH_2 and COOH termini (KIM et al. 1997). NS3 can unwind double-stranded (ds) RNA, RNA-DNA duplexes, and dsDNA in a 3′ to 5′ direction (TAI et al. 1996). This helicase activity requires divalent cations such as Mg^{2+} or Mn^{2+} and ATP, suggesting that it is coupled to ATP hydrolysis (JIN and PETERSON 1995; MORGENSTERN et al. 1997; PREUGSCHAT et al. 1996). However, certain other NTPs are able to substitute for ATP in this reaction (JIN and PETERSON 1995; MORGENSTERN et al. 1997). In contrast, ATP does not seem to be required for RNA binding (GALLINARI et al. 1998).

The crystal structure of the NS3 helicase, determined in the presence (KIM et al. 1998) or absence (CHO et al. 1998; YAO et al. 1997) of a bound oligonucleotide, has been solved to 2.1–2.3Å resolution for both the 1a (KIM et al. 1998; YAO et al. 1997) and 1b (CHO et al. 1998) genotypes of HCV. The general features of these three structures are, for the most part, quite similar, with three domains arranged in a Y configuration and separated from the other domains by distinct clefts (CHO et al. 1998; KIM et al. 1998; YAO et al. 1997). The first two domains, which contain all of the conserved NTPase/helicase motifs, each contain a parallel, six-stranded β sheet surrounded by a number of α helices. However, domain 1 contains a seventh, antiparallel β strand, and domain 2 contains an additional pair of antiparallel β strands that extend into the vicinity of domain 3. Domain 3 is completely α helical, and thus differs significantly in structure from domains 1 and 2, which may have arisen from an ancient gene duplication event. RNA binding is thought to occur in the cleft that separates domains 1 and 2 from domain 3, with the 5' end next to domain 2 and the 3' end next to domain 1 (KIM et al. 1998). The DECH motif, which has identified HCV NS3 as a member of a subfamily of the DEAD-box helicases and is thought to participate in catalysis through binding to Mg^{2+}-ATP, is located in domain 1 near the cleft that separates it from domain 2. The other helicase motifs are also situated near this cleft in domains 1 or 2. These domains are similar in overall structure to those of other well-characterized helicases, such as the *E. coli* Rep and *Bacillus stearothermophilus* PcrA DNA helicases (KOROLEV et al. 1998).

In addition to its functions in HCV polyprotein processing and RNA replication, NS3 has been proposed to regulate signal transduction by the cyclic AMP-dependent protein kinase (PKA) and influence the survival and proliferation of its cellular host. Amino acids 1487–1500 of NS3, which are located in its NTPase/helicase domain, are similar to regulatory sequences of a PKA inhibitor protein and a PKA autophosphorylation site (BOROWSKI et al. 1996). Synthetic peptides based on this sequence have been shown to inhibit PKA-catalyzed phosphorylation, as does a truncated NS3 protein consisting of amino acids 1189–1525 (BOROWSKI et al. 1996). This truncated NS3 protein is also able to interact with the catalytic subunit of PKA and inhibit its forskolin-stimulated nuclear translocation (BOROWSKI et al. 1997). However, these effects have yet to be verified for the full-length NS3 protein.

Cell modulatory functions have also been reported for the serine protease domain of NS3, including the transformation of NIH3T3 cells (SAKAMURO et al. 1995) and the suppression of actinomycin D-induced apoptosis (FUJITA et al. 1996). These effects may be related to observed correlations in the subcellular localization of NS3 and cellular p53 (ISHIDO et al. 1997; MURAMATSU et al. 1997), which has a number of antiproliferative functions. Moreover, NS3 expression apparently decreases actinomycin D-induced p53 expression. However, transfection of NIH3T3 cells with NS3 does not always lead to their transformation (FUJITA et al. 1996), and more data are needed to establish a link between NS3 and the inhibition of p53 function.

5.5 The NS4A Protein

NS4A is a hydrophobic protein of 54 amino acids, which, despite its small size, has been reported to have at least three functions: (1) activation and stabilization of the serine protease via its central domain (see elsewhere in this volume and NS3 section above), (2) anchorage of NS3 and perhaps other members of the replication complex to cellular membranes via its NH_2-terminal hydrophobic domain (HIJIKATA et al. 1993b; KIM et al. 1996), and (3) interaction with NS5A and regulation of NS5A phosphorylation (see NS5A section below).

5.6 The NS4B Protein

NS4B is a poorly characterized hydrophobic protein of approximately 30 kDa. It is a likely component of the viral replication complex, but direct evidence for this is lacking even among flaviviruses and pestiviruses.

5.7 The NS5A Protein

NS5A is a fairly hydrophilic protein that exists in at least two forms with apparent molecular masses of 56 and 58 kDa as a result of differential phosphorylation (KANEKO et al. 1994; TANJI et al. 1995b). NS5A phosphorylation occurs mostly on serine residues and, to a lesser extent, on threonine (KANEKO et al. 1994; REED et al. 1997). The primary site(s) of NS5A phosphorylation has yet to be determined, but it has been mapped by serial deletion to the region downstream of amino acid 2350 in the HCV-J isolate (TANJI et al. 1995a). This site is thought to be phosphorylated in both p56 and p58 (TANJI et al. 1995a). However, p58 may be phosphorylated by additional sites that remain unmodified in the p56 form. Analysis of the phosphorylation of various deleted forms of NS5A has suggested that these sites reside in a conserved, central region of the protein that extends from amino acid 2200 to 2250 (TANJI et al. 1995a). Site-directed mutagenesis of the nine conserved serines in this region tentatively identified these sites of p58 phosphorylation as Ser-2197, -2201, and -2204 (TANJI et al. 1995a). NS4A appears to enhance p58 production and, by inference, phosphorylation at these sites in the case of the HCV-J isolate (KANEKO et al. 1994; TANJI et al. 1995b) through direct association with NS5A (ASABE et al. 1997). However, individual NS5A phosphorylation sites and the regulation of NS5A phosphorylation by NS4A have yet to be reported for other HCV isolates, a potentially important issue, given the high percentage of Ser and

Thr residues in NS5A and the poor conservation of the COOH-terminal region proposed to contain the major site(s) of phosphorylation.

The kinase responsible for NS5A phosphorylation is thought to be of cellular origin, since: (1) NS5A contains no recognizable kinase motifs, (2) phosphorylation of NS5A expressed transiently in cultured cells occurs in the absence of other viral proteins (REED et al. 1997; TANJI et al. 1995b), and (3) phosphorylation of NS5A expressed in *E. coli* is dependent on the addition of eukaryotic cell extracts (IDE et al. 1997; Reed and Rice, unpublished data). Although the identity of the NS5A kinase is not known, some biochemical properties of an NS5A-associated kinase activity with similarities to the kinase responsible for NS5A phosphorylation in vivo have already been characterized. This NS5A-associated kinase is active in vitro over a broad pH range, has an apparent preference for $MnCl_2$ over $MgCl_2$, and is inhibited strongly by $CaCl_2$ at concentrations >0.5mM (REED et al. 1997). Furthermore, specific inhibitors of PKA protein kinase C (PKC) have little or no effect on NS5A phosphorylation in vitro or in vivo. However, both types of NS5A phosphorylation are inhibited by olomoucine, an inhibitor of certain proline-directed kinases, including extracellular signal-related kinase 1 and the cyclin-dependent kinases (CDKs) CDK2, CDK4, and CDK6. The resistance of NS5A phosphorylation to treatment with a specific PKA inhibitor casts some doubt on the suggestion that PKA is the major effector of NS5A phosphorylation (IDE et al. 1997), which was based on the observation that NS5A is phosphorylated in vitro upon the addition of purified PKA. However, NS5A contains a large number of potential phosphoacceptor residues, and the ability of other purified kinases to phosphorylate NS5A was not tested. Furthermore, it has yet to be determined whether the sites phosphorylated by PKA in vitro are utilized in vivo. For these reasons, it seems likely that a kinase other than PKA is responsible for the majority of NS5A phosphorylation.

Although the preference of the NS5A-associated kinase for Mn^{2+} over Mg^{2+} is somewhat unusual among serine/threonine kinases, as is its inhibition by Ca^{2+}, the NS5A and NS5 proteins of BVDV and yellow fever virus (YF), respectively, have been observed to associate with a kinase that has similar properties, raising the possibility that association of NS5A/NS5 proteins with the same or similar serine/threonine kinases or their phosphorylation by such kinases is a conserved feature among the *Flaviviridae* (REED et al. 1998). Consistent with this suggestion, phosphorylation has also been demonstrated for the NS5 proteins of dengue virus type 2 (KAPOOR et al. 1995) and tick-borne encephalitis virus (MOROZOVA et al. 1997). If this hypothesis is correct, analysis of the GBV NS5A proteins may reveal similar types of phosphorylation and associated serine/threonine kinase activities.

More work is needed to determine the functions of NS5A and the impact of phosphorylation on these functions in the context of viral replication. However, HCV NS5A has been proposed to function as an inhibitor of the interferon-stimulated, double-stranded RNA-dependent kinase PKR (GALE Jr. et al. 1997). PKR is a major effector of the host antiviral defense pathway which represses translation through its phosphorylation of the α subunit of the translation initiation factor eIF2. Many viruses therefore encode proteins that inhibit PKR function. Some evidence suggests that HCV NS5A interacts with PKR and interferes with its ability to phosphorylate eIF2α, perhaps by disrupting the dimerization of PKR required for its activation (GALE Jr. et al. 1998). Efforts to define the region(s) of NS5A required for its interaction with PKR have yielded conflicting results, but a conserved region of NS5A that extends from amino acid 2209 to 2248 may be involved, as well as some downstream residues (GALE Jr. et al. 1998).

Interestingly, a previous analysis of the full-length genome sequences of three HCV genotype 1b isolates obtained from different Japanese patients before and after interferon therapy had shown that the mutations which appeared during the course of this therapy were clustered primarily in E2 HVR1 and the COOH-terminal half of NS5A (ENOMOTO et al. 1995). Further analysis suggested that the amino acid sequence from position 2209 to 2248 of this region of NS5A correlated with the effectiveness of interferon treatment in these and other Japanese patients infected with genotype 1b strains of HCV; consequently, this stretch of amino acids was designated the interferon sensitivity-determining region (ISDR) (ENOMOTO et al. 1995, 1996). The sustained interferon response rate in patients with ISDR sequences that had less than or equal to three differences from that of genotype 1b reference isolates such as HCV-J was only 13%, but it was 100% in patients with four or more ISDR mutations relative to these isolates (ENOMOTO et al. 1996). With a single exception (KOMATSU et al. 1997), this result has since been more or less confirmed by other groups working with Japanese patients infected with genotype 1b, 2a, or 2b HCV strains (CHAYAMA et al. 1997; KUROSAKI et al. 1997). However, the correlation is substantially weaker or lacking in patients infected with genotype 1a HCV strains or European patients infected with HCV strains of genotype 1b, 2b, or 3a (CASATO et al. 1997; HOFGÄRTNER et al. 1997; KHORSI et al. 1997; ODEBERG et al. 1998; PAWLOTSKY et al. 1998; SAIZ et al. 1998; SQUADRITO et al. 1997; ZEUZEM et al. 1997). In one study of European patients infected with HCV-1b, an accumulation of mutations in NS5A did appear to correlate with interferon sensitivity; however, these mutations clustered between amino acids 2352 and 2361, not the ISDR (DUVERLIE et al. 1998). The reasons for these discrepancies are not clear, but differences in the interferon doses of these patients and the lower mutation rate of the ISDR in European HCV-1b patients relative to their Japanese counterparts are likely to be contributing factors. In summary, ISDR sequences may affect the response of individual HCV isolates to interferon treatment, along with other factors such as the initial quasispecies diversity and virus load (PAWLOTSKY et al. 1998), but they do not seem to have general predictive value in determining the outcome of successful interferon treatment. Furthermore, although interferon therapy may result in the selection of interferon-resistant ISDR sequences in some individuals (ENOMOTO et al. 1995), the sequence of this region seems to be fairly stable in most patients (POLYAK et al. 1998).

NH_2-terminally truncated forms of NS5A fused to the DNA-binding domain of the *Saccharomyces cerevisiae* Gal4 protein have also been shown to activate the transcription of reporter genes under the control of promoters containing Gal4 binding sites (CHUNG et al. 1997; KATO et al. 1997; TANIMOTO et al. 1997). Furthermore, this *trans*-activating ability has been linked to ISDR sequences that may be associated with increased interferon sensitivity (FUKUMA et al. 1998). The physiological relevance of these findings are questionable, since the full-length NS5A protein lacks this *trans*-activating ability and has a cytoplasmic localization. However, NS5A does contain a sequence in its COOH-terminal region that has the potential to function as a nuclear localization signal (IDE et al. 1996), raising the possibility that proteolytic removal of the NH_2-terminal region of NS5A (MARKLAND et al. 1997; MORGENSTERN et al. 1997) could result in its transport to the nucleus and activation of the transcription of certain cellular genes.

5.8 The NS5B Protein

NS5B is a 68 kDa protein with conserved sequence motifs characteristic of viral RNA-dependent RNA polymerases (RdRps), including a GDD motif believed to participate in catalysis. This RdRp activity has subsequently been demonstrated in vitro by NS5B proteins expressed in insect cells from baculovirus recombinants (BEHRENS et al. 1996; LOHMANN et al. 1997) or in *E. coli* (AL et al. 1997; YAMASHITA et al. 1998; YUAN et al. 1997). However, it is not specific for HCV templates, even in the presence of the other nonstructural proteins, suggesting that template specificity may require additional viral or cellular factors. HCV replication is thought to occur by synthesis of a full-length, minus-strand RNA. In current assays, full-length HCV RNA templates or those containing the HCV 3′ NTR are probably copied by a self-priming mechanism, in which the 3′ hydroxyl group of the template serves as the primer for RNA synthesis (BEHRENS et al. 1996; LOHMANN et al. 1997). This copy-back mechanism is probably favored by the NS5B polymerase due to the formation of a highly stable stem-loop structure at the extreme 3′ terminal of the HCV genome (BLIGHT and RICE 1997). Copy-back RNA synthesis has also been demonstrated for cellular templates with 3′ ends that are capable of base-pairing with upstream sequences (BEHRENS et al. 1996). Primers are required for RNA synthesis from homopolymeric templates and cellular RNAs that are incapable of such base-pairing (AL et al. 1997, 1998; BEHRENS et al. 1996; LOHMANN et al. 1997; YAMASHITA et al. 1998), but copying of these templates can be primed by either RNA or DNA oligonucleotides (AL et al. 1998; BEHRENS et al. 1998; LOHMANN et al. 1997; YAMASHITA et al. 1998). Divalent cations such as Mg^{2+} or Mn^{2+}, but not Zn^{2+}, are also required for NS5B RdRp activity (AL et al. 1998; BEHRENS et al. 1996; LOHMANN et al. 1997; YAMASHITA et al. 1998; YUAN et al. 1997). Maximal activity is obtained in reaction buffers that have a pH close to neutral (AL et al. 1997; BEHRENS et al. 1996; LOHMANN et al. 1997; YAMASHITA et al. 1998), a Mg^{2+} concentration between 2.5 and 20 mM (AL et al. 1997, 1998; LOHMANN et al. 1997; YAMASHITA et al. 1998; YUAN et al. 1997), and a low concentration of salt (≤100 mM KCl or NaCl) (LOHMANN et al. 1998; YAMASHITA et al. 1998). Similar to the rate observed for picornaviruses, the elongation rate of baculovirus-expressed NS5B on a genome-length HCV RNA template was estimated to be 150–200 nucleotides/min at 22°C (LOHMANN et al. 1998). This rate was independent of NS5B concentration, indicating that NS5B is highly processive (LOHMANN et al. 1998). However, the specific activity of the NS5B polymerase preparation was estimated to be approximately 10–20 times lower than that reported for the poliovirus 3D polymerase (LOHMANN et al. 1998). This suggests that a significant portion of the protein may be inactive, perhaps due to misfolding or a requirement for additional host or viral factors to achieve maximal activity.

A terminal nucleotidyl transferase activity was initially reported in a purified preparation of NS5B expressed in insect cells from a baculovirus recombinant (BEHRENS et al. 1996); however, the absence of this activity in purified preparations of NS5B expressed in *E. coli* (AL et al. 1997; YAMASHITA et al. 1998; YUAN et al. 1997) and the detection of this activity in mock-purified fractions of baculovirus-infected insect cells and purified preparations of baculovirus-expressed NS5B containing mutations that severely inhibit or abolish its RdRp activity (LOHMANN et al. 1997) indicate that it is the result of contamination by an enzyme produced by the host insect cells or baculovirus vector.

The NH_2-terminal region of NS5B appears to be important for its RdRp activity, since this activity is dramatically reduced or abrogated, respectively, by removal of 19 or 40 amino acids (LOHMANN et al. 1997). In contrast, much of the RdRp activity is retained in NS5B proteins that have been truncated at the COOH-terminus by 21 (YAMASHITA et al. 1998) or 55 (LOHMANN et al. 1997) amino acids. However, deletion of even 21 amino acids at the COOH terminus of NS5B has a dramatic effect on its localization in mammalian cells. Full-length NS5B is localized to the cytoplasm, where it is thought to associate with perinuclear membranes, whereas most of the truncated protein is transported into the nucleus (YAMASHITA et al. 1998). Since replication of HCV is thought to occur in the perinuclear region of the cell, by analogy with flaviviruses, the COOH terminus of NS5B is likely to be required for the assembly of functional replication complexes even if it is dispensable for actual RNA synthesis.

A small fraction of baculovirus-expressed NS5B has also been shown to undergo phosphorylation in insect cells (HWANG et al. 1997). This observation has yet to be confirmed in mammalian cells and its impact on the subcellular localization or RdRp activity of NS5B remains unclear.

NS5B has been shown in interact directly with NS3 and NS4A (ISHIDO et al. 1998), which in turn has been shown to form complexes with NS4B and NS5A (ASABE et al. 1997; LIN et al. 1997). These interactions, as well as the association of hydrophobic proteins such as NS4A with cellular membranes (HIJIKATA et al. 1993b), are likely to be important for the assembly of functional replication complexes. However, an efficient assay for HCV replication in vitro or in cultured cells must be developed before the contribution of these proteins to this process can be fully assessed.

6 Conclusions

Although much has been learned about HCV genome organization, polyprotein processing, protein function and structure, there are still technical hurdles to overcome and major gaps in our knowledge. We know a great deal about some of the parts but not much about how they fit together. Given the availability of full-length HCV cDNA clones functional for initiating chimpanzee infection, directed genetic analysis is now possible (KOLYKHALOV et al. 1997; YANAGI et al. 1997, 1998). With this technology, in vivo studies should yield important new information about virus evolution, immune response and pathogenesis, but they will provide only a starting point towards the understanding of replication mechanisms. Major efforts in the field should now be directed at establishing cellular and cell-free systems and additional animal models appropriate for dissecting the various steps in the HCV replication cycle and strategies for blocking them.

Acknowledgements. Our HCV work is supported by grants from the Public Health Service (CA57973 and AI40034). K.E.R. was supported in part by a predoctoral fellowship from the National Science Foundation.

References

Al RH, Xie Y, Wang Y, Hagedorn CH (1998) Expression of recombinant hepatitis C virus non-structural protein 5B in *Escherichia coli*. Virus Res 53:141–149
Al RH, Xie Y, Wang Y, Staercke CD, van Beers EH, Hagedorn CH (1997) Expression of recombinant hepatitis C virus NS5B. Nucleic Acids Symp Ser 36:197–199
Ali N, Siddiqui A (1995) Interaction of polypyrimidine tract-binding protein with the 5′ noncoding region of the hepatitis C virus RNA genome and its functional requirement in internal initiation of translation. J Virol 69:6367–6375
Ali N, Siddiqui A (1997) The La antigen binds 5′ noncoding region of the hepatitis C virus RNA in the context of the initiator AUG codon and stimulates internal ribosome entry site-mediated translation. Proc Natl Acad Sci USA 94:2249–2254
Alter HJ, Holland PV, Morrow AG, Purcell RH, Feinstone SM, Moritsugu Y (1975) Clinical and serological analysis of transfusion-associated hepatitis. Lancet 2:838–841
Alter HJ, Purcell RH, Holland PV, Popper H (1978) Transmissible agent in non-A non-B hepatitis. Lancet 1:459–463
Asabe S-I, Tanji Y, Satoh S, Kaneko T, Kimura K, Shimotohno K (1997) The N-terminal region of hepatitis C virus-encoded NS5A is important for NS4A-dependent phosphorylation. J Virol 71: 790–796
Barba G, Harper F, Harada T, Kohara M, Goulinet S, Matsuura Y, Eder G, Schaff Z, Chapman MJ, Miyamura T, Bréchot C (1997) Hepatitis C virus core protein shows a cytoplasmic localization and associates to cellular lipid storage droplets. Proc Natl Acad Sci USA 94:1200–1205
Bartenschlager R, Ahlborn-Laake L, Mous J, Jacobsen H (1993) Nonstructural protein 3 of the hepatitis C virus encodes a serine-type proteinase required for cleavage at the NS3/4 and NS4/5 junctions. J Virol 67:3835–3844
Bartenschlager R, Ahlborn-Laake L, Mous J, Jacobsen H (1994) Kinetic and structural analyses of hepatitis C virus polyprotein processing. J Virol 68:5045–5055
Bartenschlager R, Ahlborn-Laake L, Yasargil K, Mous J, Jacobsen H (1995a) Substrate determinants for cleavage in cis and in trans by the hepatitis C virus NS3 proteinase. J Virol 69:198–205
Bartenschlager R, Lohmann V, Wilkinson T, Koch JO (1995b) Complex formation between the NS3 serine-type proteinase of the hepatitis C virus and NS4A and its importance for polyprotein maturation. J Virol 69:7519–7528
Baumert TF, Ito S, Wong DT, Liang TJ (1998) Hepatitis C virus structural proteins assemble into viruslike particles in insect cells. J Virol 72:3827–3836
Behrens S-E, Grassmann CW, Thiel H-J, Meyers G, Tautz N (1998) Characterization of an autonomous subgenomic pestivirus RNA replicon. J Virol 72:2364–2372

Behrens SE, Tomei L, DeFrancesco R (1996) Identification and properties of the RNA-dependent RNA polymerase of hepatitis C virus. EMBO J 15:12–22

Blight KJ, Rice CM (1997) Secondary structure determination of the conserved 98-base sequence at the 3' terminus of hepatitis C virus genome RNA. J Virol 71:7345–7352

Borowski P, Heiland M, Oehlmann K, Becker B, Kornetzky L, Feucht H, Laufs R (1996) Non-structural protein 3 of hepatitis C virus inhibits phosphorylation mediated by cAMP-dependent protein kinase. Eur J Biochem 237:611–618

Borowski P, Oehlmann K, Heiland M, Laufs R (1997) Nonstructural protein 3 of hepatitis C virus blocks the distribution of the free catalytic subunit of cyclic AMP-dependent protein kinase. J Virol 71: 2838–2843

Bradley DW, Maynard JE, Popper H, Cook EH, Ebert JW, McCaustland KA, Schable CA, Fields HA (1983) Posttransfusion non-A non-B hepatitis: Physicochemical properties of two distinct agents. J Infect Dis 148:254–265

Bradley DW, McCaustland KA, Cook EH, Schable CA, Ebert JW, Maynard JE (1985) Posttransfusion non-A non-B hepatitis in chimpanzees: Physicochemical evidence that the tubule-forming agent is a small enveloped virus. Gastroenterology 88:773–779

Brown EA, Zhang H, Ping LH, Lemon SM (1992) Secondary structure of the 5' nontranslated regions of hepatitis C virus and pestivirus genomic RNAs. Nucleic Acids Res 20:5041–5045

Bukh J, Apgar CL (1997) Five new or recently discovered (GBV-A) virus species are indigenous to New World monkeys and may constitute a separate genus of the *Flaviviridae*. Virology 229:429–436

Bukh J, Miller RH, Purcell RH (1995) Genetic heterogeneity of hepatitis C virus: Quasispecies and genotypes. Sem Liver Dis 15:41–63

Bukh J, Purcell RH, Miller RH (1992) Sequence analysis of the 5' noncoding region of hepatitis C virus. Proc Natl Acad Sci USA 89:4942–4946

Buratti E, Baralle FE, Tisminetzky SG (1998) Localization of the different hepatitis C virus core gene products expressed in COS-1 cells. Cell Mol Biol (Noisy-Le-Grand) 44:505–512

Casato M, Agnello V, Pucillo LP, Knight GB, Leoni M, Del Vecchio S, Mazzilli C, Antonelli G, Bonomo L (1997) Predictors of long-term response to high-dose interferon therapy in type II cryoglobulinemia associated with hepatitis C virus infection. Blood 90:3865–3873

Chang J, Yang S-H, Cho Y-G, Hwang SB, Hahn YS, Sung YC (1998) Hepatitis C virus core from two different genotypes has an oncogenic potential but is not sufficient for transforming primary rat embryo fibroblasts in cooperation with the H-*ras* oncogene. J Virol 72:3060–3065

Chang SC, Yen J-H, Kang H-Y, Jang M-H, Chang M-F (1994) Nuclear localization signals in the core protein of hepatitis C virus. Biochem Biophys Res Comm 205:1284–1290

Chayama K, Tsubota A, Kobayashi M, Okamoto K, Hashimoto M, Miyano Y, Koike H, Kobayashi M, Koida I, Arase Y, Saitoh S, Suzuki Y, Murashima N, Ikeda K, Kumada H (1997) Pretreatment virus load and multiple amino acid substitutions in the interferon sensitivity-determining region predict the outcome of interferon treatment in patients with chronic genotype 1b hepatitis C virus infection. Hepatology 25:745–749

Chayama K, Tsubota A, Koida I, Arase Y, Saitoh S, Ikeda K, Kumada H (1994) Nucleotide sequence of hepatitis C virus (type 3b) isolated from a Japanese patient with chronic hepatitis C. J Gen Virol 75:3623–3628

Chen C-M, You L-R, Hwang L-H, Lee Y-H (1997) Direct interaction of hepatitis C virus core protein with the cellular lymphotoxin-β receptor. J Virol 71:9417–9426

Chen P-J, Lin M-H, Tai K-F, Liu P-C, Lin C-J, Chen D-S (1992) The Taiwanese hepatitis C virus genome: Sequence determination and mapping the 5' termini of viral genomic and antigenomic RNA. Virology 188:102–113

Cho HS, Ha NC, Kang LW, Chung KM, Back SH, Jang SK, Oh BH (1998) Crystal structure of RNA helicase from genotype 1b hepatitis C virus. A feasible mechanism of unwinding duplex RNA. J Biol Chem 273:15045–15052

Choo Q-L, Kuo G, Weiner AJ, Overby LR, Bradley DW, Houghton M (1989) Isolation of a cDNA clone derived from a blood-borne non-A non-B viral hepatitis genome. Science 244:359–362

Chung KM, Song OK, Jang SK (1997) Hepatitis C virus nonstructural protein 5A contains potential transcriptional activator domains. Mol Cells 7:661–667

Cleaves GR, Dubin DT (1979) Methylation status of intracellular dengue type 2 40S RNA. Virology 96:159–165

Cocquerel L, Meunier J-C, Pillez A, Wychowski C, Dubuisson J (1998) A retention signal necessary and sufficient for endoplasmic reticulum localization maps to the transmembrane domain of hepatitis C virus glycoprotein E2. J Virol 72:2183–2191

Dash S, Halim A-B, Tsuji H, Hiramatsu N, Gerber MA (1997) Transfection of HepG2 cells with infectious hepatitis C virus genome. Am J Pathol 151:363–373

DeFrancesco R, Urbani A, Nardi MC, Tomei L, Steinkuhler C, Tramontano A (1996) A zinc binding site in viral serine proteinases. Biochemistry 35:13282–13287

Deleersnyder V, Pillez A, Wychowski C, Blight K, Xu J, Hahn YS, Rice CM, Dubuisson J (1997) Formation of native hepatitis C virus glycoprotein complexes. J Virol 71:697–704

Dubuisson J, Hsu HH, Cheung RC, Greenberg H, Russell DR, Rice CM (1994) Formation and intracellular localization of hepatitis C virus envelope glycoprotein complexes expressed by recombinant vaccinia and Sindbis viruses. J Virol 68:6147–6160

Dubuisson J, Rice CM (1996) Hepatitis C virus glycoprotein folding: Disulfide bond formation and association with calnexin. J Virol 70:778–786

Duverlie G, Khorsi H, Castelain S, Jaillon O, Izopet J, Lunel F, Eb F, Penin F, Wychowski C (1998) Sequence analysis of the NS5A protein of European hepatitis C virus isolates and relation to interferon sensitivity. J Gen Virol 79:1373–1381

Eckart MR, Selby M, Masiarz F, Lee C, Berger K, Crawford K, Kuo C, Kuo G, Houghton M, Choo Q-L (1993) The hepatitis C virus encodes a serine protease involved in processing of the putative nonstructural proteins from the viral polyprotein precursor. Biochem Biophys Res Comm 192:399–406

Enomoto N, Sakuma I, Asahina Y, Kurosaki M, Murakami T, Yamamoto C, Izumi N, Marumo F, Sato C (1995) Comparison of full-length sequences of interferon-sensitive and resistant hepatitis C virus 1b. J. Clin. Invest. 96:224–230

Enomoto N, Sakuma I, Asahina Y, Kurosaki M, Murakami T, Yamamoto C, Ogura Y, Izumi N, Marumo F, Sato C (1996) Mutations in the nonstructural protein 5A gene and response to interferon in patients with chronic hepatitis C virus 1b infection. N Engl J Med 334:77–81

Failla C, Tomei L, DeFrancesco R (1994) Both NS3 and NS4A are required for proteolytic processing of hepatitis C virus nonstructural proteins. J Virol 68:3753–3760

Failla C, Tomei L, DeFrancesco R (1995) An amino-terminal domain of the hepatitis C virus NS3 protease is essential for interaction with NS4A. J Virol 69:1769–1777

Farci P, Alter HJ, Wong DC, Miller RH, Govindarajan S, Engle R, Shapiro M, Purcell RH (1994) Prevention of hepatitis C virus infection in chimpanzees after antibody-mediated in vitro neutralization. Proc Natl Acad Sci USA 91:7792–7796

Farci P, Shimoda A, Wong D, Cabezon T, De Gioannis D Strazzera A, Shimizu Y, Shapiro M, Alter HJ, Purcell RH (1996) Prevention of hepatitis C virus infection in chimpanzees by hyperimmune serum against the hypervariable region 1 of the envelope 2 protein. Proc Natl Acad Sci USA 93:15394–15399

Feinstone SM, Mihalik KB, Kamimura T, Alter HJ, London WT, Purcell RH (1983) Inactivation of hepatitis B virus and non-A non-B hepatitis by chloroform. Infect Immun 41:816–821

Francki RIB, Fauquet CM, Knudson DL, Brown F (1991) Classification and nomenclature of viruses: Fifth report of the international committee on taxonomy of viruses. Arch Virol Suppl 2:223

Frolov I, McBride MS, Rice CM (1998) cis-acting RNA elements required for replication of bovine viral diarrhea virus-hepatitis C virus 5′ nontranslated region chimeras. RNA 4:1418–1435

Fujita T, Ishido S, Muramatsu S, Itoh M, Hotta H (1996) Suppression of actinomycin D-induced apoptosis by the NS3 protein of hepatitis C virus. Biochem Biophys Res Comm un 229:825–831

Fukuma T, Enomoto N, Marumo F, Sato C (1998) Mutations in the interferon-sensitivity determining region of hepatitis C virus and transcriptional activity of the nonstructural region 5A protein. Hepatology 28:1147–1153

Fukushi S, Katayama K, Kurihara C, Ishiyama N, Hoshino FB, Ando T, Oya A (1994) Complete 5′ noncoding region is necessary for the efficient internal initiation of hepatitis C virus RNA. Biochem Biophys Res Comm un 199:425–432

Fukushi S, Kurihara C, Ishiyama N, Hoshino FB, Oya A, Katayama K (1997) The sequence element of the internal ribosome entry site and a 25-kilodalton cellular protein contribute to efficient internal initiation of translation of hepatitis C virus RNA. J Virol 71:1662–1666

Gale Jr. M, Blakely CM, Kwieciszewski B, Tan SL, Dossett M, Tang NM, Korth MJ, Polyak SJ, Gretch DR, Katze MG (1998) Control of PKR protein kinase by hepatitis C virus nonstructural 5A protein: molecular mechanisms of kinase regulation. Mol Cell Biol 18:5208–5218

Gale Jr. MJ, Korth MJ, Tang NM, Tan S-L, Hopkins DA, Dever TE, Polyak SJ, Gretch DR, Katze MG (1997) Evidence that hepatitis C virus resistance to interferon is mediated through repression of the PKR protein kinase by the nonstructural 5A protein. Virology 230:217–227

Gallinari P, Brennan D, Nardi C, Brunetti M, Tomei L, Steinkühler C, De Francesco R (1998) Multiple enzymatic activities associated with recombinant NS3 protein of hepatitis C virus. J Virol 72:6758–6769

Gontarek RR, Gutshall LL, Tsai J, Sathe GM, Mao JY, Prescott CD, Vecchio AM (1997) Interaction of polypyrimidine tract-binding protein with the 3' non-translated region of the hepatitis C virus genome. Nucleic Acids Symp Ser 36:146–149

Gorbalenya AE, Snijder EJ (1996) Viral cysteine proteinases. Perspectives in Drug Discovery and Design 6:64–86

Grakoui A, McCourt DW, Wychowski C, Feinstone SM, Rice CM (1993a) Characterization of the hepatitis C virus-encoded serine proteinase:determination of proteinase-dependent polyprotein cleavage sites. J Virol 67:2832–2843

Grakoui A, McCourt DW, Wychowski C, Feinstone SM, Rice CM (1993b) A second hepatitis C virus-encoded proteinase. Proc Natl Acad Sci USA 90:10583–10587

Grakoui A, Wychowski C, Lin C, Feinstone SM, Rice CM (1993c) Expression and identification of hepatitis C virus polyprotein cleavage products. J Virol 67:1385–1395

Gwack T, Kim DW, Hang JH, Choe J (1996) Characterization of RNA binding activity and RNA helicase activity of the hepatitis C virus NS3 protein. Biochem Biophys Res Comm 225:654–659

Gwack Y, Kim DW, Han JH, Choe J (1997) DNA helicase activity of the hepatitis C virus nonstructural protein 3. Eur J Biochem 250:47–54

Hahm B, Kim YK, Kim JH, Kim TY, Jang SK (1998) Heterogeneous nuclear ribonucleoprotein L interacts with the 3' border of the internal ribosomal entry site of hepatitis C virus. J Virol 72:8782–8788

Han JH, Shyamala V, Richman KH, Brauer MJ, Irvine B, Urdea MS, Tekamp-Olson P, Kuo G, Choo Q-L, Houghton M (1991) Characterization of the terminal regions of hepatitis C viral RNA: identification of conserved sequences in the 5' untranslated region and poly(A) tails at the 3' end. Proc Natl Acad Sci USA 88:1711–1715

Harada S, Watanabe Y, Takeuchi K, Suzuki T, Katayama T, Takebe Y, Saito I, Miyamura T (1991) Expression of processed core protein of hepatitis C virus in mammalian cells. J Virol 65:3015–3021

Hayashi N, Higashi H, Kaminaka K, Sugimoto H, Esumi M, Komatsu K, Hayashi K, Sugitani M, Suzuki K, Tadao O, Nozaki C, Mizuno K, Shikata T (1993) Molecular cloning and heterogeneity of the human hepatitis C virus (HCV) genome. J Hepatol 17 (Suppl. 3):S94–S107

He LF, Alling D, Popkin T, Shapiro M, Alter HJ, Purcell RH (1987) Determining the size of non-A non-B hepatitis by filtration. J Infect Dis 156:636–640

Heilek GM, Peterson MG (1997) A point mutation abolishes the helicase but not the nucleoside triphosphatase activity of hepatitis C virus NS3 protein. J Virol 71:6264–6266

Hijikata M, Kato N, Ootsuyama Y, Nakagawa M, Ohkoshi S, Shimotohno K (1991a) Hypervariable regions in the putative glycoprotein of hepatitis C virus. Biochem Biophys Res Comm 175:220–228

Hijikata M, Kato N, Ootsuyama Y, Nakagawa M, Shimotohno K (1991b) Gene mapping of the putative structural region of the hepatitis C virus genome by in vitro processing analysis. Proc Natl Acad Sci USA 88:5547–5551

Hijikata M, Mizushima H, Akagi T, Mori S, Kakiuchi N, Kato N, Tanaka T, Kimura K, Shimotohno K (1993a) Two distinct proteinase activities required for the processing of a putative nonstructural precursor protein of hepatitis C virus. J Virol 67:4665–4675

Hijikata M, Mizushima H, Tanji Y, Komoda Y, Hirowatari Y, Akagi T, Kato N, Kimura K, Shimotohno K (1993b) Proteolytic processing and membrane association of putative nonstructural proteins of hepatitis C virus. Proc Natl Acad Sci USA 90:10773–10777

Hirowatari Y, Hijikata M, Tanji Y, Nyunoya H, Mizushima H, Kimura K, Tanaka T, Kato N, Shimotohno K (1993) Two proteinase activities in HCV polypeptide expressed in insect cells using baculovirus vector. Arch Virol 133:349–356

Hofgärtner WT, Polyak SJ, Sullivan DG, Carithers Jr. RL, Gretch DR (1997) Mutations in the NS5A gene of hepatitis C virus in North American patients infected with HCV genotype 1a or 1b. J Med Virol 53:118–126

Hollinger FB, Gitnick G, Aach RD, Szmuness W, Mosley JW, Stevens CE, Peters RL, Weiner JM, Werch JB, Lander JJ (1978) Non-A non-B hepatitis transmission in chimpanzees: A project of the transfusion-transmitted viruses study group. Intervirology 10:60–68

Honda M, Brown EA, Lemon SM (1996a) Stability of a stem-loop involving the initiator AUG controls the efficiency of internal initiation of translation on hepatitis C virus RNA. RNA 2:955–968

Honda M, Ping LH, Rijnbrand RC, Amphlett E, Clarke B, Rowlands D, Lemon SM (1996b) Structural requirements for initiation of translation by internal ribosome entry within genome-length hepatitis C virus RNA. Virology 222:31–42

Hong Z, Ferrari E, Wright-Minogue J, Chase R, Risano C, Seelig G, Lee CG, Kwong AD (1996) Enzymatic characterization of hepatitis C virus NS3/4A complexes expressed in mammalian cells by using the herpes simplex virus amplicon system. J Virol 70:4261–4268

Hsu HH, Donets M, Greenberg HB, Feinstone SM (1993) Characterization of hepatitis C virus structural proteins with a recombinant baculovirus expression system. Hepatology 17:763–771

Hüssy P, Langen H, Mous J, Jacobsen H (1996a) Hepatitis C virus core protein: carboxy-terminal boundaries of two processed species suggest cleavage by a signal peptide peptidase. Virology 224:93–104

Hüssy P, Schmid G, Mous J, Jacobsen H (1996b) Purification and in vitro-phospholabeling of secretory envelope proteins E1 and E2 of hepatitis C virus expressed in insect cells. Virus Res 45:45–57

Hwang SB, Lo S-Y, Ou J-H, Lai MMC (1995) Detection of cellular proteins and viral core protein interacting with the 5′ untranslated region of hepatitis C virus RNA. J Biomed Sci 2:227–236

Hwang SB, Park K-J, Kim Y-S, Sung YC, Lai MMC (1997) Hepatitis C virus NS5B protein is a membrane-associated phosphoprotein with a predominantly perinuclear localization. Virology 227:439–446

Ide Y, Tanimoto A, Sasaguri Y, Padmanabhan R (1997) Hepatitis C virus NS5A protein is phosphorylated in vitro by a stably bound protein kinase from HeLa cells and by cAMP-dependent protein kinase A-α catalytic subunit. Gene 201:151–158

Ide Y, Zhang L, Chen M, Inchauspe G, Bahl C, Sasaguri Y, Padmanabhan R (1996) Characterization of the nuclear localization signal and subcellular distribution of hepatitis C virus nonstructural protein NS5A. Gene 182:203–211

Inudoh M, Nyunoya H, Tanaka T, Hijikata M, Kato N, Shimotohno K (1996) Antigenicity of hepatitis C virus envelope proteins expressed in Chinese hamster ovary cells. Vaccine 14:1590–1596

Ishido S, Fujita T, Hotta H (1998) Complex formation of NS5B with NS3 and NS4A proteins of hepatitis C virus. Biochem Biophy Res Comm 244:35–40

Ishido S, Muramatsu S, Fujita T, Iwanaga Y, Tong WY, Katayama Y, Itoh M, Hotta H (1997) Wild-type but not mutant-type p53 enhances nuclear accumulation of the NS3 protein of hepatitis C virus. Biochem Biophys Res Comm 230:431–436

Ito T, Lai MMC (1997) Determination of the secondary structure of and cellular protein binding to the 3′-untranslated region of the hepatitis C virus RNA genome. J Virol 71:8698–8706

Ito T, Tahara SM, Lai MMC (1998) The 3′-untranslated region of hepatitis C virus RNA enhances translation from an internal ribosomal entry site. J Virol 72:8789–8796

Jin L, Peterson DL (1995) Expression isolation and characterization of the hepatitis C virus ATPase/RNA helicase. Arch Biochem Biophys 323:47–53

Kanai A, Tanabe K, Kohara M (1995) Poly(U) binding activity of hepatitis C virus NS3 protein, a putative RNA helicase. FEBS Lett 376:221–224

Kaneko T, Tanji Y, Satoh S, Hijikata M, Asabe S, Kimura K, Shimotohno K (1994) Production of two phosphoproteins from the NS5A region of the hepatitis C viral genome. Biochem Biophy Res Comm 205:320–326

Kapoor M, Zhang L, Ramachandra M, Kusukawa J, Ebner KE, Padmanabhan R (1995) Association between NS3 and NS5 proteins of dengue virus type 2 in the putative RNA replicase is linked to differential phosphorylation of NS5. J Biol Chem 270:19100–19106

Kato N, Hijikata M, Nakagawa M, Ootsuyama Y, Muraiso K, Ohkoshi S, Shimotohno K (1991) Molecular structure of the Japanese hepatitis C viral genome. FEBS Lett 280:325–328

Kato N, Lan K-H, Ono-Nita SK, Shiratori Y, Omata M (1997) Hepatitis C virus nonstructural region 5A protein is a potent transcriptional activator. J Virol 71:8856–8859

Kato N, Ootsuyama Y, Ohkoshi S, Nakazawa T, Sekiya H, Hijikata M, Shimotohno K (1992) Characterization of hypervariable regions in the putative envelope protein of hepatitis C virus. Biochem Biophys Res Comm 189:119–127

Kato N, Sekiya H, Ootsuyama Y, Nakazawa T, Hijikata M, Ohkoshi S, Shimotohno K (1993) Humoral immune response to hypervariable region 1 of the putative envelope glycoprotein (gp70) of hepatitis C virus. J Virol 67:3923–3930

Kawamura T, Furusaka A, Koziel MJ, Chung RT, Wang TC, Schmidt EV, Liang TJ (1997) Transgenic expression of hepatitis C virus structural proteins in the mouse. Hepatology 25:1014–1021

Khorsi H, Castelain S, Wyseur A, Izopet J, Canva V, Rombout A, Capron D, Capron J-P, Lunel F, Stuyver L, Duverlie G (1997) Mutations of hepatitis C virus 1b NS5A 2209–2248 amino acid sequence do not predict the response to recombinant interferon-alfa therapy in French patients. J Hepatol 27:72–77

Kim DW, Gwack Y, Han JH, Choe J (1995) C-terminal domain of the hepatitis C virus NS3 protein contains an RNA helicase activity. Biochem Biophy Res Comm 215:160–166

Kim DW, Gwack Y, Han JH, Choe J (1997) Towards defining a minimal functional domain for NTPase and RNA helicase activities of the hepatitis C virus NS3 protein. Virus Res 49:17–25

Kim DW, Suzuki R, Harada T, Saito I, Miyamura T (1994) *Trans*-suppression of gene expression by hepatitis C viral core protein. Jpn J Med Sci Biol 47:211–220

Kim JL, Morgenstern KA, Griffith JP, Dwyer MD, Thomson JA, Murcko MA, Lin C, Caron PR (1998) Hepatitis C virus NS3 RNA helicase domain with a bound oligonucleotide: the crystal structure provides insights into the mode of unwinding. Structure 6:89–100

Kim JL, Morgenstern KA, Lin C, Fox T, Dwyer MD, Landro JA, Chambers SP, Markland W, Lepre CA, O'Malley ET, Harbeson SL, Rice CM, Murcko MA, Caron PR, Thomson JA (1996) Crystal structure of the hepatitis C virus NS3 protease domain complexed with a synthetic NS4A cofactor peptide. Cell 87:343–355

Kohara M, Tsukiyama-Kohara K, Maki N, Asano K, Yamaguchi K, Miki K, Tanaka S, Hattori N, Matsuura Y, Saito I, Miyamura T, Nomoto A (1992) Expression and characterization of glycoprotein gp35 of hepatitis C virus using recombinant vaccinia virus. J Gen Virol 73:2313–2318

Kolykhalov AA, Agapov EV, Blight KJ, Mihalik K, Feinstone SM, Rice CM (1997) Transmission of hepatitis C by intrahepatic inoculation with transcribed RNA. Science 277:570–574

Kolykhalov AA, Agapov EV, Rice CM (1994) Specificity of the hepatitis C virus serine proteinase: Effects of substitutions at the 3/4A 4A/4B 4B/5A and 5A/5B cleavage sites on polyprotein processing. J Virol 68:7525–7533

Kolykhalov AA, Feinstone SM, Rice CM (1996) Identification of a highly conserved sequence element at the 3' terminus of hepatitis C virus genome RNA. J Virol 70:3363–3371

Komatsu H, Fujisawa T, Inui A, Miyagawa Y, Onoue M (1997) Mutations in the nonstructural protein 5A gene and response to interferon therapy in young patients with chronic hepatitis C virus 1b infection. J Med Virol 53:361–365

Korolev S, Yao N, Lohman TM, Weber PC, Waksman G (1998) Comparisons between the structures of HCV and Rep helicases reveal structural similarities between SF1 and SF2 super-families of helicases. Prot Sci 7:605–610

Kurosaki M, Enomoto N, Marumo F, Sato C (1993) Rapid sequence variation of the hypervariable region of hepatitis C virus during the course of chronic infection. Hepatology 18:1293–1299

Kurosaki M, Enomoto N, Murakami T, Sakuma I, Asahina Y, Yamamoto C, Ikeda T, Tozuka S, Izumi N, Marumo F, Sato C (1997) Analysis of genotypes and amino acid residues 2209 to 2248 of the NS5A region of hepatitis C virus in relation to response to interferon-β therapy. Hepatology 25:750–753

Kyono K, Miyashiro M, Taguchi I (1998) Detection of hepatitis C virus helicase activity using the scintillation proximity assay system. Anal Biochem 257:120–126

Lanford RE, Notvall L, Chavez D, White R, Frenzel G, Simonsen C, Kim J (1993) Analysis of hepatitis C virus capsid E1 and E2/NS1 proteins expressed in insect cells. Virology 197:225–235

Leary TP, Muerhoff AS, Simons JN, Pilot-Matias TJ, Erker JC, Chalmers ML, Schlauder GG, Dawson GJ, Desai SM, Mushahwar IK (1996) Sequence and genomic organization of GBV-C: a novel member of the Flaviviridae associated with human non-A-E hepatitis. J Med Virol 48:60–67

Leinbach SS, Bhat RA, Xia S-M, Hum W-T, Stauffer B, Davis AR, Hung PP, Mizutani S (1994) Substrate specificity of the NS3 serine proteinase of hepatitis C virus as determined by mutagenesis at the NS3/NS4A junction. Virology 204:163–169

Lesniewski R, Okasinski G, Carrick R, Van Sant C, Desai S, Johnson R, Scheffel J, Moore B, Mushahwar I (1995) Antibody to hepatitis C virus second envelope (HCV-E2) glycoprotein: a new marker of HCV infection closely associated with viremia. J Med Virol 45:415–22

Lin C, Lindenbach BD, Prágai B, McCourt DW, Rice CM (1994a) Processing of the hepatitis C virus E2-NS2 region: Identification of p7 and two distinct E2-specific products with different C termini. J Virol 68:5063–5073

Lin C, Prágai B, Grakoui A, Xu J, Rice CM (1994b) Hepatitis C virus NS3 serine proteinase: *trans*-cleavage requirements and processing kinetics. J Virol 68:8147–8157

Lin C, Thomson JA, Rice CM (1995) A central region in the hepatitis C virus NS4A protein allows formation of an active NS3-NS4A serine proteinase complex in vivo and in vitro. J Virol 69:4373–4380

Lin C, Wu JW, Hsiao, Su MS (1997) The hepatitis C virus NS4A protein: interactions with the NS4B and NS5A proteins. J Virol 71:6465–6471

Linnen J, Wages J, Zhangkeck ZY, Fry KE, Krawczynski KZ, Alter H, Koonin E, Gallagher M, Alter M, Hadziyannis S, Karayiannis P, Fung K, Nakatsuji Y, Shih J, Young L, Piatak M, Hoover C, Fernandez J, Chen S, Zou JC, Morris T, Hyams KC, Ismay S, Lifson JD, Hess G, Kim JP (1996) Molecular cloning and disease association of hepatitis G virus: a transfusion-transmissible agent. Science 271:505–508

Liu Q, Tackney C, Bhat RA, Prince AM, Zhang P (1997) Regulated processing of hepatitis C virus core protein is linked to subcellular localization. J Virol 71:657–662

Lo S-Y, Masiarz F, Hwang SB, Lai MMC, Ou J-H (1995) Differential subcellular localization of hepatitis C virus core gene products. Virology 213:455–461

Lo S-Y, Selby M, Tong M, Ou J-H (1994) Comparative studies of the core gene products of two different hepatitis C virus isolates: Two alternative forms determined by a single amino acid substitution. Virology 199:124–131

Lo S-Y, Selby MJ, Ou J-H (1996) Interaction between hepatitis C virus core protein and E1 envelope protein. J Virol 70:5177–5182

Lohmann V, Körner F, Herian U, Bartenschlager R (1997) Biochemical properties of hepatitis C virus NS5B RNA-dependent RNA polymerase and identification of amino acid sequence motifs essential for enzymatic activity. J Virol 71:8416–8428

Lohmann V, Roos A, Korner F, Koch JO, Bartenschlager R (1998) Biochemical and kinetic analyses of NS5B RNA-dependent RNA polymerase of the hepatitis C virus. Virology 249:108–118

Love RA, Parge H, Wickersham JA, Hostomsky Z, Habuka N, Moomaw EW, Adachi T, Hostomska Z (1996) The crystal structure of hepatitis C virus NS3 proteinase reveals a trypsin-like fold and a structural zinc binding site. Cell 87:331–342

Lu HH, Wimmer E (1996) Poliovirus chimeras replicating under the translational control of genetic elements of hepatitis C virus reveal unusual properties of the internal ribosomal entry site of hepatitis C virus. Proc Natl Acad Sci USA 93:1412–1417

Manabe S, Fuke I, Tanishita O, Kaji C, Gomi Y, Yoshida S, Mori C, Takamizawa A, Yoshida I, Okayama H (1994) Production of nonstructural proteins of hepatitis C virus requires a putative viral protease encoded by NS3. Virology 198:636–644

Markland W, Petrillo RA, Fitzgibbon M, Fox T, McCarrick R, McQuaid T, Fulghum JR, Chen W, Fleming MA, Thomson JA, Chambers SP (1997) Purification and characterization of the NS3 serine protease domain of hepatitis C virus expressed in *Saccharomyces cerevisiae*. J Gen Virol 78:39–43

Matsumoto M, Hsieh T-Y, Zhu N, VanArsdale T, Hwang SB, Jeng K-S, Gorbalenya AE, Lo S-Y, Ou J-H, Ware CF, Lai MMC (1997) Hepatitis C virus core protein interacts with the cytoplasmic tail of lymphotoxin-β receptor. J Virol 71:1301–1309

Matsumoto M, Hwang SB, Jeng K-S, Zhu N, Lai MMC (1996) Homotypic interaction and multimerization of hepatitis C virus core protein. Virology 218:43–51

Matsuura Y, Harada S, Suzuki R, Watanabe Y, Inoue Y, Saito I, Miyamura T (1992) Expression of processed envelope protein of hepatitis C virus in mammalian and insect cells. J Virol 66:1425–1431

Matsuura Y, Suzuki T, Suzuki R, Sato M, Aizaki H, Saito I, Miyamura T (1994) Processing of E1 and E2 glycoproteins of hepatitis C virus expressed in mammalian and insect cells. Virology 205:141–150

Michalak JP, Wychowski C, Choukhi A, Meunier JC, Ung S, Rice CM, Dubuisson J (1997) Characterization of truncated forms of the hepatitis C virus glycoproteins. J Gen Virol 78:2299–2306

Miller RH, Purcell RH (1990) Hepatitis C virus shares amino acid sequence similarity with pestiviruses and flaviviruses as well as members of two plant virus supergroups. Proc Natl Acad Sci USA 87:2057–2061

Mizuno M, Yamada G, Tanaka T, Shimotohno K, Takatani M, Tsuji T (1995) Virion-like structures in Hela G cells transfected with the full-length sequence of the hepatitis C virus genome. Gastroenterology 109:1933–1940

Mizushima H, Hijikata H, Asabe S-I, Hirota M, Kimura K, Shimotohno K (1994a) Two hepatitis C virus glycoprotein E2 products with different C termini. J Virol 68:6215–6222

Mizushima H, Hijikata M, Tanji Y, Kimura K, Shimotohno K (1994b) Analysis of N-terminal processing of hepatitis C virus nonstructural protein 2. J Virol 68:2731–2734

Moradpour D, Englert C, Wakita T, Wands JR (1996) Characterization of cell lines allowing tightly regulated expression of hepatitis C virus core protein. Virology 222:51–63

Moradpour D, Wakita T, Wands JR, Blum HE (1998) Tightly regulated expression of the entire hepatitis C virus structural region in continuous human cell lines. Biochem Biophys Res Comm 246:920–924

Morgenstern KA, Landro JA, Hsiao K, Lin C, Gu Y, Su MS-S, Thomson JA (1997) Polynucleotide modulation of the protease nucleoside triphosphatase and helicase activities of a hepatitis C virus NS3-NS4A complex isolated from transfected COS cells. J Virol 71:3767–3775

Moriya K, Fujie H, Shintani Y, Yotsuyanagi H, Tsutsumi T, Ishibashi K, Matsuura Y, Kimura S, Miyamura T, Koike K (1998) The core protein of hepatitis C virus induces hepatocellular carcinoma in transgenic mice. Nature Medicine 4:1065–1067

Moriya K, Yotsuyanagi Y, Shintani Y, Fujie H, Ishibashi K, Matsuura Y, Miyamura T, Koike K (1997) Hepatitis C virus core protein induces hepatic steatosis in transgenic mice. J Gen Virol 78:1527–1531

Morozova OV, Tsekhanovskaya NA, Maksimova TG, Bachvalova TN, Matveeva VA, Kit YY (1997) Phosphorylation of tick-borne encephalitis virus NS5 protein. Virus Res 49:9–15

Muerhoff AS, Leary TP, Simons JN, Pilotmatias TJ, Dawson GJ, Erker JC, Chalmers ML, Schlauder GG, Desai SM, Mushahwar IK (1995) Genomic organization of GB viruses A and B: two new members of the *Flaviviridae* associated with GB agent hepatitis. J Virol 69:5621–5630

Muramatsu S, Ishido S, Fujita T, Itoh M, Hotta H (1997) Nuclear localization of the NS3 protein of hepatitis C virus and factors affecting the localization. J Virol 71:4954–4961

Nishihara T, Nozaki C, Nakatake H, Hoshiko K, Esumi M, Hayashi N, Hino K, Hamada F, Mizuno K, Shikata T (1993) Secretion and purification of hepatitis C virus NS1 glycoprotein produced by recombinant baculovirus-infected insect cells. Gene 129:207–214

Nolandt O, Kern V, Muller H, Pfaff E, Theilmann L, Welker R, Krausslich HG (1997) Analysis of hepatitis C virus core protein interaction domains. J Gen Virol 78:1331–1340

Odeberg J, Yun Z, Sonnerborg A, Weiland O, Lundeberg J (1998) Variation in the hepatitis C virus NS5a region in relation to hypervariable region 1 heterogeneity during interferon treatment. J Med Virol 56:33–38

Ogata N, Alter HJ, Miller RH, Purcell RH (1991) Nucleotide sequence and mutation rate of the H strain of hepatitis C virus. Proc Natl Acad Sci USA 88:3392–3396

Okamoto H, Kurai K, Okada S-I, Yamamoto K, Iizuka H, Tanaka T, Fukuda S, Tsuda F, Mishiro S (1992) Full-length sequence of a hepatitis C virus genome having poor homology to reported isolates: Comparative study of four distinct genotypes. Virology 188:331–341

Okamoto H, Okada S, Sugiyama Y, Kurai K, Iizuka H, Machida A, Miyakawa Y, Mayumi M (1991) Nucleotide sequence of the genomic RNA of hepatitis C virus isolated from a human carrier: comparison with reported isolates for conserved and divergent regions. J Gen Virol 72:2697–2704

Pasquinelli C, Shoenberger JM, Chung J, Chang KM, Guidotti LG, Selby M, Berger K, Lesniewski R, Houghton M, Chisari FV (1997) Hepatitis C virus core and E2 protein expression in transgenic mice. Hepatology 25:719–727

Pawlotsky J-M, Germanidis G, Neumann AU, Pellerin M, Frainais P-O, Dhumeaux D (1998) Interferon resistance of hepatitis C virus genotype 1b: relationship to nonstructural 5A gene quasispecies mutations. J Virol 72:2795–2805

Pestova TV, Shatsky IN, Fletcher SP, Jackson RJ, Hellen CUT (1998) A prokaryotic-like mode of cytoplasmic eukaryotic ribosome binding to the initiation codon during internal translation initiation of hepatitis C and classical swine fever virus RNAs. Genes Dev 12:67–83

Pieroni L, Santolini E, Fipaldini C, Pacini L, Migliaccio G, La Monica N (1997) In vitro study of the NS2-3 protease of hepatitis C virus. J Virol 71:6373–6380

Pileri P, Uematsu Y, Compagnoli S, Galli G, Falugi F, Petracca R, Weiner AJ, Houghton M, Rosa D, Grandi G, Abrignani S (1998) Binding of hepatitis C virus to CD81. Science 282:938–941

Pizzi E, Tramontano A, Tomei L, La Monica N, Failla C, Sardana M, Wood T, DeFrancesco R (1994) Molecular-model of the specificity pocket of the hepatitis C virus protease: Implications for substrate recognition. Proc Natl Acad Sci USA 91:888–892

Polyak SJ, McArdle S, Liu S-L, Sullivan DG, Chung M, Hofgärtner WT, Carithers Jr. RL, McMahon BJ, Mullins JI, Corey L, Gretch DR (1998) Evolution of hepatitis C virus quasispecies in hypervariable region 1 and the putative interferon sensitivity-determining region during interferon therapy and natural infection. J Virol 72:4288–4296

Poole TL, Wang C, Popp RA, Potgieter LN, Siddiqui A, Collett MS (1995) Pestivirus translation initiation occurs by internal ribosome entry. Virology 206:750–754

Preugschat F, Averett DR, Clarke BE, Porter DJT (1996) A steady-state and pre-steady-state kinetic analysis of the NTPase activity associated with the hepatitis C virus NS3 helicase domain. J Biol Chem 271:24449–24457

Prince AM, Brotman B, Grady GF, Kuhns WJ, Hazzi C, Levine RW, Millian SJ (1974) Long-incubation post-transfusion hepatitis without serological evidence of exposure to hepatitis B virus. Lancet 2: 241–246

Ralston R, Thudium K, Berger K, Kuo C, Gervase B, Hall J, Selby M, Kuo G, Houghton M, Choo Q-L (1993) Characterization of hepatitis C virus envelope glycoprotein complexes expressed by recombinant vaccinia viruses. J Virol 67:6753–6761

Ravaggi A, Natoli G, Primi D, Albertini A, Levrero M, Cariani E (1994) Intracellular localization of full-length and truncated hepatitis C virus core protein expressed in mammalian cells. J Hepatol 20: 833–836

Ray RB, Lagging LM, Meyer K, Ray R (1996a) Hepatitis C virus core protein cooperates with *ras* and transforms primary rat embryo fibroblasts to tumorigenic phenotype. J Virol 70:4438–4443

Ray RB, Lagging LM, Meyer K, Steele R, Ray R (1995) Transcriptional regulation of cellular and viral promoters by the hepatitis C virus core protein. Virus Res 37:209–220

Ray RB, Meyer K, Ray R (1996b) Suppression of apoptotic cell death by hepatitis C virus core protein. Virology 226:176–182

Ray RB, Meyer K, Steele R, Shrivastava A, Aggarwal BB, Ray R (1998a) Inhibition of tumor necrosis factor (TNF-α)-mediated apoptosis by hepatitis C virus core protein. J Biol Chem 273:2256–2259

Ray RB, Steele R, Meyer K, Ray R (1997) Transcriptional repression of p53 promoter by hepatitis C virus core protein. J Biol Chem 272:10983–10986

Ray RB, Steele R, Meyer K, Ray R (1998b) Hepatitis C virus core protein represses $p21^{WAF1/Cip1/Sid1}$ promoter activity. Gene 208:331–336

Reed KE, Gorbalenya AE, Rice CM (1998) The NS5A/NS5 proteins of viruses from three genera of the family *Flaviviridae* are phosphorylated by associated serine/threonine kinases. J Virol 72:6199–6206

Reed KE, Grakoui A, Rice CM (1995) The hepatitis C virus NS2-3 autoproteinase: cleavage site mutagenesis and requirements for bimolecular cleavage. J Virol 69:4127–4136

Reed KE, Xu J, Rice CM (1997) Phosphorylation of the hepatitis C virus NS5A protein in vitro and in vivo: properties of the NS5A-associated kinase. J Virol 71:7187–7197

Reynolds JE, Kaminski A, Carroll AR, Clarke BE, Rowlands DJ, Jackson RJ (1996) Internal initiation of translation of hepatitis C virus RNA: the ribosome entry site is at the authentic initiation codon. RNA 2:867–878

Reynolds JE, Kaminski A, Kettinen HJ, Grace K, Clarke BE, Carroll AR, Rowlands DJ, Jackson RJ (1995) Unique features of internal initiation of hepatitis C virus RNA translation. EMBO J 14: 6010–6020

Rijnbrand R, Bredenbeek PJ, van der Straaten T, Whetter L, Inchauspe G, Lemon S, Spaan W (1995) Almost the entire 5′ non-translated region of hepatitis C virus is required for cap-independent translation. FEBS Lett 365:115–119

Rijnbrand RC, Abbink TE, Haasnoot PC, Spaan WJ, Bredenbeek PJ (1996) The influence of AUG codons in the hepatitis C virus 5′ nontranslated region on translation and mapping of the translation initiation window. Virology 226:47–56

Ruggieri A, Harada T, Matsuura Y, Miyamura T (1997) Sensitization to Fas-mediated apoptosis by hepatitis C virus core protein. Virology 229:68–76

Ryu W-S, Choi D-Y, Yang J-Y, Kim C-H, Kwon Y-S, So H-S, Cho JM (1995) Characterization of the putative E2 envelope glycoprotein of hepatitis C virus expressed in stably transformed Chinese hamster ovary cells. Molecules and Cells 5:563–568

Saiz JC, Lopez-Labrador FX, Ampurdanes S, Dopazo J, Forns X, Sanchez-Tapias JM, Rodes J (1998) The prognostic relevance of the nonstructural 5A gene interferon sensitivity determining region is different in infections with genotype 1b and 3a isolates of hepatitis C virus. J Infect Dis 177:839–847

Sakamuro D, Furukawa T, Takegami T (1995) Hepatitis C virus nonstructural protein NS3 transforms NIH 3T3 cells. J Virol 69:3893–3896

Santolini E, Migliaccio G, La Monica N (1994) Biosynthesis and biochemical properties of the hepatitis C virus core protein. J Virol 68:3631–3641

Santolini E, Pacini L, Fipaldini C, Migliaccio G, Monica N (1995) The NS2 protein of hepatitis C virus is a transmembrane polypeptide. J Virol 69:7461–7471

Satoh S, Tanji Y, Hijikata M, Kimura K, Shimotohno K (1995) The N-terminal region of hepatitis C virus nonstructural protein 3 (NS3) is essential for stable complex formation with NS4A. J Virol 69:4255–4260

Sekiya H, Kato N, Ootsuyama Y, Nakazawa T, Yamauchi K, Shimotohno K (1994) Genetic alterations of the putative envelope proteins encoding region of the hepatitis C virus in the progression to relapsed phase from acute hepatitis: humoral immune response to hypervariable region 1. Int J Cancer 57:664–670

Selby MJ, Glazer E, Masiarz F, Houghton M (1994) Complex processing and protein: protein interactions in the E2:NS2 region of HCV. Virology 204:114–122

Shih C-M, Chen C-M, Chen SY, Lee Y-HW (1995) Modulation of the *trans*-suppression activity of hepatitis C virus core protein by phosphorylation. J Virol 69:1160–1171

Shih CM, Lo SJ, Miyamura T, Chen SY, Lee YH (1993) Suppression of hepatitis B virus expression and replication by hepatitis C virus core protein in HuH-7 cells. J Virol 67:5823–5832

Shimizu YK, Igarashi H, Kiyohara T, Cabezon T, Farci P, Purcell RH, Yoshikura H (1996) A hyperimmune serum against a synthetic peptide corresponding to the hypervariable region 1 of hepatitis C virus can prevent viral infection in cell cultures. Virology 223:409–412

Simmonds P (1994) Variability of hepatitis C virus genome. Curr Stud Hematol Blood Transfus 61:12–35

Simons JN, Leary TP, Dawson GJ, Pilotmatias TJ, Muerhoff AS, Schlauder GG, Desai SM, Mushahwar IK (1995a) Isolation of novel virus-like sequences associated with human hepatitis. Nature Medicine 1:564–569

Simons JN, Pilotmatias TJ, Leary TP, Dawson GJ, Desai SM, Schlauder GG, Muerhoff AS, Erker JC, Buijk SL, Chalmers ML, Vansant CL, Mushahwar IK (1995b) Identification of two flavivirus-like genomes in the GB hepatitis agent. Proc Natl Acad Sci USA 92:3401–3405

Sizova DV, Kolupaeva VG, Pestova TV, Shatsky IN, Hellen CUT (1998) Specific interaction of eukaryotic translation initiation factor 3 with the 5' nontranslated regions of hepatitis C virus and classical swine fever virus RNAs. J Virol 72:4775–4782

Spaete RR, Alexander D, Rugroden ME, Choo Q-L, Berger K, Crawford K, Kuo C, Leng S, Lee C, Ralston R, Thudium K, Tung JW, Kuo G, Houghton M (1992) Characterization of the hepatitis E2/NS1 gene product expressed in mammalian cells. Virology 188:819–830

Squadrito G, Leone F, Sartori M, Nalpas B, Berthelot P, Raimondo G, Pol S, Brechot C (1997) Mutations in the nonstructural 5A region of hepatitis C virus and response of chronic hepatitis C to interferon alfa. Gastroenterology 113:567–572

Srinivas RV, Ray RB, Meyer K, Ray R (1996) Hepatitis C virus core protein inhibits human immunodeficiency virus type 1 replication. Virus Res 45:87–92

Stempniak M, Hostomska Z, Nodes BR, Hostomsky Z (1997) The NS3 proteinase domain of hepatitis C virus is a zinc-containing enzyme. J Virol 71:2881–2886

Suzich JA, Tamura JK, Palmer-Hill F, Warrener P, Grakoui A, Rice CM Feinstone SM, Collett MS (1993) Hepatitis C virus NS3 protein polynucleotide-stimulated nucleoside triphosphatase and comparison with the related pestivirus and flavivirus enzymes. J Virol 67:6152–6158

Suzuki R, Matsuura Y, Susuki T, Ando A, Chiba J, Harada S, Saito I, Miyamura T (1995) Nuclear localization of the truncated hepatitis C virus core protein with its hydrophobic C terminus deleted. J Gen Virol 76:53–61

Tabor E, Garety RJ, Drucker JA, Seeff LB, Hoofnagle JF, Jackson DR, April M, Barker LF, Pineda-Tamondong G (1978) Transmission of non-A non-B hepatitis from man to chimpanzee. Lancet 1:463–466

Tai C-L, Chi W-K, Chen D-S, Hwang L-H (1996) The helicase activity associated with hepatitis C virus nonstructural protein 3 (NS3). J Virol 70:8477–8484

Tanaka T, Kato N, Cho M-J, Shimotohno K (1995) A novel sequence found at the 3' terminus of hepatitis C virus genome. Biochem Biophys Res Comm 215:744–749

Tanaka T, Kato N, Cho M-J, Sugiyama K, Shimotohno K (1996) Structure of the 3' terminus of the hepatitis C virus genome. J Virol 70:3307–3312

Tanimoto A, Ide Y, Arima N, Sasaguri Y, Padmanabhan R (1997) The amino terminal deletion mutants of hepatitis C virus nonstructural protein NS5A function as transcriptional activators in yeast. Biochem Biophys Res Comm 236:360–364

Tanji Y, Hijikata M, Hirowatari Y, Shimotohno K (1994a) Hepatitis C virus polyprotein processing: kinetics and mutagenic analysis of serine proteinase-dependent cleavage. J Virol 68:8418–8422

Tanji Y, Hijikata M, Hirowatari Y, Shimotohno K (1994b) Identification of the domain required for trans-cleavage activity of hepatitis C viral serine proteinase. Gene 145:215–219

Tanji Y, Hijikata M, Satoh S, Kaneko T, Shimotohno K (1995a) Hepatitis C virus-encoded nonstructural protein NS4A has versatile functions in viral protein processing. J Virol 69:1575–1581

Tanji Y, Kaneko T, Satoh S, Shimotohno K (1995b) Phosphorylation of hepatitis C virus-encoded nonstructural protein NS5A. J Virol 69:3980–3986

Thomssen R, Bonk S, Propfe C, Heermann KH, Kochel HG, Uy A (1992) Association of hepatitis C virus in human sera with beta-lipoprotein. Med. Microbiol. Immunol. 181:293–300

Thomssen R, Bonk S, Thiele A (1993) Density heterogeneities of hepatitis C virus in human sera due to the binding of beta-lipoproteins and immunoglobulins. Med Microbiol Immunol 182:329–334

Tomei L, Failla C, Santolini E, DeFrancesco R, La Monica N (1993) NS3 is a serine protease required for processing of hepatitis C virus polyprotein. J Virol 67:4017–4026

Trowbridge R, Gowans EJ (1998) Identification of novel sequences at the 5' terminus of the hepatitis C virus genome. J Viral Hepat 5:95–98

Tsuchihara K, Tanaka T, Hijikata M, Kuge S, Toyoda H, Nomoto A, Yamamoto N, Shimotohno K (1997) Specific interaction of polypyrimidine tract-binding protein with the extreme 3'-terminal structure of the hepatitis C virus genome the 3'X. J Virol 71:6720–6726

Tsukiyama-Kohara K, Iizuka N, Kohara M, Nomoto A (1992) Internal ribosome entry site within hepatitis C virus RNA. J Virol 66:1476–1483

Wakita T, Taya C, Katsume A, Kato J, Yonekawa H, Kanegae Y, Saito I, Hayashi Y, Koike M, Kohara M (1998) Efficient conditional transgene expression in hepatitis C virus cDNA transgenic mice mediated by the Cre/loxP system. J Biol Chem 273:9001–9006

Wang C, Sarnow P, Siddiqui A (1993) Translation of human hepatitis C virus RNA in cultured cells is mediated by an internal ribosome-binding mechanism. J Virol 67:3338–3344

Weiner AJ, Brauer MJ, Rosenblatt J, Richman KH, Tung J, Crawford K, Bonino F, Saracco G, Choo Q-L, Houghton M, Han JH (1991) Variable and hypervariable domains are found in the regions of HCV corresponding to the flavivirus envelope and NS1 proteins and the pestivirus envelope glycoproteins. Virology 180:842–848

Weiner AJ, Geysen HM, Christopherson C, Hall JE, Mason TJ, Saracco G, Bonino F, Crawford K, Marion CD, Crawford KA, Brunetto M, Barr PJ, Miyamura T, McHutchinson J, Houghton M (1992) Evidence for immune selection of hepatitis C virus (HCV) putative envelope glycoprotein variants: potential role in chronic HCV infections. Proc Natl Acad Sci USA 89:3468–3472

Wengler G, Wengler G, Gross HJ (1978) Studies on virus-specific nucleic acids synthesized in vertebrate and mosquito cells infected with flaviviruses. Virology 89:423–437

Yamada N, Tanihara K, Takada A, Yorihuzi T, Tsutsumi M, Shimomura H, Tsuji T, Date T (1996) Genetic organization and diversity of the 3' noncoding region of the hepatitis C virus genome. Virology 223:255–261

Yamashita T, Kaneko S, Shirota Y, Qin W, Nomura T, Kobayashi K, Murakami S (1998) RNA-dependent RNA polymerase activity of the soluble recombinant hepatitis C virus NS5B protein truncated at the C-terminal region. J Biol Chem 273:15479–15486

Yan Y, Li Y, Munshi S, Sardana V, Cole JL, Sardana M, Steinkuehler C, Tomei L, De Francesco R, Kuo LC, Chen Z (1998) Complex of NS3 protease and NS4A peptide of BK strain hepatitis C virus: a 2.2 Å resolution structure in a hexagonal crystal form. Protein Sci 7:837–847

Yanagi M, Purcell RH, Emerson SU, Bukh J (1997) Transcripts from a single full-length cDNA clone of hepatitis C virus are infectious when directly transfected into the liver of a chimpanzee. Proc Natl Acad Sci USA 94:8738–8743

Yanagi M, St. Claire M, Shapiro M, Emerson SU, Purcell RH, Bukh J (1998) Transcripts of a chimeric cDNA clone of hepatitis C virus genotype 1b are infectious in vivo. Virology 244:161–172

Yao N, Hesson T, Cable M, Hong Z, Kwong AD, Le HV, Weber PC (1997) Structure of the hepatitis C virus RNA helicase domain. Nat Struct Biol 4:463–467

Yasui K, Wakita T, Tsukiyama-Kohara K, Funahashi SI, Ichikawa M, Kajita T, Moradpour D, Wands JR, Kohara M (1998) The native form and maturation process of hepatitis C virus core protein. J Virol 72:6048–6055

Yen J-H, Chang SC, Hu C-R, Chu S-C, Lin S-S, Hsieh Y-S, Chang M-F (1995) Cellular proteins specifically bind to the 5'-noncoding region of hepatitis C virus RNA. Virology 208:723–732

Yi M, Nakamoto Y, Kaneko S, Yamashita T, Murakami S (1997) Delineation of regions important for heteromeric association of hepatitis C virus E1 and E2. Virology 231:119–129

Yoo BJ, Selby M, Choe J, Suh BS, Choi SH, Joh JS, Nuovo GJ, Lee H-S, Houghton M, Han JH (1995) Transfection of a differentiated human hepatoma cell line (Huh7) with in vitro-transcribed hepatitis C virus (HCV) RNA and establishment of a long-term culture persistently infected with HCV. J Virol 69:32–38

Yuan Z-H, Kumar U, Thomas HC, Wen Y-M, Monjardino J (1997) Expression purification and partial characterization of HCV RNA polymerase. Biochem Biophy Res Comm 232:231–235

Zeuzem S, Lee J-H, Roth WK (1997) Mutations in the nonstructural 5A gene of European hepatitis C virus isolates and response to interferon alfa. Hepatology 25:740–744

Zhu N, Khoshnan A, Schneider R, Matsumoto M, Dennert G, Ware C, Lai MMC (1998) Hepatitis C virus core protein binds to the cytoplasmic domain of tumor necrosis factor (TNF) receptor 1 and enhances TNF-induced apoptosis. J Virol 72:3691–3697

Internal Ribosome Entry Site-Mediated Translation in Hepatitis C Virus Replication

R.C.A. Rijnbrand and S.M. Lemon

1	Introduction	85
2	Primary, Secondary, and Tertiary Structure of the Hepatitis C Virus 5' Nontranslated RNA	90
3	Mapping the Hepatitis C Virus Internal Ribosome Entry Site	94
3.1	The 5' Limit of the Hepatitis C Virus Internal Ribosome Entry Site	94
3.2	The 3' Border: Is There a Requirement for Sequence from the Coding Region?	96
4	Hepatitis C Virus Internal Ribosome Entry Site Mediated Ribosome Entry	98
4.1	Ribosome Entry Takes Place at the Polyprotein Initiation Site	98
4.2	Interactions Between the 40S Ribosome Subunit and the Viral RNA	101
4.3	Involvement of Canonical and Noncanonical Host Translation Factors in Hepatitis C Virus Translation	101
4.4	Assembly of the Translation Complex on the Viral RNA	104
5	Involvement of the 3' Nontranslated RNA in Cap-Independent Translation	105
6	Changes in the 5' Nontranslated RNA Affect Translation	106
6.1	The Hepatitis C Virus Internal Ribosome Entry Site and Viral Tropism	106
6.2	Genetic Variation and Differences in Internal Ribosome Entry Site Activity	108
7	Similarities Between Hepatitis C Virus and Prokaryotic Translation Initiation	109
8	Summary: Unique Features of Flaviviral Internal Ribosome Entry Site Elements	109
	References	111

1 Introduction

The initiation of protein translation on eukaryotic messenger RNAs predominantly follows the first AUG rule, as described by Kozak (1989b). This states that the ribosome begins scanning an RNA molecule from its extreme 5' end until it encounters an AUG codon, at which point translation is initiated. In eukaryotes, mRNA molecules usually carry an ^{m7}GpppG cap structure at their 5' terminus. This 5' cap-structure strongly enhances translation, as it facilitates binding of translation initiation factors and the 40S ribosome subunit to the mRNA (reviewed

---

Department of Microbiology and Immunology, The University of Texas Medical Branch at Galveston, 4.104 Medical Research Building, 301 University Boulevard, Galveston, TX 77555-1019, USA

by PAIN 1996; SACHS et al. 1997). The scanning process, or 3' movement of the 40S ribosome subunit along the RNA in search of an AUG codon, is inhibited by very stable RNA structures. Also, AUG codons positioned between the 5' terminus and the AUG codon initiating a specific coding region reduce the efficiency of translation.

When picornaviral RNA sequences were determined, they appeared to defy the classical rules for eukaryotic translation initiation. Firstly, the 5' non-translated RNA (NTR) of picornaviruses, such as poliovirus and encephalomyocarditis virus (EMCV), were found to be relatively lengthy, representing up to 10% of the entire viral genome. They also contained multiple AUG codons between the 5' terminus and the site at which translation of the long open reading frame is initiated. Moreover, subsequent studies demonstrated that the picornavirus 5'NTR is highly structured and thus relatively resistant to scanning by the 40S ribosome subunit. Finally, picornavirus RNAs had been shown many years before to lack a 5' cap-structure and to have in its place a small viral protein covalently linked to the 5' end of the RNA (LEE et al. 1977). All of these characteristics conflict with the optimal conditions for cap-dependent translation of eukaryotic mRNAs. Thus, it is not surprising that picornaviruses were recognized in the late 1980s to initiate translation by an alternative mechanism involving internal ribosome entry (JANG et al. 1988; PELLETIER et al. 1988; reviewed in EHRENFELD and SEMLER 1995; HELLEN and WIMMER 1995; JACKSON and KAMINSKI 1995). The initiation of translation by this mechanism does not require a 5' cap-structure, but is critically dependent on a lengthy RNA segment upstream of the initiation codon. This highly structured RNA segment acts to direct the 40S ribosome subunit to the site of translation initiation, usually an AUG placed hundreds of nucleotides (nts) downstream of the 5' end of the RNA. The controlling segment of the 5'NTR has been referred to as the "ribosome landing pad" or, more commonly, the "internal ribosome entry site" (IRES). In the case of picornaviruses, the IRES is 3' (and functionally independent) of RNA structures located at the extreme 5' end of the genome that are required for RNA replication (ALEXANDER et al. 1994; JIA et al. 1996; LU and WIMMER 1996; MOLLA et al. 1992; ROHLL et al. 1994). IRES function depends on primary, secondary and tertiary RNA structure. Since their discovery in picornaviruses, such IRES elements have also been identified within a number of other viral and cellular mRNAs (BERLIOZ et al. 1995; BERNSTEIN et al. 1997; GAN and RHOADS 1996; IIZUKA et al. 1995; LIU and INGLIS 1992; MACEJAK and SARNOW 1991; MAGA et al. 1995; NANBRU et al. 1997; OH et al. 1992; PRATS et al. 1992; SIMONS et al. 1996; STONELEY et al. 1998; TEERINK et al. 1995; THIEL and SIDDELL 1994; THOMAS et al. 1991; VAGNER et al. 1995a,b; VERVER et al. 1991; YE et al. 1997).

When the sequence of the hepatitis C virus (HCV) genome became available, several characteristics were reminiscent of the picornaviral 5'NTR and IRES. Although the HCV 5'NTR (341 nts) was found to be shorter than that of the picornaviruses (600–1400 nts), it is still lengthier than the 5'NTR of most cellular mRNAs (KOZAK 1987) and considerably longer than the 5'NTR of yellow fever virus and other "classical" flaviviruses. The sequence upstream of the large open-reading frame within the HCV genome also contains several AUG codons (3–5, depending on the virus strain). In addition, the 5′NTR was recognized to form several stable RNA structures based on the presence of covariant nucleotide substitutions in different strains of the virus (BROWN et al. 1992; TSUKIYAMA-KOHARA et al. 1992). All of these features suggested that HCV translation might be initiated by a cap-independent, IRES-directed process, a fact that has since been confirmed by a number of investigators (TSUKIYAMA-KOHARA et al. 1992; WANG et al. 1993). In addition, translation driven by the HCV 5′NTR has been shown to be resistant to the shut-off of cap-dependent translation in cells expressing the polio- or coxsackievirus 2A proteinases (BORMAN et al. 1995, 1997; RIJNBRAND et al. 1995; TSUKIYAMA-KOHARA et al. 1992; WANG et al. 1993).

Among other members of the family *Flaviviridae*, HCV (genus *Hepacivirus*) is most closely related to members of the genus *Pestivirus* (for review see RICE 1997) and the novel, unclassified flavivirus, GB virus B (GBV-B) (MUERHOFF et al. 1995 ; OHBA et al. 1996). Short runs of sequence identity and common folding patterns have been recognized in the 5′NTRs of the pestiviruses bovine viral diarrhea virus (BVDV), border disease virus (BDV) and hog cholera virus (HoCV, also known as classical swine fever virus; BROWN et al. 1992; HONDA et al. 1998a) (Fig. 1). In contrast, as alluded to above, members of the genus *Flavivirus*, have 5′NTRs of 80–120 nts, which is close to the optimal leader length for cap-dependent translation. Unlike HCV and pestiviral RNAs, these latter flaviviral RNAs are thought to possess a 5′ cap-structure, and their polyproteins contain an apparent methyltransferase domain (capping enzyme) (KOONIN 1993). The HCV and pestivirus polyproteins do not contain such a sequence motif. However, the exact structure of the 5′ terminus of the HCV, pestivirus and GBV-B genomic RNAs is not known. There are no data available that distinguish between the presence of a 5′ cap-structure (presumed to be very unlikely), a picornavirus-like genome-linked protein, or an unprotected 5′ end (BROCK et al. 1992).

Very recently, PESTOVA et al. (1998) demonstrated that the 5′NTRs of HCV and HoCV are capable of binding specifically to the 40S ribosome subunit in the absence of any additional canonical or noncanonical translation factors. This makes these viral RNAs completely unique among all eukaryotic RNAs, as even the picornaviral IRES elements do not form binary complexes with the 40S subunit in the absence of a number of translational factors. Some canonical translation factors (such as eukaryotic initiation factor 3, or eIF3; SIZOVA et al. 1998) are likely to play an important role in HCV translation. However, the lack of a need for such factors in the specific interaction that occurs between the HCV IRES and the 40S subunit is very reminiscent of the interaction between the Shine-Dalgarno site of bacterial RNAs and the prokaryotic 30S ribosome subunit. This is a central feature of the HCV IRES, and it is likely to be dependent both on primary and higher order RNA structures.

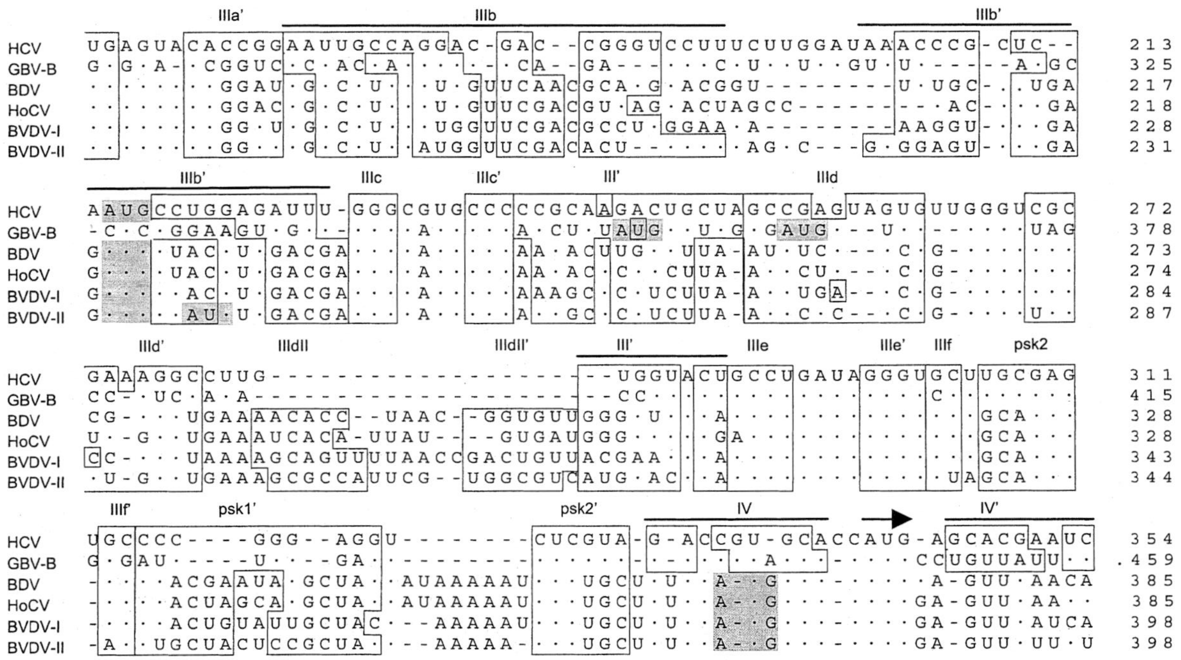

Fig. 1. Sequence alignment of the 5' NTR sequences of hepatitis C virus (HCV) (1a isolate; KOLYKHALOV et al. 1997), GBV-B (MUERHOFF et al. 1995), BDV (isolate X818; BECHER et al. 1998), HoCV (isolate alfort; MEYERS et al. 1996), BVDV type I (isolate NADL; COLLETT et al. 1988), BVDV type II (isolate 890; RIDPATH et al. 1995). Base-paired elements are divided into four structural domains (I, II, III and IV) based on local arrangements. Lines above the sequence indicate multiple base-paired regions that contribute to a specific structural element. Individual hairpins are labeled according to HONDA et al. (1996a), with the 3' base-pairing element marked by (¹). Stems involved in the formation of the pseudoknot are indicated as *psk1* and *psk2*. *Open boxes* indicate proposed helical structures, while *shaded boxes* indicate the location of AUG codons within the 5'NTRs. Gaps introduced to optimize alignment are indicated by (-); positions at which nucleotides are identical to the HCV sequence are shown with a (.).

2 Primary, Secondary, and Tertiary Structure of the Hepatitis C Virus 5' Nontranslated RNA

The 5'NTR is one of the most conserved regions of the HCV genome, reflecting its importance in both viral replication and translation. The small amount of sequence variation observed within this element is limited to specific regions (BUKH et al. 1992; SMITH et al. 1995). Overall, 5'NTRs from different HCV strains share over 85% nucleotide sequence identity. However, even though there is an impressive conservation of primary nucleotide sequence within the genus *Hepacivirus*, the level of identity between the 5'NTRs of HCV, pestiviruses and GBV-B is relatively low (Fig. 1). There are only a few short segments possessing high identity (Figs. 1, 2). Despite this, secondary and tertiary RNA structures appear to be largely conserved among these different viruses (BROWN et al. 1992; HONDA et al. 1996a, 1999b) (Fig. 2). Nonetheless, there is considerable variability in the length of these 5'NTRs. The shortest 5'NTR is that of HCV (341 nts), followed by pestiviruses (374–385 nts), and GBV-B (445 nts). The additional length of the GBV-B 5'NTR appears to be due to the presence of two additional stem-loop structures that are not found in other flaviviral IRESs. The structures of the HCV, pestiviral and GBV-B IRESs are, however, very different from the RNA structures that comprise the type 1 and type 2 IRES elements of picornaviruses (WIMMER et al. 1993). This has led to their proposed classification as type 3 viral IRES elements (LEMON and HONDA 1997). IRES elements have also been identified within the 5'NTRs of the unclassified flaviviruses, GB virus C (GBV-C, or "hepatitis G virus") and GB virus A (GBV-A) (SIMONS et al. 1996). These latter IRESs have a very different structure than that found in the other flaviviruses, however, and they cannot be considered to be type 3 elements (SIMONS et al. 1996).

Models of the secondary (and to some extent tertiary) structure of the 5'NTRs of HCV, pestiviruses and GBV-B have been developed based on a combination of phylogenetic comparative sequence analysis and computer assisted folding (BROWN et al. 1992; DENG and BROCK 1993; HONDA et al. 1996a, 1999a; LE et al. 1995; SMITH et al. 1995; WANG et al. 1995). For HCV and the pestivirus HoCV, the resulting models have been partially confirmed by analyses of the nuclease sensitivity of synthetic RNAs (BROWN et al. 1992; HONDA et al. 1996a, 1999a; SIZOVA et al. 1998). In addition, several conserved structures that are important for translation have been confirmed by mutational analysis (BURATTI et al. 1997; HONDA et al. 1996a, 1999a; REYNOLDS et al. 1996; RIJNBRAND et al. 1997; WANG et al. 1994, 1995) (Fig. 2).

All of these 5'NTRs have four major structural domains (Fig. 3), of which the most 5', domain I, is least conserved in sequence. Domains II and III of these viruses share considerable structural similarities, as well as several short tracts in which there is a high level of primary nucleotide sequence identity. The latter segments are generally predicted to be single stranded (LEMON and HONDA 1997) (Figs. 1, 2, 4). Domain IV spans the initiator AUG codon and appears to be

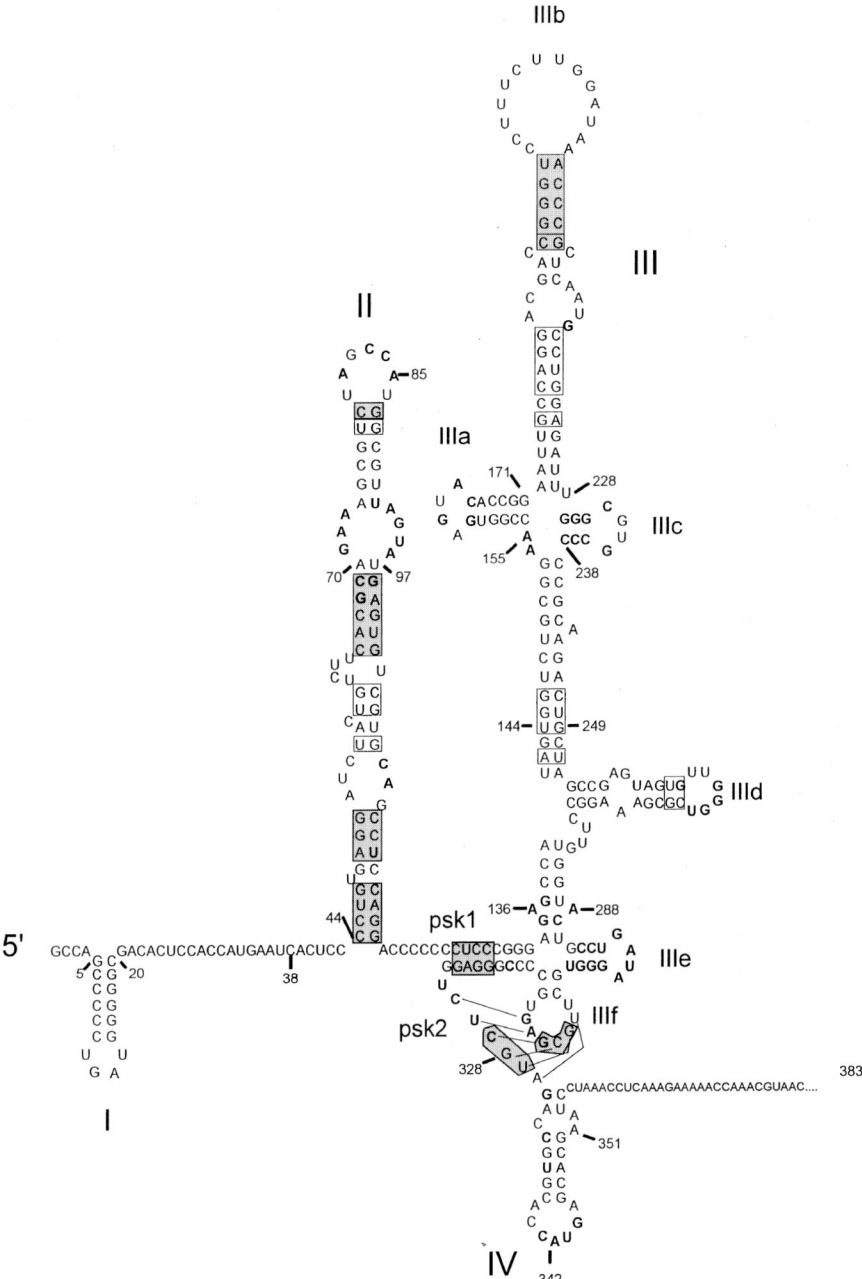

Fig. 2. Secondary and tertiary structure of the hepatitis C virus (HCV) 5'NTR (HCV-H strain). Structural domains are labeled I, II, III and IV, with predicted hairpin loops indicated by *letters*. Short nucleotide sequences that are conserved in HCV as well as pestiviruses and GBV-B are indicated in *bold font*. Base-pairings that have been validated by genetic analysis are indicated by *shaded boxes*, while *open boxes* indicate base pair interactions confirmed by natural sequence covariation

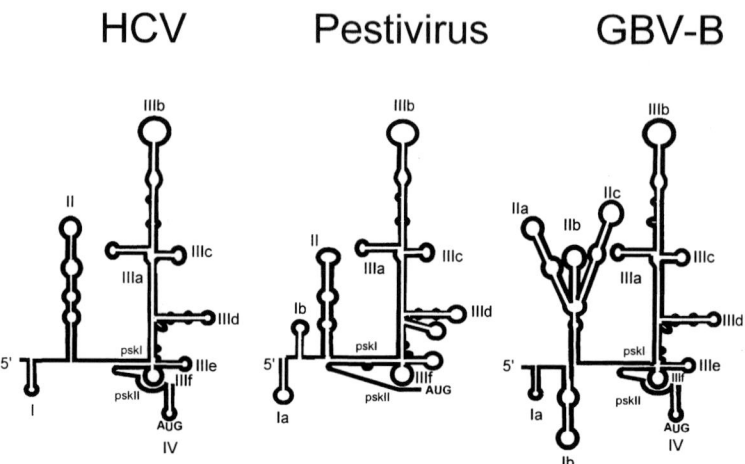

Fig. 3. The RNA structures predicted for the hepatitis C virus (HCV), pestivirus, and GB virus B (GBV-B) 5'NTRs. The polyprotein initiation site is indicated by AUG. Major structural domains are labeled as in Figs. 1 and 2

uniquely structured in HCV and GBV-B (HONDA et al. 1996a). These structural domains are considered in greater detail in the following paragraphs.

The most 5' structure in each of these flaviviral RNAs appears to be a small stem-loop (Fig. 3; stem-loop Ia). This segment of the 5'NTR has the lowest level of sequence identity between HCV, the pestiviruses, and GBV-B. In contrast to the HCV 5'NTR, the 5'NTRs of GBV-B and the pestiviruses are predicted to form an additional stem-loop upstream of their respective IRES elements (Fig. 3, stem-loop Ib). These structures are not part of the IRES (which is composed of domains II, III and part of IV, see below) and they most likely function as signals for RNA replication. However, there are as yet no experimental data supporting the existence of these stem-loops or their putative role in replication of the viral RNA.

Domain II is complex and consists of multiple stems and bulge loops (Fig. 3). Many of the nucleotides in the loop regions of this structure are identical in each of these viruses (HCV, pestiviruses, GBV-B), in contrast to the majority of nucleotides present in base-paired regions (HONDA et al. 1999a) (Fig. 4). The apical loop contains a conserved AGCCA sequence (AACCA in BDV X818), while two bulge loops in this stem have GA and AGUA sequences that are absolutely conserved among the genera. The strong conservation of the loop nucleotides is likely to reflect their involvement in an RNA/RNA or possibly RNA-protein interaction involving conserved host cell elements during viral translation or possibly RNA replication. Despite conservation of many features of domain II, however, there are also a number of very distinct differences between the genera. The most notable difference is in the GBV-B RNA, which appears to form two additional stem-loops (IIc and IId) that are not present in hepaciviruses or pestiviruses (HONDA et al. 1996a; LEMON and HONDA 1997) (Fig. 3). As indicated above, these insertions largely account for the greater length of the GBV-B 5'NTR.

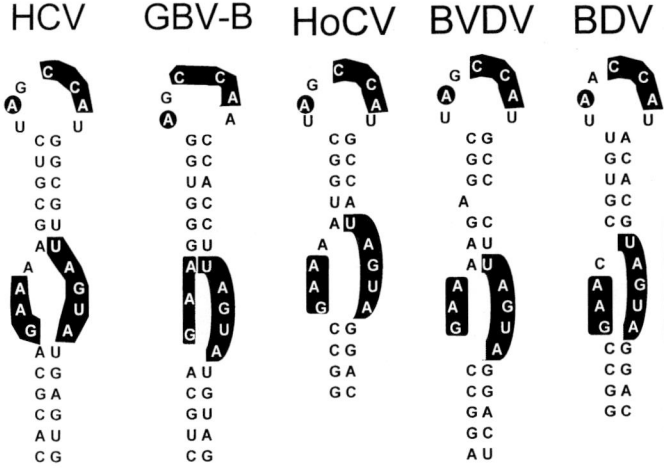

Fig. 4. Primary nucleotide sequence of the apical segments of domain II in hepatitis C virus (HCV), GB virus B (GBV-B) (IIa), hog cholera virus (HoCV), bovine viral diarrhea virus (BVDV) and border disease virus (BDV). Conserved nucleotide sequences are *boxed*. (Modified from HONDA et al. 1998a)

Domain III is the largest RNA structure within the 5′NTR, and in several ways it demonstrates the highest degree of structural conservation among these different viruses. It comprises the core of the flaviviral IRES and has at least weak translation initiating activity even in the absence of domain II (TSUKIYAMA-KOHARA et al. 1992). One very striking feature of domain III is an RNA pseudoknot (Figs. 2, 3), which is positioned immediately upstream of the initiation site. The distance between the 3′ end of this pseudoknot and the initiator AUG codon is fixed at 11 nts in HCV and GBV-B, and 12 nts in the pestiviruses (HONDA et al. 1996a; LE et al. 1995; WANG et al. 1995). RNA pseudoknots have also been predicted in picornaviral (LE et al. 1992, 1993), coronaviral (LE et al. 1994) and the GBV-A and GBV–C virus (SIMONS et al. 1996) IRES elements. However, unlike the case with the type 3 IRESs (RIJNBRAND et al. 1997; WANG et al. 1995), there are as yet no experimental data to support such structures in the IRESs of these other viruses. The sequences of the unpaired loop segments of stem-loops IIId and IIIe, and the base-paired stems of IIIc and IIIe, are relatively well conserved among the type 3 flaviviral IRESs. A remarkable feature is the conservation of four unpaired A residues (positions 136, 154, 155 and 288 in HCV) that are located in small bulge-loops in the domain III structures of each of these viruses (Figs. 1, 2).

Domain IV contains a stem-loop that spans the initiator AUG. This stem-loop structure has been predicted to exist only in HCV and GBV-B (HONDA et al. 1996a; SMITH et al. 1995). No comparable structural element appears to be present in the pestivirus 5′NTR (HONDA et al. 1996a; SIZOVA et al. 1998). Although the AUG codon at which translation is initiated is located within the loop region of this structure, the stem-loop is not required for either HCV or GBV-B IRES-mediated translation (HONDA et al. 1996a; Rijnbrand and Lemon, in preparation; TSUKI-YAMA-KOHARA et al. 1992; WANG et al. 1993). The stem-loop may nonetheless play a key role in regulating the initiation of translation on these viral RNAs (see below).

It is important to remember that these predictions of IRES structure represent at best only crude approximations of the higher order structures assumed by these viral RNAs. Recent studies with other RNAs suggest that these IRES segments are likely to fold into compact structures with interiors from within which water molecules are relatively excluded. A better understanding of the tertiary structure assumed by the type 3 flaviviral IRES is likely to be gained from studies employing biophysical approaches that have been used previously to solve complex protein structures, such as NMR and X-ray crystallography.

3 Mapping the Hepatitis C Virus Internal Ribosome Entry Site

Several approaches have been taken to determine the minimal sequence required for HCV and pestivirus IRES-dependent translation. To a large extent the techniques used have been similar to those employed in earlier studies of picornaviral IRES elements. A common approach has been the analysis of bicistronic RNAs in which two reporter protein-coding sequences are separated by an IRES sequence. Translation of the upstream reading frame occurs in a 5' end-dependent fashion and involves scanning of the 40S subunit, while translation of the downstream reading frame is driven by the IRES element. Easily quantifiable reporter proteins, like luciferase (WANG et al. 1993; TSUKIYAMA-KOHARA et al. 1992) or chloramphenicol acetyl transferase (CAT; RIJNBRAND et al. 1995) have been used by a number of laboratories. However, others have used either the HCV core protein-coding region (TSUKIYAMA-KOHARA et al. 1992; FUKUSHI et al. 1994) or even the entire HCV open reading frame as the reporter sequence in an effort to study the IRES in a more natural context (HONDA et al. 1996b). Using these approaches, several laboratories have attempted to map the 5' and 3' borders of the HCV and pestivirus IRES elements.

3.1 The 5' Limit of the Hepatitis C Virus Internal Ribosome Entry Site

With a few exceptions, these studies paint a relatively consistent picture of the 5' limits of the IRES. Data related to the 5' border of the HCV IRES have been provided by TSUKIYAMA-KOHARA et al. (1992), WANG et al. (1993), KETTINEN et al. (1994), FUKUSHI et al. (1994), RIJNBRAND et al. (1995), HONDA et al. (1996b, 1999a), REYNOLDS et al. (1996), and KAMOSHITA et al. (1997). These results are summarized in Fig. 5. Early reports by TSUKIYAMA-KOHARA et al. (1992) and KETTINEN et al. (1993), reported the 5' border to be located between nts 110 and 156. In contrast FUKUSHI et al. (1994) found the entire 5'NTR to be required for IRES activity. Despite these reports, most studies have shown the 5' border of the IRES to be

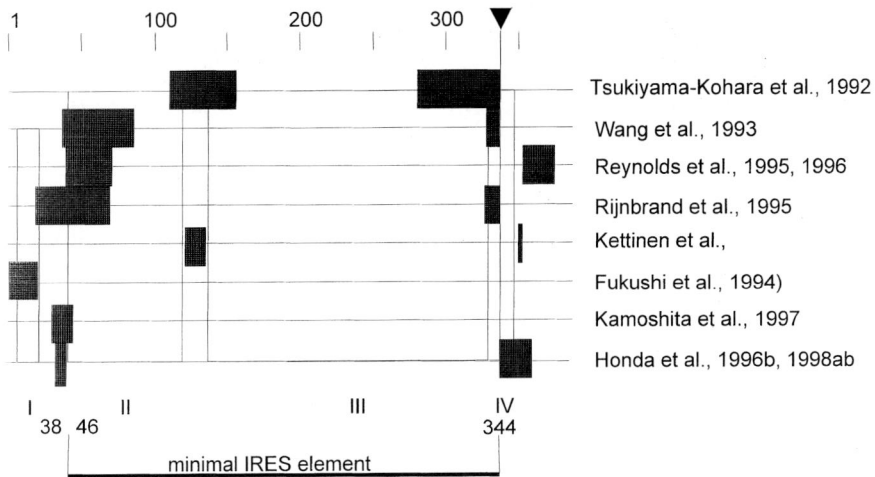

Fig. 5. Summary of published efforts to define the 5' and 3' borders of the hepatitis C virus (HCV) IRES element. Solid boxes indicate the area to which the 5' or 3' border was mapped. The solid line at the bottom of the diagram indicates the consensus minimal IRES element

located at or near the 5' end of domain II (nucleotide 44; HONDA et al. 1996b; KAMOSHITA et al. 1997; REYNOLDS et al. 1996; RIJNBRAND et al. 1995; WANG et al. 1993). We have found that deletion of nts 32–37 does not interfere with IRES function, while substitutions at nts 45–46 have a strong, negative effect on translation (HONDA et al. 1999a,b). These results place the 5' border of the HCV IRES between nts 38 and 46.

A similar situation exists for the pestivirus IRES element. In an early study, POOLE et al. (1995) mapped the 5' border of the BVDV IRES in an in vitro translation system and found it to be located between nts 139 and 154, downstream of domain II. However, RIJNBRAND et al. (1995), studying translation in transfected cells, obtained results suggesting that the 5' border of the HoCV IRES is located between nts 28 and 66. This would place the 5' limit of the HoCV IRES between stem-loop Ia and domain II, similar to the majority of studies related to HCV.

There are several possible explanations for the discrepancies that exist between different studies that have attempted to map the 5' limits of these IRES elements. REYNOLDS et al. (1996) suggested that translation directed by the HCV IRES might be less dependent on 5'NTR sequences (requiring only nts 120–341) when the downstream, core protein-coding sequence is present in RNA transcripts. This seems unlikely, however, since HONDA et al. (1996b) and FUKUSHI et al. (1994) both found a requirement for the upstream 5'NTR sequence when studying translation on RNAs containing the HCV core protein-coding sequence. It is more likely that varying conditions for translation may explain these results. For example, KAMOSHITA et al. (1997) found the 5' limit of the HCV 5' IRES to be located between nts 28 and 45 under physiological salt conditions, while TSUKIYAMA-KOHARA et al. (1992) mapped the 5' limit to between nts 110 and 156 when studying similar transcripts at low salt concentrations.

Most studies agree that stem-loop I is not required for IRES activity. In contrast, its inclusion in RNA transcripts has generally been found to exert a slightly negative influence on translation. Translation has been shown to be enhanced upon removal of the structure from both HCV and HoCV RNAs (HONDA et al. 1996b; KAMOSHITA et al. 1997; RIJNBRAND et al. 1995, 1997). The inhibitory effect of stem-loop I on HCV translation may be cell-type specific, as KAMOSHITA et al. (1997) found that RNA transcripts lacking nts 1–22 were translated slightly more efficiently than the full length 5'NTR in a HeLa S10 lysate, equally efficiently in HeLa cells, but less efficiently in African green monkey kidney cells. Such differences suggest the possibility that a host cell-specific factor(s) may interact with stem-loop I, resulting in a reduction in translation efficiency. However, deletion of the most 5' 28 nts resulted in an increase in translation in all cases (KAMOSHITA et al. 1997). The retention of stem-loop I in the HCV genome in the face of its apparent negative effect on translation probably reflects an essential role for this structure in RNA replication (most likely positive-strand initiation).

3.2 The 3' Border: Is There a Requirement for Sequence from the Coding Region?

The IRES extends in a 3' direction at least as far as the initiator AUG codon. The intervening sequence, including the RNA pseudoknot involving stem-loop IIIf (Figs. 2, 3), is essential for translation on both HCV and HoCV RNAs. However, as long as the general structure of the pseudoknot is preserved, the actual nucleotide sequence appears to be of less importance (RIJNBRAND et al. 1997; WANG et al. 1994, 1995). Each of the upstream hairpin structures in domains II and III appear to be required for IRES-mediated translation (HONDA et al. 1996b; RIJNBRAND et al. 1995; WANG et al. 1994). However, nucleotide substitutions made within base-paired segments of these domains are generally well tolerated, provided that base-pairing is preserved. This explains why the considerable variation in the primary sequence of the base-paired segments of domain III that exists among different HCV genotypes. Nonetheless, compensatory substitutions within stem-loop II and III may reduce translational efficiency by two-fold (BURRATTI et al. 1997; HONDA et al. 1999a). Nucleotide substitutions within the unpaired loop sequences generally appear to have a much greater impact on translation.

The most controversial aspect of the HCV IRES has been the suggestion that the extreme 5' capsid coding sequence may be required for IRES activity (REYNOLDS et al. 1995). In early studies, significant translational activity was observed with dicistronic transcripts that contained the HCV IRES fused directly to sequence encoding either CAT (TSUKIYAMA-KOHARA et al. 1992) or luciferase (WANG et al. 1993) at the natural HCV polyprotein initiation site. However, more recent studies indicate that IRES activity is dependent on the nature of the sequence immediately downstream of the initiator AUG codon. First, REYNOLDS et al. (1995) noted a significant increase in translational activity if the most 5' core protein-coding sequences were included in the RNA transcript. RNAs containing 0–10 nts of the core coding region translated at near background levels, while transcripts containing 34 nts of the core coding region efficiently translated either of two heterologous reporter proteins (influenza NS or secretory alkaline phosphatase).

In addition to these observations, HONDA et al. (1996b) observed that the introduction of 13 silent nucleotide substitutions between nts 8 and 41 of the core protein-coding sequence substantially reduced the translation of nearly genome-length HCV RNAs in a cell-free, in vitro translation system. However, these silent mutations in the core coding sequence had no apparent influence on the level of translation when these RNAs were transcribed in vivo in an hepatocyte derived cell line using the hybrid vaccinia-T7 expression system.

Finally, LU and WIMMER (1996) found that chimeric polioviruses in which the native picornaviral IRES was replaced by the HCV IRES required at least 24 nts of the core protein-coding sequence to be viable. A chimeric virus containing 370 nts of the core protein-coding region demonstrated the most efficient replication. The authors suggested that the integrity of the first 123 amino acids of the HCV core protein was essential for the replication of the chimeric virus (presumably to ensure sufficient translation), since both a frameshift mutation and an internal deletion in this region severely impaired replication. However, although a nonviable chimera in which the HCV initiator codon was the first codon of the poliovirus polyprotein translated with notably poor activity in a HeLa cell lysate, the translational efficiency of the viable chimeric RNAs did not closely parallel the plaque sizes. Thus, it seems more likely that polyprotein processing or some other aspect of picornavirus replication may have been limiting. This interpretation is consistent with recent studies in our laboratory, in which we observed no transactivation of IRES activity following high level expression of the core protein in Huh-7 cells (T.-H. Wang et al., manuscript in preparation).

It is important to note that the first several hundred nucleotides of the HCV open reading frame are highly conserved with strong evidence for codon bias (BUKH et al. 1994; SMITH et al. 1997). The presence of conserved secondary RNA structure within this segment of the genome is supported by the existence of covariant nucleotide substitutions (SMITH et al. 1997). However, there is no evidence that the two stem-loop structures that have been proposed in this segment play any kind of a role in cap-independent translation.

In conflict with the hypothesis that IRES activity is critically dependent on the presence on the HCV core coding region are other studies that demonstrate relatively efficient IRES-mediated translation in the absence of such sequence (RIJNBRAND et al. 1995; TSUKIYAMA-KOHARA et al. 1992; WANG et al. 1993). Two different explanations have been proposed for the discrepancy between these studies and the work of REYNOLDS et al. (1995). First, REYNOLDS et al. (1995) suggested that a fortuitous (albeit weak) sequence homology between the luciferase and HCV capsid sequences may have accounted for the IRES activity observed by WANG et al. (1993) in transcripts containing the luciferase sequence fused directly to the HCV initiator AUG. Such homology was not evident in the influenza NS' reporter sequence used by Reynolds and coworkers. As an alternative explanation, HONDA et al. (1996a) proposed that the fusion of some reporter gene sequences with the

HCV 5'NTR may result in local RNA structure that is unfavorable for translation initiation. This view is supported by strong evidence that the 40S ribosome interacts with the RNA directly at the site of initiator AUG codon, and that structure surrounding the AUG reduces translation activity (HONDA et al. 1996a; REYNOLDS et al. 1996; RIJNBRAND et al. 1996) (see the following section). If this is the case, then the inclusion of the 5' core protein-coding sequence may be beneficial simply because it ensures the absence of unfavorable base pair formation in this region. Although we favor the latter explanation, the available evidence cannot distinguish completely between these two possibilities, and it should be noted that they are not necessarily mutually exclusive. However, an important piece of information that may be relevant to this controversy is the observation that RNA transcripts representing only the 5'NTR, and terminating just upstream of domain IV, are capable of forming binary complexes with purified 40S ribosome subunits (PESTOVA et al. 1998). This suggests that the downstream sequence may not be an essential component of the IRES.

4 Hepatitis C Virus Internal Ribosome Entry Site Mediated Ribosome Entry

4.1 Ribosome Entry Takes Place at the Polyprotein Initiation Site

In the case of poliovirus RNAs, the 40S ribosome subunit appears to make an initial contact in the region of an AUG codon that is located near the 3' end of the IRES (JACKSON et al. 1990). Subsequent to this contact, the 40S subunit scans in a 3' direction towards the authentic initiator AUG. In contrast, in other picornaviruses (the cardioviruses and aphthoviruses, and probably also HAV), the 40S subunit forms a primary contact with the viral RNA within a few nucleotides of the initiator AUG codon (reviewed by STEWART and SEMLER 1997). Translation directed by the HCV IRES appears to occur by a process that more closely resembles the cardiovirus scheme, but there are significant differences with even these picornaviral IRES elements. With all picornavirus IRESs, ribosome entry is thought to be dependent upon a conserved pYxxxAUG motif (PILIPENKO et al. 1992). A sequence similar to this picornavirus motif, with appropriate spacing between the pyrimidine rich (pY) region and the downstream AUG codon, is present within stem-loop IIIb of the HCV IRES (BROWN et al. 1992) (Figs. 1, 2). However, this motif is lacking in most of the pestiviruses and GBV-B (Fig. 1), and mutational analyses of the HCV motif do not support a role in translation similar to that of the pYxxxAUG motif of picornaviruses. Mutations in the AUG codon itself do not affect translation efficiency (REYNOLDS et al. 1996; RIJNBRAND et al. 1996; WANG et al. 1994; YEN et al. 1995). Furthermore, small changes in the pyrimidine-rich loop sequence of stem-loop IIIb have rather minimal effects on translation efficiency (WANG et al. 1994; YEN et al. 1995), although a five-fold decrease in translation was observed following substitutions of six of the pyrimidine residues in this loop sequence (BURATTI et al. 1997). Thus, in the case of the HCV IRES, translation does not appear to be dependent upon a pYxxxAUG motif. A possible explanation for the reduction in IRES activity observed with the IIIb loop mutants (BURATTI et al. 1997) is provided by the observation that this sequence is complementary to an unpaired loop segment of 18S ribosome RNA (BROWN et al. 1992; DENG and BROCK 1993). Thus, base pair interactions between these segments may play a role in the binding of the viral RNA to the ribosome subunit. Alternatively, PESTOVA et al. (1998) and SIZOVA et al. (1998) have shown that this stem-loop interacts with eIF3 (discussed below), and this protein-RNA interaction may be important in the formation of binary complexes between the 5'NTR and the 40S subunit.

As mentioned previously, the HCV, pestivirus and GBV-B 5'NTRs all contain multiple AUG codons upstream of the initiator AUG (Fig. 1). Although some of these are conserved between different HCV genotypes and even between different viral genera, there are no AUG codons that are absolutely conserved within the 5'NTRs of all of these viruses (Fig. 1). This suggests that internally located AUG codons are not required, per se, for IRES mediated translation in these flaviviruses. Mutations involving the AUG codons at nts 85 and 96 in the HCV 5'NTR did result in a significant reduction in translation efficiency (REYNOLDS et al. 1996; RIJNBRAND et al. 1996). Substitutions of the third base of the AUG-85 codon had little effect on translation provided that the predicted base pair interaction involving nt 87 was preserved by an appropriate compensatory substitution at nt 79 (REYNOLDS et al. 1996).

Several lines of evidence suggest that the 40S ribosome subunit does not scan for the polyprotein initiation site on HCV RNA, but rather enters on the RNA directly at the site of translation initiation. Translation is not initiated at AUG codons placed either upstream (REYNOLDS et al. 1996; RIJNBRAND et al. 1996) or downstream of the authentic initiator codon (RIJNBRAND et al. 1996). A similar observation was made for the pestivirus HoCV (RIJNBRAND et al. 1997). These experiments indicate that ribosome scanning, if it occurs at all, is limited to a narrow window upstream of the initiation site. These data also indicate that the location of the initiator AUG codon is critical for efficient IRES function, thereby making it an intrinsic part of the IRES. Parenthetically, these observations explain the lack of IRES activity in the RNA transcripts studied by Yoo et al. (1992), who failed to identify evidence for internal initiation of translation on HCV RNA. The transcripts studied by Yoo et al. (1992) contained a 30 nts spacer sequence placed between the 5'NTR sequence and the initiator AUG, which opened the reading frame of a downstream CAT reporter sequence.

Equally strong evidence for direct positioning of the 40S subunit over the initiator AUG has been presented by HONDA et al. (1996a). Nuclease mapping studies and phylogenetic analyses suggest that the HCV initiator AUG codon is located within the unpaired loop sequence of a small stem-loop structure (domain IV) (HONDA et al. 1996a) (Fig. 2). This structure must be melted to accommodate the proper positioning of the 40S subunit over the AUG codon. HONDA et al.

(1996a) determined the translational activities of a number of mutant RNAs containing substitutions in this stem-loop. Both in vitro and in vivo, mutations that even minimally enhanced the stability of this putative stem-loop resulted in significant decrements in translation. However, these mutations had no effect on translation when it was initiated by a 5′ end-dependent scanning mechanism on RNAs with large 5′ deletions (HONDA et al. 1996a). These data clearly indicate that the 40S subunit does not approach the initiator AUG in a scanning mode. The normal predicted free energy of stem-loop IV is only −6.2kcal/mol. Scanning ribosomes are easily capable of melting stem-loops as stable as −30kcal/mol (KOZAK 1989c). The fact that mutations that minimally stabilize stem-loop IV have such dramatic effects on translation can only be explained by the absence of scanning. However, a mutation that significantly decreased the stability of stem-loop IV resulted in only a marginal increase in translation (HONDA et al. 1996a). This indicates that the stem-loop is not required for IRES activity and that it does not (by itself) significantly hinder the docking of the 40S subunit to the RNA. The presence of IRES activity in transcripts lacking the 3′ stem sequence (TSUKIYAMA-KOHARA et al. 1992; WANG et al. 1993) provides further evidence that stem-loop IV is not essential for cap-independent translation.

As indicated above, these data provide a potential explanation for the poor translational activity of some RNA transcripts in which a reporter protein sequence is fused directly to the HCV initiator codon (REYNOLDS et al. 1995). Should the fusion result in even minimally stable RNA structure in the region surrounding the AUG, translation would be adversely affected (HONDA et al. 1996a). Why the domain IV structure is conserved in HCV and GBV-B (but not the pestiviruses) remains a mystery. Its function remains unknown, although it is tempting to consider a role in regulating translation. For example, the specific binding of a protein to this structure could favorably influence the stability of the stem-loop, and this would likely have an important negative impact on translation. However, no such protein has yet been identified. It is relevant to note that similar translational regulatory mechanisms have been identified within prokaryotic viruses (reviewed by HONDA et al. 1996a).

Direct interaction of the 40S subunit with the viral RNA at the site of translation initiation is also consistent with the fact that some non-AUG codons, like AUU or CUG, may effectively substitute for the authentic initiator AUG. These mutants demonstrate only minimal changes in translation efficiency in a cell-free translation system (REYNOLDS et al. 1995), although AUU functions rather inefficiently in transfected cells (HONDA et al. 1996b). Other non-AUG codons, such as ACG, do not function very efficiently as translation initiation sites (REYNOLDS et al. 1995; RIJNBRAND et al. 1996). It appears that functional non-AUG codons must contain at least two of the three nucleotides normally present in the AUG initiator, perhaps to allow effective base-pairing with the Met-tRNA anti-codon. These results are at odds with most studies of cellular RNAs that follow the scanning model, although initiation may occur with efficiencies up to 67% efficiency when AUU codons are placed in an optimal initiation context (KOZAK 1989a; PEABODY 1987). The greater recognition of non-AUG initiator codons within the context of the HCV IRES is likely to reflect the importance of upstream RNA sequences and structures that facilitate the binding of the RNA to the 40S subunit and that act to properly position the initiator site at the Met-tRNA anti-codon.

4.2 Interactions Between the 40S Ribosome Subunit and the Viral RNA

Each of these different lines of evidence suggests that the 40S ribosome subunit becomes positioned directly over the translation initiation codon. However, the interaction between the AUG codon and the Met-tRNA anti-codon is not likely to be the first contact to occur. The initial contacts are more likely to involve the conserved upstream stem-loop structures in domain II and III (HONDA et al. 1996a). Purified 40S subunits are able to bind to transcripts lacking stem-loop IV (PESTOVA et al. 1998), clearly indicating that the initial contact of the 40S subunit with the RNA is not dependent on sequence downstream of the AUG codon. In addition, 40S ribosome subunits are able to bind to the RNA in the absence of the Met-tRNA-GTP ternary complex, or any other cellular translation factor. Although the order is not known, the Met-tRNA-GTP ternary complex may interact in a subsequent step with the RNA-40S subunit complex. The translation initiation factor 3 (eIF3) binds to both the HCV IRES and the 40S subunit (PESTOVA et al. 1998; SIZOVA et al. 1998; TOLAN et al. 1983), effecting a conformational change in the complex that places the AUG codon opposite the Met-tRNA anti-codon. As the distance between the AUG initiator codon and the most 3' base pairs of the pseudoknot is 11–12 nts, melting of the pseudoknot would not seem to be required to achieve these complexes (HONDA et al. 1996a). The footprint of an 80S ribosome covers approximately 30 nts, and the length of footprint upstream of the initiator AUG is on the order of about 11 nts (KOZAK 1977).

It may be that ribosome entry is directed at least in part by base pairing between the viral RNA and 18S rRNA. Certainly, several potential interactions between 18S rRNA and HCV and pestivirus NTRs have been identified (BROWN et al. 1992; DENG and BROCK 1993; LE et al. 1995). However, as probable as this hypothesis seems, particularly in light of the binary complexes formed by HCV RNA and 40S subunits (PESTOVA et al. 1998), there are thus far no experimental data that directly support base pairing between viral and cellular RNAs.

4.3 Involvement of Canonical and Noncanonical Host Translation Factors in Hepatitis C Virus Translation

Several cellular proteins appear to interact specifically with the HCV 5'NTR (Fig. 6), and some of these have been suggested to function as noncanonical translation initiation factors. Most notable among these are the polypyrimidine tract-binding protein (PTB, also known as "heterogeneous nuclear ribonucleoprotein I" or hnRNP I) (ALI and SIDDIQUI 1995) and the La autoantigen (ALI and

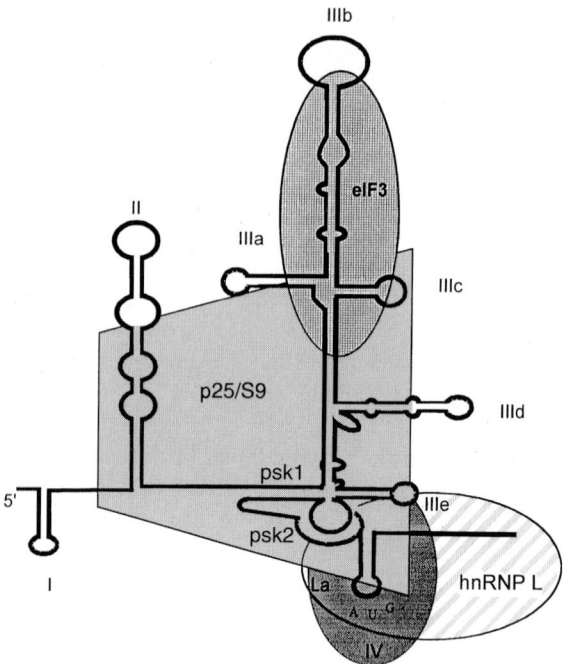

Fig. 6. Location of binding sites as they have been mapped for polypyrimidine tract-binding protein (PTB) (hnRNP I) (ALI and SIDDIQUI 1995); p25/S9 (FUKUSHI et al. 1997, PESTOVA et al. 1998), p120/eIF3 (BURATTI et al. 1998; PESTOVA et al. 1998; SIZOVA et al. 1998; YEN et al. 1995), La (ALI and SIDDIQUI 1997) and hnRNP L (HAHM et al. 1998)

SIDDIQUI 1997). However, an essential role for any noncanonical translation factor in HCV translation seems less likely in light of the observation that the viral 5′NTR forms a binary complex with the 40S ribosome subunit in the absence of any canonical or noncanonical translation initiation factor (PESTOVA et al. 1998). While the interaction of eIF3 with stem-loops IIIb and IIIc may play an important role in mediating the correct alignment of the viral RNA on the ribosome subunit (PESTOVA et al. 1998; SIZOVA et al. 1998), the role of other cellular proteins in translation initiation is uncertain. Although it cannot be excluded that additional factors may stimulate internal initiation of translation following assembly of the RNA-40S complex, the minimal set of components required for this process appear to be GTP, eIF2, and eIF3 in concert with the 40S and 60S ribosome subunits (PESTOVA et al. 1998). This distinguishes the HCV IRES from the picornaviral IRES elements that appear to require almost all canonical initiation factors (PESTOVA et al. 1996).

Both PTB and La have been postulated to play important roles in picornaviral translation (reviewed in BELSHAM et al. 1995, 1996; EHRENFELD and SEMLER 1995; HELLEN and WIMMER 1995). PTB is a 57kDa nuclear protein that binds to polypyrimidine tracts typically found at the 3′ end of introns in pre-mRNAs. Although its cellular functions are incompletely defined, it may play a role in RNA splicing. The involvement of PTB in translation mediated by the EMCV and FMDV IRESs was suggested by the results of depletion analyses (BORMAN et al. 1993; BOROVJAGIN et al. 1994, HELLEN et al. 1993; KAMINSKI et al. 1995). Thus, it is interesting that a bacterially produced GST-PTB fusion protein was reported to interact with RNA fragments derived from the 5' 122 nts of the HCV 5'NTR (ALI and SIDDIQUI 1995). These results are controversial, however. YEN et al. (1995) were unable to detect complex formation with a probe representing the 5' 130 nts of the HCV 5'NTR. Since ALI and SIDDIQUI (1995) studied the interaction between purified GST-PTB (or PTB-enriched fractions of cell lysates) and RNA probes synthesized with photoreactive 4-thio-UTP, there is concern about the specificity of the RNA binding activity they observed.

In searching for a functional role for PTB in HCV translation, ALI and SIDDIQUI (1995) found that translationally active lysates that were depleted of PTB by an immunoadsorption process no longer supported efficient IRES activity. However, reconstitution of the lysate with GST-PTB did not restore translation. This suggests that some other essential translation factor(s) may have been removed with PTB during the depletion step, or that GST-PTB is not able to substitute for native PTB in this system. In contrast, KAMINSKI et al. (1995) found that IRES activity was retained following the removal of PTB from cell lysates by an alternative RNA adsorption strategy. Lysates that no longer supported translation directed by the EMCV IRES retained the ability to support HCV IRES activity. Thus, it seems certain that the HCV IRES has at least a much reduced requirement for PTB compared with the EMCV IRES. Although it is difficult to prove absolutely, the results reported by KAMINSKI et al. (1995) suggest that PTB is not essential for translation mediated by the HCV IRES.

The La autoantigen (also a nuclear protein) binds to RNA fragments containing AUG codons (MCBRATNEY and SARNOW 1996), as well as a wide variety of viral 5'NTRs, including those of poliovirus and other picornaviruses, the human immunodeficiency virus, and influenza virus (SVITKIN et al. 1994; PARK and KATZE 1995). La protein appears to enhance the activity of the poliovirus IRES, and to facilitate the correct initiation of cap-independent translation on poliovirus RNAs (MEEROVITCH et al. 1993). Thus, it is of interest that ALI and SIDDIQUI (1997) found La to bind to a site overlapping the polyprotein initiation site and stem-loop IV of HCV RNA. The binding of La to the HCV 5'NTR was extraordinarily dependent on the presence of the initiator AUG codon. Moreover, the addition of purified La protein to rabbit reticulocyte lysate (which generally has a low abundance of La) enhanced HCV IRES-directed translation up to 60-fold. Since La possesses RNA unwinding activity (HUHN et al. 1997) and also binds to the 40S subunit (PEEK et al. 1996), it is intriguing to speculate that La might mediate the unwinding of stem-loop IV in a way that facilitates entry of the ribosome onto the viral RNA. However, further studies are needed to confirm these results. Given the importance of the initiator AUG in the binding of La to the HCV IRES (ALI and SIDDIQUI 1997), it would be particularly interesting to study the binding of La to translationally active RNA transcripts that contain noncanonical initiator codons such as AUU or CUG (see above).

HAHM et al. (1998) reported recently that yet a third nuclear protein, hnRNP L, binds specifically to HCV RNA in the vicinity of the initiation codon. Greater binding activity was observed with RNA probes containing increasing lengths of the capsid protein-coding sequence. These RNAs also demonstrated greater translational activity, suggesting a possible correlation between hnRNPL binding and activity of the IRES.

An interaction between domain II of the HCV 5′NTR and an as yet unidentified 25kDa protein present in HeLa lysate was reported by FUKUSHI et al. (1997). Translation was reduced by mutations in domain II that altered binding of the protein. However, these mutations also destabilized base pair interactions within the structure of the IRES (HONDA et al. 1998a). Thus, it is not possible to draw any conclusions concerning the role of this 25kDa protein in translation. PESTOVA et al. (1998) also reported the interaction of a 25kDa protein (ribosome protein S9) with the HCV and HoCV IRES elements. The binding of this protein to the viral RNA was dependent on structure in domains II and III (hairpin IIIc, pseudoknot), as well as sequence downstream of the RNA pseudoknot. Its role in HCV translation, if any, is not known.

YEN et al. (1995) described the interaction of two additional cytoplasmic proteins (87 and 120kDa, respectively) with an RNA fragment representing stem-loops IIIa, IIIb, and IIIc of the HCV 5′NTR (nts 131–253). Although the binding of 87kDa protein was dependent on the sequence of the IIIb loop, a mutational analysis suggested a role for only the 120kDa protein in IRES-mediated translation. However, BURATTI et al. (1998) also observed an interaction between hairpin IIIb and cellular proteins of 120 and 170kDa. These proteins are likely to be subunits of eIF3, which appears to play an important role in mediating the interaction of HCV RNA with the 40S subunit as described in the preceding section (PESTOVA et al. 1998; SIZOVA et al. 1998).

4.4 Assembly of the Translation Complex on the Viral RNA

Coupled with the data described above, a rudimentary understanding of the distribution of eIF2, eIF3 and ribosomal protein S9 on the 40S ribosome subunit (BOMMER et al. 1991) suggests a possible model for the assembly of the HCV translation initiation complex. According to this model, the 40S ribosome subunit, containing eIF2, Met-tRNA and S9, interacts with the folded viral 5′NTR at multiple contact points, such that the AUG codon is placed near the anticodon of the Met-tRNA (Fig. 7A). The lateral side of the 40S subunit particle is likely to become bound to the top of domain III through interactions between eIF3 and both the 40S particle and the viral RNA. The binding of eIF2 results in further positioning of the AUG codon so that it is in close proximity to the anticodon on the Met-tRNA. This seems likely to occur in association with the melting of stem-loop IV to a single-stranded form, allowing the RNA to become correctly positioned within the RNA groove of the 40S subunit and setting the stage for initiation of translation (Fig. 7B). Interestingly, eIF2 which appears to be required for directing the ribosome towards the AUG initiator codon on the HCV IRES (PESTOVA et al. 1998), has been implicated in AUG codon recognition in both IRES and cap-dependent translation (DASSO et al. 1990; DONAHUE et al. 1988; THOMAS et al.

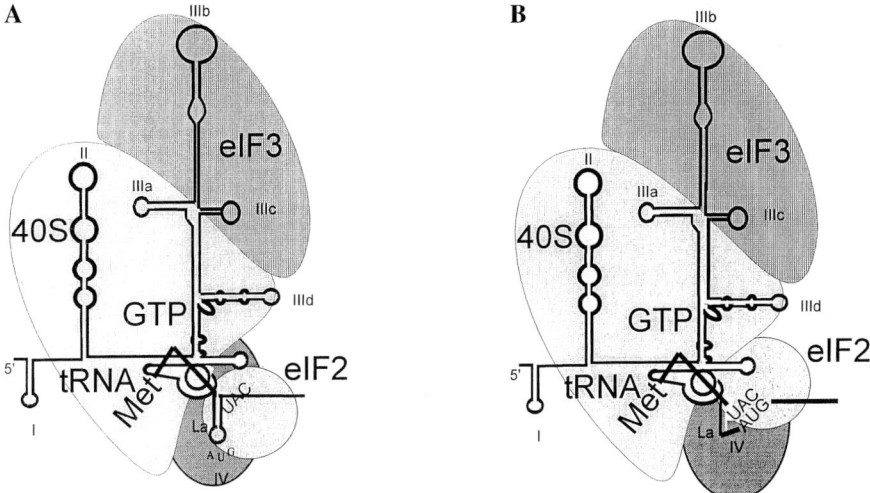

Fig. 7A,B. Model for the assembly of a 48S preinitiation complex on the hepatitis C virus internal ribosome entry site (HCV IHRES). The positioning of eIF3 and the 40S ribosome subunit is based on the interaction of subunits of these elements with the HCV IRES. After the formation of the initial complex **A**, a conformational change occurs in the RNA that places the anti-codon over the AUG codon and results in the unfolding of stem-loop IV **B**. eIF2 may play an essential role in this process

1996). Subsequent binding of the 60S subunit results in the assembly of a functional 80S ribosome, which then begins to translate the HCV polyprotein.

Although there are data to support each of these interactions, the precise order in which these events occur remains uncertain and several different sequences for assembly of the initiation complex can be envisioned (Fig. 8). Both canonical and noncanonical translation initiation factors may facilitate the latter stages of this process. The assembly process is likely to be very similar to other flaviviral IRESs.

5 Involvement of the 3′ Nontranslated RNA in Cap-Independent Translation

One feature that distinguishes the RNAs of HCV, GBV-B and pestiviruses from other RNAs containing IRESs is the lack of a poly-(A) tail at the extreme 3′ end of the RNA. There is, however, some evidence that the 3′ end of the HCV genomic RNA, like the poly-(A) tail of eukaryotic mRNAs, may have a positive influence on translational activity. RNA transcripts containing the HCV IRES and the highly structured 3′ terminus of the viral RNA were several-fold more active in a translation assay than RNAs that lacked the 3′ terminal sequence (ITO et al. 1998). There were no apparent differences in the stability of the RNAs that could account for this difference. PTB, which binds to pyrimidine-rich sequences within the 3′NTR

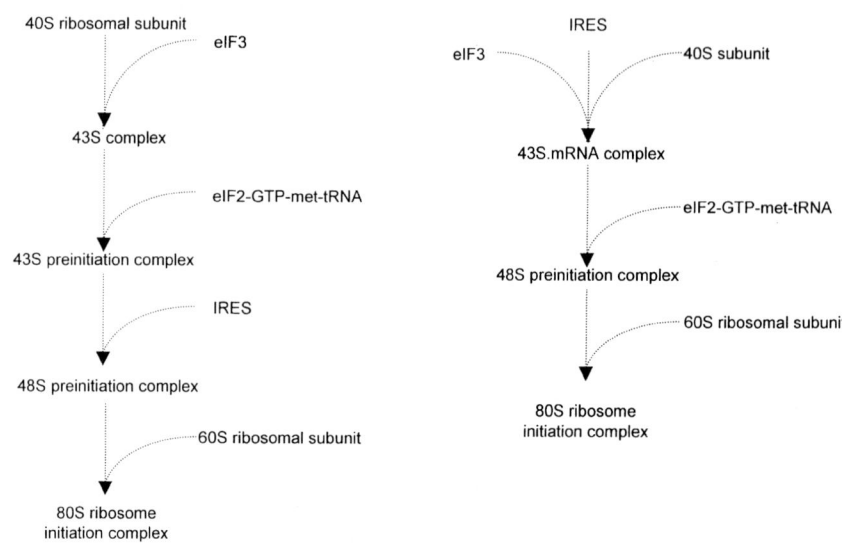

Fig. 8A,B. Two possible scenarios leading to formation of the 48S preinitiation complex on the hepatitis C virus internal ribosome entry site (HCV IHRES). **A** The 40S ribosome subunit binds both eIF3 and the ternary complex (eIF2-GTP-Met-tRNA) to form a 43S preinitiation complex (as normally occurs in cap-dependent translation). The 43S complex subsequently binds to the HCV RNA. In a subsequent step, the 48S preinitiation complex is formed by a conformational change in the HCV IRES that places the AUG codon against the Met-tRNA anti-codon. **B** An alternative scenario in which eIF3 and the 40S ribosome subunit bind are bound by the IRES. In a subsequent step, the ternary complex is incorporated into the 40S-RNA complex, resulting in a conformational change in the viral RNA that places codon and anti-codon opposite each other to form a 48S initiation complex. Subsequent interactions with the 60S ribosome subunit in either scenario lead to formation of a translationally active ribosome

and is active as a dimer, could potentially facilitate an interaction between 5′ and 3′ nontranslated sequences of the virus. However, the nature of this interaction and the mechanism for translational enhancement remain unknown.

6 Changes in the 5′ Nontranslated RNA Affect Translation

6.1 The Hepatitis C Virus Internal Ribosome Entry Site and Viral Tropism

In addition to the liver, HCV RNA has been detected in circulating peripheral blood mononuclear cells (PBMCs) (LERAT et al. 1996). Although the significance of these observations remains a subject of debate, the possibility that HCV might infect such cells in vivo is strengthened by evidence for replication of the virus in cultured lymphoblastoid cells. Following the inoculation of two different lymphoblastoid cell lines with wild-type virus, new quasispecies emerged which were characterized by three unique nucleotide substitutions within the IRES (NAKAJIMA et al. 1996). The novel quasispecies were not present within the original viral inoculum. However, the nucleotide substitutions within the 5′NTR of these viral quasispecies were also identified in viral sequences amplified directly from the PBMCs of infected chimpanzees (SHIMIZU et al. 1997). These data, coupled with the emergence of identical nucleotide changes in virus populations infecting two different lymphoblastoid cell lines (NAKAJIMA et al. 1996), strongly suggest the selection of quasispecies with greater "fitness" for replication in cells of lymphoid origin. One interesting possibility is that the nucleotide substitutions within the 5′NTR of these quasispecies may enhance the activity of the IRES in a cell type-specific fashion. This would not be without precedent, as a similar, cell type-specific adaptation has been demonstrated within the IRES of hepatitis A virus (HAV) during passage of this virus in monkey kidney cells (SCHULTZ et al. 1996). These HAV mutations, as well as mutations in the IRES elements of other picornaviruses, have been shown to affect viral translation in a cell-type specific fashion (SCHULTZ et al. 1996; LA MONICA and RACANIELLO 1989; SHIROKI et al. 1997).

To determine whether the HCV mutations noted by NAKAJIMA et al. (1996) might have a cell-type specific influence on translation, Lerat et al. (in preparation) compared the activity of the IRES present in the quasispecies that was dominant in the serum inoculum (NC-1) with that recovered from the infected lymphoblastoid cell lines (NC-7). Interestingly, the translational activity of the B cell "adapted" NC-7 IRES was enhanced over that of the NC-1 IRES by a factor of 2- to 2.5-fold in two different transfected B cell lines (Raji and Bjab cells) and in one T cell derived cell line (Molt4). In contrast, there was no difference in the relative activities of these IRESs in an in vitro transcription-translation system, in hepatocyte-derived Huh7 cells, or in hematopoietic cell lines of nonlymphoid lineage. These results suggest that the mutations in the NC-7 sequence may represent a specific adaptation of the HCV IRES that enhances translation in lymphoblastoid cells. If so, the selection of these mutations during passage of the virus suggests that translation might be a limiting factor in HCV replication in such cells. Moreover, these results suggest that certain cell type-specific factors, possibly proteins, must be important to IRES activity, despite the results described in the previous section. Unfortunately, however, there is at present no way to directly assess the impact of these mutations on replication of the virus in lymphoblastoid cells. Although infectious HCV RNA can be transcribed from genome-length cDNA clones (YANAGI et al. 1997; KOLYKHALOV et al. 1997), it has not yet proven possible to rescue virus from this RNA in cultured cells.

6.2 Genetic Variation and Differences in Internal Ribosome Entry Site Activity

Based on information from several studies, it appears that there may be slight differences in the activities of the IRES sequences in different genotypes of HCV (BURATTI et al. 1997; COLLIER et al. 1998; HONDA et al. 1999b; TSUKIYAMA-KOHARA et al. 1992). However, it is not clear whether such differences are of biological significance or otherwise influence the clinical course of disease. Unfortunately, it is not possible to determine the influence of IRES activity on the pathogenicity of HCV in the absence of effective small animal models of hepatitis C. Nonetheless, these studies are of interest because of their relevance to structure-function relationships within the IRES.

BURATTI et al. (1997) found translational activity to be comparable for RNAs containing either the 1b or 2a genotype 5'NTR, while RNAs containing the 5'NTR sequence of genotype 3 virus directed translation with approximately twofold lower efficiency. In contrast, TSUKIYAMA-KOHARA et al. (1992) reported that the genotype 2b IRES was slightly more active than the 1b IRES. In an effort to pinpoint the nucleotide differences responsible for this difference, KAMOSHITA et al. (1997) constructed a number of chimeric 5' 1b/2b NTRs. Although they were able to confirm their original observation that the 2b genotype 5'NTR was two- to fivefold more active in directing translation than the comparable 1b sequence, they were not able to identify specific nucleotide sequence differences responsible for this variation in activity. Their results suggest that the difference in IRES activity may be due to the influence of multiple nucleotide substitutions on the conformation of the folded RNA.

HONDA et al. (1996b) noted that the 5'NTR sequence of a genotype 1a virus was approximately twofold more efficient in directing translation of the downstream core protein-coding sequence than the comparable 1b sequence. Unlike KAMOSHITA et al. (1997), HONDA et al. (1999b) were able to show that this difference was due to a specific dinucleotide substitution involving nts 34–35 of the HCV genome. Interesting, this site is upstream of the IRES. The deletion of nts 32–37 from the 1b sequence restored its translational activity to that of the 1a virus, indicating that the dinucleotide was inhibitory in the 1b virus. Further analysis suggested that the difference in translational activity was due to an RNA-RNA interaction involving the dinucleotide sequence and unidentified bases within the capsid coding sequence (HONDA et al. 1999b). Significant differences were found in the translational activities of these genotypes only with RNAs containing the entire core protein-coding sequence. It is not known whether similar long-range RNA interactions occur in all genotype 1b viruses or in other HCV genotypes. However, it is important to note that RNA transcribed from a genome-length cDNA clone containing the genotype 1b sequence studied by HONDA et al. (1999b) was recently shown to be infectious in a chimpanzee (M. Beard and S.M. Lemon, In press).

Finally, COLLIER et al. (1998) tested the translational strengths of IRES sequences from seven different HCV genotypes (1a, 1b, 2b, 3a, 4a, 5a and 6a) in four different cell lines. They found that the genotype 2b IRES had the highest activity in all four cell lines, while the genotype 6a IRES generally had the lowest activity. The difference observed in the activities of the genotype 2b and 1b IRES elements (40%–60%) was similar to that found by TSUKIYAMA-KOHARA et al. (1992). However, no differences were observed between the 1a and 1b genotypes. This is consistent with the results of HONDA et al. (1996b), since the constructs evaluated by COLLIER et al. (1998) contained only 15 nts of the core protein-coding sequence. This indicates the dangers inherent in studying IRES activity in the context of RNA transcripts containing heterologous reporter protein-coding sequences.

7 Similarities between Hepatitis C Virus and Prokaryotic Translation Initiation

In many ways, the translation mechanism utilized by HCV, GBV-B and pestiviruses appears similar to that utilized by prokaryotes (PESTOVA et al. 1998). Translation initiation factors with functions analogous to eIF2 and eIF3 have been identified in prokaryotes, but no protein homologues of the other canonical eukaryotic translation initiation factors have yet been described. Moreover, the initiation of translation occurs by a very different process. In contrast to eukaryotes, the initiation of translation in prokaryotes is not dependent on the 5′ end of the RNA. A conserved Shine-Dalgarno sequence, located just upstream of the AUG initiator codon, interacts with the 16S rRNA and serves to direct ribosome entry onto the RNA. This mechanism places the ribosome subunit at the AUG initiator codon. Thus, prokaryotic translation initiation is very similar to that which occurs with the flaviviral IRES. In both cases, the messenger RNA has the ability to interact directly with the ribosome subunit (PESTOVA et al. 1998), and there is complete absence of any ribosome scanning (HONDA et al. 1996a; RIJNBRAND et al. 1996). These two features of translation initiation are not found in any other eukaryotic RNA. This is likely to be significant not only for the number of initiation factors involved in cap-independent translation of flaviviral RNAs, but also the possible mechanisms by which the process is regulated.

8 Summary: Unique Features of Flaviviral Internal Ribosome Entry Site Elements

Although internal ribosome entry results in the cap-independent initiation of translation on a number of different viral and cellular RNAs, the extent to which the responsible IRES elements share a common mechanism of translation initiation remains to be defined. The flaviviral IRES superficially resembles the type 2 picornavirus IRES in that there does not appear to be any scanning of the 40S subunit for the initiation codon (LEMON and HONDA 1997; WIMMER et al. 1993). Nonetheless, there are a number of differences between these IRESs that suggest significantly different mechanisms of translation initiation: (1) The flavivirus IRES is generally only half the length of the picornavirus IRES elements, and it assumes a completely different secondary and tertiary structure. (2) There is no conserved pYxxxAUG element in the flaviviral IRES, in contrast to the picornavirus IRES (EHRENFELD and SEMLER 1995; HELLEN and WIMMER 1995). (3) Despite the evidence described above for cell-specific differences in HCV IRES activity (Lerat et al., in preparation), the flaviviral IRES is remarkably versatile in its ability to drive efficient cap-independent translation in a variety of different cell lines. In contrast, picornaviral IRES elements appear to have much greater cell specificity (BORMAN et al. 1997). (4) The HCV IRES has no strict requirement for PTB, La, or any of the canonical translation initiation factors other than eIF2 and eIF3. This is in stark contrast to picornavirus IRES elements (PESTOVA et al. 1996). (5) Translation initiation occurs in the absence of a canonical AUG codon in some mutated HCV RNAs, a feature that has not been described for either type I or II picornavirus IRES elements. (6) The 40S ribosome subunit is directly positioned on top of the AUG codon after forming a complex with the flaviviral IRES, and there is little or no ability of the 40S subunit to scan for distally located AUG codons. At least with the type I picornavirus IRES elements, there is good evidence for scanning (BELSHAM 1992; HELLEN et al. 1994; PILIPENKO et al. 1994). These differences make it clear that the HCV IRES and related IRES elements in pestiviruses and GBV-B should be classified as a separate group (type III) of virus IRES elements as proposed by LEMON and HONDA (1997).

There are also similarities and differences with cellular IRES elements. Unlike picornaviral IRESs, initiation of translation occurs at non-AUG codons in some cellular IRES elements. Cellular IRESs control translation on mRNAs encoding both c-myc and fibroblast growth factor (FGF-2), and translation is initiated at CUG codons (NANBRU et al. 1997; VAGNER et al. 1995). It is not known how the ribosome selects for the translation initiation site, but the recognition of CUG codons is efficient (VAGNER et al. 1995). Scanning as well as a second IRES element may be involved in the recognition of the four initiation sites in the FGF-2 message (VAGNER et al. 1995). Scanning also appears to be involved in recognition of the initiation sites on eIF4E and c-myc mRNAs as well (GAN et al. 1998; NANBU et al. 1997; VAGNER et al. 1995). Thus, the mechanism of internal entry on HCV RNA may be significantly different from that occurring on these cellular RNAs.

A great deal of effort has gone into the analysis of the flaviviral IRES element. However, there is much that remains to be done in order to arrive at a satisfactory understanding of the structure and function of this RNA element. Biophysical approaches to determining the structure of the 5'NTR, including NMR and X-ray crystallography, are very likely to provide a new view of the three dimensional structure of this unique RNA element. The interactions of viral and cellular proteins with the IRES require further investigation, as a better understanding of these interactions is certain to provide valuable insights into the translation mechanism and its regulation. The process of ribosome entry on the viral RNA is an attractive target for the development of effective anti-viral compounds.

Acknowledgements. This work was supported in part by grants from the National Institute of Allergy and Infectious Diseases (U19-AI40035-01) and the Texas Advanced Technology Program (004952-025).

References

Alexander L, Lu H-H, Wimmer E (1994) Polioviruses containing picornavirus type 1 and/or type 2 internal ribosomal entry site elements: genetic hybrids and the expression of a foreign gene. Proc Natl Acad Sci USA 91:1406–1410
Ali N, Siddiqui A (1995) Interaction of polypyrimidine tract-binding protein with the 5′ noncoding region of the hepatitis C virus RNA genome and its functional requirement in internal initiation of translation. J Virol 69:6367–6375
Ali N, Siddiqui A (1997) The La antigen binds 5′ noncoding region of the hepatitis C virus RNA in the context of the initiator AUG codon and stimulates internal ribosome entry site-mediated translation. Proc Natl Acad Sci USA 94:2249–2254
Becher P, Orlich M, Thiel HJ (1998) Complete genomic sequence of border disease virus, a pestivirus from sheep. J Virol 72:5165–5173
Belsham GJ (1992) Dual initiation sites of protein synthesis on foot-and-mouth disease virus RNA are selected following internal entry and scanning of ribosomes in vivo. EMBO J 11:1105–1110
Belsham GJ, Sonenberg, N, Svitkin YV (1995) The role of the La autoantigen in internal initiation. Curr Top Microbiol 203:85–98
Belsham GJ, Sonenberg N (1996) RNA-protein interactions in regulation of picornavirus RNA translation. Micrbiol. Rev. 60:499–511
Berlioz C, Torrent C, Darlix J-L (1995) An internal ribosomal entry signal in the rat VL30 region of the Harvey murine sarcoma virus leader and its use in dicistronic retroviral vectors. J Virol 69:6400–6407
Bernstein J, Sella O, Le SY, Elroy-Stein O (1997) PDGF2/c-sis mRNA leader contains a differentiation-linked internal ribosomal entry site (D-IRES). J Biol Chem 272:9356–9362
Bommer UA, Lutsch G, Stahl J, Bielka H (1991) Eukaryotic initiation factors eIF-2 and eIF-3: interactions, structure and localization in ribosomal initiation complexes. Biochemie 73:1007–1019
Borman A, Howell MT, Patton JG, Jackson RJ (1993) The involvement of a spliceosome component in internal initiation of human rhinovirus RNA translation. J Gen Virol 74:1775–1788
Borman AM, Bailly JL, Girard M, Kean KM (1995) Picornavirus internal ribosome entry segments: comparison of translation efficiency and the requirements for optimal internal initiation of translation in vitro. Nucleic Acids Res 23:3656–3663
Borman AM, Le Mercier P, Girard M, Kean KM (1997) Comparison of picornaviral IRES-driven internal initiation of translation in cultured cells of different origins. Nucleic Acids Res 25:925–932
Borovjagin A, Pestova TV, Shatsky IN (1994) Pyrimidine tract binding protein strongly stimulates in vitro encephalomyocarditis virus RNA translation at the level of preinitiation complex formation. FEBS 351:299–302
Brock KV, Deng R, Riblet SM (1992) Nucleotide sequencing of 5′ and 3′ termini of bovine viral diarrhea virus by RNA ligation and PCR. J Virol Meth 38:39–46
Brown EA, Zhang H, Ping L, Lemon SM (1992) Secondary structure of the 5′ nontranslated regions of hepatitis C virus and pestiviruses genomic RNAs. Nucleic Acids Res 20:5041–5045
Bukh J, Purcell RH, Miller RH (1992) Sequence analysis of the 5′ noncoding region of hepatitis C virus. Proc Natl Acad Sci USA 89:4942–4946
Bukh J, Purcell RH, Miller RH (1994) Sequence analysis of the core gene of 14 hepatitis C virus genotypes. Proc Natl Acad Sci USA 91:8239–8243
Buratti E, Gerotto M, Pontisso P, Alberti A, Tisminetzky SG, Baralle FE (1997) In vivo translational efficiency of different hepatitis C virus 5′-UTRs. FEBS 411:275–280
Buratti E, Tisminetzky SG, Zotti M, Baralle FE (1998) Functional analysis of the interaction between HCV 5′ UTR and putative subunits of eukaryotic translation initiation factor iF3. Nucleic Acids Res 26:3179–3187

Collett MS, Larson R, Gold C, Strick D, Anderson DK, Purchio AF (1988) Molecular cloning and nucleotide sequence of the pestivirus bovine viral diarrhea virus. Virology 165:191–199

Collier AJ, Tang S, Elliot RM (1998) Translational efficiencies of the 5' untranslated region from representatives of the six major genotypes of hepatitis C virus using a novel bicistronic reporter assay system. J Gen Virol 79:2359–2366

Dasso MC, Milburn SC, Hershey JW, Jackson RJ (1990) selection of the 5'-proximal translation initiation site is influenced by mRNA and eIF-2 concentrations. Eur J Biochem 187: 361–371

Deng R, Brock KV (1993) 5' and 3' untranslated regions of pestivirus genome: primary and secondary structure analyses. Nucleic Acids Res 21:1949–1957

Donahue TF, Cigan AM, Pabich EK, Valavicius BC (1988) Mutations at a Zn(II) finger motif in the yeast eIF2B gene alter ribosomal start-site selection during the scanning process. Cell 54:621–632

Ehrenfeld E, Semler BL (1995) Anatomy of the poliovirus internal ribosome entry site. Curr Top Microbiol 203:65–83

Fukushi S, Katayama K, Kurihara C, Ishiyama N, Hoshino FB, Ando T, Oya A (1994) Complete 5' noncoding region is necessary for the efficient internal initiation of hepatitis C virus RNA. Biochem Biophys Res Comm 199:425–432

Fukushi S, Kurihara C, Ishiyama N, Hoshino FB, Oya A, Katayama K (1997) The sequence element of the internal ribosome entry site and a 25-kilodalton cellular protein contribute to efficient internal initiation of translation of hepatitis C virus RNA. J Virol 71:1662–1666

Gan W, Rhoads RE (1996) Internal initiation of translation directed by the 5'-untranslated region of the mRNA for eIF4G, a factor involved in the picornavirus-induced switch from cap-dependent to internal initiation. J Biol Chem 271:623–626

Gan W, Celle ML, Rhoads RE (1998) Functional characterization of the internal ribosome entry site of eIF4G mRNA. J Biol Chem 273:5006–5012

Hahm B, Kim YK, Kim JH, Kim TY, Jang SK (1998) HnRNP L interacts with the 3' border of internal ribosomal entry site of hepatitis C virus. J Virol 72:8782–8788

Hellen CU, Witherell GW, Schmid M, Shin SH, Pestova TV, Gil A, Wimmer E (1993) A cytoplasmic 57-kDa protein that is required for translation of picornavirus RNA by internal ribosomal entry is identical to the nuclear pyrimidine tract-binding protein. Proc Natl Acad Sci USA 90:7642–7646

Hellen CU, Wimmer E (1995) Translation of encephalomyocarditis virus RNA by internal ribosomal entry. Curr Top Microbiol 203:31–63

Hellen CUT, Pestova TV, Wimmer E (1994) Effect of mutations downstream of the internal ribosome entry site on initiation of poliovirus protein synthesis. J Virol 68:6312–6322

Honda M, Brown EA, Lemon SM (1996a) Stability of a stem-loop involving the initiator AUG controls the efficiency of internal initiation of translation on hepatitis C virus RNA. RNA 2:955–968

Honda M, Ping LH, Rijnbrand RCA, Amphlett E, Clarke B, Rowlands D, Lemon SM (1996b) Structural requirements for initiation of translation by internal ribosome entry within genome-length hepatitis C virus RNA. Virology 222:31–42

Honda M, Beard M, Ping L, Lemon SM (1999a) A phylogenetically conserved stem-loop structure at the 5' border of the internal ribosome entry site of hepatitis C virus is required for cap-independent viral translation. J Virol 73:1165–1174

Honda M, Rijnbrand R, Abell G, Kim D, Lemon SM (1999b) Natural variation in translation activities of the 5' nontranslated RNAs of genotypes 1a and 1b hepatitis C virus: Evidence for a long range RNA-RNA interaction outside of the internal ribosomal entry site. J Virol 73:4941–4951

Huhn P, Pruijn GJ, van Venrooij WJ, Bachmann M (1997) Characterization of the autoantigen La (SS-B) as a dsRNA unwinding enzyme. Nucleic Acids Res 25:410–416

Iizuka N, Chen C, Yang Q, Johannes G, Sarnow P (1995) Cap-independent translation and internal initiation of translation in eukaryotic cellular mRNA molecules. Curr Top Microbiol 203:155–177

Ito T, Tahara SM, Lai MMC (1998) The 3' untranslated region of hepatitis C virus RNA enhances translation from an internal ribosomal entry site. J Virol 72:8789–8796

Jackson RJ, Howell MT, Kaminski A (1990) The novel mechanism of initiation of picornavirus RNA translation. Trends Biochem Sci 15:477–483

Jackson RJ, Kaminski A (1995) Internal initiation of translation in eukaryotes: The picornavirus paradigm and beyond. RNA 1:985–1000

Jang SK, Krausslich MJH, Nicklin GM, Duke AC, Palmenberg AC, Wimmer E (1988) A segment of the 5' nontranslated region of encephalomyocarditis virus RNA directs internal entry of ribosomes during in vitro translation. J Virol 62:2636–2643

Jia XY, Tesar M, Summers DF, Ehrenfeld E (1996) Replication of hepatitis A viruses with chimeric 5' nontranslated regions. J Virol 70:2861–2868

Kaminski A, Hunt SL, Patton JG, Jackson RJ (1995) Direct evidence that the polypyrimidine tract binding protein (PTB) is essential for internal initiation of translation of encephalomyocarditis virus RNA. RNA 1:924–938

Kamoshita N, Tsukiyama-Kohara K, Kohara M, Nomoto A (1997) Genetic analysis of internal ribosomal entry site on hepatitis C virus RNA: implication for involvement of the high ordered structure and cell type-specific transacting factors. Virology 233:9–18

Kettinen HK, Grace K, Grunert S, Klarke B, Rowlands D, Jackson R (1994) Mapping of the internal ribosome entry site at the 5′ end of the hepatitis C virus genome. In: Nishioka K, Suzuki H, Mishihiro S, Oda T (eds) Proceedings of the international symposium on viral hepatitis and liver disease. Tokyo, pp 125–131

Kolykhalov AA, Agapov EV, Blight KJ, Mihalik K, Feinstone SM, Rice CM (1997) Transmission of hepatitis C by intrahepatic inoculation with transcribed RNA. Science 277:570–574

Koonin EV (1993) Computer-assisted identification of a putative methyltransferase domain in NS5 protein of flaviviruses and lambda 2 protein of reovirus. J Gen Virol 74:733–740

Kozak M (1977) Nucleotide sequences of 5′-terminal ribosome-protected initiation regions from two reovirus messages. Nature 269:391–394

Kozak M (1987) An analysis of 5′-noncoding sequences from 699 vertebrate messenger RNAs. Nucleic Acids Res 15:8125–8148

Kozak M (1989a) Context effects and inefficient initiation at non-AUG codons in eucaryotic cell-free translation systems. Mol Cell Biol 9:5073–5080

Kozak M (1989b) The scanning model for translation: An update. J Cell Biol 108:229–241

Kozak M (1989c) Circumstances and mechanisms of inhibition of translation by secondary structure in eucaryotics mRNA. Mol Cell Biol 9:5134–5142

La Monica N, Racaniello VR (1989) Differences in replication of attenuated and neurovirulent polioviruses in human neuroblastome cell line SH-SY5Y. J Virol 63:2357–2360

Le SY, Chen JH, Sonenberg N, Maizel JV (1992) Conserved tertiary structure elements in the 5′ untranslated region of human enteroviruses and rhinoviruses. Virology 191:858–866

Le SY, Chen JH, Sonenberg N, Maizel JV, Jr. (1993) Conserved tertiary structural elements in the 5′ nontranslated region of cardiovirus, aphthovirus and hepatitis A virus RNAs. Nucleic Acids Res 21:2445–2451

Le SY, Chen JH, Sonenberg N, Maizel JV, Jr. (1994) Distinct structural elements and internal entry of ribosomes in mRNA3 encoded by infectious bronchitis virus. Virology 198:405–411

Le SY, Chen JH, Sonenberg N, Maizel JV, Jr. (1995) Unusual folding regions and ribosome landing pad within hepatitis C virus and pestivirus RNAs. Gene 154:137–143

Lee YF, Nomoto A, Detjen BM, Wimmer E (1977) A protein covalently linked to poliovirus genome RNA. Proc Natl Acad Sci USA 74:59–63

Lemon SM, Honda M (1997) Internal ribosome entry sites within the RNA genomes of hepatitis C virus and other Flaviviruses. Seminars in Virology 8:274–288

Lerat H, Berby F, Trabaud MA, Vidalin O, Major M, Trepo C, Inchauspe G (1996) Specific detection of hepatitis C virus minus strand RNA in hematopoietic cells. J Clin Invest 97:845–851

Liu DX, Inglis SC (1992) Internal entry of ribosomes on a tricistronic mRNA encoded by infectious bronchitis virus. J Virol 66:6143–6154

Lu H-H, Wimmer E (1996) Poliovirus chimeras replicating under the translational control of genetic elements of hepatitis C virus reveal unusual properties of the internal ribosomal entry site of hepatitis C virus. Proc Natl Acad Sci USA 93:1412–1417

Macejak DG, Sarnow P (1991) Internal initiation of translation mediated by the 5′ leader of a cellular mRNA. Nature 353:90–94

Maga JA, Widmer G, LeBowitz JH (1995) Leishmania RNA virus 1-mediated cap-independent translation. Mol Cell Biol 15:4884–4889

McBratney S, Sarnow P (1996) Evidence for involvement of trans-acting factors in selection of the AUG start codon during eukaryotic translational initiation. Mol Cell Biol 16:3523–3534

Meerovitch K, Svitkin YV, Lee HS, Lejbkowicz F, Kenan DJ, Chan EK, Agol VI, Keene JD, Sonenberg N (1993) La autoantigen enhances and corrects aberrant translation of poliovirus RNA in reticulocyte lysate. J Virol 67:3798–3807

Meyers G, Thiel HJ, Rumenapf T (1996) Classical swine fever virus: recovery of infectious viruses from cDNA constructs and generation of recombinant cytopathogenic defective interfering particles. J Virol 70:1588–1595

Molla A, Jang SK, Paul AV, Reuer Q, Wimmer E (1992) Cardioviral internal ribosomal entry site is functional in a genetically engineered dicistronic poliovirus. Nature 356:255–257

Muerhoff AS, Leary TP, Simons JN, Pilot-Matias TJ, Dawson GJ, Erker JC, Chalmers ML, Schlauder GG, Desai SM, Mushahwar IK (1995) Genomic organisation of GB viruses A and B: two new members of the Flaviviridae associated with GB agent hepatitis. J Virol 69:5621–5630

Nakajima N, Hijikata M, Yoshikura H, Shimizu YK (1996) Characterization of long-term cultures of hepatitis C virus. J Virol 70:3325–3329

Nanbru C, Lafon I, Audigier S, Gensac MC, Vagner S, Huez G, Prats AC (1997) Alternative translation of the proto-oncogene c-myc by an internal ribosome entry site. J Biol Chem 272:32061–32066

Oh SK, Scott MP, Sarnow P (1992) Homeotic gene Antennapedia mRNA contains 5'-noncoding sequences that confer translational initiation by internal ribosome binding. Genes Dev 6:1643–1653

Ohba K, Mizokami M, Lau JY, Orito E, Ikeo K, Gojobori T (1996) Evolutionary relationship of hepatitis C, pesti-, flavi, plantviruses, and newly discovered GB hepatitis agents. FEBS 378:232–234

Pain VM (1996) Initiation of protein synthesis in eukaryotic cells. Eur J Biochem 236:747–771

Park YW, Katze MG (1995) Translational control by influenza virus. Identification of cis-acting sequences and trans-acting factors which may regulate selective viral mRNA translation. J Biol Chem 270:28433–28439

Peabody DS (1987) Translation initiation at an ACG triplet in mammalian cells. J Biol Chem 262:11847–11851

Peek R, Pruijn GJ, van Venrooij WJ (1996) Interaction of the La (SS-B) autoantigen with small ribosomal subunits. Eur J Biochem 236:649–655

Pelletier J, Kaplan G, Racaniello V, Sonenberg N (1988) Cap-independent translation of poliovirus mRNA is conferred by sequence elements within the 5' noncoding region. Mol Cell Biol 8:1103–1112

Pestova TV, Hellen CU, Shatsky IN (1996) Canonical eukaryotic initiation factors determine initiation of translation by internal ribosomal entry. Mol Cell Biol 16:6859–6869

Pestova TV, Shatsky IN, Fletcher SP, Jackson RJ, Hellen CUT (1998) A prokaryotic-like mode of cytoplasmatic eukaryotic ribosome binding to the initiation codon during internal translation of hepatitis C virus and classical swine fever virus RNAs. Genes Dev 12:67–83

Pilipenko EV, Gmyl AP, Maslova SV, Svitkin YV, Sinyakov AN, Agol VI (1992) Prokaryotic-like cis elements in the cap-independent internal initiation of translation on picornavirus RNA. Cell 68:119–131

Pilipenko EV, Gmyl AP, Maslova SV, Belov GA, Sinyakov AN, Huang M, Brown TDK, Agol VI (1994) Starting window, a distinct element in the cap-independent internal initiation of translation of picornaviral RNA. J Mol Biol 241:398–414

Poole TL, Wang C, Popp RA, Potgieter LND, Siddiqui A, Collet MS (1995) pestivirus translation initiation occurs by internal ribosome entry. Virology 206:750–754

Prats AC, Vagner S, Prats H, Amalric F (1992) cis-acting elements involved in the alternative translation initiation process of human basic fibroblast growth factor mRNA. Mol Cell Biol 12:4796–4805

Reynolds JE, Kaminski A, Kettinen HJ, Grace K, Clarke BE, Carroll AR, Rowlands DJ, Jackson RJ (1995) Unique features of internal initiation of hepatitis C virus RNA translation. EMBO J 14:6010–6020

Reynolds JE, Kaminski A, Carroll AR, Clarke BE, Rowlands DJ, Jackson RJ (1996) Internal initiation of translation of hepatitis C virus RNA: the ribosome entry site is at the authentic initiation codon. RNA 2:867–878

Rice MC (1997) The Flaviviridae (931–959) In: Fields BN, Knipe DM, Howley PM (eds) Virology. Lippincott-Raven, Philadelphia, p 3

Ridpath, JF, Bolin SR (1995) The genomic sequence of a virulent bovine viral diarrhea virus (BVDV) from the type 2 genotype: detection of a large genomic insertion in a noncytopathic BVDV. Virology 212:39–46

Rijnbrand R, Bredenbeek P, van der Straaten T, Whetter L, Inchauspe G, Lemon S, Spaan W (1995) Almost the entire 5' non-translated region of hepatitis C virus is required for cap-independent translation. FEBS 365:115–119

Rijnbrand R, van der Straaten T, van Rijn P, Spaan WJM, Bredenbeek PJ (1997) Internal entry of ribosomes is directed by the 5' noncoding region of classical swine fever virus and is dependent on the presence of an RNA pseudoknot upstream of the initiation codon. J Virol 71:451–457

Rijnbrand RCA, Abbink TEM, Haasnoot PC, Spaan WJM, Bredenbeek PJ (1996) The influence of AUG codons in the hepatitis C virus 5' nontranslated region on translation and mapping of the translation initiation window. Virology 226:47–56

Rohll JB, Percy N, Ley R, Evans DJ, Almond JW, Barclay WS (1994) The 5'-untranslated regions of picornavirus RNAs contain independent functional domains essential for RNA replication and translation. J Virol 68:4384–4391

Sachs AB, Sarnow P, Hentze MW (1997) Starting at the beginning, middle, and end: translation initiation in eukaryotes. Cell 89:831–838

Schultz DE, Honda M, Whetter L, McKnight KL, Lemon SM (1996) Mutations within the 5' non-translated RNA of cell culture-adapted hepatitis A virus which enhance cap-independent translation in cultured African green monkey kidney cells. J Virol 70:1041–1049

Shimizu YK, Igarashi H, Kanematu T, Fujiwara K, Wong DC, Purcell RH, Yoshikura H (1997) Sequence analysis of the hepatitis C virus genome recovered from serum, liver, and peripheral blood mononuclear cells of infected chimpanzees. J Virol 5769–5773

Shiroki K, Ishii T, Aoki T, Ota Y, Wang WX, Komatsu T, Ami Y, Arita M, Abe S, Hashizume S, Nomoto A (1997) Host range phenotype induced by mutations in the internal ribosomal entry site of poliovirus RNA. J Virol 71:1–8

Simons JN, Desai SM, Schultz DE, Lemon SM, Mushawar IK (1996) Translation initiation in GB viruses A and C: evidence for internal ribosome entry and implications for genome organisation. J Virol 70:6126–6135

Sizova DV, Kolupaeva VG, Pestova TV, Shatsky IN, Hellen CUT (1998) Specific interaction of eukaryotic translation initiation factor 3 with the 5' nontranslated regions of hepatitis C virus and classical swine fever virus RNAs. J Virol 72:4775–4782

Smith DB, Mellor J, Jarvis LM, Davidson F, Kolberg J, Urdea MS, Yap PL, Simmonds P (1995) Variations of the hepatitis C virus 5' non-coding region: implications for the secondary structure, virus detection and typing The international HCV collaborative study group. J Gen Virol 76:1749–1761

Smith DB, Simmonds P (1997) Characteristics of nucleotide substitution in the hepatitis C virus genome: constraints on sequence change in coding regions at both ends of the genome. J Mol Evol 45:238–246

Stewart SR, Semler BL (1997) RNA determinants of picornavirus cap-independent translation initiation. Seminars in Virology 8:242–255

Stoneley M, Paulin FE, Le Quesne JP, Chappell SA (1998) C-Myc 5' untranslated region contains an internal ribosome entry segment. Oncogene 16:423–428

Svitkin YV, Pause A, Sonenberg N (1994) La autoantigen alleviates translational repression by the 5' leader sequence of the human immunodeficiency virus type 1 mRNA. J Virol 68:7001–7007

Teerink H, Voorma HO, Thomas AA (1995) The human insulin-like growth factor II leader 1 contains an internal ribosomal entry site. Biochem Biophys Acta 1264:403–408

Thiel V, Siddell SG (1994) Internal ribosome entry in the coding region of murine hepatitis virus mRNA 5. J Gen Virol 75:3041–3046

Thomas AA, ter Haar E, Wellink J, Voorma HO (1991) Cowpea mosaic virus middle component RNA contains a sequence that allows internal binding of ribosomes and that requires eukaryotic initiation factor 4F for optimal translation. J Virol 65:2953–2959

Thomas AA, Rijnbrand R, Voorma HO (1996) Recognition of the initiation codon for protein synthesis in foot-and-mouth disease virus RNA. J Gen Virol 77:265–272

Tolan DR, Hershey JW, Traut RT (1983) Crosslinking of eukaryotic initiation factor eIF3 to the 40S ribosomal subunit from rabbit reticulocytes. Biochimie 65:427–436

Tsukiyama-Kohara K, Iizuka N, Kohara M, Nomoto A (1992) Internal ribosome entry site within hepatitis C virus RNA. J Virol 66:1476–1483

Vagner S, Waysbort A, Marenda M, Gensac MC, Amalric F, Prats AC (1995a) Alternative translation initiation of the Moloney murine leukemia virus mRNA controlled by internal ribosome entry involving the p57/PTB splicing factor. J Biol Chem 270:20376–20383

Vagner S, Gensac MC, Maret A, Bayard F, Amalric F, Prats H, Prats AC (1995b) Alternative translation of human fibroblast growth factor 2 mRNA occurs by internal entry of ribosomes. Mol Cell Biol 15:35–44

Verver J, Le Gall O, van Kammen A, Wellink J (1991) The sequence between nucleotides 161 and 512 of cowpea mosaic virus M RNA is able to support internal initiation of translation in vitro. J Gen Virol 72:2339–2345

Wang C, Sarnow P, Siddiqui A (1993) Translation of human hepatitis C virus RNA in cultured cells is mediated by an internal ribosome binding mechanism. J Virol 67:3338–3344

Wang C, Sarnow P, Siddiqui A (1994) A conserved helical element is essential for internal initiation of translation of hepatitis C virus RNA. J Virol 68:7301–7307

Wang C, Le S-Y, Ali N, Siddiqui A (1995) An RNA pseudoknot is an essential structural element of the internal ribosome entry site located within the hepatitis C virus 5' noncoding region. RNA 1:526–537

Wimmer E, Hellen CU, Cao X (1993) Genetics of poliovirus. Annu Rev Genet 27:353–436

Yanagi M, Purcell RH, Emerson SU, Bukh J (1997) Transcripts from a single full-length cDNA clone of hepatitis C virus are infectious when directly transfected into the liver of a chimpanzee. Proc Natl Acad Sci USA 94:8738–8743

Ye X, Fong P, Iizuka N, Choate D, Cavener DR (1997) Ultrabithorax and Antennapedia 5′ untranslated regions promote developmentally regulated internal translation initiation. Mol Cell Biol 17: 1714–1721

Yen JH, Cahng SC, Hu CR, Chu SC, Lin SS, Hsieh YS, Chang MF (1995) Cellular proteins specifically bind to the 5′ noncoding region of hepatitis C virus RNA. Virology 208:723–732

Yoo BJ, Spaete RR, Geballe AP, Selby M, Houghton M, Han JH (1992) 5′ end dependent translation initiation of hepatitis C viral RNA and the presence of putative positive and negative translational control elements within the 5′ untranslated region. Virology 191:889–899

Hepatitis C Virus Core Protein: Possible Roles in Viral Pathogenesis

M.M.C. Lai[1] and C.F. Ware[2]

1	Introduction 117
2	Biochemical Properties and Biosynthesis of the Core Protein 118
3	Possible Pathogenic Roles of the Core Protein 120
3.1	Direct Cytotoxicity of the Core Protein 120
3.1.1	Modulation of Cellular Gene Expression 120
3.1.2	Transforming Activity of the Core Protein 121
3.1.3	Modulation of Apoptosis 122
3.2	Interactions of the Core Protein with Host's Immune Systems 122
3.2.1	Binding to Lymphotoxin β-Receptor 123
3.2.2	Binding to Tumor Necrosis Factor Receptor 123
3.3.3	Binding to Other Cytokine Receptors 124
4	Implications of the Interactions Between the Core Protein and Tumor Necrosis Factor Receptor Family 125
5	Perspectives 130
References	 130

1 Introduction

One of the hallmarks of hepatitis C virus (HCV) infections is its propensity to cause chronic infections despite the presence of HCV-specific cellular and humoral immunity. Thus, HCV appears to have an ability to evade the host's immunity, similar to many other viruses. Various viruses employ different mechanisms, most often mediated by viral nonstructural proteins, to evade the host's defense. Some of the viral nonstructural proteins are important for viral replication, whereas others may perform functions strictly to counter the host's defense, thus ensuring viral survival. HCV encodes several nonstructural proteins, which include proteins for RNA replication (ns5b and probably ns3) and protein processing (ns2, ns3 and ns4a); the functions of the remaining nonstructural proteins are not yet clear. They may

---

[1] Howard Hughes Medical Institute and Department of Molecular Microbiology & Immunology, University of Southern California School of Medicine, 2011 Zonal Avenue, HMR-401, Los Angeles, CA 90033-1054, USA
[2] Division of Molecular Immunology, La Jolla Institute for Allergy and Immunology, San Diego, CA 92121, USA

perform functions to disrupt host's defense, thus contributing to the persistence of HCV infection. For example, the ns5a protein has a potential ability to counter the effects of interferon (GALE et al. 1997). Ironically, however, one of the HCV proteins that have been demonstrated to play a role in disrupting the host's response to viral infection so far is a structural protein, namely, the core protein. This chapter is devoted to the discussion of the core protein in this respect.

All of the structural and nonstructural proteins of HCV are synthesized as a polyprotein, which is processed into multiple proteins. The NH_2-terminal portion of the polyprotein encodes structural proteins, the first of which is the core (C) protein. The core protein forms the viral nucleocapsid, which is enveloped by two envelope proteins, E1 and E2. An electron micrograph of the HCV nucleocapsid released from the virion by detergent treatment showed a putatively icosahedral particle of 33nm (TAKAHASHI et al. 1992). It contains a protein reactive with the anti-core antibody, but the apparent molecular weight (26kDa) of the protein is larger than the known core protein (p21) of HCV. More recently, HCV virion-like particles have been purified from insect cells expressing the C, E1 and E2 proteins, establishing the identity of the core protein as an internal nucleocapsid protein of the virus particles (BAUMERT et al. 1998). The core protein can oligomerize (MATSUMOTO et al. 1996), consistent with its ability to form nucleocapsid. In the HCV-infected hepatocytes and when expressed as an individual protein, the core protein is localized in the cytoplasm; however, some truncated forms of the core protein are found in the nucleus (LANFORD et al. 1993; LIU et al. 1997; Lo et al. 1995; SHIH et al. 1993; SUZUKI et al. 1995), which is likely not the site of viral replication, and may thus perform functions unrelated to viral replication. Indeed, the core protein has been shown to have the capability to activate or suppress the promoters of many cellular genes tested (see Sect. 3). Furthermore, it can interact with a variety of cellular proteins, some of which may impact on the host's defense. Therefore, the core protein likely plays an important role in the pathogenesis or establishment of persistent infection of HCV.

2 Biochemical Properties and Biosynthesis of the Core Protein

The mature core protein of HCV consists of 191 amino acids, the last 20 or so residues serving as the signal peptide of the adjoining downstream protein E1. The core protein is cleaved from E1 by cellular signal peptidases in the endoplasmic reticulum (ER); thus, it is associated with E1 and E2 in the membrane fraction of the cell (SANTOLINI et al. 1994), specifically located at the cytoplasmic side of the ER (SANTOLINI et al. 1994). Various other proteolytic events (most likely mediated by cellular proteases as well) further cleave the core protein into several truncated forms of the protein, including cleavages at amino acid (aa) residues 182, 178 (HUSSY et al. 1996), 171 (LIU et al. 1997) and 153 (Lo et al. 1994) under various conditions. These cleavages occur not only during in vitro translation in rabbit reticulocyte lysates but also during in vivo expression of the protein in various cell types. The regulation of core protein cleavage and its functional significance is not clear. Since most of the truncated proteins (shorter than 179 amino acids) are located in the nucleus, they likely have different functions from those of the full-length core protein, which is localized in the cytoplasm.

The 191-amino acid protein consists of two distinct domains: an NH_2-terminal two-third domain of highly charged amino acids and a COOH-terminal one-third domain (aa 120–191) of hydrophobic residues (Fig. 1). The hydrophobic domain is responsible for the binding of the core protein to the membrane, particularly the ER, of the cells. In the NH_2-terminal domain, there are three stretches of highly Arg- and Lys-rich sequences within amino acid residues 6–23, 39–74 and 101–121 (SHIH et al. 1993), the significance of which is not clear. There are several stretches of potential nuclear localization signals (aa 1–25, 38–43, 58–64 and 66–71; CHANG et al. 1994; SHIH et al. 1993); thus, when the COOH-terminal domain is removed, the truncated proteins are translocated into the nucleus. The NH_2-terminal portion also enables the core protein to bind to ribosomes either in rabbit reticulocyte lysates or in the cells and can bind to cellular DNA and RNA (SANTOLINI et al. 1994). The NH_2-terminal hydrophilic domain also is responsible for the homotypic interactions of the core protein to form dimers and multimers (MATSUMOTO et al. 1996). It is not clear whether the functional unit of the core protein is a monomer or multimers.

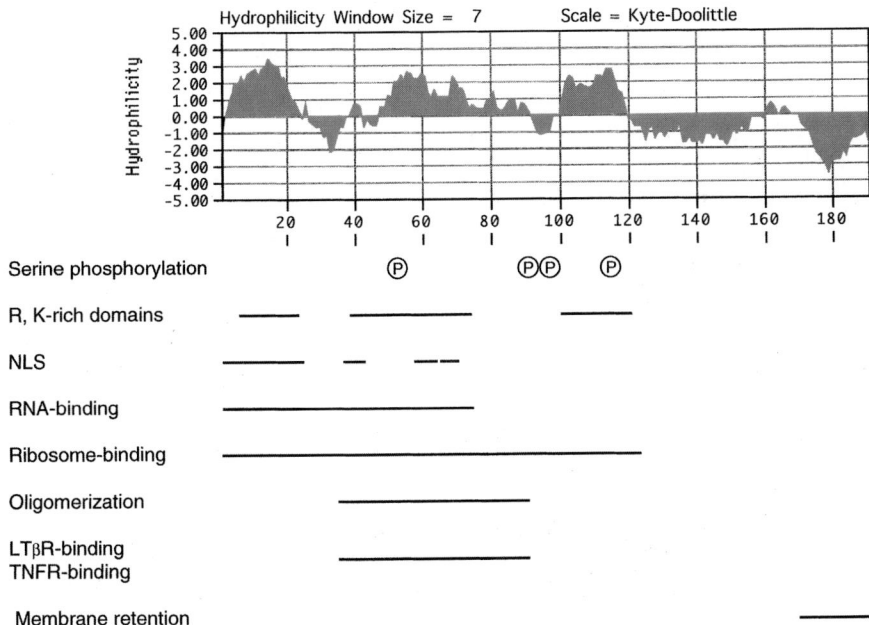

Fig. 1. Hydropathy plot and functional domains of the hepatitis C virus (HCV) core protein. NLS: nuclear localization signal

The core protein is phosphorylated. The main phosphorylated residues are SER-53, 93, 96, and 116 (SHIH et al. 1995). Phosphorylation is probably carried out by cellular protein kinases A and C. Phosphorylation of the protein is required for some of its biological activities, e.g. inhibition of replication and gene expression of hepatitis B virus (SHIH et al. 1995; see Sect. 3). These diverse biochemical properties and variable subcellular localizations suggest that the core protein may have multiple functions in the life cycle of HCV.

In natural infections in humans or chimpanzees, HCV core protein has been detected only in the cytoplasm of infected cells. Only rarely has the core protein been detected in the nucleus of hepatocytes, for example, in a core protein transgenic mouse (KAWAMURA et al. 1997). Thus, it is not clear whether the nuclear localization of the truncated core proteins has biological significance. Nevertheless, the nuclear localization of the truncated core proteins is consistent with its potential functions in regulating cellular gene activities.

3 Possible Pathogenic Roles of the Core Protein

The primary function of the core protein is the formation of the viral nucleocapsid, but its various biochemical properties suggest that it may also interfere with host functions in the virus-infected cells. These activities may directly perturb cellular functions, thereby causing cytotoxicity. Alternatively, some of the activities may affect the host's specific or innate immune response or other defense mechanisms.

3.1 Direct Cytotoxicity of the Core Protein

The core protein possesses several functions that can potentially cause direct cytotoxicity.

3.1.1 Modulation of Cellular Gene Expression

The core protein has been reported to modulate the promoter activity of several cellular genes, including c-*myc*, c-*fos*, p53 and p21(WAF1), and viral genes, including the long terminal repeat (LTR) of Rous sarcoma virus and human immunodeficiency virus (HIV), SV40 early promoter, and the endogenous promoters of hepatitis B virus (HBV) (CHANG et al. 1998; RAY et al. 1995, 1997, 1998; SHIH et al. 1993). Some of the promoters are activated by the core protein, whereas the others are suppressed. As a consequence, the core protein may disrupt the normal functions of the cells. However, all of these effects were demonstrable only in transient transfection of the reporter constructs containing these promoters; so far, the core protein has not been directly demonstrated to alter the activities of the endogenous cellular genes. Nevertheless, the replication and gene expression of HBV and HIV are both inhibited by the HCV core protein in several different cell lines (SHIH et al. 1993; SRINIVAS et al. 1996). This finding is consistent with clinical studies showing that the replication of HBV is usually suppressed by the coinfection with HCV (LIAW et al. 1998). The mechanism of these inhibitions is not clear, but the phosphorylation of the core protein appears to be essential for the inhibition of HBV gene expression and replication by HCV (SHIH et al. 1995). It is not clear whether the core protein binds to these promoter elements directly or to transcription factors; at least in the case of p21, the core protein does not bind directly to the promoter region (RAY et al. 1998). In this regard, it is interesting to note that HCV core protein has been shown to bind heterogeneous nuclear ribonucleoprotein K (hnRNP K; HSIEH et al. 1998). HnRNP K is involved in cellular pre-mRNA splicing and nuclear RNA transport but may also enhance or suppress promoter activities of several cellular genes, including c-*myc* and c-*fos* (TOMONAGA and LEVENS 1995) and thymidine kinase (TK) (HSIEH et al. 1998). The binding of the core protein to hnRNP K resulted in the reversion of the suppressive effects of hnRNP K on the TK promoter (HSIEH et al. 1998). Although the biological significance of this finding in HCV infection is still not clear, it suggests a potential mechanism for the core protein to alter cellular gene expression and thus derange the normal cellular functions. Conceivably, the reported modulating functions of the core protein on other promoters may also be mediated by its binding to various transcription factors. One such factor may be RNA helicase (YOU et al. 1999b).

These observations suggest that the core protein itself could cause direct cytotoxicity. Indeed, it has been difficult to establish cell lines stably expressing the HCV core protein. Furthermore, a transgenic mouse expressing the core protein has been shown to develop liver steatosis, suggesting that this protein alone has the potential to cause liver injury (MORIYA et al. 1997). However, several other similar transgenic mouse lines did not develop any liver pathology (KAWAMURA et al. 1997; PASQUINELLI et al. 1997). Another transgenic mouse line expressing HCV structural proteins under inducible conditions developed hepatitis mediated by CD4- and CD8-positive cells (WAKITA et al. 1998). Also, several permanent cell lines expressing the core protein either constitutively (ZHU et al. 1998) or under inducible conditions (MORADPOUR et al. 1996) did not have gross abnormality. Thus, the potential cytotoxicity of the core protein may require other factors as well. Regardless, the ability of the core protein to modulate the promoter activities may play a role in HCV pathogenesis.

The core protein also binds to lipid droplets containing apolipoprotein A2 in the cells (BARBA et al. 1997). This property may account for HCV-induced steatosis.

3.1.2 Transforming Activity of the Core Protein

The expression of the core protein has been reported to cooperate with the *ras* oncogene to transform rat primary embryo fibroblasts (RAY et al. 1996). Although this reported transforming property was not confirmed in a recent study (CHANG et al. 1998), it was shown that Rat-1 cells stably expressing the core protein did exhibit various parameters of transformed cells, such as focus formation, anchorage-independent growth and tumor formation in nude mice (CHANG et al. 1998).

Similar observations have been made on a mouse cell line expressing the core protein C (TSUCHIHARA et al. 1999). Thus, the core protein appears to have oncogenic potential under some conditions. This transforming activity may be related to the ability of the core protein to suppress the promoter activity of the p53 gene (RAY et al. 1997) and one of its target genes, p21 (WAF1; RAY et al. 1998), which inhibits cyclin-dependent kinases and regulates cell cycle progression. The HCV core protein responsive element in the p21 promoter was mapped downstream of the p53-binding site, but the core protein did not bind to the p21 promoter directly. Thus, the effects of the core protein on p53 and p21 appear to be independent, but may cooperate to promote cell growth through repression of p21 transcription. These transforming properties may contribute to the increased risks of hepatocellular carcinoma in HCV infections. Indeed, a core protein transgenic mouse line developed hepatocellular carcinoma (MORIYA et al. 1998).

3.1.3 Modulation of Apoptosis

The core protein has also been reported to modulate the apoptosis induced by various agents. Depending on the experimental conditions, the core protein can either inhibit or enhance apoptosis; for example, it has been shown to inhibit c-myc- and cisplatin-induced, p53-independent apoptosis in HeLa and Chinese hamster cell lines stably expressing the core protein, whereas the UV-induced apoptosis was not affected (RAY et al. 1996). This inhibition may be the result of alteration of cellular gene expression by the core protein, since, in this cell line, the core protein is expressed in the nucleus. The inhibition of apoptosis may contribute to the persistence of HCV infections, as well as cellular transformation. On the other hand, the core protein can enhance the apoptosis induced by lymphotoxin-$\alpha\beta$, tumor necrosis factor or Fas ligand in various cell lines (CHEN et al. 1997; RUGGIERI et al. 1997; ZHU et al. 1998). Some of these effects may be caused by the binding of the core protein to these cytokine receptors (see Sect. 3.2). The increased cellular sensitivity to cytokines will explain the occurrence of hepatitis in HCV infections. The precise effects of the core protein likely depend on the exact cell types and are regulated by other undefined factors.

3.2 Interactions of the Core Protein with Host's Immune Systems

Another interesting property of the core protein is its ability to bind to the cytoplasmic domain of several members of tumor necrosis factor receptor (TNFR) family. This property confers to the core protein a unique ability to disrupt the cytokine and immune functions of the host cells, thus potentially contributing to the immune-mediated mechanism of the HCV pathogenesis or its persistent infection.

3.2.1 Binding to Lymphotoxin β-Receptor

By using the yeast two-hybrid screening approach, several cellular proteins have been detected as the possible interacting partners of the HCV core protein. The first protein identified is lymphotoxin β receptor (LTβR; CHEN et al. 1997; MATSUMOTO et al. 1997), which is a member of the TNFR family. The binding site was mapped to the cytoplasmic domain of the receptor, which interacts with the NH_2-terminal hydrophilic portion of the HCV core protein. Although the binding affinity has not been directly measured, this binding appears to be weak, as the core protein failed to compete for the dimerization of the LTβR itself (CHEN et al. 1997). Nevertheless, the core protein-binding site corresponds to the binding site of a signal transduction molecule of the receptor, i.e., TRAF-3 (TNFR-associated factor-3; VAN ARSDALE et al. 1997); thus, the binding of the core protein is expected to compete for the binding of TRAF-3, potentially resulting in the disruption of the signal transduction of the receptor. A stable HeLa cell line constitutively expressing the core protein has been shown to be more sensitive than the parental cell line to killing induced by lymphotoxin-αβ complex plus γ-interferon (CHEN et al. 1997). However, this effect was seen only with the core protein-expressing HeLa cells but not with the HepG2 cells, which would have been more relevant to viral hepatitis. Lymphotoxin has been shown to play an important role in the development of lymph nodes and may also play a role in other immune functions, including formation of germinal centers, inasmuch as the LTβR is expressed in most of the cell types, except lymphocytes, whereas its ligand is expressed only on activated T and B cells (WARE et al. 1995). The possible biological significance of this interaction will be discussed below in Sect. 4.

3.2.2 Binding to Tumor Necrosis Factor Receptor

By GST-fusion protein pull-down assay, the core protein was found to bind to the TNFR-1 (also known as TNFR-60), which was the prototype member of the TNFR family (ZHU et al. 1998). The binding domain is mapped to the COOH-terminal end of the cytoplasmic domain of the receptor in a region termed the "death domain", which is responsible for the apoptosis signaling triggered by the binding of TNF. The binding of the core protein to TNFR reduced the binding of the TNFR signaling molecules, including TRADD and TRAF-2 (Zhu et al., unpublished), which also bind directly or indirectly to the death domain of TNFR. Furthermore, in some cell lines expressing the HCV core protein, e.g. BC10ME cells (a mouse fibroblast cell line), the TNF-induced NFκB activation, which is partially mediated by TRADD and TRAF-2, is inhibited (ZHU et al. 1998). These findings suggest that the core protein can disrupt the signal transduction cascade of TNFR. However, in other cell lines, such as HepG2 and HeLa cells, TNF-induced NFκB activation was not affected by the presence of core protein, indicating that the TNFR signaling pathway is different between different cell types, and that the effect of the core protein may differ between different cells (ZHU et al. 1998).

Cells expressing the core protein exhibited an altered sensitivity to TNF. However, the responses appear to vary with cell type: (1) enhanced sensitivity has been reported for murine BC10ME cells, and human HeLa and HepG2 cells. The core-expressing cells exhibit an approximately two-fold enhanced apoptotic response to TNF (ZHU et al. 1998). This effect in BC10ME cells could be the result of inhibition of NFκB activation, since NFκB is known to inhibit TNF-induced apoptosis (BEG and BALTIMORE 1996; VAN ANTWERP et al. 1996; WANG et al. 1996). However, this mechanism could not explain the enhanced sensitivity of core-expressing HeLa and HepG2 cells, since NFκB activation was not affected by the core protein in these cells (ZHU et al. 1998). Thus, the core protein may also affect other signaling pathways of TNFR, such as FADD or JNK pathways. (2) No alteration in TNF sensitivity has been reported for two other HeLa and HepG2 cell lines stably expressing the core protein (CHEN et al. 1997). (3) Decreased TNF sensitivity has been reported for a core-protein-expressing human breast cancer cell line, MCF 7 cells (RAY et al. 1998). The last two reports did not examine whether the core protein binds to the TNFR in their respective cell lines. The variability of the effects of the core protein on TNF sensitivity suggests that the effects of the core protein may be modulated by other factors, such as the genetic background and production of other cytokines. Interestingly, MCF 7 cells do not express caspase 3, which is a key component of apoptosis pathway (JANICKE et al. 1998), suggesting that these cells undergo apoptosis by a distinct mechanism. Furthermore, the state of the core protein likely plays a role as well: if the core protein (as a truncated core protein molecule) is localized in the nucleus, it will be less likely to bind to the TNFR, but, instead, will be more likely to participate in the regulation of cellular gene functions. The activated or suppressed cellular genes may modulate any effects resulting from the binding of the core protein to TNFR.

It is not clear whether the expression of the core protein in animals will have similar effects to what was seen in tissue culture. Several transgenic mice expressing the HCV core protein in the liver have been reported (KAWAMURA et al. 1997; PASQUINELLI et al. 1997). These mice did not show any detectable phenotype. However, the effects of TNF treatment on these transgenic animals have not been directly examined. Recently a transgenic mouse line expressing the core protein in every tissue, including liver, lymph node and spleen, has been developed and showed a reduced sensitivity to TNF (R. Schneider, G. Dennert and M.M. Lai, unpublished observation). This effect was precisely the opposite to that seen in tissue culture. This transgenic mouse line mimics natural HCV infection because this line also includes the possible effects of HCV infection of lymphocytes. These results further indicate that the potential effects of the core protein in HCV infections in humans likely depend on many cellular factors.

3.3.3 Binding to Other Cytokine Receptors

The HCV core protein may also bind to other members of TNFR family. Preliminary data suggest that the core protein binds to Fas antigen but not to CD40 (Zhu et al., unpublished). The binding of the core protein to Fas antigen is consistent with the finding that the core protein sensitizes cells to Fas-mediated apoptosis (RUGGIERI et al. 1997). Recently, several other receptors (e.g., TRAIL receptors) have been identified in this family that contain a death domain (see Sec. 4), suggesting the possibility that core protein may interact with additional TNFR-related receptors. Furthermore, the signal transducers of several of these receptors also contain a death domain; it is possible that the core protein binds to still more members of the TNFR and their signal transducers that contain the death domain. Thus, the overall effects of the core protein in HCV infection may be much more complex than demonstrated so far. The core protein also has been shown to suppress the generation of virus specific cytotoxic T cells by an unknown mechanism (LARGE et al. 1999).

4 Implication of the Interactions Between the Core Protein and Tumor Necrosis Factor Receptor Family

The discussion above showed that the core protein is capable of binding to multiple members of TNFR family and, as a result, alter the sensitivity of the cells to ligands of these receptors. Does this property play any role in HCV infection? A brief review of the biological effects of TNFR family suggests that the core protein, via its interactions with these receptors, will have profound consequences on the host. Significantly, in chronic HCV patients, there is also increased expression of TNF and Fas ligand (LARREA et al. 1996; MAURI et al. 1998; TILG et al. 1992), indicating that whatever effects the core protein might have on the response of TNFRs will be amplified during HCV infection.

The TNFR family has recently experienced a population explosion and now includes over 20 receptor-ligand pairs that are involved in the development, homeostasis, and effector functions of the immune system (Fig. 2; SMITH et al. 1994; WARE et al. 1998). Members of this receptor family can be categorized into two major groups based upon their interactions with distinct families of signaling proteins, either the death domain (DD) family or the TRAF family of zinc RING finger proteins. HCV core protein binds to members of the both groups. Both groups of receptors initiate cell death in different cellular contexts, typically tumor cells, although the DD receptor family can initiate apoptosis of normal, non-transformed cells. Additionally, these receptors activate the NFκB family of transcription factors that control expression of many proinflammatory and anti-apoptotic genes, and the c-Jun transcription factor family that are involved in responses to stress. The signaling pathways that control apoptosis and NFκB are simultaneously activated, and yet, either pathway can be blocked without disrupting the other (LIU et al. 1996). In the human and mouse cell lines expressing the HCV core protein as described above, the NFκB activation is blocked only in the mouse cell line, but not in human cell lines (ZHU et al. 1998), and yet all the cell lines have increased sensitivity to TNF, consistent with the multiple and divergent pathways for the TNFR. Variable effects of the core protein on NFκB activation have also been reported in other cell lines (SHRIVASTAVA et al. 1998; YOU et al. 1999a), suggesting that these effects are modulated by cellular conditions.

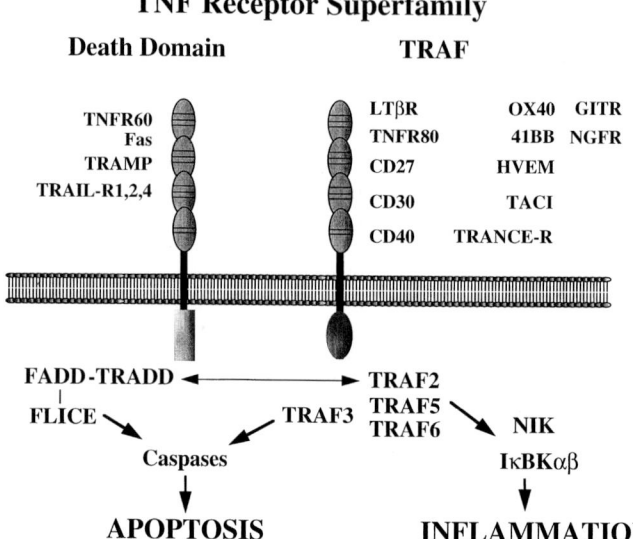

Fig. 2. The tumor necrosis factor (TNF) receptor superfamily and signaling proteins. Receptors are categorized into two major groups, death domain (DD) or TNFR-associated factors (TRAF)-binding receptors. CD, cluster of differentiation; FADD/TRADD, Fas/TNFR-associated DD proteins; FLICE (MACH), caspase 8; GITR, glucocorticoid-induced TNF receptor-related gene; HVEM, herpesvirus entry mediator; LT, lymphotoxin; NIK, TRAF2-associated kinase; IκBK, inhibitor of κB kinase; NGFR, p75 nerve growth factor (neurotrophin) receptor; TNF, tumor necrosis factor; TRAIL, TNF-related apoptosis-inducing ligand; TRAMP, TNF receptor-associated membrane protein (DR3/WSL-1); TACI, T cell activation CAML-interacting protein

For both receptor types, their signaling pathways have been shown to bifurcate at the earliest step in signaling by recruiting distinct signaling proteins to the cytoplasmic domains of the receptor. For TNFR1, this involves recruitment of TRADD and FADD, the latter of which provides an adapter link to FLICE (caspase 8) containing a death effector domain that activates other caspases leading to apoptosis (Hsu et al. 1996). TRADD recruits TRAF2, which, in turn, activates NFκB and JNK-dependent pathways. Similarly, the LTβR utilizes TRAF3 to propagate signals for cell death (van Arsdale et al. 1997), and TRAF5 (Nakano et al. 1996; or TRAF2) to activate NFκB/JNK pathways. For at least some receptors, additional signaling cascades can be activated independently of DD or TRAFs, through proteins such as FAN (involved in sphingomyelinase activation; Adam-Klages et al. 1996) or DAXX, a Fas-binding protein involved in death signaling independently from FADD (Yang et al. 1997). This list is unlikely exhaustive of the signaling pathways activated by these receptors, and we should probably assume that the signaling domains act as manifolds for initiating signals that lead to diverse cellular responses. Thus, it is not surprising that the biological response elicited by these receptors may depend on the state of cellular differentiation, sometimes resulting in opposite effects in different cell types.

Table 1. Virus homologues and modulators of the tumor necrosis factor (TNF) superfamily

Virus	Virus gene product	Effect	Reference
Hepatitis C virus	Core protein	Binds LTβR/TNFR; increases apoptosis to TNF	(Matsumoto et al. 1997) (Zhu et al. 1998)
Adenovirus	E-1B	Inhibits Fas apoptosis	(Hashimoto et al. 1991)
	E3–14.7K	Inhibits TNF apoptosis	(Horton et al. 1991)
	E3–10.4K/ 14.5K	Down regulates surface Fas & inhibits apoptosis	(Shisler et al. 1997)
Epstein-Barr virus	LMP-1	Sequesters TRAFs	(Mosialos et al. 1995; Devergne et al. 1996)
Herpes simplex virus1/2	Envelope gD	Entry protein via HVEM; LIGHT virokine	(Montgomery et al. 1996; Mauri et al. 1998)
Equine herpes Virus-2	E8	Inhibits Fas and TNF apoptosis	(Bertin et al. 1997)
Poxvirus	T2	Binds TNF/LTα; blocks T cell apoptosis	(Upton et al. 1991)
	A53R	Binds TNF, not LTα	(Smith et al. 1996)
	crmA	Caspase inhibitor of apoptosis	(Ray et al. 1992)
	MC159	Inhibits Fas and TNF apoptosis	(Bertin et al. 1997)
Avian leukosis virus	ALV-B Env SU	Entry factor via CAR-1 (TRAIL-R2)(tv-b^{s3}); induces apoptosis	(Brojatsch et al. 1996)

LTβR, lymphotoxin β receptor; HVEM, herpesvirus entry mediator; LTα, lymphotoxin-α; LIGHT, Lymphotoxin-like ligand that competes with envelope glycoprotein D of herpes simplex virus for binding to HVEM, a receptor expressed on T lymphocytes; CAR-1, cellular avian leukosis virus receptor (the avian homologue of TRAIL-R2).

The finding that HCV core protein directly binds LTβR (Chen et al. 1997; Matsumoto et al. 1997) and TNFR1 (Zhu et al. 1998) at the domains where their signal transducers bind suggests that the signaling pathways of these receptors are likely altered by the core protein. As discussed above, overexpression of HCV core protein in mouse and human tumor cells leads to an enhanced apoptotic response to TNF or LTαβ in at least some cell lines (Chen et al. 1997; Zhu et al. 1998). These effects are similar to the action of adenovirus E1A oncoprotein, which sensitizes cells to apoptosis and concurrently increases sensitivity to TNF (Duerksen-Hughes et al. 1989; Table 1), although E1A will convert a resistant cell type to a sensitive phenotype, a feature not associated with HCV core protein. This effect of E1A is attenuated by adenovirus E3 proteins that block sensitivity to apoptosis signaled by TNFR (Gooding, 1992; Wold and Gooding 1991). Conceivably, the apoptosis-modulating effect of the HCV core protein may similarly be countered by other viral proteins. Because of the presence of multiple and interacting signal transduction pathways of these receptors, it is not surprising that different cells expressing the core protein may respond differently to TNF (Ray et al. 1998; Zhu et al. 1998).

What is the consequence of the altered sensitivity of core-expressing cells to TNF or LT? This may depend on the cell types affected. Theoretically, enhancement of the apoptosis of leukocytes might accelerate their early death and diminish the vitality of the immune response, which in turn may contribute to virus persistence. This is a possible scenario in HCV infections since leukocytes are potential targets for HCV infection (BLIGHT et al. 1994; LERAT et al. 1996; MULLER et al. 1993; SHIMIZU et al. 1997). The enhanced sensitivity of hepatocytes, however, may accelerate the development of hepatitis, or facilitate the spread of virus by formation of apoptotic bodies, which are engulfed by neighboring cells, without being exposed to antibodies. On the other hand, if the core protein reduces cell sensitivity to TNF (RAY et al. 1998), it will provide a mechanism for virus-infected cells to escape the host's defense mechanism, thus ensuring viral persistence.

The mechanism by which HCV core protein causes enhanced sensitivity to cytokines is not entirely clear. It could nullify receptor signaling by blocking recruitment or activation of signaling proteins in either a global or selective fashion. In support of the latter case, HCV core protein enhances apoptosis without affecting NFκB activation in human cell lines following stimulation with TNF (ZHU et al. 1998). However, the lack of a tissue culture model for HCV infection hampers identification of the specific effects of core and other HCV proteins. Animal models with transgenic expression of HCV proteins may provide some new perspectives.

Modulating members of the TNF family in vivo by transgenic or gene deletion methods has been shown to result in profound alteration of innate defenses, as well as cellular and humoral immunity (VON BOEHMER 1997; Table 2). The phenotype of TNFR1-knockout mice suggests that TNFR1-specific signaling is critical for innate defenses, particularly for resistance to intracellular parasites such as *Listeria monocytogenes*, although for some viruses, such as mousepox or lymphocytic choriomeningitis virus (LCMV), TNFR1 signaling does not appear to be essential for the host's defense (PFEFFER et al. 1993; ROTHE et al. 1993). LTαβ and TNF are

Table 2. Phenotypes of mice deficient in tumor necrosis factor (TNF) and lymphotoxins (LTs)

Ligand or receptor	Spleen				Lymph node	Peyer's patches	Thymus	Organ infiltrates
	Marginal zone	1° lymphoid follicles	2° follicles germinal centers	FDC network				
TNF	Increased	Absent	None	None	Present	Present	Normal	None
TNFR-1	–	Absent	None	None	Present	Present	Normal	None
TNFR-2	Normal	Normal	Normal	Normal	Normal	Normal	Normal	None
LTα[b]	Absent	Absent	Absent	None	Absent	Absent	Normal	Present in tg[c]
LTβ	Absent	Absent	Absent	None	Partially absent[a]	Absent	Normal	Masssive[c]
LTβR	Absent	Absent	Absent	None	Absent	Absent	Normal	Absent

[a] Presence of cervical, mesenteric cadual nodes.
[b] LTα expressed as transgene (tg) under insulin promotor infiltrates in pancreas and kidney.
[c] CD4+ T cell and B cells in liver and lung.

also required for specific immunologic defenses. Mice deficient in LTα (Fu et al. 1997; MATSUMOTO et al. 1996), LTβ (ALIMZHANOV et al. 1997; KONI et al. 1997) or TNF (PASPARAKIS et al. 1996) lack primary follicles in the spleen and fail to form germinal centers during antigen-specific immune responses. Here, LTβR and TNFR1 signaling pathways act at discreet steps of stromal cell differentiation required for tissue microenvironments that promote efficient Ig class switching by B lymphocytes (Fu et al. 1997). In addition, deletion of either LTα, LTβ or LTβR genes, but not TNF, disrupts signals essential for development of peripheral lymphoid organs (including most lymph nodes and Peyer's patches; ALIMZHANOV et al. 1997; DE TOGNI et al. 1994; KONI et al. 1997; FÜTTERER et al. 1998; Table 2). These results indicate that LTβR and TNFR1 signaling is necessary for the formation of tissue microenvironments that allow lymphocytes to interact with antigen-presenting cells, and for B lymphocyte migration and differentiation into antibody-secreting cells. Thus, a reasonable prediction is that disruption of LTβR or TNFR signaling by HCV core protein could lead to a poor-quality immune response that may allow virus to persist. A poor-quality antibody response may result if HCV infects cells of the marginal zone or antigen-presenting cells in lymphoid organs, thus disrupting the ability of B cells to switch Ig subclasses. In the liver, faulty LTβR or TNFR1 signaling caused by HCV core protein might alter trafficking and infiltration of infected liver tissue by activated T cells, thus preventing efficient virus clearance. This speculation is based in part on the observations of LTα transgenic mice (driven by insulin promoter) that show ectopic infiltration of T cells into the pancreas and kidney in a pattern resembling organized lymphoid tissue (KRATZ et al. 1996; SACCA et al. 1997). Curiously, the lungs and liver of mice deficient in LTα or LTβ exhibit a significant infiltration with CD4+ T cells and B cells that is not due to infection, but appears as a result of altered trafficking of lymphocytes (ALIMZHANOV et al. 1997; KONI et al. 1997). In view of the fact that HCV core affects TNF or LTαβ signaling, are similar phenotypes observed in mice carrying an HCV core transgene? So far, several transgenic mice that express HCV core protein either ubiquitously (R. Schneider et al. unpublished) or in liver-specific mode (KAWAMURA et al. 1997; PASQUINELLI et al. 1997), have no apparent phenotype that affects the development of lymph nodes or organization of splenic germinal centers. This deficiency of phenotype probably was due to late expression of the core protein during embryonic development. More subtle changes in lymphoid tissue architecture should be examined, specifically at the level of organization of germinal centers and follicles following administration of antigen, as well as the processes of tissue remodeling in the liver that occur during chronic infection.

Chronically inflamed tissue typically develops pseudo-organized lymphoid tissue with follicle-like foci containing T and B cells. However, the immunologic picture of the liver chronically infected with HCV shows some lymphocytic infiltration, with evidence for cytotoxic T lymphocytes in the circulation, but a lack of organized lymphoid tissue with plasma cells (that is found frequently, for example, in autoimmune hepatitis). Is the lack of organized lymphoid follicles in the liver evidence of altered signaling of LTβR or TNFR caused by HCV? Currently, very little is understood about the roles of TNFR and LTβR in chronic inflammation of non-lymphoid tissue, an area that may be important for understanding the molecular basis of persistent HCV infections.

5 Perspectives

The ability of the HCV core protein to modulate cellular gene expression and to interact with cellular proteins, particularly the members of TNFR family, suggests that the core protein plays an important role in HCV pathogenesis. Because of these diverse interactions, the effects of the core protein are likely determined by the combined effects of multiple, sometimes counteracting, factors. Because of the sequence and structural relatedness of the members of TNFR family, it is conceivable that the core protein may interact with many other members of the family as well. These interactions will undoubtedly have profound effects on the pathogenesis and persistence of HCV infection and may also contribute to the occurrence of autoimmune diseases which have been found to be frequently associated with HCV infections (ALMASIO et al. 1992; HADDAD et al. 1992; JOHNSON et al. 1993; TRAN et al. 1992). This is particularly true if the core protein binds to Fas antigen. Further identification of the interacting partners of the core protein will facilitate our understanding of the HCV pathogenesis. The core protein may also be a potential target of antiviral agents.

References

Adam-Klages S, Adam D, Wiegmann K, Struve S, Kolanus W, Schneider-Mergener J, Kronke M (1996) FAN, a novel WD-repeat protein, couples the p55 TNF-receptor to neutral sphingomyelinase. Cell 86:937–947
Alimzhanov MB, Kuprash DV, Kosco-Vilbois MH, Luz A, Turetskaya RL, Tarakhovsky A, Rajewsky K, Nedospasov SA, Pfeffer K (1997) Abnormal development of secondary lymphoid tissues in lymphotoxin β-deficient mice. Proc Natl Acad Sci USA 94:9302–9307
Almasio P, Provenzano G, Scimemi M, Cascio G, Craxi A, Pagliaro L (1992) Hepatitis C virus and Sjogren's syndrome. Lancet 339:989–990
Barba G, Harper F, Harada T, Kohara M, Goulinet S, Matsuura Y, Eder G, Schaff Z, Chapman MJ, Miyamura T, Brechot C (1997) Hepatitis C virus core protein shows a cytoplasmic localization and associates to cellular lipid storage droplets. Proc Natl Acad Sci USA 94:1200–1205
Baumert TF, Ito S, Wong DT, Liang JT (1998) Hepatitis C virus structural proteins assemble into virus-like particles in insect cells. J Virol 72:3827–3836
Beg AA, Baltimore D (1996) An essential role for NF-κB in preventing TNF-α-induced cell death. Science 274:782–784
Bertin J, Armstrong RC, Ottilie S, Martin DA, Wang Y, Banks S, Wang, G.-H, Senkevich TG, Alnemri ES, Moss B, Lenardo MJ, Tomaselli KJ, Cohen JI (1997) Death effector domain-containing herpesvirus and poxvirus proteins inhibit both Fas- and TNFR1-induced apoptosis. Proc Natl Acad Sci USA 94:1172–1176
Blight K, Lesniewski RR, Labrooy JT, Gowans EJ (1994) Detection and distribution of hepatitis C-specific antigens in naturally infected liver. Hepatology 20:553–557

Brojatsch J, Naughton J, Rolls MM, Zingler K, Young JAT (1996) CAR1, a TNFR-related protein, is a cellular receptor for cytopathic avian leukosis-sarcoma viruses and mediates apoptosis. Cell 87:845–855

Chang J, Yang SH, Cho YG, Hwang S B, Hahn YS, Sung YC (1998) Hepatitis C virus core from two different genotypes has an oncogenic potential but is not sufficient for transforming primary rat embryo fibroblast in cooperation with the H-ras oncogene. J Virol 72:3060–3065

Chang SC, Yen JH, Kang HY, Jang MH, Chang MF (1994) Nuclear localization signals in the core protein of hepatitis C virus. Biochem Biophys Res Commun 205:1284–1290

Chen C-M, You L-R, Hwang L-H, Lee Y-HW. (1997) Direct interaction of hepatitis C virus core protein with the cellular lymphotoxin-β receptor modulates the signal pathway of the lymphotoxin-β receptor. J. Virol. 71:9417–9426

De Togni P, Goellner J, Ruddle NH, Streeter PR, Fick A, Mariathasan S, Smith SC, Carlson R, Shornick LP, Strauss-Schoenberger J, Russell JH, Karr R, Chaplin DD (1994) Abnormal development of peripheral lymphoid organs in mice deficient in lymphotoxin. Science 264:703–706

Devergne O, Hatzivassiliou E, Izumi KM, Kaye KM, Kleijnen MF, Kieff E, Mosialos G (1996) Association of TRAF1, TRAF2, and TRAF3 with an Epstein-Barr virus LMP1 domain important for B-lymphocyte transformation: role in NF-κB activation. Mol Cell Biol 16:7098–7108

Duerksen-Hughes P, Wold WSM, Gooding LR (1989) Adenovirus E1 A renders infected cells sensitive to cytolysis by tumor necrosis factor. J Immunol 143:4193–4200

Fu Y-X, Molina H, Matsumoto M, Huang G, Min J, Chaplin DD (1997) Lymphotoxin-α (LTα) supports development of splenic follicular structure that is required for IgG responses. J Exp Med 185:2111–2120

Fütterer A, Mink K, Luz A, Kosco-Vilbois MH, Pfeffer K (1998) The lymphototoxin beta receptor controls organogenesis and affinity maturation in peripheral lymphoid tissues. Immunity 9:59–70

Gale MJJ, Korth MJ, Tang NM, Tan SL, Hopkins DA, Dever TE, Polyak SJ, Gretch DR, Katze MG (1997) Evidence that hepatitis C virus resistance to interferon is mediated through repression of the PKR protein kinase by the nonstructural 5 A protein. Virology 230:217–227

Gooding LR (1992) Virus proteins that counteract host immune defenses. Cell 71:5–7

Haddad J, Deny P, Munz-Gotheil C, Ambrosini JC, Trinchet JC, Paterson D, Mal F, Callard P, Beaugrand M (1992) Lymphocytic sialadenitis of Sjogren's syndrome associated with chronic hepatitis C virus liver disease. Lancet 339:321–323

Hashimoto S, Ishii A, Yonehara S (1991) The E1b oncogene of adenovirus confers cellular resistance to cytotoxicity of tumor necrosis factor and monoclonal anti-Fas antibody. Int Immunol 3:343–351

Horton TM, Ranheim TS, Aquino L, Kusher DI, Saha SK, Ware CF, Wold WSM, Gooding LR (1991) Adenovirus E3 14.7 K protein functions in the absence of other adenovirus proteins to protect transfected cells from tumor necrosis factor cytolysis. J Virol 65:2629–2639

Hsieh T-Y, Matsumoto M, Chou H-C, Schneider R, Hwang SB, Lee AS, Lai MMC (1998) Hepatitis C virus core protein interacts with heterogeneous nuclear ribonucleoprotein K. J Biol Chem 273:17651–17659

Hsu H, Shu HB, Pan MG, Goeddel DV (1996) TRADD-TRAF2 and TRADD-FADD interactions define two distinct TNF receptor 1 signal transduction pathways. Cell 84:299–308

Hüssy P, Langen H, Mous J, Jacobsen H (1996) Hepatitis C virus core protein: carboxy-terminal boundaries of two processed species suggest cleavage by a signal peptide peptidase. Virology 224:93–104

Janicke RU, Sprengart ML, Wati MR, Porter AG (1998) Caspase-3 is required for DNA fragmentation and morphological changes associated with apoptosis. J Biol Chem 273:9357–9360

Johnson RJ, Gretch DR, Yamabe H, Hart J, Bacchi CE, Hartwell P, Couser WG, Corey L, Wener MH, Alpers CE, Wilson R (1993) Membranoproliferative glomerulonephritis associated with hepatitis C virus infection. N Eng J Med 328:465–470

Kawamura T, Furusaka A, Koziel MJ, Chung RT, Wang TC, Schmidt EV, Liang TJ (1997) Transgenic expression of hepatitis C virus structural proteins in the mouse. Hepatology 25:1014–1021

Koni PA, Sacca R, Lawton P, Browning JL, Ruddle NH, Flavell RA (1997) Distinct roles in lymphoid organogenesis for lymphotoxins α and β revealed in lymphotoxin β-deficient mice. Immunity 6:491–500

Kratz A, Campos-Neto A, Hanson MS, Ruddle NH (1996) Chronic inflammation is caused by lymphotoxin in lymphoid neogenesis. J Exp Med 183:1461–1472

Lanford RE, Notvall L, Chavez D, White R, Frenzel G, Simonsen C, Kim J (1993) Analysis of hepatitis C virus capsid, E1, and E2/NS1 proteins expressed in insect cells. Virology 197:225–235

Large MK, Kittlesen DJ, Hahn YS (1999) Suppression of host immune response by the core protein of hepatitis C virus: possible implications for hepatitis C virus persistence. J Immunol 162:931–938

Larrea E, Garcia N, Qian C, Civeira MP, Prieto J (1996) Tumor necrosis factor α gene expression and the response to interferon in chronic hepatitis C. Hepatology 23:210–217

Lerat H, Berby F, Trabaud MA, Vidalin O, Major M, Trepo C, Inchauspe G (1996) Specific detection of hepatitis C virus minus strand RNA in hematopoietic cells. J Clin Invest 97:845–851

Liaw YF, Tsai SL, Sheen IS, Chao M, Yeh CT, Hsieh SY, Chu CM (1998) Clinical and virological course of chronic hepatitis B virus infection with hepatitis C and D virus markers. Am J Gastroenterol 93:354–359

Liu Q, Tackney C, Bhat RA, Prince AM, Zhang P (1997) Regulated processing of hepatitis C virus core protein is linked to subcellular localization. J Virol 71:657–662

Liu ZG, Hsu H, Goeddel DV, Karin M (1996) Dissection of TNF receptor 1 effector functions: JNK activation is not linked to apoptosis while NF-κB activation prevents cell death. Cell 87:565–576

Lo S-Y, Masiarz F, Hwang SB, Lai MMC, Ou J-H (1995) Differential subcellular localization of hepatitis C virus core gene products. Virology 213:455–461

Lo S-Y, Selby M, Tong M, Ou J-H (1994) Comparative studies of the core gene products of two different hepatitis C virus isolates: two alternative forms determined by a single amino acid substitution. Virology 199:124–131

Matsumoto M, Hsieh T-Y, Zhu N, VanArsdale T, Hwang SB, Jeng K-S, Gorbalenya AE, Lo S-Y, Ou J-H, Ware CF, Lai MMC (1997) Hepatitis C virus core protein interacts with the cytoplasmic tail of lymphotoxin-β receptor. J Virol 71:1301–1309

Matsumoto M, Hwang SB, Jeng K-S, Zhu N, Lai MMC (1996) Homotypic interaction and multimerization of hepatitis C virus core protein. Virology 218:43–51

Mauri DN, Ebner R, Montgomery RI, Kochel KD, Cheung TC, Yu G-L, Ruben S, Murphy M, Eisenbery RJ, Cohen GH, Spear PG, Ware CF (1998) LIGHT, a new member of the TNF superfamily and lymphotoxin α are ligands for herpesvirus entry mediator. Immunity 8:21–30

Montgomery RI, Warner MS, Lum BJ, Spear PG (1996) Herpes simplex virus-1 entry into cells mediated by a novel member of the TNF/NGF receptor family. Cell 87:427–436

Moradpour D, Englert C, Wakita T, Wands JR (1996) Characterization of cell lines allowing tightly regulated expression of hepatitis C virus core protein. Virology 222:51–63

Moriya K, Fujie H, Shintani Y, Yotsuyanagi H, Tsutsumi T, Ishibashi K, Matsuura Y, Kimura S, Miyamura T, Koike K (1998) The core protein of hepatitis C virus induces hepatocellular carcinoma in transgenic mice. Nature Med 4:1065–1067

Moriya K, Yotsuyanagi H, Shintani Y, Fujie H, Ishibashi K, Matsuura Y, Miyamura T, Koike K (1997) Hepatitis C virus core protein induces hepatic steatosis in transgenic mice. J Gen Virol 78:1527–1531

Mosialos G, Birkenbach M, Yalamanchili R, VanArsdale T, Ware C, Kieff E (1995) The Epstein-Barr virus transforming protein LMP1 engages signaling proteins for the tumor necrosis factor receptor family. Cell 80:389–399

Muller HM, Pfaff E, Goeser T, Kallinowski B, Solbach C, Theilmann L (1993) Peripheral blood leukocytes serve as a possible extrahepatic site for hepatitis C virus replication. J Gen Virol 74:669–676

Nakano H, Oshima H, Chung W, Williams-Abbott L, Ware C, Yagita H, Okumura K (1996) TRAF5, an activator of NF-κB and putative signal transducer for the lymphotoxin-β receptor. J Biol Chem 271:14661–14664

Pasparakis M, Alexopoulou L, Episkopou V, Kollias G (1996) Immune and inflammatory responses in TNFα-deficient mice: a critical requirement for TNFα in the formation of primary B cell follicles, follicular dendritic cell networks and germinal centers, and in the maturation of the humoral immune response. J Exp Med 184:1397–1411

Pasquinelli C, Schoenberger JM, Chung J, Chang K-M, Guidotti LG, Selby M, Berger K, Lesniewski R, Houghton M, Chisari FV (1997) Hepatitis C virus core and E2 protein expression in transgenic mice. Hepatology 25:719–727

Pfeffer K, Matsuyama T, Kundig TM, Wakeham A, Kishihara K, Shahinian A, Wiegmann K, Ohashi PS, Kronke M, Mak T-W (1993) Mice deficient for the 55kd tumor necrosis factor receptor are resistant to endotoxic shock, yet succumb to L. monocytogenes infection. Cell 73:457–467

Ray CA, Black RA, Kronheim SR, Greenstreet TA, Sleath PR, Salvesen GS, Pickup DJ (1992) Viral inhibition of inflammation: cowpox virus encodes an inhibitor of the interleukin-1β converting enzyme. Cell 69:597–604

Ray RB, Lagging LM, Meyer K, Ray R (1996) Hepatitis C virus core protein cooperates with ras and transforms primary rat embryo fibroblasts to tumorigenic phenotype. J Virol 70:4438–4443

Ray RB, Lagging LM, Meyer K, Steele R, Ray R (1995) Transcriptional regulation of cellular and viral promoters by the hepatitis C virus core protein. Virus Res 37:209–220

Ray RB, Meyer K, Ray R (1996) Suppression of apoptotic cell death by hepatitis C virus core protein. Virology 226:176–182

Ray RB, Meyer K, Steele R, Shrivastava A, Aggarwal BB, Ray R (1998) Inhibition of tumor necrosis factor (TNF-alpha)-mediated apoptosis by hepatitis C virus core protein. J Biol Chem 273:2256–2259

Ray RB, Steele R, Meyer K, Ray R (1997) Transcriptional repression of p53 promoter by hepatitis C virus core protein. J Biol Chem 272:10983–10986

Ray RB, Steele R, Meyer K, Ray R (1998) Hepatitis C virus core protein represses p21WAF/Cip1/Sid1 promoter activity. Gene 208:331–336

Rothe J, Lesslauer W, Lotscher H, Lang Y, Koebel P, Kontgen F, Althage A, Zinkernagel R, Steinmetz M, Bluethmann H (1993) Mice lacking the tumour necrosis factor receptor 1 are resistant to TNF-mediated toxicity but highly susceptible to infection by Listeria monocytogenes. Nature 364: 798–802

Ruggieri A, Harada T, Matsuura Y, Miyamura T (1997) Sensitization to Fas-mediated apoptosis by hepatitis C virus core protein. Virology 229:68–76

Sacca R, Turley S, Soong L, Mellman I, Ruddle NH (1997) Transgenic expression of lymphotoxin restores lymph nodes to lymphotoxin-α-deficient mice. J Immunol 159:4252–4260

Santolini E, Migliaccio G, La Monica N (1994) Biosynthesis and biochemical properties of the hepatitis C virus core protein. J Virol 68:3631–3641

Shih C-M, Chen C-M, Chen S-Y, Lee Y-HW (1995) Modulation of the trans-suppression activity of hepatitis C virus core protein by phosphorylation. J Virol 69:1160–1171

Shih C-M, Lo SJ, Miyamura T, Chen S-Y, Lee Y-HW (1993) Suppression of hepatitis B virus expression and replication by hepatitis C virus core protein in HuH-7 cells. J Virol 67:5823–5832

Shimizu YK, Igarashi H, Kanematu T, Fujiwara K, Wong DC, Purcell RH, Yoshikura H (1997) Sequence analysis of the hepatitis C virus genome recovered from serum, liver, and peripheral blood mononuclear cells of infected chimpanzees. J Virol 71:5769–5773

Shisler J, Yang C, Walter B, Ware CF, Gooding LR (1997) The adenovirus E3–10.4K/14.5K complex mediates loss of cell surface Fas (CD95) and resistance to Fas-induced apoptosis. J Virol 71:8299–8306

Shrivastava A, Manna SK, Ray R, Aggarwal BB (1998) Ectopic expression of hepatitis C virus core protein differentially regulates nuclear transcription factors. J Virol 72:9722–9728

Smith CA, Farrah T, Goodwin RG (1994) The TNF receptor superfamily of cellular and viral proteins: activation, costimulation and death. Cell 76:959–962

Smith CA, Hu F-Q, Smith TD, Richards CL, Smolak P, Goodwin RG, Pickup DJ (1996) Cowpox virus genome encodes a second soluble homologue of cellular TNF receptors, distinct from CrmB, that binds TNF but not LTα. Virology 223:132–147

Srinivas RV, Ray RB, Meyer K, Ray R (1996) Hepatitis C virus core protein inhibits human immunodeficiency virus type 1 replication. Virus Res 45:87–92

Suzuki R, Matsuura Y, Suzuki T, Ando A, Chiba J, Harada S, Saito I, Miyamura T (1995) Nuclear localization of the truncated hepatitis C virus core protein with its hydrophobic C terminus deleted. J Gen Virol 76:53–61

Takahashi K, Kishimoto S, Yoshizawa H, Okamoto H, Yoshikawa A, Mishiro S (1992) p26 protein and 33-nm particle associated with nucleocapsid of hepatitis C virus recovered from the circulation of infected hosts. Virology 191:413–434

Tilg H, Wilmer A, Vogel W, Herold M, Nolchen B, Judmaier G, Huber C (1992) Serum levels of cytokines in chronic liver diseases. Gastroenterology 103:264–274

Tomonaga T, Levens D (1995) Heterogeneous nuclear ribonucleoprotein K is a DNA-binding transactivator. J Biol Chem 270:4875–4881

Tran A, Quaranta J-F, Benzaken S, Thiers V, Chau HT, Hastier P, Regnier D, Dreyfus G, Pradier C, Sadoul J-L, Hebutern X, Rampal P (1992) High prevalence of thyroid autoantibodies in a prospective series of patients with chronic hepatitis C before interferon therapy. Hepatology 18:253–257

Tsuchihara K, Hajikata M, Fukuda K, Kuroki T, Yamamoto N, Shimotohno K (1999) Hepatitis C virus core protein regulates cell growth and signal transduction pathway transmitting growth stimuli. Virology 258:100–107

Upton C, Macen JL, Schreiber MM, McFadden G (1991) Myxoma virus expresses a secreted protein with homology to the tumor necrosis factor receptor gene family that contributes to viral virulence. Virology 184:370–382

Van Antwerp DJ, Martin SJ, Kafri T, Green DR, Verma IM (1996) Suppression of TNF-α-induced apoptosis by NF-κB. Science 274:787–789
Van Arsdale TL, VanArsdale SL, Force WR, Walter BN, Mosialos G, Kieff E, Reed JC, Ware CF (1997) Lymphotoxin-b receptor signaling complex: role of tumor necrosis factor receptor-associated factor 3 recruitment in cell death and activation of nuclear factor κB. Proc Natl Acad Sci USA 94:2460–2465
von Boehmer H (1997) Lymphotoxins: from cytotoxicity to lymphoid organogenesis. Proc Natl Acad Sci USA 94:8926–8927
Wakita T, Taya C, Katsume A, Kato J, Yonekawa H, Kanegae Y, Saito I, Hayashi Y, Koike M, Kohara M (1998) Efficient conditional transgene expression in hepatitis C virus cDNA transgenic mice mediated by the Cre/loxP system. J Biol Chem 273:9001–9006
Wang CY, Mayo MW, Baldwin ASJ (1996) TNF- and cancer therapy-induced apoptosis: potentiation by inhibition of NF-κB. Science 274:784–787
Ware CF, Santee SM, Glass A (1998) Tumor necrosis factor-related ligands and receptors. 3rd edn. In: Thomson AW (ed) The cytokine handbook. Academic, London
Ware CT, VanArsdale TL, Crowe PD, Browning JL (1995) The ligands and receptors of the lymphotoxin system. In: Griffiths GM, Tschopp J (eds) Pathways for cytolysis. Springer, Berlin Heidelberg New York, pp 175–218
Wold WS, Gooding LR (1991) Region E3 of adenovirus: a cassette of genes involved in host immunosurveillance and virus-cell interactions. Virology 184:1–8
Yang X, Khosravi-Far R, Chang HY, Baltimore D (1997) Daxx, a novel FAS-binding protein that activates JNK and apoptosis. Cell 89:1067–1076
You LR, Chen CM, Lee YHW (1999a) Hepatitis C virus core protein enhances NFκB signal pathway triggering by lymphotoxin-beta receptor ligand and tumor necrosis factor alpha. J Virol 73:1672–1681
You LR, Chen CM, Yeh TS, Tsai TY, Mai RT, Lin CH, Lee YH (1999b) Hepatitis C virus core protein interacts with cellular RNA helicase. J Virol 73:2841–2853
Zhu N, Khoshnan A, Schneider R, Ware CF, Lai MMC (1998) Hepatitis C virus core protein binds to the cytoplasmic domain of tumor necrosis factor (TNF) receptor 1 and enhances TNF-induced apoptosis. J Virol 72:3691–3697

Folding, Assembly and Subcellular Localization of Hepatitis C Virus Glycoproteins

J. DUBUISSON

1	Introduction	135
2	Formation of Hepatitis C Virus Glycoprotein Complexes	136
2.1	Native E1E2 Complexes	136
2.2	Disulfide Bond Aggregates	137
3	Folding of Hepatitis C Virus Glycoproteins	139
3.1	Intramolecular Disulfide Bond Formation	139
3.2	Folding of Subdomains	140
3.3	Role of E2 in the Folding of E1	140
3.4	Role of Endoplasmic Reticulum Chaperones	142
4	Sequences Involved in Hepatitis C Virus Glycoprotein Interactions	143
5	Subcellular Localization of Hepatitis C Virus Glycoproteins	144
6	Concluding Remarks	145
References		146

1 Introduction

Hepatitis C Virus (HCV) glycoproteins E1 and E2 are produced by proteolytic cleavage of the HCV polyprotein (RICE 1996). Cotranslational cleavages at the C/E1, E1/E2 and NS2/NS3 sites produce E1 and a short-lived precursor of E2, E2-NS2. This precursor is cleaved to produce E2, E2-p7 and NS2 (GRAKOUI et al. 1993; LIN et al. 1994; MIZUSHIMA et al. 1994; SELBY et al. 1994). For some HCV strains, processing at the E2/p7 site is inefficient, leading to the production of a reasonably stable E2-p7 species (DUBUISSON et al. 1994; LIN et al. 1994; MIZUSHIMA et al. 1994; SELBY et al. 1994). HCV glycoproteins E1 and E2 are heavily modified by N-linked glycosylation and are believed to be type I transmembrane glycoproteins with a NH_2-terminal ectodomain and a COOH-terminal hydrophobic anchor. For E2, COOH-terminal deletions removing its hydrophobic region result in secretion of the ectodomain (HUSSY et al. 1996; INUDOH et al. 1996; LESNIEWSKI et al. 1995; MATSUURA et al. 1994; MICHALAK et al. 1997; NISHIHARA et al. 1993;

Equipe Hépatite C, CNRS-UMR 8526, Institut de Biologie de Lille et Institut Pasteur de Lille, BP447, 59021 Lille cédex, France

SELBY et al. 1994; SPAETE et al. 1992). This is in accordance with other data proposing that the hydrophobic anchor domain begins at amino acid 718 (position on the polyprotein; MIZUSHIMA et al. 1994). The situation appears to be more confusing for E1, since a truncated form ending at amino acid 340 is secreted only if it contains an internal deletion between amino acids 262 and 290, suggesting that a second membrane anchor might exist (MATSUURA et al. 1994). However, truncated forms ending at amino acid 311 or 334 and containing this internal sequence can also be secreted (HUSSY et al. 1996; MICHALAK et al. 1997). Since E1 and its truncated forms do not fold properly in the absence of E2 (see below), interpretation of these results is difficult because the absence of secretion of some truncated forms of E1 can be due to retention in the endoplasmic reticulum (ER) by interaction with chaperones. E1, like E2, is probably anchored by its COOH-terminal hydrophobic sequence and the NH_2-terminal limit of the potential transmembrane domain of E1 remains to be established. Studies using transient viral and non-viral expression systems have shown that HCV glycoproteins interact to form complexes which have been proposed as functional subunits of the HCV particle (DELEERSNYDER et al. 1997; DUBUISSON et al. 1994; GRAKOUI et al. 1993; LANFORD et al. 1993; RALSTON et al. 1993). However, the low levels of HCV particles in patient samples and lack of a cell culture system supporting efficient HCV replication or particle assembly have hampered the characterization of virion glycoprotein complexes.

2 Formation of Hepatitis C Virus Glycoprotein Complexes

Studies using transient expression systems have shown that E2 and/or its related products (E2-NS2 and E2-p7) interact with E1 to form complexes. Purified HCV glycoprotein complexes expressed by using vaccinia virus recombinants have been shown to be noncovalently associated (RALSTON et al. 1993). In contrast, a fraction of E1 and E2 present in lysates of cells infected with vaccinia-HCV recombinants has been reported to be associated via disulfide linkages (GRAKOUI et al. 1993). Other studies on the formation of intracellular complexes have shown that, in the presence of nonionic detergents, two forms of E1E2 complexes are detected: a heterodimer of E1 and E2 stabilized by noncovalent interactions, and heterogeneous disulfide-linked aggregates (DUBUISSON et al. 1994; DUBUISSON and RICE 1996). Additional studies indicate that the noncovalent heterodimer is composed of native HCV glycoproteins whereas the disulfide-linked aggregates is formed by misfolded proteins (DELEERSNYDER et al. 1997).

2.1 Native E1E2 Complexes

Due to lack of appropriate immune reagents for HCV glycoproteins, it has been difficult to distinguish between glycoprotein molecules that undergo productive folding and assembly from those which follow a nonproductive pathway leading to misfolding and aggregation. However the isolation and characterization of a conformation-sensitive E2-reactive monoclonal antibody (MAb) (H2) has recently been reported (DELEERSNYDER et al. 1997). This MAb selectively recognizes slowly maturing E1E2 heterodimers which are noncovalently linked, protease-resistant, and no longer associated with the ER chaperone calnexin. This heterodimer probably represents the native prebudding form of the HCV glycoprotein complex. Both E2 and unprocessed E2-p7 are present in HCV glycoprotein complexes recognized by MAb H2. At present, it is not known whether one or both of these E2 forms is present in mature HCV particles. However, for the BK strain of HCV, processing at the E2/p7 site is efficient and leads to the cleavage of E2-p7 species (DUBUISSON et al. 1994; LIN et al. 1994), suggesting that, at least for this strain, unprocessed E2-p7 is not incorporated into HCV particles. In the case of the pestivirus classical swine fever virus, processing in this region of the polyprotein is similar but E2-p7 is either not present in mature virus or at levels too low to be detected (ELBERS et al. 1996).

In addition to forming E1E2 complexes, HCV glycoproteins have been shown to interact with other HCV proteins. In immunoprecipitation studies, E1 has been shown to coprecipitate with the core protein (LO et al. 1996) and NS2 has been detected after immunoprecipitation with anti-E2 antibodies (SELBY et al. 1994). However, neither NS2 nor the core protein could be detected in association with native E1E2 complexes (J. Dubuisson, unpublished results), suggesting that these interactions have probably no implications for HCV virion morphogenesis.

2.2 Disulfide Bond Aggregates

Studies of reactivity with conformation-sensitive MAbs and protease sensitivity indicate that the production of properly assembled E1E2 oligomers is inefficient (DELEERSNYDER et al. 1997). This does not seem to be due to mutations introduced during cDNA synthesis or PCR amplification of the original clone used in these studies. Indeed, similar results are obtained with a vaccinia recombinant expressing the sequence of the structural proteins of a recently characterized infectious cDNA clone (KOLYKHALOV et al. 1997; J. Dubuisson and C.M. Rice, unpublised data). Alternatively, this tendency towards aggregation could be due to abnormally high level production driven by the viral expression systems used. However, analysis of HCV glycoprotein assembly in a non-viral expression system showed similar results (J. Dubuisson and D. Moradpour, unpublished data) suggesting that this tendency towards aggregation could be an intrinsic property of these glycoproteins. Indeed, slow folding of HCV glycoproteins may increase the fraction of these proteins shunted into competing nonproductive pathways, such as aggregation and aberrant disulfide bond formation (FISCHER and SCHMID 1990). An attractive hypothesis is that such nonproductive pathways may occur in authentic HCV-infected cells. In the context of virus replication, inefficient folding of the HCV glycoproteins would down-regulate particle formation and virus replication to minimize exposure of viral antigens to the immune system and/or reduce pathogenicity. Together, these results suggest that HCV glycoproteins can follow two different pathways: a productive pathway leading to the formation of noncovalent native E1E2 complexes and a nonproductive pathway leading to the formation of large aggregates (Fig. 1). A consequence of these observations is that the quality of these proteins should be carefully monitored when they are produced to evaluate their role as potential

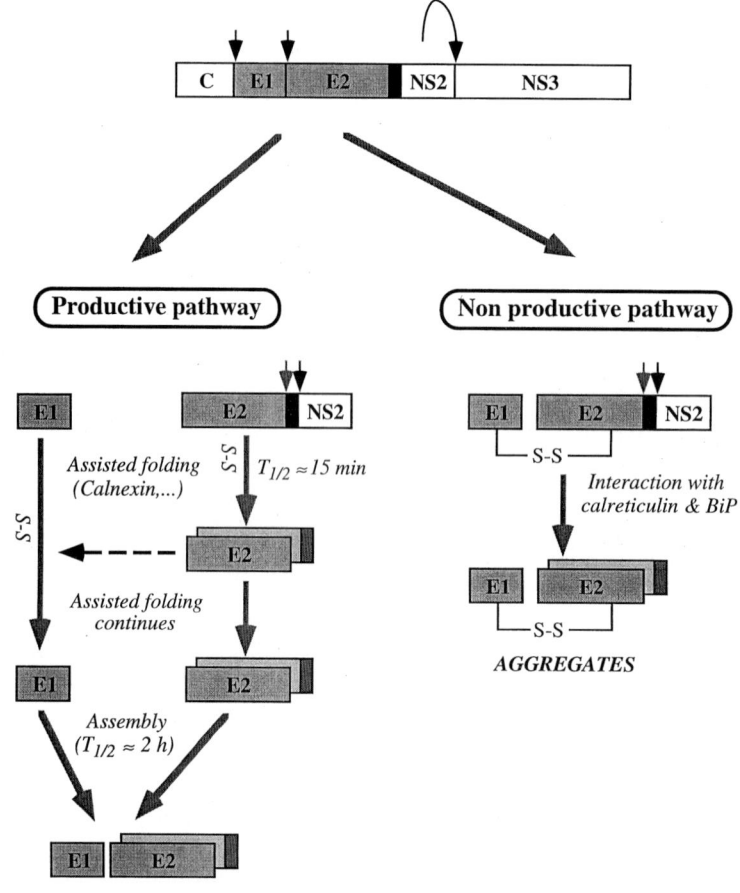

Fig. 1. A model for hepatitis C virus (HCV) glycoprotein assembly. HCV glycoproteins E1 and E2 are produced by cleavage of the polyprotein. After synthesis, these glycoproteins can follow two different pathways. In the nonproductive pathway, E1 and E2 form heterogeneous aggregates stabilized by disulfide bonds. These aggregates can interact with the molecular chaperones BiP and calreticulin. In the productive pathway, E1 and E2 fold slowly and interact to form a noncovalent heterodimer. During their folding, HCV glycoproteins interact with calnexin, and E2 plays a chaperone-like role for E1. Intermolecular disulfide bonds and acquisition of intramolecular disulfide bonds are indicated by S-S. See text for more details

vaccine or diagnostic tools. In addition, fundamental studies analyzing HCV envelope formation or the interaction of HCV glycoproteins with a potential receptor should take this phenomenon of aggregation into account.

3 Folding of Hepatitis C Virus Glycoproteins

Studies of HCV glycoprotein assembly have shown that properly folded E1 and E2 interact to form a heterodimer stabilized by noncovalent interactions. The kinetics of association between E1 and E2 indicate that the formation of stable E1E2 complexes is slow ($T_{1/2} \approx 2h$) (DELEERSNYDER et al. 1997). To identify the limiting steps in HCV glycoprotein assembly, the folding of E1 and E2 has been monitored in pulse-chase experiments by analyzing the formation of their intramolecular disulfide bonds (DUBUISSON and RICE 1996; MICHALAK et al. 1997), the formation of conformation-dependent epitopes (DELEERSNYDER et al. 1997; HABERSETZER et al. 1998), and their interactions with ER chaperones (CHOUKHI et al. 1998; DUBUISSON and RICE 1996).

3.1 Intramolecular Disulfide Bond Formation

The lumen of the ER provides an environment optimized for the oxidation of proteins targeted to the secretory pathway. Indeed, the ER lumen, in contrast to the cytosol, has a redox potential sufficiently oxidizing to allow disulfide bond formation. In addition, it contains a folding enzyme, protein disulfide isomerase (PDI) which, depending on the redox conditions, can catalyze reduction, isomerization, or oxidation of disulfide bonds (FREEDMAN et al. 1994). Its role in the ER is the isomerization of disulfide bonds which helps proteins acquire correct cysteine pairing during the folding process. Like most cellular glycoproteins, HCV glycoproteins contain highly conserved cysteine residues which are potentially involved in disulfide bond formation. SDS-gel electrophoresis under nonreducing conditions can be used to monitor disulfide bond formation in viral glycoproteins (DOMS et al. 1993). This method takes advantage of an increase in mobility as a protein acquires more compact conformations stabilized by the formation of disulfide bonds and often reflects progressive folding towards the native state. When analyzed under nonreducing conditions, HCV glycoprotein E1 co-migrates with its reduced form during a 5-min pulse and the first 30min of chase (DUBUISSON and RICE 1996). It takes at least 60min to start detecting a second form of E1 with a faster electrophoretic mobility, indicating that disulfide bond formation is slow for E1. In contrast to E1, disulfide bond formation in E2 and E2-p7 appears to be complete by the time of E2-NS2 cleavage ($T_{1/2} \approx 15min$). These results have led to the conclusion that disulfide bond formation is a limiting step for the folding of E1. It is likely that, due to prolonged interaction with ER chaperone(s) (e.g., calnexin) in the productive pathway (Fig. 1), cysteine residues in E1 are slowly accessible to form intramolecular disulfide bonds.

3.2 Folding of Subdomains

Disulfide bond formation in E2 is faster than in E1 indicating that this step is not limiting for the folding of E2. However, in pulse-chase experiments, the kinetics of detection of E2 with the conformation-sensitive MAb H2 indicate that the folding of E2 is slow (DELEERSNYDER et al. 1997; MICHALAK et al. 1997). These observations suggest that there is a limiting step in the folding of E2 after it has acquired its intramolecular disulfide bonds. Proline isomerization is important for protein folding because prolines are usually free to isomerize when polypeptide chains are in the unfolded state, but are constrained in properly folded chains (DELEERSNYDER et al. 1997; MICHALAK et al. 1997; NILSSON and ANDERSON 1991). Due to the large number of prolines in E2 (24 residues for genotype 1a), proline isomerization may be rate-limiting for E2 folding, however, there is currently no experimental evidence for that. Human MAbs which recognize conformation-dependent epitopes on E2 have also been obtained (HABERSETZER et al. 1998). The kinetics of recognition of E2 by these MAbs is faster than with MAb H2 and they are similar to the kinetics of disulfide bond formation in E2, indicating that some state of folding can rapidly be observed for this glycoprotein. In addition, these human MAbs recognize aggregates as well as native complexes suggesting that some state of folding can also be observed in the nonproductive pathway. Together, these data indicate that some subdomain(s) of E2 can be formed early after synthesis.

3.3 Role of E2 in the Folding of E1

The effect of coexpression of E1 and E2 glycoproteins on each other's folding has recently been evaluated. Kinetics of folding of E2, analyzed with the conformation-sensitive MAb H2 (DELEERSNYDER et al. 1997), show similar results in the presence or absence of E1, suggesting that the presence of E1 is not necessary for the folding of E2 (MICHALAK et al. 1997). Since a MAb which recognizes a properly folded E1 glycoprotein is not available, disulfide bond formation of E1 in the presence or absence of E2 has been monitored by SDS-PAGE under nonreducing conditions. An oxidized form of E1, which appears slowly, can clearly be detected when E2 is coexpressed with E1 (DUBUISSON and RICE 1996). However, in the absence of E2, no oxidized form of E1 is detected, indicating that E2 affects the folding of E1 (MICHALAK et al. 1997). It is very likely that E2 plays this chaperone-like role by interacting directly with E1. Indeed, noncovalent HCV glycoprotein complexes are detected in association with calnexin long before native complexes are formed (DUBUISSON and RICE 1996). It is therefore possible that E2 plays its chaperone role in a complex formed of E1, E2 and calnexin (Fig. 2; see below). However, noncovalent interactions between E1 and E2 are barely detected immediately after

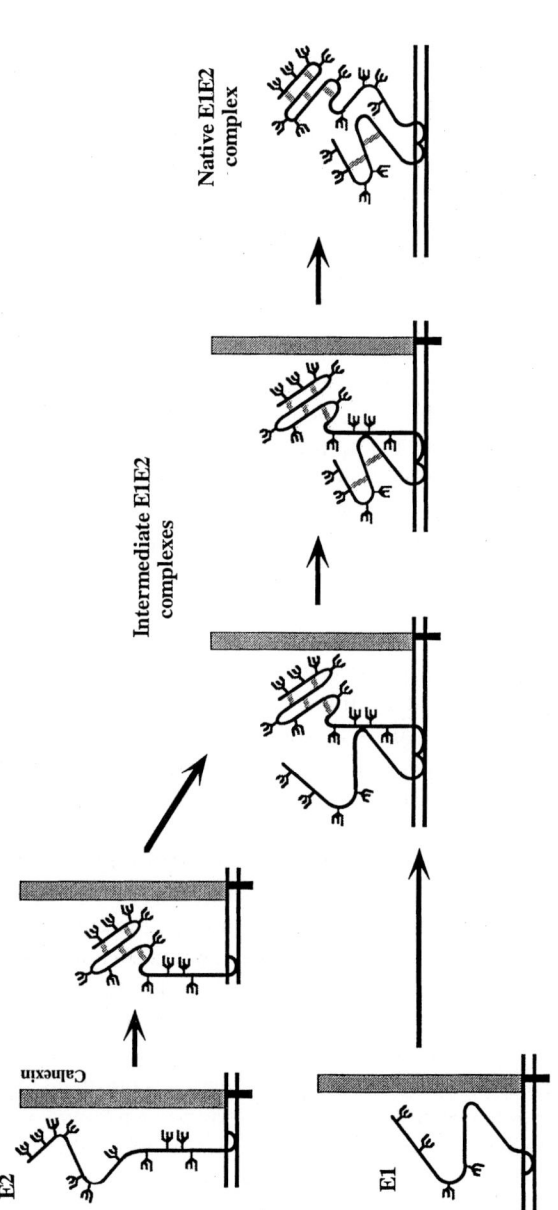

Fig. 2. A model for hepatitis C virus (HCV) glycoprotein folding. Based on the accumulation of direct and indirect observations, a model identifying some steps of HCV glycoprotein folding can be drawn. Immediately after their synthesis, monomeric forms of E1 and E2 interact with calnexin. Intramolecular disulfide bonds form rapidly in E2 leading to the folding of at least one subdomain. This partially folded form of E2 interacts with the non-oxidized form of E1, forming an intermediate E1E2 complex which is associated with calnexin. Extensive folding of E1 occurs in the context of the intermediate complex. Finally, E1 and E2 acquire their final state of folding and form a native E1E2 complex which is retained in the ER by at least one specific sequence (transmembrane domain of E2)

HCV glycoprotein synthesis, suggesting that E2 needs to be partially folded to potentially protect E1 from misfolding. This could also explain the delay observed for intramolecular disulfide bond formation in E1. This chaperone-like activity of E2 is very similar to the role of prM of the Japanese encephalitis virus (KONISHI and MASON 1993). Cosynthesis of prM is indeed required for proper folding, membrane association, and assembly of the E protein of this virus.

3.4 Role of Endoplasmic Reticulum Chaperones

Along with folding enzymes, such as PDI and prolyl *cis-trans* isomerases (GETHING and SAMBROOK 1992), the ER contains molecular chaperones including immunoglobulin heavy-chain binding protein (BiP or GRP78; HARTL et al. 1994), GRP94 (JAKOB and BUCHNER 1994), calnexin (BERGERON et al. 1994) and calreticulin (NAUSEEF et al. 1995; PETERSON et al. 1995; WADA et al. 1995). Molecular chaperones are proteins that associate specifically with incompletely folded or unassembled proteins, and increase the efficiency by which they acquire their correct three-dimensional structure. Like folding enzymes, chaperones do not specify folding and assembly information but rather help protein attain their native folded structures by limiting nonproductive interactions and potentially dead-end pathways. In the ER, chaperones have the additional task of helping to retain incorrectly and incompletely folded molecules until they have folded and assembled properly (HAMMOND and HELENIUS 1995).

ER chaperones calnexin, calreticulin and BiP but not GRP94 interact with HCV glycoproteins (CHOUKHI et al. 1998). Interactions with calnexin and/or calreticulin has been shown for several viral glycoproteins (GAUDIN 1997; HAMMOND et al. 1994; OTTEKEN and MOSS 1996; PETERSON et al. 1995; YAMASHITA et al. 1996). Their preference for interaction with glycoproteins is based on a lectin-like affinity for monoglucosylated N-linked oligosaccharides (HAMMOND et al. 1994; HEBERT et al. 1995; PETERSON et al. 1995; WARE et al. 1995). Binding of substrate glycoproteins to and release from calnexin and calreticulin depends on trimming and reglucosylation of the N-linked glycans (HAMMOND et al. 1994; HEBERT et al. 1995). The absence of interaction between HCV glycoproteins, and calreticulin or calnexin after tunicamycin treatment (CHOUKHI et al. 1998) is consistent with the view that these chaperones act as lectins.

It has been shown that calreticulin and BiP interact preferentially with aggregates of HCV glycoproteins, whereas calnexin associates preferentially with noncovalent E1E2 complexes (CHOUKHI et al. 1998). This indicates that these chaperones recognize HCV proteins in a different state of folding. Since the kinetics of association of HCV glycoproteins with calnexin and calreticulin are similar, it is likely that, instead of interacting sequentially with these glycoproteins, calreticulin is involved in the nonproductive pathway of HCV glycoprotein complex formation and calnexin in the productive pathway (Fig 1). In addition, calnexin has been shown to associate with the oxidized form of HCV glycoproteins (DUBUISSON and RICE 1996). This reinforces the idea that proteins composing the native complex have been in contact with calnexin during their folding (Fig. 2). The aggregates observed in the dead-end pathway have probably preferentially interacted with calreticulin and/or BiP. However, these aggregates are stable in cells expressing HCV glycoproteins (DELEERSNYDER et al. 1997), whereas the complexes formed between HCV glycoproteins and calreticulin are transient (CHOUKHI et al. 1998). The dissociation of such aggregates from calreticulin could be due to partial folding of the proteins involved in the aggregates. Indeed, it has been shown that a subdomain of E2 can be folded in such aggregates (HABERSETZER et al. 1998).

The major role of chaperones is to help protein attain their native folded structures by limiting nonproductive interactions and dead-end pathways. Since a large portion of HCV glycoproteins is involved in nonproductive interactions, it is possible that, due to the slow folding of HCV glycoproteins, the ER chaperones cannot fully play their role at physiological concentrations. Although molecular chaperones are abundant in the ER (MARQUARDT et al. 1993; ROWLING et al. 1994), a majority of these molecules could be involved in protein–protein interactions, leaving only a fraction of them free to interact with newly synthesized proteins. It has indeed been observed that the efficiency of HCV glycoprotein folding is reduced in rapidly dividing cells (Dubuisson, unpublished data). However, overexpression of calnexin, calreticulin or BiP with HCV glycoproteins has not shown any improvement in the assembly of native HCV glycoprotein complexes (CHOUKHI et al. 1998). This suggests that a proper balance of chaperone activities could be required for optimal folding. In addition, other chaperone(s) and/or foldase(s) might be necessary to assist in HCV glycoprotein folding.

4 Sequences Involved in Hepatitis C Virus Glycoprotein Interactions

When they are coexpressed, truncated forms of E1 and E2 can interact to form complexes which are released in the supernatant of culture medium suggesting that sequences important for HCV glycoprotein interactions are located in their ectodomain (HUSSY et al. 1996; LANFORD et al. 1993; MATSUURA et al. 1994; MICHALAK et al. 1997). In addition, regions important for interactions between E1 and E2 have been mapped by far-western blotting using bacterial recombinant proteins or by pull-down assays using proteins expressed in mammalian cells (YI et al. 1997). NH_2-terminal sequences in E2 (amino acids 415–500 of the polyprotein) and also in E1 have been shown to be important for HCV glycoprotein interactions. However, since native complexes are formed by assembly of extensively folded proteins, it cannot be concluded that the sequences identified as important for HCV glycoprotein interactions are accessible for protein–protein interactions in properly folded proteins. Indeed, E1 and E2 have a tendency to form aggregates when they are coexpressed (DUBUISSON et al. 1994) and might therefore contain sequences for interactions leading to their assembly in the nonproductive pathway. In addition, analysis of secreted complexes composed of truncated forms of E1 and E2 indicate that these proteins are aggregated (MICHALAK et al. 1997).

Deletion of the COOH-terminal transmembrane domain of E2 has been shown to abolish the formation E1E2 heterodimers and aggregates (MICHALAK et al. 1997; SELBY et al. 1994) and similar results have been observed when the transmembrane domain of E2 is replaced by the anchor signal of CD4 or a glycosyl phosphatidylinositol moiety (COCQUEREL et al. 1998). These data suggest that the transmembrane domains of HCV glycoproteins could interact directly to stabilize E1E2 heterodimers. Alternatively, regions in both transmembrane domains and ectodomains of HCV glycoproteins could be important to stabilize E1E2 heterodimers and disruption in one domain (e.g. transmembrane domain) could impede the formation of properly folded heterodimers.

5 Subcellular Localization of Hepatitis C Virus Glycoproteins

Prolonged interactions of HCV glycoprotein complexes with ER chaperones has been observed (CHOUKHI et al. 1998). It has been suggested that association of HCV oligomers with calnexin could be responsible for their retention in the ER (DUBUISSON and RICE 1996). However, recent data indicate that native E1E2 complexes do not interact with calnexin and are still retained in the ER (DELEERSNYDER et al. 1997). HCV glycoprotein complexes may therefore contain a retention signal to allow budding into an intracellular compartment. For the flaviviruses, virions appear in intracellular vesicles (probably modified ER) and are released from cells via the exocytosis pathway (reviewed in GRIFFITHS and ROTTIER 1992; PETTERSSON 1991). Efficient HCV particle formation has not been observed in transient expression assays, suggesting that essential viral or host factors are missing or blocked. The native E1E2 complexes, identified in the ER by the conformation-sensitive MAb H2, most probably represent a prebudding form of HCV glycoprotein oligomer.

To determine if E1 and/or E2 contain an ER-targeting signal for ER retention of E1E2 complexes, these proteins have been expressed alone and their intracellular localization studied. Due to misfolding of E1 in the absence of E2 (MICHALAK et al. 1997), no conclusion, on the localization of its native form can be drawn from the expression of E1 alone. However, E2 expressed in the absence of E1 can fold properly and is retained in the ER, as shown by the lack of complex glycans, its intracellular distribution and the absence of its expression on the cell surface (COCQUEREL et al. 1998). An ER retention signal has been mapped on E2. Replacement of the transmembrane domain of E2 with the anchor sequence of CD4 or a glycosyl phosphatidylinositol moiety has been shown to be sufficient for export to the cell surface. In addition, a chimeric protein containing the ectodomain of CD4 fused to the transmembrane domain of E2 is retained in the ER. These experiments indicate that the COOH-terminal 29 amino acids of E2 contain the information for

ER retention. However, the ratio of chimeric E2 proteins leaving the ER is improved after removal of the COOH-terminal 56 amino acids of its ectodomain, suggesting that this sequence can reinforce E2 retention in the ER. Alternatively, as recently shown (MICHALAK et al. 1997), the presence of these 56 residues could reduce the efficiency of E2 folding, which, in turn, could lead to less efficient export out of the ER.

HCV glycoprotein E2 is retained in the ER by a targeting signal which does not show the characteristics of a classical ER retention motif. Several specific signals have been identified for the retention/retrieval of ER proteins. They include a tetra-amino acid sequence, KDEL, for soluble proteins, a di-lysine motif at the COOH-terminal of type I proteins, or a di-arginine NH_2-terminal motif for type II proteins (reviewed in NILSSON and WARREN 1994; PELHAM 1995). However, ER retention by a transmembrane domain has been shown for some other proteins (AHN et al. 1993; BONIFACINO et al. 1991; YANG et al. 1997). Recently, it has been proposed that in the absence of dominant luminal or cytosolic associations, proteins distribute based on interactions between their transmembrane domain and the surrounding lipid environment (BRETCHER and MUNRO 1993; YANG et al. 1997).

The ER-retention signal present in E2 could be sufficient to retain E1E2 complexes in the ER, but the presence of another signal in E1 cannot be excluded. Since coexpression with E2 is required for the proper folding of E1 (MICHALAK et al. 1997), an ER localization of E1 due to a specific retention signal cannot be descriminated from a retention due to misfolding of the protein (HAMMOND and HELENIUS 1995). In addition, coexpression of E1 with an E2 that has the transmembrane domain deleted or replaced abolishes complex formation (COCQUEREL et al. 1998; MICHALAK et al. 1997; SELBY et al. 1994), and does not allow an indirect study of E1 localization.

The transmembrane domain of HCV glycoprotein E2 is multifunctional. Besides anchoring E2 in cell membranes and potentially in HCV envelope (MIZUSHIMA et al. 1994), the E2 transmembrane domain has other functions. Its COOH-terminal half is the signal sequence for the p7 polypeptide (LIN et al. 1994). It plays an important role in the interaction between HCV glycoproteins to form native E1E2 complexes, and it is also responsible for the retention of E2 in the ER.

6 Concluding Remarks

Studies using transient viral and non-viral expression systems have shown that HCV glycoproteins can follow two different folding pathways leading to the formation of a noncovalent heterodimer or disulfide bond aggregates (Fig. 1). As described in this review, there is evidence suggesting that the noncovalent heterodimer could be the prebudding form of the HCV glycoprotein complex. However, it has now become clear that there is no single mature conformation or quaternary structure for viral glycoproteins since oligomeric structures change at different stages of assembly, release, and entry. The characterization of the noncovalent E1E2 heterodimer gives only a partial picture of the HCV glycoprotein complex. Further studies will be needed to characterize the quaternary structure of the E1E2 complex of the virion and the modifications it acquires during the entry process, when the viral envelope initiates its contact with the host cell membrane. These modifications will be difficult to study in the absence of a cell culture system supporting efficient HCV particle assembly.

Acknowledgements. This work was supported by the following grants: an ATIPE from the CNRS, a grant from the ARC (N° 1039) and a grant from the INSERM/AFS (INSERM N° 5FS10).

References

Ahn K, Szczesna-Skorupa E, Kemper B (1993) The amino-terminal 29 amino acids of cytochrome P450 2C1 are sufficient for retention in the endoplasmic reticulum. J Biol Chem 268:18726–18733
Bergeron JJM, Brenner MB, Thomas DY, Williams DB (1994) Calnexin: a membrane-bound chaperone of the endoplasmic reticulum. Trends Biochem Sci 19:124–129
Bonifacino JS, Cosson P, Shah N, Klausner RD (1991) Role of potentially charged transmembrane residues in targeting proteins for retention and degradation within the endoplasmic reticulum. EMBO J 10:2783–2793
Bretcher MS, Munro S (1993) Cholesterol and Golgi Apparatus. Science 261:1280–1281
Choukhi A, Ung S, Wychowski C, Dubuisson J (1998) Involvement of endoplasmic reticulum chaperones in folding of hepatitis C virus glycoproteins. J Virol 72:3851–3858
Cocquerel L, Meunier J-C, Pillez A, Wychowski C, Dubuisson J (1998) A retention signal necessary and sufficient for endoplasmic reticulum localization maps to the transmembrane domain of hepatitis C virus glycoprotein E2. J Virol 72:2183–2191
Deleersnyder V, Pillez A, Wychowski C, Blight K, Xu J, Hahn YS, Rice CM, Dubuisson J (1997) Formation of native hepatitis C virus glycoprotein complexes. J Virol 71:697–704
Doms RW, Lamb RA, Rose JK, Helenius A (1993) Folding and assembly of viral membrane proteins. Virology 193:545–562
Dubuisson J, Hsu HH, Cheung RC, Greenberg HB, Russell DG, Rice CM (1994) Formation and intracellular localization of hepatitis C virus envelope glycoprotein complexes expressed by recombinant vaccinia and Sindbis viruses. J Virol 68:6147–6160
Dubuisson J, Rice CM (1996) Hepatitis C virus glycoprotein folding: disulfide bond formation and association with calnexin. J Virol 70:778–786
Elbers K, Tautz N, Becher P, Stoll D, Rümenapf T, Thiel H-J (1996) Processing in the pestivirus E2-NS2 region: identification of proteins p7 and E2p7. J Virol 70:4131–4135
Fischer G, Schmid FX (1990) The mechanism of protein folding. Implications of in vitro refolding models for de novo protein folding and translocation in the cell. Biochemistry 29:2205–2212
Freedman RB, Hirst TR, Tuite MF (1994) Protein disulphide isomerase: building bridges in protein folding. Trends Biochem Sci 19:331–336
Gaudin Y (1997) Folding of rabies virus glycoprotein: epitope acquisition and interaction with endoplasmic reticulum chaperones. J Virol 71:3742–3750
Gething M-J, Sambrook J (1992) Protein folding in the cell. Nature 355:33–45
Grakoui A, Wychowski C, Lin C, Feinstone SM, Rice CM (1993) Expression and identification of hepatitis C virus polyprotein cleavage products. J Virol 67:1385–1395
Griffiths G, Rottier P (1992) Cell biology of viruses that assemble along the biosynthetic pathway. Semin Cell Biol 3:367–381
Habersetzer F, Fournillier A, Dubuisson J, Rosa D, Abrigniani S, Wychowski C, Nakano I, Trépo C, Desgranges C, Inchauspé G (1998) Characterization of human monoclonal antibodies specific of the hepatitis C virus glycoprotein E2 with in vitro binding neutralization properties. Virology 249: 32–41

Hammond C, Braakman I, Helenius A (1994) Role of N-linked oligosaccharides, glucose trimming and calnexin during glycoprotein folding in the endoplasmic reticulum. Proc Natl Acad Sci USA 91: 913–917

Hammond C, Helenius A (1995) Quality control in the secretory pathway. Curr Opin Cell Biol 7:523–529

Hartl F-U, Hlodan R, Langer T (1994) Molecular chaperones in protein folding: the art of avoiding sticky situations. Trends Biochem Sci 19:20–25

Hebert DN, Foellmer B, Helenius A (1995) Glucose trimming and reglucosylation determine glycoprotein association with calnexin in the endoplasmic reticulum. Cell 81:425–433

Hussy P, Schmid G, Mous J, Jacobsen H (1996) Purification and in vitro-phospholabeling of secretory envelope proteins E1 and E2 of hepatitis C virus expressed in insect cells. Virus Res 45:45–57

Inudoh M, Nyunoya H, Tanaka T, Hijikata M, Kato N, Shimotohno K (1996) Antigenicity of hepatitis C virus envelope proteins expressed in hamster ovary cells. Vaccine 14:1590–1596

Jakob U, Buchner J (1994) Assisting spontaneity: the role of Hsp90 and small Hsps as molecular chaperones. Trends Biochem Sci 19:205–211

Kolykhalov AA, Agapov EV, Blight K, Mihalik K, Feinstone SM, Rice CM (1997) Transmission of hepatitis C by intrahepatic inoculation with transcribed RNA. Science 277:570–574

Konishi E, Mason PW (1993) Proper maturation of the Japanese Encephalitis Virus envelope glycoprotein requires cosynthesis with the premembrane protein. J Virol 67:1672–1675

Lanford RE, Notvall L, Chavez D, White R, Frenzel G, Simonsen C, Kim J (1993) Analysis of hepatitis C virus capsid, E1, and E2/NS1 proteins expressed in insect cells. Virology 197:225–235

Lesniewski R, Okasinski G, Carrick R, Van Sant C, Desai S, Johnson R, Scheffel J, Moore B, Mushahwar I (1995) Antibody to hepatitis C virus second envelope (HCV-E2) glycoprotein: a new marker of HCV infection closely associated with viremia. J Med Virol 45:415–422

Lin C, Lindenbach BD, Pragai B, McCourt DW, Rice CM (1994) Processing of the hepatitis C virus E2-NS2 region: identification of p7 and two distinct E2-specific products with different C termini. J Virol 68:5063–5073

Lo S-Y, Selby MJ, Ou J-H (1996) Interaction between hepatitis C virus core protein and E1 envelope protein. J Virol 70: 5177–5182

Marquardt M, Hebert D, Helenius A (1993) Post-translational folding of influenza hemagglutinin in isolated endoplasmic reticulum-derived microsomes. J Biol Chem 268:19618–19625

Matsuura Y, Suzuki T, Suzuki R, Sato M, Aizaki H, Saito I, Miyamura T (1994) Processing of E1 and E2 glycoproteins of hepatitis C virus expressed in mammalian and insect cells. Virology 205:141–150

Michalak J-P, Wychowski C, Choukhi A, Meunier J-C, Ung S, Rice CM, Dubuisson J (1997) Characterization of truncated forms of hepatitis C virus glycoproteins. J Gen Virol 78:2299–2306

Mizushima H, Hijikata M, Asabe S-I, Hirota M, Kimura K, Shimotohno K (1994) Two hepatitis C virus glycoprotein E2 products with different C termini. J Virol 68:6215–6222

Nauseef WM, McCormick SJ, Clark RA (1995) Calreticulin functions as a molecular chaperone in the biosynthesis of myeloperoxidase. J Biol Chem 270:4741–4747

Nilsson B, Anderson S (1991) Proper and improper folding of proteins in the cellular environment. Annu Rev Microbiol 45:607–635

Nilsson T, Warren G (1994) Retention and retrieval in the endoplasmic reticulum and the Golgi apparatus. Curr Opin Cell Biol 6:517–521

Nishihara T, Nozaki C, Nakatake H, Hoshiko K, Esumi M, Hayashi N, Hino K, Hamada F, Mizuno K, Shikata T (1993) Secretion and purification of hepatitis C virus NS1 glycoprotein produced by recombinant baculovirus-infected insect cells. Gene 129:207–214

Otteken A, Moss B (1996) Calreticulin interacts with newly synthesized human immunodeficiency virus type 1 envelope glycoprotein, suggesting a chaperone function similar to that of calnexin. J Biol Chem 271:97–103

Pelham HRB (1995) Sorting and retrieval between the endoplasmic reticulum and Golgi apparatus. Curr Opin Cell Biol 7:530–535

Peterson JR, Ora A, Nguyen Van P, Helenius A (1995) Transient, lectin-like association of calreticulin with folding intermediates of cellular and viral glycoprotein. Mol Biol 6:1173–1184

Pettersson RF (1991) Protein localization and virus assembly at intracellular membranes. Curr Top Microbiol Immunol 170:67–104

Ralston R, Thudium K, Berger K, Kuo C, Gervase B, Hall J, Selby M, Kuo G, Houghton M, Choo Q-L (1993) Characterization of hepatitis C virus envelope glycoprotein complexes expressed by recombinant vaccinia viruses. J Virol 67:6753–6761

Rice CM (1996) Flaviviridae: the viruses and their replication. In: Fields BN, Knipe DM, Howley PM (eds) Fields Virology. Lippincott-Raven, Philadelphia, pp 931–959

Rowling P, Mclaughlin SH, Pollock GS, Freedman RB (1994) A single purification procedure for the major resident proteins of the ER lumen: endoplasmin, BiP, calreticulin and protein disulfide isomerase. Protein Expr Purif 5:331–336

Selby MJ, Glazer E, Masiarz F, Houghton M (1994) Complex processing and protein: protein interactions in the E2:NS2 region of HCV. Virology 204:114–122

Spaete RR, Alexander D, Rugroden ME, Choo Q-L, Berger K, Crawford K, Kuo C, Leng S, Lee C, Ralston R, Thudium K, Tung JW, Kuo G, Houghton M (1992) Characterization of the hepatitis E2/NS1 gene product expressed in mammalian cells. Virology 188:819–830

Wada I, Imai S, Kai M, Sakane F, Kanoh H (1995) Chaperone function of calreticulin when expressed in the endoplasmic reticulum as the membrane-anchored and soluble forms. J Biol Chem 270:20298–20304

Ware FE, Vassilakos A, Peterson PA, Jackson MR, Lehrman MA, Williams DB (1995) The molecular chaperone calnexin binds Glc1Man9GlcNAc2 oligosaccharide as an initial step in recognizing unfolded glycoproteins. J Biol Chem 270:4697–4704

Yamashita Y, Shimokata K, Mizuno S, Daikoku T, Tsurumi T, Nishiyama Y (1996) Calnexin acts as a molecular chaperone during the folding of glycoprotein B of human cytomegalovirus. J Virol 70:2237–2246

Yang M, Ellenberg J, Bonifacino JS, Weissman AM (1997) The transmembrane domain of a carboxy-terminal anchored protein determines localization to the endoplasmic reticulum. J Biol Chem 272:1970–1975

Yi M, Nakamoto Y, Kaneko S, Yamashita T, Murakami S (1997) Delineation of regions important for heteromeric association of hepatitis C virus E1 and E2. Virology 231:119–129

Structure and Function of the Hepatitis C Virus NS3-NS4A Serine Proteinase

R. De Francesco and C. Steinkühler

1	Introduction	149
2	NS3-NS4A: A Heterodimeric Serine Proteinase	152
3	Three-Dimensional Structure of the NS3-NS4A Proteinase	153
3.1	NS3-NS4A Is a Chymotrypsin-Like Serine Proteinase	153
3.2	Mechanism of NS3 Proteinase Activation by NS4A	157
3.3	Stability of the NS3/NS4A Complex	160
3.4	Substrate Specificity of the NS3-NS4A Serine Proteinase	161
3.4.1	Primary Specificity	162
3.4.2	Distal Subsites	164
3.5	The Metal-Binding Site of the NS3-NS4A Proteinases	165
References		167

1 Introduction

Hepatitis C virus (HCV) is a relatively young member of the Flaviviridae family that is nowadays recognized as the major etiological agent of both blood-borne and sporadic non-A-non-B hepatitis (Houghton 1996). It is estimated that about 0.5%–1.5% of the total world population is infected with this virus. In many cases, infection with HCV leads to life-threatening disease, such as cirrhosis of the liver and hepatocarcinoma (Simmonds et al. 1998). A vaccine against HCV has not yet been discovered, and treatment with interferon is effective only in about 20% of the patients. There is thus a compelling need for a deeper understanding of the HCV life cycle and identification of targets for the development of more effective antiviral treatments.

The HCV (+)-stranded RNA genome encodes a precursor polyprotein of ~3,000kDa that is processed cotranslationally and post-translationally in order to release the viral structural and nonstructural proteins (Bartenschlager 1997). Some of the viral nonstructural proteins have been associated with enzymatic

---

Istituto di Ricerche di Biologia Molecolare (IRBM) "P. Angeletti", Via Pontina Km 30.600, 00040 Pomezia (Rome), Italy

activities that are thought to be essential for viral replication. Among these, nonstructural protein 3 (NS3) has been shown to contain a serine proteinase domain responsible for the most of the proteolytic maturation events within the nonstructural portion of the viral polyprotein (NEDDERMANN et al. 1997). Studies carried out with other members of the Flaviviridae family support the hypothesis that inactivation of the serine proteinase activity associated with NS3 leads to the production of noninfectious viral particles (CHAMBERS et al. 1990). Thus, the NS3 serine proteinase of HCV has become one of the major targets for the discovery of novel anti-HCV drugs.

HCV NS3 is a multidomain, 631-amino acids (aas) long protein that contains, in addition to the serine proteinase domain, an RNA helicase and an RNA-stimulated ATPase (see the chapter by A.D. Kwong et al., this volume). Despite the high degree of sequence variability among different HCV isolates, it was possible to predict the existence of a serine proteinase domain within the NH_2-terminal third of NS3 on the basis of the conserved sequence patterns common to all viral and cellular serine proteinases (MILLER and PURCELL 1990; BAZAN and FLETTERICK 1989). The NS3 proteinase domain has thus been mapped by deletion mutagenesis to the NH_2-terminal 180 amino acids of NS3 (BARTENSCHLAGER et al. 1994; FAILLA et al. 1995; TANJI et al. 1994b; HAHM et al. 1995; HAN et al. 1995; LIN et al. 1994). Within this region, residues His-57, Asp-81 and Ser-139 constitute the catalytic triad of the enzyme and mutagenesis of each of these residues was found to abolish proteolytic activity (BARTENSCHLAGER et al. 1993; ECKART et al. 1993; GRAKOUI et al. 1993; TOMEI et al. 1993; MANABE et al. 1994; HIJIKATA et al. 1993b). The role of NS3 in the maturation of the viral polyprotein has been elucidated by transient transfection and cell-free translation studies. From these studies it emerged that the serine proteinase contained within NS3 is required for the processing of all the polyprotein junctions that are COOH-terminal of NS3 itself, i.e. at the NS3/NS4A, NS4A/NS4B, NS4B/NS5A and NS5A/NS5B boundaries. Several lines of evidence, such as insensitivity to dilution and failure to observe *trans*-cleavage at the NS3-NS4A site, have suggested that processing at the latter junction occurs exclusively in *cis*, i.e., within the same polyprotein molecule. Conversely, the remaining cleavage sites were found to be processed also in *trans* (BARTENSCHLAGER et al. 1993; GRAKOUI et al. 1993; LIN et al. 1994; LIN and RICE 1995; HIJIKATA et al. 1993b; TOMEI et al. 1993).

Kinetic studies of the NS3-dependent polyprotein processing in cells transiently expressing HCV polyprotein revealed a preferential but not obligated order of cleavage (BARTENSCHLAGER et al. 1994; FAILLA et al. 1995; LIN et al. 1994; TANJI et al. 1994a). Proteolytic processing at the NS3-NS4A junction is rapid and precedes all other NS3-mediated cleavages. This has led to the suggestion that cleavage at this site might be a cotranslational event. However, cleavage at the NS3-NS4A site was demonstrated to rely on the previous interaction of the NS3 proteinase with the cofactor NS4A (see below), implying that this processing event occurs post-translationally (FAILLA et al. 1994). Proteolytic processing at the NS3-NS4A site is followed by cleavage at the NS5A-NS5B site. This cleavage event generates mature NS5B, the viral RNA-dependent RNA polymerase, and an NS4A-NS5A precursor that is processed more slowly down to the final mature proteins, namely NS4A, NS4B and NS5A. Notably, only cleavage of the NS4A-NS4B precursor requires the presence of membranes for efficient processing in cell-free translation experiments (LIN and RICE 1995; KOCH et al. 1996).

The location of the sites cleaved by the NS3 proteinase within the HCV polyprotein was obtained by sequencing the NH_2-terminals of mature NS4A, NS4B, NS5A and NS5B (GRAKOUI et al. 1993; PIZZI et al. 1994). Comparison of the sequences flanking the peptide bonds cleaved by NS3 in the polyprotein of the different HCV genotypes yielded the substrate consensus sequence Asp/GluXaa$_4$Cys/Thr-Ser/Ala (Fig. 1A). Cleavage was thus demonstrated to occur after a cysteine residue in all *trans*-cleavage sites, whereas the intramolecular site between NS3 and NS4A was shown to be unique in this respect having a threonine residue in the P1 position. The only other features of all the cleavage sites are a conserved negatively charged residue in the P6 positions, and a serine or alanine residue in the P1' positions.

The specificity of cleavage by the NS3-NS4 proteinase was ultimately confirmed using different forms of the recombinant enzyme on synthetic peptide substrates (LANDRO et al. 1997; STEINKÜHLER et al. 1996b; URBANI et al. 1997; ZHANG et al. 1997). Cleavage kinetics were measured using peptides corresponding to all the natural cleavage sites of the HCV polyprotein. The order of cleavage

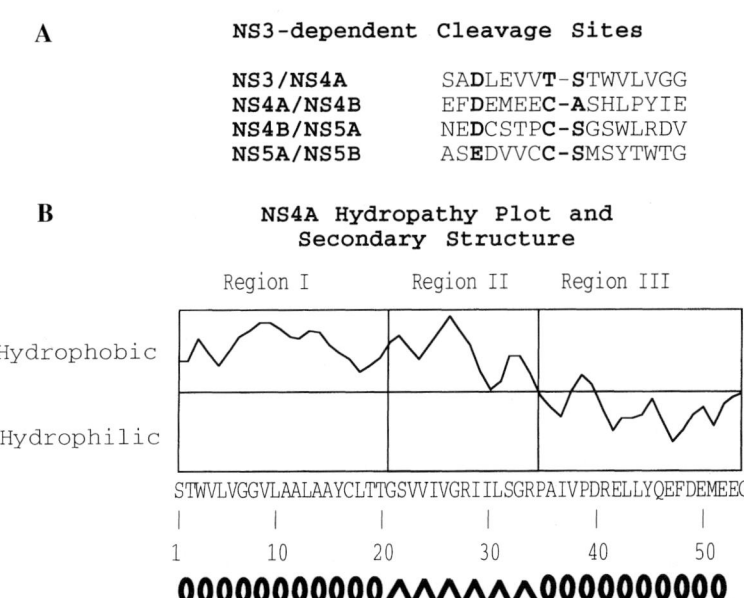

Fig. 1. A Sequence of the NS3-dependent cleavage sites within the viral polyprotein (HCV genotype 1b). B Hydropathy plot and secondary structure prediction of the NS4A protein. The hydropathy was calculated using the algorithm of Kite-Doolittle. The secondary structure prediction was carried out using the peptide structure program of the GCG-Winsconsin software package

efficiency, expressed as k_{cat}/K_m, was determined as follows: NS5A-NS5B > NS4A-NS4B ≫ NS4B-NS5A. Peptides harboring the sequence of the NS3-NS4A site were not cleaved, in keeping with the notion that the latter site is recognized in an exclusively intramolecular fashion. Since the relative kinetics on the synthetic peptides representing the *trans*-cleavage sites parallels the efficiency of cleavage observed at the respective sites in the polyprotein, it was possible to conclude that the primary structure is an important determinant of the efficiency with which each *trans*-site is cleaved during polyprotein processing.

2 NS3-NS4A: A Heterodimeric Serine Proteinase

Although the NH_2-terminal serine proteinase domain of NS3 shows some intrinsic enzymatic activity, NS4A, a second viral protein, is an essential NS3 proteinase cofactor for efficient proteolytic processing of the HCV polyprotein (FAILLA et al. 1994; BARTENSCHLAGER et al. 1994; LIN et al. 1994; TANJI et al. 1995a). The presence of the NS4A protein is in fact strictly required for cleavage at the NS3/NS4A, NS4A/4B and NS4B/NS5A sites, and it increases the extent of cleavage at the NS5A/NS5B junction. NS4A can exert its cofactor function in *cis* (i.e., when expressed as an NS3-NS4A precursor protein) as well as in *trans* (i.e., when expressed as a separate molecule or as part of the polyprotein substrate). In transfected cells, NS3 and NS4A form a tight complex that can be detected by co-immunoprecipitation of the two proteins with antibodies directed either against NS3 or against NS4A (HIJIKATA et al. 1993; BARTENSCHLAGER et al. 1995b; FAILLA et al. 1995; LIN et al. 1995; SATOH et al. 1995). In addition, complex formation with NS4A significantly increases the stability of NS3 in the cytoplasm of mammalian cultured cells and targets the NS3 protein to the membranes of the endoplasmic reticulum (TANJI et al. 1995a). The conclusion drawn from these findings is that NS4A is a functional analog of Flavivirus NS2B and Pestivirus p10 proteins and that a heterodimeric NS3-NS4A species is likely to be the biologically relevant form of the HCV serine proteinase.

NS4A is a protein of 54 residues and is predicted to be a membrane-bound protein. Analysis of hydropathy plots and secondary structure prediction highlight the presence of three regions within NS4A (Fig. 1B). Region I, residues 1–20, is highly hydrophobic and has been predicted by multiple sequence alignment to form a *trans*-membrane α helix (ROST et al. 1995). Region II, residues 21–34, is also hydrophobic but is rather predicted to form a β-strand structure (TOMEI et al. 1996). Finally, region III (i.e. the 20 COOH-terminal residues of NS4A) is more hydrophilic with predicted preference for α-helical conformation. Deletion mutagenesis experiments have shown that the integrity of NS4A region II is indeed required and sufficient to fully activate the NS3 serine-proteinase activity (LIN et al. 1995; SHIMIZU et al. 1996; TOMEI et al. 1996; KOCH et al. 1996; BUTKIEWICZ et al. 1996). Synthetic peptides with the amino acid sequence of this region have been found to bind to the NS3 enzyme with a 1:1 stoichiometry and to activate its proteolytic activity in vitro (LIN et al. 1995; SHIMIZU et al. 1996; STEINKÜHLER et al. 1996a,b; TOMEI et al. 1996; BIANCHI et al. 1997).

Deletion mutagenesis experiments performed on the NS3 protein have pinpointed the region of the protein that is involved in the NS3-NS4A interaction (FAILLA et al. 1995; SATOH et al. 1995; KOCH et al. 1996). This region has been mapped within the minimal serine proteinase domain: more precisely, the 22 NH_2-terminal amino acids of NS3 were found to be indispensable both for the NS3-NS4A complex formation and for the modulation of the proteolytic activity by NS4A. Interestingly, the same region did not appear to be required for the NS4A-independent residual serine proteinase activity of NS3, measured on the NS5A/NS5B site. It was further proposed that the same region of NS3 forms an autonomous NS4A-binding domain (SATOH et al. 1995).

Many of the experiments carried out using transient expression in cell culture and cell-free in vitro translation have indicated that both basal and NS4A-stimulated proteinase activities of truncated NS3 proteins, essentially lacking the COOH-terminal helicase domain, are virtually identical to the full-length 70kDa NS3 protein. Following this observation, several groups adopted a "minimalist" approach and studied a simplified system composed of a recombinant protein encompassing amino acids 1–180, and a synthetic peptide based on the sequence of region II of NS4A as a co-factor mimic. Purification of the truncated enzyme using several different heterologous expression systems has been reported (SUZUKI et al. 1995; SHOJI et al. 1995; MORI et al. 1996; STEINKÜHLER et al. 1996a,b; MARKLAND et al. 1997; VISHNUVARDHAN et al. 1997) and the effort has culminated in the crystallization of the free NS3 proteinase domain (HCV 1b genotype, LOVE et al. 1996) and the complex with a co-factor peptide (HCV-1a genotype, KIM et al. 1996; HCV 1b genotype, YAN et al. 1998).

3 Three-Dimensional Structure of the NS3-NS4A Proteinase

3.1 NS3-NS4A Is a Chymotrypsin-Like Serine Proteinase

Serine proteinases of the chymotrypsin family contain two domains of similar structure, consisting each of a six-stranded β-barrel (LESK and FORDHAM 1996). These two domains are believed to have arisen by gene duplication and divergence. The two domains pack together asymmetrically, with the residues of the so-called catalytic triad lying between them. The catalytic triad in chymotrypsin is formed by residues Asp102/His-57/Ser-195. These residues are spatially arranged to form a charge relay system that polarizes the side-chain of Ser-195 for nucleophilic attack on the C atom of the scissile bond. During the proteolysis reaction, an intermediate forms in which this carbon atom is tetrahedral, and a negative charge is developed on the former carbonyl oxygen of the peptide bond being cleaved ("oxyanion").

This negative charge is stabilized by the proteinase "oxyanion-binding hole", in which there are hydrogen bonds from the negatively charged oxygen atom of the substrate to the NH groups of residues 193 and 195 (chymotrypsin numbering).

Analysis of the X-ray structures of the truncated NS3 protein complexed with the NS4A-derived peptide (KIM et al. 1996; YAN et al. 1998) revealed that the serine proteinase domain adopts a canonical chymotrypsin-like fold, consisting of two β-barrels, each containing a "Greek key" motif (Fig. 2A, B). The COOH-terminal domain (residues 94–180) contains a conventional six-stranded β-barrel, which ends with a structurally conserved COOH-terminal helix. Conversely, the NH_2-terminal domain (NS3 residues 1–93 and NS4A residues 21–34) contains eight β-strands. As discussed below, two additional strands, A_0 and D'_1, are contributed by the proteinase extreme NH_2-terminal and by the NS4A peptide, respectively. The loops connecting the various secondary structural elements are relatively short in the NS3 proteinase, compared to cellular serine proteinases. Implication of this feature with respect to substrate recognition is discussed below (Sect. 3.4.2).

The NS3 proteinase active site is similar to the active site of the other chymotrypsin-like enzymes. The nucleophilic Ser-139, together with the general acid/base catalyst His-57 and Asp-81, form the NS3 catalytic triad. As expected, these residues are located in a crevice at the interface of the two domains (Fig. 2A): His-57 and Asp-81 are located at the beginning of helix α_1 and β-strand F_1, respectively. Both residues are found in the NH_2-terminal domain. The catalytic residue, Ser-139, and the oxyanion binding loop (residues 137–139) follows helix α_2 in the COOH-terminal domain. The polypeptide backbone conformation of this region is virtually identical to that observed in other chymotrypsin-like proteinases.

As anticipated on the basis of secondary structure predictions and studies with model peptides (FAILLA et al. 1995; TOMEI et al. 1996), the NS4 A cofactor assumes an extended structure and forms β-strand D_1 (Figs. 2, 3). This secondary structure element interacts exclusively with the NH_2-terminal barrel of the enzyme. Consistent with the early mutagenesis experiments (see above), the region of the NS3 protein that is in direct contact with the NS4A peptide includes the NH_2-terminal 22 residues of NS3, corresponding to A_0 and α_0. The β-strand contributed by NS4A, D'_1, intercalates between the two β-strands A_0 and A_1 in an antiparallel fashion. Helix α_0 is juxtaposed against the NH_2-terminal of the NS4A peptide (Fig. 2A, Fig. 3). Nearly all of the main chain carbonyl and amide groups comprised between NS4A residues 23 ad 31 are engaged in hydrogen bonds with this region of NS3, including hydrogen bonds to the side chains of Arg-11 and Glu-32. With the exception of the most terminal residues, namely Gly-21, Gly-33 and Arg-

Fig. 2A,B. The NS3-NS4A serine proteinase adopts a chymotrypsin-like fold. **A** The three-dimensional structure of the HCV NS3 proteinase domain complexed with an NS4A-derived peptide. The side-chains of the amino acids of the catalytic triad (His-57, Asp-81, Ser-139) and of the amino acids involved in metal coordination (Cys-97, Cys-99, Cys-145 and His-149) are shown. The coordinates used to generate the picture are deposited in the Protein Data Base (PDB id: 1JXP) and refer to the HCV-1b NS3-NS4A complex (YAN et al. 1998). **B** Topology of the secondary structure of the NS3-NS4A serine proteinase. β-strands are shown as *arrows*. The strand contributed by the NS4A peptide is *shaded*. The NH_2-terminal domain is at the *top*, the COOH-terminal domain is at the *bottom*

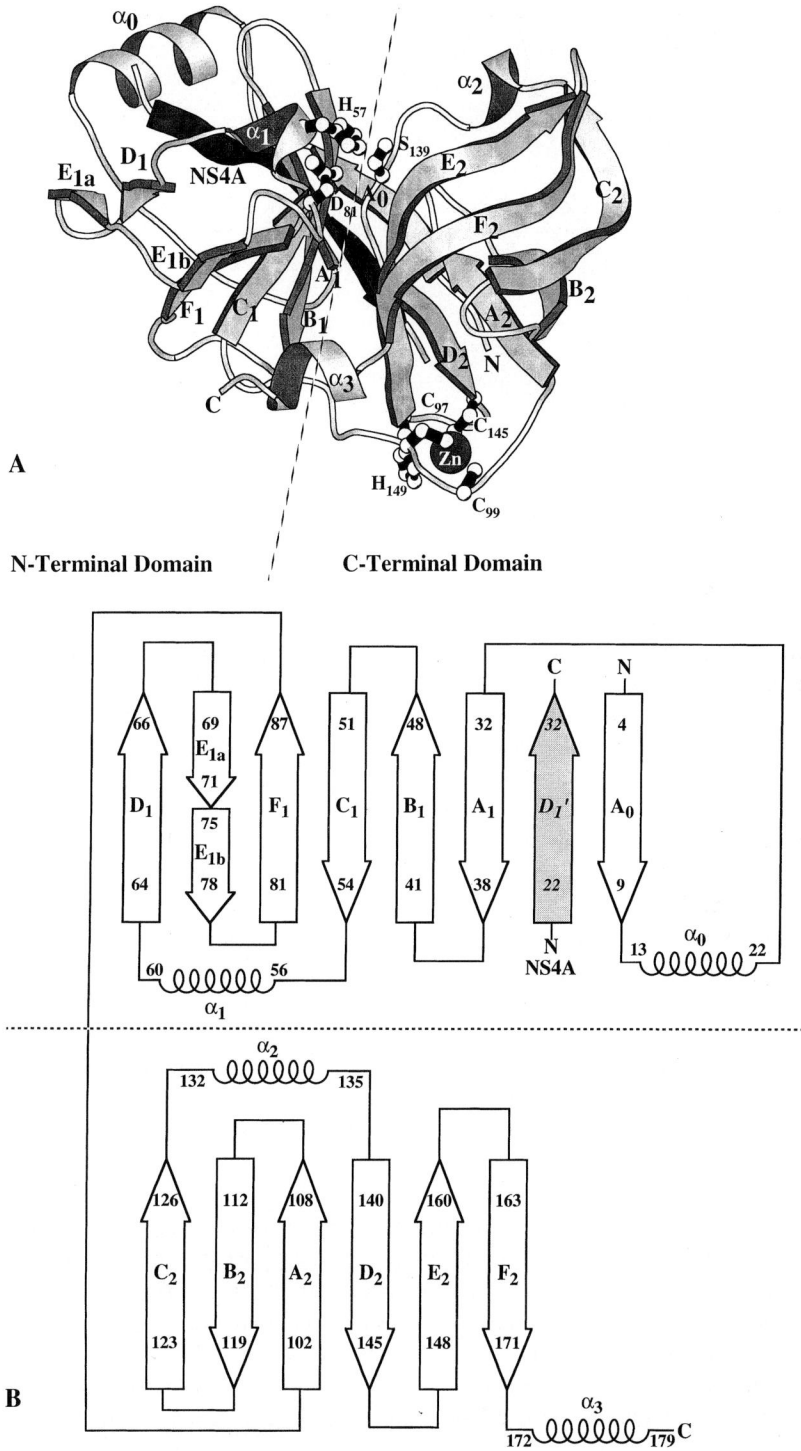

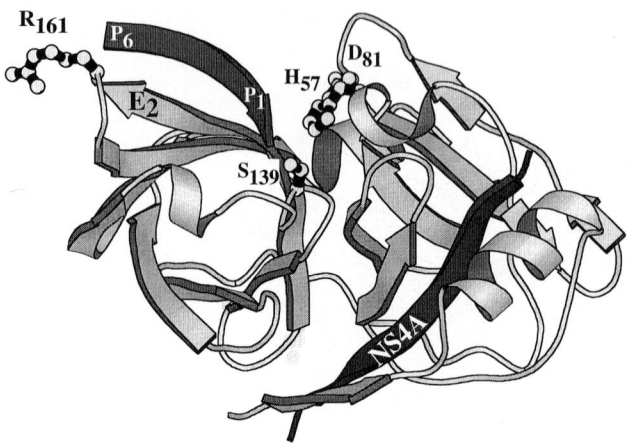

Fig. 3. NS3-NS4A proteinase-substrate interactions. The three-dimensional structure of the NS3-NS4A serine proteinase. The side-chains of the amino acids of the catalytic triad (His-57, Asp-81, Ser-139) and of Arg161 are shown. The P-side of the substrate (P6-P1) is modelled as a β-strand making contacts with the protein strand βE2

34 in the 1b genotype, the NS4A peptide is completely buried within the complex. Several side chains emanating from NS4A bind into in hydrophobic pockets formed by NS3, thus contributing to the formation of the hydrophobic core of the NS3-NS4A complex. The interaction of NS4A pepide with the core of NS3 buries ~1650Å^2 of surface area, while the interaction with the NH$_2$-terminal region buries an additional 750Å^2. Overall, about 2,400Å of surface area is buried by the interaction between the enzyme and the cofactor. For this reason, NS4A has been suggested to be an integral structural component of the proteinase (KIM et al. 1996).

In the crystal structure without NS4A, the COOH-terminal β-barrel of the proteinase domain adopts the same fold observed in the complex, except for the lack of the COOH-terminal α-helix (LOVE et al. 1996). Conversely, the NH$_2$-terminal of the protein assumes a significantly different conformation: the NH$_2$-terminal 30 amino acids of NS3 extend away from the protein, forming four consecutive β-strands that have been named A_0, B_0, C_0, and D_0, respectively. In order to avoid confusion, it is important to point out that strand A_0 of the uncomplexed NS3 proteinase does not match exactly strand A_0 found in the complex. The asymmetric unit in this crystal structure is an trimer, whereby the NH$_2$-terminal extended region of each monomer interacts with neighboring molecules, binding in a long, shallow hydrophobic valley, located mainly on the COOH-terminal β-barrel and composed of residues from strands A_2, B_2, and D_2, but also some residues from A_1 (LOVE et al. 1996). These interactions are likely to be peculiar features of the enzyme crystallized in the absence of the NS4A cofactor. In solution, the NH$_2$-terminal region of NS3 is probably disordered when not engaged in the interaction with NS4A, thus providing an explanation for the enhanced metabolic stability of the NS3-NS4A complex with respect to the uncomplexed enzyme (TANJI et al. 1995a).

3.2 Mechanism of NS3 Proteinase Activation by NS4A

Binding of NS4A has several structural consequences on the conformation of the NS3 proteinase. The most dramatic conformational change impacts the NH_2-terminal 22 amino acids of the proteinase. These residues are believed to be essentially misfolded in the uncomplexed form of NS3 (see above). Upon NS4A binding, this region forms strand A_0 and α_0. These two structural elements that have no counterpart in other serine proteinases and it is, a posteriori, not surprising that their deletion results in an enzyme that can no longer be activated by NS4A, but which retains the basal proteolytic activity (Fig. 4A).

A visual comparison between the structures of the complexed and the uncomplexed NS3 proteinase domain has suggested that a second, more subtle but possibly crucial, conformational change, i.e., displacement of strand D_1 of the proteinase (residues 64–66), might occur upon complex formation (YAN et al. 1998). In the uncomplexed enzyme, NS3 strand D_1 appears to be located in position seen for the corresponding D_1 strand in chymotrypsin. According to the analysis of Yan and coworkers (1998), upon complex formation the NH_2-terminal portion of the D'_1 β-strand provided by NS4A assumes the location previously occupied by D_1. In this way, strand D'_1 of NS4A ends up being adjacent to helix α_1. In particular, the side chain of residue 23 of NS4A is now positioned to make a hydrophobic contact with Ala59 of helix α_1. The resulting hydrophobic interaction is proposed to rigidify helix α_1, in which the catalytic histidine resides. In fact helix $\alpha 1$ is very well defined in the complex crystal structure (YAN et al. 1998), whereas the same region (residues 57–63) are described as being very mobile in the uncomplexed NS3 structure. As a result, the imidazole of His-57, expected to extract a proton from nucleophile Ser-139 during the enzymatic reaction, is within hydrogen bonding distance from the side chain of the catalytic serine in the NS3-NS4A complex (KIM et al. 1997; YAN et al. 1998). Conversely, in the uncomplexed NS3 structure, the side chain of His-57 points toward the catalytic serine but is too far away to form the hydrogen bond observed in serine proteinases structures.

The side chain of Asp-81 is the third member of the catalytic triad. In the NS3-NS4A complex structure, this residue is properly hydrogen bonded to the His-57 imidazole in order to provide stabilization to the charge relay that occurs during catalysis. In the structure of the uncomplexed NS3 proteinase, the carboxyl group of Asp-81 points away from His-57 and forms a hydrogen bond with the guanidium moiety of Arg-155, a residue that is strikingly conserved in all HCV isolates (a serine residue is commonly found at this position in other serine proteinases). These observations offer an attractive explanation of how activation of the NS3 proteinase activity is brought about by the NS4A cofactor. It should be taken in account, however, that relatively minor readjustments would be sufficient to cor-

A

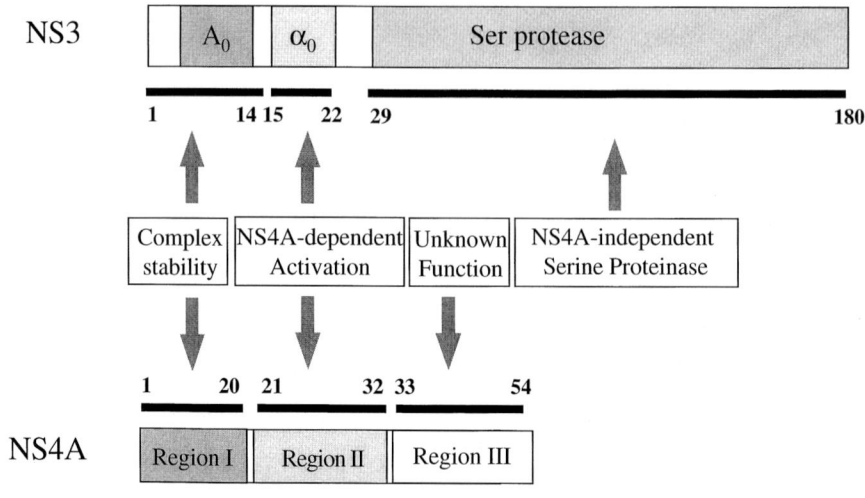

B

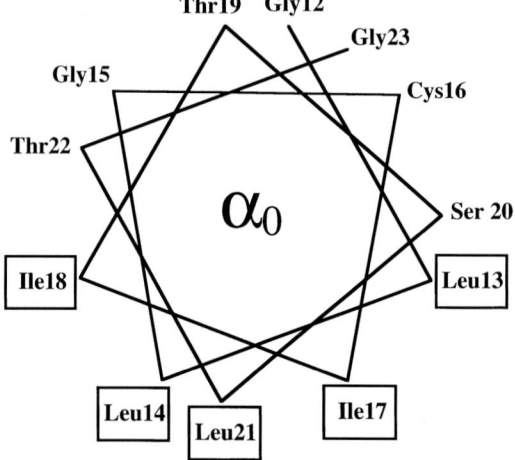

Fig. 4A,B. The region of NS3 and NS4A involved in the stabilization of the proteinase complex. **A** The region of NS3 and NS4A involves in NS4A-dependent activation of the proteinase and in the stabilization of the NS3-NS4A complex. The boundary of the various regions were defined by deletion mapping. **B** Helical wheel representation of helix α_0. All the hydrophobic amino acids (*boxed*) are found on the outer face of the helix (*bottom*) rather than in the protein interior

rectly position the side chain of Asp-81 in the absence of the co-factor, suggesting that also in this case a classic catalytic triad configuration could probably exist during catalysis (LOVE et al. 1996). This hypothesis is in line with the finding that uncomplexed NS3 retains some proteinase activity and that complex formation with NS4A does not alter the pK_a value of His-57, as determined both by activity titration and by NMR (LANDRO et al. 1997; URBANI et al. 1998).

A different mechanism has also been invoked to explain activation of the NS3 proteinase by its NS4A cofactor. According to the hypothesis presented by Kim and co-workers (1996), incorporation of NS4A in the NH_2-terminal β-barrel of NS3 could "lead to alterations in the active site by providing a more rigid and precise framework for residues that form the prime-side of the substrate-binding channel. In addition, activation via direct interaction of NS4A with the prime-side residues of the substrate is possible". According to this view, binding of NS4A could directly or indirectly promote the formation of novel binding site(s) for the recognition of the prime-side of the substrate. To substantiate this hypothesis, LANDRO et al. (1997) propose a kinetic scheme according to which an ordered, sequential binding of NS4A and substrate occur: according to the proposed mechanism, NS4A binds obligatorily to NS3 prior to substrate binding, thus contributing to the formation of the enzyme substrate-binding site. They further reported evidence for the influence of NS4A on binding of synthetic peptides to prime-side subsites. Two classes of peptide inhibitors are reported. The first class does not extend in the P' region and inhibits the NS3 proteinase in an NS4A-independent fashion. The second class of inhibitors includes P' residues and displays a large increase in NS3 binding affinity when NS4A is also bound to the enzyme. Furthermore, sequential truncation of P' residues pinpoint the P1' and P4' residues as most important for the NS4A-dependent affinity increase displayed by these inhibitors. In this respect, it is worthy of notice that molecular modeling of a decapeptide into the enzyme substrate-binding cleft has revealed the possibility that the P4' residue side chain be in direct contact with NS4A. Overall, this study suggests that the correct formation of the P' subsites is indeed driven by the NS4A co-factor.

The different mechanisms of NS4A activation outlined above, however, are not to be considered mutually exclusive. The kinetic consequences of complex formation have been investigated in detailed using recombinant serine proteinase domain in conjunction with NS4A-derived synthetic peptides. The fashion in which NS4A peptide mimics the enzymatic activity of NS3 depends on both the assay conditions and the substrate peptide used. Thus, either an enhancement of the enzyme catalytic constant (k_{cat}) or a simultaneous decrease in K_m and a concomitant increase in k_{cat} values was observed (SHIMIZU et al. 1996; STEINKÜHLER et al. 1996a,b; BIANCHI et al. 1997; LANDRO et al. 1997; URBANI et al. 1997). These findings are suggestive of both an activation of the catalytic machinery through a conformational rearrangement and altered substrate binding modes, possibly mediated through the interaction of the proteinase S' site with the P' portion of the substrate in the NS3-NS4A complex.

3.3 Stability of the NS3/NS4A Complex

The structural changes induced by the formation of a complex between the truncated NS3 proteinase and a peptide spanning NS4A residues 21–34 can be monitored by changes in the protein near-UV CD spectrum and in its tryptophan fluorescence spectrum (BIANCHI et al. 1997). These spectroscopic changes were interpreted in terms of changes in the environment of Trp-85, which is engaged in the interaction with the side chain of Val-23 of NS4A. The results of these studies have permitted calculation of complex dissociation constants in the low micromolar range and a complex half-life of 3.5 min (BIANCHI et al. 1997). Conversely, the native NS3-NS4A complex was observed to be stable for several hours in cultured human cells transiently transfected with HCV cDNA (BARTENSCHLAGER et al. 1995b). Therefore, the complex with the peptide analog of NS4A appears to be significantly less stable than what can be extrapolated for the native complex with the full-length co-factor. Systematic deletion experiments in transfected cells (BARTENSCHLAGER et al. 1995b; LIN et al. 1995) showed that, while the central domain of NS4A is required and sufficient in the activation of the proteinase, truncations affecting the NH_2-terminal hydrophobic sequence of NS4A (region I) impair the efficiency of co-immunoprecipitation of NS3 and NS4A (Fig. 4A). Since the same region of NS4A is likely to be responsible for the membrane anchoring of NS3-NS4A complexes, tight complex formation might involve membrane association (see also below). These observations suggest that the NH_2-terminal, hydrophobic domain of NS4A is involved in the formation of a stable NS3-NS4A complex, but also that formation of such a long-lived complex is not strictly required for activation of the NS3 serine proteinase. The tight association observed between NS3 and NS4A might serve a function different from polyprotein processing. Given that NS3 contains an RNA-helicase domain and is therefore postulated to have a role in the replication of the viral genome, the formation of a stable, membrane-bound NS3-NS4A complex might be required for the formation of a membrane-associated replication complex.

Deletion mapping experiments were also used to define how regions within the NS3 NH_2-terminal were involved in the interaction with NS4A, both with respect to proteinase activation and stable complex formation. The results of these experiments are summarized in Fig. 4A. The 180 NH_2-terminal residues of NS3 retain full proteolytic activity and are sufficient to form a stable complex with the NS4A cofactor (BARTENSCHLAGER et al. 1995b; FAILLA et al. 1995; LIN et al. 1995; SATOH et al. 1995). Further COOH-terminal deletions result in the complete loss of activity. Conversely, deletion from the NH_2-terminal side of the protein led to a gradual and characteristic inactivation of the NS3 proteinase. Constructs lacking up to 14 NH_2-terminal residues show a normal processing pattern and they can still be activated by NS4A (FAILLA et al. 1995; BARTENSCHLAGER et al. 1995b). Interestingly, however, NS3 constructs bearing the same NH_2-terminal deletions are no longer able to achieve stable complex formation, as judged by the lack of an immunoprecipitable NS3-NS4A complex (BARTENSCHLAGER et al. 1995b; KOCH et al. 1996). These findings imply that residues at the extreme NH_2-terminal of NS3 are not essential for NS4A-dependent proteinase activation, but important for the stabilization of the NS3-NS4A complex. These residues encompass β-strand A_0. This secondary structure element resembles, in the NS3-NS4A crystal structure, a "molecular clamp" that locks the NS4A cofactor onto the NS3 proteinase domain (Fig. 3).

NS3 constructs further lacking residues 15–22 can no longer be activated by NS4A and are appreciably less stable (BARTENSCHLAGER et al. 1995b; FAILLA et al. 1995; SATOH et al. 1995). These residues correspond to helix α_0 (Fig. 4A). The minimal region required for basal, nonactivated activity of NS3 proteinase begins with residue 29, only four residues upstream of the first amino acid of strand A_1 (BARTENSCHLAGER et al. 1995b; FAILLA et al. 1995). The region comprised residues 23–29, connecting A_0 to A_1, does not seem to play any role with regard to either NS3-NS4A complex formation or NS3 proteinase basal activity.

Altogether, these data suggest that α_0, although not sufficient to promote the formation of a stable complex, is crucial for activation of the NS3 proteolytic activity by the NS4A cofactor. This conclusion is further supported by the finding that substitution of NS3 α_0 residues with helix-breaking amino acids severely compromises NS4A-dependent proteinase activation of NS3 (KOCH et al. 1996). A surprising, structural feature of helix α_0 is that all the residues facing outwards and away from the rest of the protein are bulky hydrophobic amino acids (Fig. 4B). The side chains of these amino acids create an unusually large hydrophobic patch on the NS3-NS4A complex surface, and it has been suggested that they might constitute a second domain of attachment to the phospholipid bilayer of the endoplasmic reticulum membrane (YAN et al. 1998). Exposing such a large hydrophobic surface to an aqueos solvent would be, from a thermodynamic point of view, extremely costly in terms of free energy. Therefore, the formation of helix α_0 may be largely unfavoured in aqueos solution, whereas it would promptly form in a membranous environment. This observation might explain why the complex between the isolated NS3 proteinase domain and the NS4A peptide is only marginally stable in solution (BIANCHI et al. 1997), despite a buried surface area in excess of 2,000Å^2 revealed in the crystal structure. It has, in fact, been proposed that the crystal structure of the NS3-NS4A complex may represent the membrane-bound form of the proteinase (YAN et al. 1998). In this frame of thinking, binding of an NS4A peptide corresponding to region II to NS3 may be necessary but not sufficient to properly fold the extreme NH_2-terminal of NS3, namely A_0 and α_0, when the complex is present in solution. Conversely, membrane attachment, driven by region I of the NS4A co-factor, might provide the hydrophobic surface necessary to stabilize helix α_0, thus allowing the formation of a complex as seen in the crystal structure.

3.4 Substrate Specificity of the NS3-NS4A Serine Proteinase

Substrate recognition by all serine proteinases involves binding of the substrate into a cleft on the surface of the enzyme. The substrate binding cleft of chymotrypsin-like proteinases is normally found at the interface between the NH_2-terminal and the COOH-terminal β-barrel domains. Substrates and peptide inhibitors bind to serine proteinases through the formation of an antiparallel β-sheet between a region of the substrate adjacent to the scissile bond (normally P3-P1) and β-strand E_2 of the enzyme, which forms the base of the interdomain cleft (LESK and FORDHAM 1996). Formation of this β sheet-like interaction between the enzyme and the substrate direct the substrate (or inhibitor) side chains into the S4-S1 binding pockets. Assuming that HCV NS3 binds peptide substrate in a fashion equivalent to chymotrypsin-like proteinase complexes with protein or peptide inhibitors, we can explain the observed substrate specificity.

3.4.1 Primary Specificity

The primary specificity of a proteinase is defined by the side chain of the amino acid that preceeds the scissile bond, i.e. the P1 position. The specificity for the P1 amino acid is imposed by the shape of the S1 pocket on the enzyme, often referred to as the proteinase specificity pocket. As stated above, the NS3 cleavage sites have the consensus sequence Asp/GluXaa$_4$Cys/Thr-Ser/Ala (Fig. 1A), with cleavage occuring after cysteine in all *trans* cleavage sites (i.e. NS4A/NS4B, NS4B/NS5A and NS5A/NS5B) or after threonine in the intramolecular cleavage site between NS3 and NS4A. Preference for cysteine residues in the P1 positions of NS3 substrates can be rationalized on the basis of the peculiar structure of the S1 pocket of the enzyme. The structure shows a specificity pocket that is shallow and hydrophobic, being occluded on the botton by the aromatic ring of Phe-154. The side chains of Ala-157 and Leu-135 line the side of the S1 pocket (PIZZI et al. 1994; KIM et al. 1996; LOVE et al. 1996; YAN et al. 1998). The side chain of cysteine found in the substrate P1 position has been suggested to engage in a favourable sulfhydryl-aromatic interactions with the electron-rich π clouds on the ring system of phenylalanine. The role of Phe-154, Ala-157 and Leu-135 in the recognition of the P1 residue was further substantiated by site-directed mutagenesis experiments. In these studies, the residues predicted to be part of the specificity pocket were mutagenized in order to engineer proteinases with altered substrate specificities (FAILLA et al. 1996; Koch and BARTENSCHLAGER 1997). Substitution at positions 135 or 157 resulted in reduced processing efficiencies, but in no case was the substrate specificity was altered. Conversely, NS3 proteinase constructs in which Phe-154 was replaced by threonine resulted in a dramatic change of enzyme specificity: large, hydrophobic residues such as leucine or phenylanine were selectively recognized by the mutant enzyme, indicating that the mutations gave rise to the a much larger and deeper S1 pocket.

The finding that a threonine residue is always found in the P1 position of the *cis*-cleavable bond deserves some special attention. The side-chain of this residue cannot be optimally fitted into the S1 pocket and was shown to decrease cleavage efficiency when introduced into the P1 positions of the *trans* cleavage sites (BARTENSCHLAGER et al. 1995a; KOLYKHALOV et al. 1994). Conversely, a cysteine in the P1 position of the NS3-NS4A site is cleaved with equal or greater efficiency than observed for threonine (BARTENSCHLAGER et al. 1995a; LEINBACH et al. 1994).

In addition, Komoda and co-workers (1994b) where able to demonstrate trans cleavage of a chimeric protein harboring the NS3/NS4A cleavage site expressed in E. coli when the wild-type threonine residue was replaced by cysteine. All these observations suggest that threonine is suboptimal, even in the context of the NS3/NS4A precursor, and raise the question why should a threonine be strictly conserved at the *cis*-cleavage site site throughout evolution of the virus.

Using transient expression in eukaryotic cells or in vitro translation of HCV polyprotein precursors, the NS3/NS4A cleavage was shown to be remarkably tolerant to substitutions of the P1 threonine (BARTENSCHLAGER et al. 1995a; LEINBACH et al. 1994; KOLYKHALOV et al. 1994; TANJI et al. 1994). Replacement of threonine residues with amino acids having a small side chain (i.e., glycine, valine, alanine, serine) was well tolerated, as was substitution of the residue in P1 with larger polar or negatively charged amino acids (asparagine, aspartate). Only amino acids with a basic (arginine) or bulky hydrophobic side chain (tyrosine, phenylalanine, isoleucine) led to severely decreased processing efficiency. The extreme tolerance of the P1 residue of the *cis*-cleavage site to substitution with so many diverse amino acid has led to the conclusion that processing at this site is determined primarily by polyprotein folding, rather than by primary sequence (BARTENSCHLAGER et al. 1995a).

Conversely, all the *trans*-cleavage sites were shown to be affected to some extent by substitution of the P1 residue. Interestingly, susceptibility to the reported mutations of these sites was shown to depend on the sequence context, with a gradient of increasing sensitivity following the order NS4A/NS4B < NS4B/NS5A < NS5A/NS5B (BARTENSCHLAGER et al. 1995a; KOLYKHALOV et al. 1994). Although some residues other than cysteine were accepted in this position, the P1 residue of the *trans*-cleavage sites was nevertheless identified by these studies as the major determinant of substrate specificity.

In the context of synthetic peptide substrates, a more pronounced sensitivity of the NS3 proteinase to P1 substitutions was observed than using polyprotein substrates in transient transfection experiments (LANDRO et al. 1997; URBANI et al. 1997; ZHANG et al. 1997). Substitution of the P1 cysteine with natural and nonnatural amino acids in the context of an NS4A/NS4B-based substrate gave the following order of decreasing efficiencies: cysteine > homocysteine > allylglycine > 2-amino-butyric acid > threonine > norvaline > valine. Serine, alanine, glycine or leucine in the P1 position yielded uncleavable substrates (URBANI et al. 1997). Similar results were obtained introducing modifications in the P1 position of substrates based on the sequence of the NS5A/NS5B junction (LANDRO et al. 1997; ZHANG et al. 1997). These findings have to be compared with the results from mutagenesis experiments of the P1 residues in the polyprotein NS4A/NS4B junction, where only arginine and aspartate resulted in complete abolishment of cleavage (KOLYKHALOV et al. 1994), while processing was still observed with residues such as alanine, glycine, serine or leucine (BARTENSCHLAGER et al. 1995a; KOLYKHALOV et al. 1994). These differences may reflect an increased activity of the NS3 proteinase on polyprotein substrates that render it less discriminating towards suboptimal P1 residues.

3.4.2 Distal Subsites

Analysis of the consensus sequence for cleavage by the NS3-NS4A proteinase of HCV (Fig. 1A) indicate that the P1 specificity is important but obviously not the only determinant of cleavage. An acidic amino acid at the P6 position and a serine or an alanine residues at the P1' position seem also to be important for substrate recognition by this enzyme. Consistent with the lack of apparent specificity over the P2-P5 region, the structure of the NS3 proteinase reveals a substrate binding cleft that is rather flat and solvent exposed. No pockets corresponding to the S2-S5 subsites are easily identified on the enzyme surface. Several loops involved in substrate recognition by cellular serine proteinases are in fact shorter or absent in the NS3 proteinase. Most notably, in chymotrypsin and elastase the loop connecting β-strands E_1 and F_1 is positioned such that it can interact with residues on the P side of the substrate and plays a pivotal role in substrate recognition. In the picornaviral and bacterial chymotrypsin-like proteinases the loop connecting β-strands B_2 and C_2 has an analogous function. It has been pointed out that the lack of corresponding loops in the NS3 proteinase domain goes along with the apparent lack of substrate specificity displayed by this enzyme over P2-P5 (LOVE et al. 1996).

In the case of other chymotrypsin-like proteinases, β-strand E_2 forms β-sheet-like hydrogen bonds with the substrate over P2-P3. In the NS3 serine proteinase, this β-strand is longer than usually observed in the homologous enzymes (Fig. 3). It has been suggested that the peptide backbone β-like interaction between the NS3 proteinase strand E2 and the substrates continues for three more residues, until P6, and that the continuous main chain interaction might compensate for the lack of P2-P5 side-chain to enzyme interactions (LOVE et al. 1996). A basic residue, Arg-161, follows invariantly the end of β-strand E_2 of NS3 (Fig. 3). The side chain of this residue is likely to form a stabilizing ion pair with the side chain of the conserved acidic residue in the substrate P6 position.

Contrary to what one might have expected on the basis of the observed cleavage consensus sequence, both the P1' and P6 positions of NS3 cleavage sites were tolerant to extensive substitutions (BARTENSCHLAGER et al. 1995a; KOLYK-HALOV et al. 1994). In particular, the acidic residue conserved in the P6 position of all cleavage sites was shown not to be essential in any of the cleavage sites. It was noticed however, that additional acidic residues are always present in the P-region of all cleavage sites. It has been thus suggested that an overall negative charge in the P-region of the substrate might be important for cleavage efficiency. In line with this hypothesis, it has been found that the simultaneous substitution of both P6 and P5 acidic residues present in the NS5A/NS5B junction completely abolishes cleavage (KOMODA et al. 1994b).

The studies just described relied on the expression of mutated polyproteins in cellular or cell-free in vitro translation systems. The substrate specificity of the NS3-NS4A proteinase was also studied in detail using purified recombinant enzyme and synthetic peptide substrates. For efficient activity on peptide substrates the NS3-NS4A proteinase requires at least decamer peptide substrates spanning P6-P4' (STEINKÜHLER et al. 1996b), suggesting recognition of the P1'-P4' region by the enzyme. Alanine scanning experiments performed on peptide substrates based either on the NS4A/NS4B (URBANI et al. 1997) or on the NS5A/NS5B (ZHANG et al. 1997) cleavage sites led to the conclusion that, besides P1, also P6, P3 and P4' residues contribute to substrate recognition and cleavage by the NS3-NS4A proteinase.

Based on these kinetic studies, it can be concluded that binding of substrates to the NS3-NS4A proteinase active site is mediated by multiple, possibly cooperative, weak interactions involving both the peptide main chain and the side chains of residues spanning from P6 through P4'. Most of the substrate binding energy is provided by the interaction of the enzyme with the P side of the substrate. Interaction with the P' portion of the substrate, although less important in quantitative terms, is also crucial for recognition. The subsites of the enzyme that are presumed to interact with this part of the substrate may form only in the presence of the NS4A cofactor, explaining, at least in part, the need for a cofactor by the NS3 serine proteinase. Ultimately, the efficiency with which the bound substrate will proceed through the transition state is strongly influenced by the nature of the residue in the P1 position (URBANI et al. 1997). This unique substrate recognition mechanism, mediated by an extended network of interaction, is likely to make the development of low molecular weight inhibitors of the NS3 proteinase a very difficult task.

3.5 The Metal Binding Site of the NS3-NS4A Proteinases

Sequence alignments of the different HCV genotypes and of the related GB viruses A, B and C identified three strictly conserved cysteines and one histidine residue in the NS3 proteinase domain. In a homology model of the HCV NS3 proteinase domain the conserved residues were found to cluster in space, leading to the hypothesis that they could serve as ligands of a metal binding site (DE FRANCESCO et al. 1996). This prediction was confirmed experimentally by the finding that the purified NS3 proteinase domain contains stoichiometric amounts of zinc (De FRANCESCO et al. 1996; STEMPNIAK et al. 1997). Although the metal binding site is remote from the enzyme active site and it is unlikely to play a role in catalysis, zinc was found to be instrumental to the enzyme's function. Thus, zinc addition was shown to enhance the proteinase activity of an in vitro translated NS3 protein (STEMPNIAK et al. 1997). Furthermore, zinc addition was required for the production of soluble NS3 proteinase in *E. coli* grown in defined media (DE FRANCESCO et al. 1996). Also, in vitro refolding of the purified enzyme was dependent on the presence of this metal (URBANI et al. 1998). Both Co(II) and Cd(II) could be incorporated in the zinc site and a spectroscopic characterization of the resulting Co- and Cd-substituted enzymes was consistent with the prediction of tetracoordinate ligation of the metal (DE FRANCESCO et al. 1996). X-ray crystallography later confirmed the assignment of the metal ligands: the zinc ion was shown to be tetrahedrally coordinated by Cys-97, Cys-99 and Cys-145 and, through a bridging water molecule, by His-149 (LOVE et al. 1996; KIM et al. 1996; YAN et al. 1998).

This indirect role of His-149 is consistent with the relatively weak effects of mutations in this position. Whereas substitutions of either cysteine are incompatible with enzymatic activity (HIJIKATA et al. 1993), mutagenesis of His-149 into alanine leads to an active enzyme, even though with impaired catalytic properties and characterized by decreased kinetics of metal incorporation (STEMPNIAK et al. 1997; URBANI et al. 1998). An NMR study of the metal binding site has shown that His-149 is rather flexible, shuttling in a pH-dependent way between a "closed" conformation, in which it participates in metal chelation, and an "open" conformation, leading to solvent exposure of the zinc (URBANI et al. 1998). This flexibility also finds support in one of the published crystal structures (LOVE et al. 1996) in which His-149 participates in metal chelation in only two of the three monomers in the asymmetric unit of the crystal, moving away in the third. Rearrangements of the metal coordination geometry have also been detected spectroscopically upon binding of the enzyme to its NS4A co-factor (URBANI et al. 1998). Anions such as cyanide or azide have been found to replace the water ligand, thereby perturbing the metal coordination. These perturbations go along with alterations of the specific activity of the NS3 proteinase (URBANI et al. 1998). This is suggestive of an indirect influence of the metal coordination geometry on the active site conformation. This influence can be rationalized by considering the topological location of the zinc binding site. The residues participating in metal coordination are positioned on a long loop connecting the two domains of NS3 and in a hairpin loop situated in the second domain. Conformational perturbations in this region can be expected to affect the relative orientation of the two domains and, consequently, the positioning of the residues constituting the catalytic machinery of the proteinase. In fact, many extracellular proteinases possess disulfide bridges that are believed to stabilize the relative orientation of the residues involved in catalysis. Disulfide bridges are unlikely to be stable in the reducing intracellular environment and have been substituted, in the serine proteinases of HCV-related viruses, but also in the picornavirus 2A proteinases, by metal centers presumably serving the same purpose.

The finding of a metal binding site in the HCV NS3 proteinase domain has raised the question of its involvement in the proteolytic processing of the NS2/NS3 cleavage site. This processing event is catalyzed by a viral-encoded enzymatic function that requires most of the NS2 protein and the NH_2-terminal portion of NS3. Cleavage is not dependent on the NS3 serine proteinase activity but requires the presence, within the NS2/NS3 precursor, of the NS3 proteinase domain, and cannot be substituted by other parts of the HCV polyprotein (GRAKOUI et al. 1993; HIJIKATA et al. 1993; REED et al. 1995; SANTOLINI et al. 1995). Processing at the NS2/NS3 site was inhibited by metal-chelating agents and restored by subsequent addition of either Zn or Cd (HIJIKATA et al. 1993; PIERONI et al. 1997), leading to the suggestion that this junction was processed by a zinc-dependent autoproteinase. Extensive mutagenesis studies have shown that both the residues participating in zinc chelation within NS3 as well as a cysteine and a histidine residue in NS2 were required for this cleavage activity (HIJIKATA et al. 1993). The metal-dependence of the processing at the NS2/NS3 junction may therefore reflect either the requirement of a native NS3 proteinase domain and hence of the structural integrity of its zinc binding site, or the necessity for a metal ion directly involved in the mechanism of catalysis. In the latter case, the question arises whether the putative catalytic metal is distinct from the metal contained in NS3. The metal binding site of NS3 is positioned in the vicinity of the NH_2-terminal of the protein, which arises by processing of the NS2/NS3 junction. It could be speculated that this location, the peculiar ligand array around the metal and its structural flexibility may serve the dual role of both participating in the mechanism of processing of the NS2/NS3 junction and of providing conformational stability to the mature NS3 protein.

References

Bartenschlager R, Ahlborn-Laake L, Mous J, Jacobsen H (1993) Nonstructural protein 3 of the hepatitis C virus encodes a serine-type proteinase required for cleavage at the NS3/4 and NS4/5 junctions. J Virol 67:3835–3844
Bartenschlager R, Ahlborn Laake L, Mous J, Jacobsen H (1994) Kinetic and structural analyses of hepatitis C virus polyprotein processing. J Virol 68:5045–5055
Bartenschlager R, Ahlborn-Laake L, Yasargil K, Mous J, Jacobsen H (1995a) Substrate determinants for cleavage in cis and in trans by the hepatitis C virus NS3 proteinase. J Virol 69:198–205
Bartenschlager R, Lohman V, Wilkinson T, Koch JO (1995b) Complex formation between the NS3 serine-type proteinase of the hepatitis C virus and NS4A and its importance for polyprotein maturation. J Virol 69:7519–7528
Bartenschlager R (1997) Molecular targets in inhibition of the hepatitis C virus replication. Antiviral Chem Chemother 8:281–301
Bazan FJ, Fletterick RJ (1989) Detection of a trypsin-like serine protease domain in Flavivirus and Pestivirus. Virology 171:637–639
Bianchi E, Urbani A, Biasiol G, Brunetti M, Pessi A, De Francesco R, Steinkühler C (1997) Complex formation between the hepatitis C virus serine proteinase and a synthetic NS4A cofactor peptide. Biochemistry 36:7890–7897
Butkiewicz NJ, Wendel M, Zhang R, Jubin R, Pichardo J, Smith EB, Hart AM, Ingram R, Durkin J, Mui PW, Murray MG, Ramanathan L, Dasmahapatra B (1996) Enhancement of hepatitis C virus NS3 proteinase activity by association with NS4A-specific synthetic peptides: identification of sequence and critical residues of NS4A for the cofactor activity. Virology 225:328–338
Chambers TJ, Weir RC, Grakoui A, McCourt DW, Bazan JF, Fletterick RJ, Rice CM (1990) Evidence that the N-terminal domain of nonstructural protein NS3 from yellow fever virus is a serine protease responsible for site-specific cleavages in the viral polyprotein. Proc Natl Acad Sci USA 87:8898–8902
De Francesco R, Urbani A, Nardi MC, Tomei L, Steinkühler C, Tramontano A (1996) A zinc binding site in viral serine proteinases. Biochemistry 35:13282–13287
Eckart MR, Selby M, Masiarz F, Lee C, Berger K, Crawford K, Kuo G, Houghton M, Choo QL (1993) The hepatitis C virus encodes a serine proteinase involved in processing of the putative nonstructural proteins from the viral polyprotein precursor. Biochem Biophys Res Commun 192:399–406
Failla C, Tomei L, De Francesco R (1994) Both NS3 and NS4A are required for proteolytic processing of hepatitis C virus nonstructural proteins. J Virol 68:3753–3760
Failla C, Tomei L, De Francesco R (1995) An amino-terminal domain of the hepatitis C virus NS3 proteinase is essential for interaction with NS4A. J Virol 69:1769–1777
Failla C, Pizzi E, De Francesco R, Tramontano A (1996) Redesigning the substrate specificity of the hepatitis C virus proteinase. Folding Design 1:35–42
Grakoui A, McCourt DW, Wychowski C, Feinstone SM, Rice C (1993) Characterization of the hepatitis C virus-encoded serine proteinase: determination of proteinase-dependent polyprotein cleavage sites. J Virol 67:2832–2843
Hahm B, Han DS, Back SH, Song OK, Cho MJ, Kim CJ, Shimotohno K, Jang SK (1995) NS3–4A of hepatitis C virus is a chymotrypsin-like proteinase. J Virol 69:2534–2539
Han DS, Hahm B, Rho HM, Jang SK (1995) Identification of the serine proteinase domain in NS3 of the hepatitis C virus. J Gen Virol 76:985–993

Hijikata M, Mizushima H, Tanji Y, Komoda Y, Hirowatari Y, Akagi T, Kato N, Kimura K, Shimotohno K (1993a) Proteolytic processing and membrane association of putative nonstructural proteins of hepatitis C virus. Proc Natl Acad Sci USA 90:10733–10737

Hijikata M, Mizushima H, Akagi T, Mori S, Kakiuchi N, Kato N, Tanaka T, Kimura K, Shimotohno K (1993b) Two distinct proteinase activities required for the processing of a putative nonstructural precursor protein of hepatitis C virus. J Virol 67:4665–4675

Houghton M (1996) Hepatitis C Viruses. In: Fields BN, Knipe DM, Howley PM (eds) Fields' virology, 3rd edn. Lippincott-Raven, Philadelphia, pp 1035–1058

Kim JL, Morgenstern KA, Lin C, Fox T, Dwyer MD, Landro JA, Chambers SP, Markland W, Lepre CA, O'Malley ET, Harbeson SL, Rice CM, Murcko MA, Caron PR, Thomson JA (1996) Crystal structure of the hepatitis virus NS3 proteinase domain complexed with a synthetic NS4A cofactor peptide. Cell 87:343–355

Koch JO, Lohmann V, Herian U, Bartenschlager R (1996) In vitro studies on the activation of the hepatitis C virus NS3 proteinase by the NS4A cofactor. Virology 221:54–66

Koch JO, Bartenschlager R (1997) Determinants of substrate specificity in the NS3 serine proteinase of the hepatitis C virus. Virology 237:78–88

Kolykhalov AA, Agapov EV, Rice CM (1994) Specificity of the hepatitis C virus NS3 serine proteinase: effects of substitutions at the 3/4A, 4A/4B, 4B/5A, and 5A/5B cleavage sites on polyprotein processing. J Virol 68:7525–7533

Komoda Y, Hijikata M, Sato S, Asabe SI, Kimura K, Shimotohno K (1994b) Substrate requirements of hepatitis C virus serine proteinase for intermolecular polypeptide cleavage in *Escherichia coli*. J Virol 68:7351–7357

Landro JA, Raybuck SA, Luong YPC, O'Malley ET, Harbeson SL, Morgenstern KA, Rao G, Livingston DJ (1997) Mechanistic role of an NS4A peptide cofactor with the truncated NS3 proteinase of hepatitis C virus: elucidation of the NS4A stimulatory effect via kinetic analysis and inhibitor mapping. Biochemistry 36:9340–9348

Leinbach SS, Bhat RA, Xia SM, Hum WT, Stauffer B, Davis A, Hung PP, Mizutani S (1994) Substrate specificity of the NS3 serine proteinase of hepatitis C virus as determined by mutagenesis at the NS3/NS4A junction. Virology 204:163–169

Lesk AM, Fordham WD (1996) Conservation and variability in the structure of serine proteinases of the chymotrypsin family. J Mol Biol 258:501–537

Lin C, Pragai BM, Grakoui A, Xu J, Rice CM (1994b) Hepatitis C virus NS3 serine proteinase: trans-cleavage requirements and processing kinetics. J Virol 68:8147–8157

Lin C, Rice CM (1995) The hepatitis C virus NS3 proteinase and NS4A co-factor: establishment of a cell-free trans-processing assay. Proc Natl Acad Sci USA 92:7622–7626

Lin C, Thomson JA, Rice CM (1995) A central region in the hepatitis C virus NS4A protein allows formation of an active NS3-NS4A serine proteinase complex in vivo and in vitro. J Virol 69:4373–4380

Love RA, Parge HE, Wickersham JA, Hostomsky Z, Habuka N, Moomaw EW, Adachi T, Homstomska Z (1996) The crystal structure of hepatitis C virus NS3 proteinase reveals a trypsin-like fold and a structural zinc binding site. Cell 87:331–342

Manabe S, Fuke I, Tanishita O, Kaji C, Gomi Y, Yoshida S, Mori C, Takamizawa A, Yosida I, Okayama H (1994) Production of nonstructural proteins of hepatitis C virus requires a putative viral proteinase encoded by NS3. Virology 198:636–644

Markland W, Petrillo RA, Fitzgibbon M, Fox T, McCarrick R, McQuaid T, Fulghum JR, Chen W, Fleming MA, Thomson JA, Chambers SP (1997) Purification and characterization of the NS3 serine proteinase domain of hepatitis C virus expressed in Saccharomyces cerevisiae. J Gen Virol 78:39–43

Miller RH, Purcell RH (1990) Hepatitis C virus shares amino acid sequence similarity with pestiviruses and flaviviruses as well as members of two plant virus supergroups. Proc Natl Acad Sci USA 87:2057–2061

Mori A, Yamada K, Kimura J, Koide T, Yuasa S, Yamada E, Miyamura T (1996) Enzymatic characterization of purified NS3 serine proteinase of hepatitis C virus expressed in *Escherichia coli*. FEBS Lett 378:37–42

Neddermann P, Tomei L, Steinkühler C, Gallinari P, Tramontano A, De Francesco R (1997) The nonstructural proteins of the hepatitis C virus: structure and functions. Biol Chem 378:469–476

Pieroni L, Santolini E, Fipaldini C, Pacini L, Migliaccio G, La Monica N (1997) In vitro study of the NS2-3 proteinase of hepatitis C virus. J Virol 71:6373–6380

Pizzi E, Tramontano A, Tomei L, La Monica N, Failla C, Sardana M, Wood T, De Francesco R (1994) Molecular model of the specificity pocket of the hepatitis C virus proteinase: implications for substrate recognition. Proc Natl Acad Sci USA 91:888–892

Reed KE, Grakoui A, Rice CM (1995) Hepatitis C virus-encoded NS2–3 proteinase: cleavage-site mutagenesis and requirements for bimolecular cleavage. J Virol 69:4127–4136

Rost B, Casadio R, Fariselli P, Sander C (1995) Transmembrane helices predicted at 95% accuracy. Protein Sci 4:521–533

Santolini E, Pacini L, Fipaldini C, Migliaccio G, Monica N (1995) The NS2 protein of hepatitis C virus is a transmembrane polypeptide. J Virol 69:7461–7471

Satoh S, Tanji Y, Hijikata M, Kimura K, Shimotohno K (1995) The N-terminal region of hepatitis C virus nonstructural protein 3 (NS3) is essential for stable complex formation with NS4A. J Virol 69:4255–4260

Shimizu Y, Yamaji K, Masuho Y, Yokota T, Inoue H, Sudo K, Satoh S, Shimotohno K (1996) Identification of the sequence of NS4A required for enhanced cleavage of the NS5A/5B site by hepatitis C virus NS3 proteinase. J Virol 70:127–132

Shoji I, Suzuki T, Chieda S, Sato M, Harada T, Chiba T, Matsuura Y, Miyamura T (1995) Proteolytic activity of NS3 serine proteinase of hepatitis C virus efficiently expressed in *Escherichia coli*. Hepatology 22:1648–1655

Simmonds P, Mutimer D, Follet EAC (1998) Hepatitis C. In: Collier L, Balows A, Sussman M (eds) Microbiology and microbial infections. Oxford University Press, New York

Steinkühler C, Tomei L, De Francesco R (1996a) In vitro activity of hepatitis C virus proteinase NS3 purified from recombinant baculovirus-infected Sf9 cells. J Biol Chem 271:6367–6373

Steinkühler C, Urbani A, Tomei L, Biasiol G, Sardana M, Bianchi E, Pessi A, De Francesco R (1996b) Activity of purified hepatitis C virus proteinase NS3 on peptide substrates. J Virol 70:6694–6700

Stempniak M, Hostomska Z, Nodes BR, Hostomsky Z (1997) The NS3 proteinase domain of hepatitis C virus is a zinc-containing enzyme. J Virol 71:2881–2886

Suzuki T, Sato M, Chieda S, Shoji I, Harada T, Yamakawa Y, Watabe S, Matsuura Y, Miyamura T (1995) In vivo and in vitro trans-cleavage activity of hepatitis C virus serine proteinase expressed by recombinant baculoviruses. J Gen Virol 76:3021–3029

Tanji Y, Hijikata M, Hirowatari Y, Shimotohno K (1994a) Hepatitis C virus polyprotein processing: kinetics and mutagenic analysis of serine proteinase-dependent cleavage. J Virol 68:8418–8422

Tanji Y, Hijikata M, Hirowatari Y, Shimotohno K (1994b) Identification of the domain required for trans-cleavage activity of hepatitis C viral serine proteinase. Gene 145:215–219

Tanji Y, Hijikata M, Satoh S, Kaneko T, Shimotohno K (1995a) Hepatitis C virus-encoded nonstructural protein NS4A has versatile functions in viral protein processing. J Virol 69:1575–1581

Tomei L, Failla C, Santolini E, De Francesco R, La Monica N (1993) NS3 is a serine proteinase required for processing of hepatitis C virus polyprotein. J Virol 67:4017–4026

Tomei L, Failla C, Vitale RL, Bianchi E, De Francesco R (1996) A central hydrophobic domain of the hepatitis C virus NS4A protein is necessary and sufficient for the activation of the NS3 proteinase. J Gen Virol 77:1065–1070

Urbani A, Bianchi E, Narjes F, Tramontano A, De Francesco R, Steinkühler C, Pessi A (1997) Substrate specificity of the hepatitis C virus serine proteinase NS3. J Biol Chem 272:9204–9209

Urbani A, Bazzo R, Nardi MC, Cicero D, De Francesco R, Steinkühler C, Barbato G (1998) The metal binding site of the hepatitis C virus NS3 proteinase. A spectroscopic study. J Biol Chem 273:18760–18769

Vishnuvardan D, Kakiuchi N, Urvil PT, Shimotohno K, Kumar PKR, Nishikawa S (1997) Expression of highly active recombinant NS3 proteinase domain of hepatitis C virus in *E. coli*. FEBS Lett 402:209–212

Yan Y, Li Y, Munshi S, Sardana V, Cole J, Sardana M, Steinkühler C, Tomei L, De Francesco R, Kuo L, Chen Z (1998) Complex of NS3 proteinase and NS4A peptide of BK strain hepatitis C virus: a 2.2Å resolution structure in a hexagonal crystal form. Protein Sci 7:837–847

Zhang R, Durkin J, Windsor WT, McNemar C, Ramanathan L, Le HV (1997) Probing the substrate specificity of hepatitis C virus NS3 serine proteinase by using synthetic peptides. J Virol 71:6208–6213

Structure and Function of Hepatitis C Virus NS3 Helicase

A.D. Kwong, J.L. Kim, and C. Lin

1	Introduction	171
2	Genomic Organization of Helicase Domain in NS3/4A	173
2.1	Classification of NS3 Helicase	173
2.2	Mapping of the NS3 Helicase Domain	173
3	Biochemical Properties of NS3 Helicase	175
3.1	Basal and Nucleic Acid Stimulated-NTPase Activity	175
3.2	RNA Binding Activity	176
3.3	Unwinding Activity	177
3.4	Salt Sensitivity	178
4	Function of the Conserved Motifs in NS3 Helicase	178
4.1	Motif I	178
4.2	Motif II	179
4.3	Motif III	180
4.4	Motif VI	180
5	Three-Dimensional Structure of NS3 Helicase	180
5.1	Structure of Helicase Conserved Sequence Motifs	182
5.2	Structure of the ATP Binding Site	184
5.3	Structure of the Nucleic Acid Binding Site	184
6	Structure-Directed Mutagenesis of Hepatitis C Virus Helicase Motif VI	185
7	Structure-Based Mutagenesis of the Hepatitis C Virus Helicase Oligonucleotide-Binding Site	187
8	Models of Unwinding	187
9	Hepatitis C Virus Helicase Antiviral Drug Development	189
9.1	Helicase Targets for Small Molecule Inhibitors	190
9.2	High Throughput Helicase Assays	190
9.3	Herpes Simplex Virus Helicase Inhibitors as a Proof of Concept	191
10	Summary	191
References		191

1 Introduction

Hepatitis C virus (HCV) is a positive-stranded RNA virus with a linear RNA genome approximately 9.6kb in size (Choo et al. 1989). This genome encodes a

---

Vertex Pharmaceuticals, Inc., 130 Waverly Street, Cambridge, MA 02139, USA

single polyprotein of approximately 3010 amino acids and at least ten viral proteins: NH$_2$-C-E1-E2-p7-NS2-NS3-NS4A-NS4B-NS5A-NS5B-COOH (CHOO et al. 1991; KAITO et al. 1994; RICE 1996; TAKAMIZAWA et al. 1991). The individual proteins are released from the polyprotein by both host signal peptidases and viral proteases (GRAKOUI et al. 1993; HIJIKATA et al. 1991; LIN et al. 1994 and reviewed by Reed and Rice elsewhere in this volume (pp 55–84)). Some or all of the nonstructural proteins (NS2, NS3, NS4A, NS4B, NS5A and NS5B) are believed to interact to form the viral replication machinery (HOUGHTON 1996; RICE 1996).

HCV NS3 is a 631 residue multi-functional enzyme that consists of a serine protease domain in the NH$_2$-terminal 181 amino acids, and a nucleic acid-stimulated NTPase and helicase domain in the COOH-terminal 450 amino acids. NS3 protease cleaves at the NS3/NS4A junction and forms a tight, non-covalent complex with NS4A which is necessary for efficient processing of the remaining polyprotein (BARTENSCHLAGER et al. 1995; BUTKIEWICZ et al. 1996; FAILLA et al. 1994; TANJI et al. 1995; and reviewed by DeFrancesco and Steinkuhler elsewhere in this volume (pp 149–170)). The structure of the serine protease domain has been determined in the absence (LOVE et al. 1996) and presence of a synthetic NS4A cofactor peptide (KIM et al. 1996). These X-ray crystallographic analyses have shown that NS4A is deeply buried in the core of NS3, where it stabilizes the active conformation of the NS3/4A protease (KIM et al. 1996). No evidence exists to suggest that the serine protease domain and helicase domains are separated by proteolytic processing in vivo. The two enzymes may be permanently linked fortuitously or because of a functional interdependence between the two domains. The effect of poly(U) on NS3 protease activity has been contradictory. MORGANSTERN et al. (1997) observed a five-fold stimulation using NS3/4A, whereas GALLINARI et al. (1998) observed inhibition of full length NS3 protease activity.

All pesti- and flaviviruses contain conserved helicase motifs in their homologous NS3 proteins, suggesting that helicases play an important role in the life cycle of these viruses (KADARÉ and HAENNI 1997; MILLER and PURCELL 1990). Helicases are enzymes which can unwind double-stranded regions of DNA or RNA in an NTP (usually ATP)-dependent manner. Although a genetic knockout experiment to test the essentiality of NS3 helicase in the background of an infectious HCV cDNA clone (KOLYKHALOV et al. 1997; YANAGI et al. 1997, 1998) has not been reported, HCV RNA helicase activity is thought to be essential for virus growth. In order for HCV to replicate, negative-stranded RNA must be synthesized using the incoming positive-stranded RNA as the template. The negative-stranded replicative intermediate is then used as a template to synthesize positive-stranded progeny RNA, which is packaged into viral capsids. Because the positive and negative RNA strands are complementary, NS3 RNA helicase is thought to be required for strand separation. Additional functions of the NS3 RNA helicase in HCV replication might include the melting of secondary structures of positive-strand RNA in order to increase translational efficiency of the polyprotein or to allow the access of viral replicase template to RNA in highly stable secondary structures such as the internal ribosome entry site (IRES) element and the 98-bp 3′X element RNA (KOLYKHALOV et al. 1996; TANAKA et al. 1996).

2 Genomic Organization of Helicase Domain in NS3/4A

2.1 Classification of NS3 Helicase

Based on sequence comparisons, three superfamilies (SF) of helicase have been defined, each of which include cellular and viral DNA and RNA helicases (GORBALENYA and KOONIN 1993; GORBALENYA et al. 1989). HCV NS3 helicase and related NS3-like proteins in the poty-, flavi-, and pestivirus families have been classified in helicase superfamily-2 (SF2) on the basis of sequence analyses of seven highly conserved helicase amino acid motifs (GORBALENYA and KOONIN 1993; KADARÉ and HAENNI 1997). Cellular helicases in SF2 are involved in many processes including translation (eIF-4A), RNA processing and transcription (human helicase A), and DNA repair (UvrB, Rad-3 and Ercc-3). As elegantly reviewed by KADARÉ and HAENNI (1997), viruses which encode SF2 helicases include vaccinia virus (SHUMAN 1993), herpes simplex virus (HSV) (MARTINEZ et al. 1992), plum pox potyvirus (PPV) (LAIN et al. 1991a,b), tamarillo mosaic virus (TaMV) and potyvirus (EAGLES et al. 1994). In addition to HCV NS3 helicase, other RNA helicases of the SF2 family were predicted in the homologous NS3 proteins of the pestivirus such as bovine viral diarrhea virus (BVDV) (WARRENER and COLLETT 1995), or in flaviviruses such as West Nile virus (WNV) (WENGLER and WENGLER 1991), yellow fever virus (YFV) (WARRENER et al. 1993), hepatitis G virus (HGV) (LAXTON et al. 1998), and Japanese encephalitis (JE) virus (KUO et al. 1996). HGV is the most closely related to HCV and encodes a helicase which has been shown to have poly(U)-stimulated ATPase activity and DNA helicase activity (LAXTON et al. 1998).

2.2 Mapping of the NS3 Helicase Domain

Both the NS3/4A complex and the full length NS3 protein have been demonstrated to unwind double-stranded nucleic acid (GALLINARI et al. 1998; HONG et al. 1996; MORGENSTERN et al. 1997). As shown in Table I, using a variety of assay conditions, investigators have mapped the NS3 nucleic acid-stimulated NTPase and duplex nucleic acid unwinding activity to the COOH-terminal portion of NS3, downstream of the serine protease domain. The minimal sequence of NS3 required for helicase activity starts adjacent to the protease domain and is about 400 amino acids long (KIM et al. 1997a).

Many studies have demonstrated that the serine protease (LANDRO et al. 1997; MARKLAND et al. 1997; STEINKÜHLER et al. 1996) and RNA helicase domains (GALLINARI et al. 1998; KIM et al. 1997a; TAI et al. 1996) of NS3 can be expressed independently as catalytically active species. Some evidence suggests that the NS3 protease domain may have a slight modulating effect on NS3 helicase activity. Examples include differences in pH optima of ATPase and RNA unwinding activities between the native NS3/4A protein complex (HONG et al. 1996; MORGEN-

Table 1. Comparison of various HCV helicase proteins reported in the literature

HCV strain	Polyprotein (NS3) position	Number of residues	HCV helicase assay conditions	Reference
1a	1027–1657 (1–631)+4A	631	100mM MES (pH 6.0); 2.5mM ATP; 1.0mM MgCl$_2$; 2.0mM DTT; 1U prime RNase inhibitor; 100μg/ml BSA; 10–30fmol NS34A; 50–250fmol ds RNA substrate Incubate 30min at 37°C	Hong et al. 1996
1a	1027–1657 (1–631)+4A	631	25mM HEPES (pH 7.5); 5.0mM ATP; 3.0mM MgCl$_2$; 1.0mM DTT; 10U RNasin; 100μg/ml BSA; ~100fmol NS34A; ~1.0pmol dsRNA substrate Incubate 10min at 37°C	Morgenstern et al. 1997
1b	1027–1657 (1–631)	631	25mM MOPS (pH 7.0); 3.0mM ATP; 3.0mM MnCl$_2$; 2.5mM DTT; 2.5U RNasin; 100μg/ml BSA; 1.25–80nM NS3; 1nM dsRNA substrate Incubate 30min at 37°C	Gallinari et al. 1998
1b	1176–1657 (150–631)	482	20mM HEPES (pH 7.0); 2.5mM ATP; 1.5mM MnCl$_2$; 2.0mM DTT; 2.5U RNasin; 100μg/ml BSA; ~20pmol NS3; 0.66pmol dsRNA substrate Incubate 30min at 37°C	Tai et al. 1996
1a	1192–1657 (165–631)	466	25mM MOPS (pH 6.5); 5.0mM ATP; 3.0mM MnCl$_2$; 2.0mM DTT; 2.5U RNasin 100ug/ml BSA; 1.0pmol NS3; 0.5pmol dsRNA substrate Incubate 30min at 37°C	Kim et al. 1995; Gwack et al. 1996; Kim et al. 1997b
1b	1193–1657 (166–630)	465	50mM MOPS (pH 7.0); 5mM ATP; 3mM MgCl$_2$ 200nmol NS3; 2nmol dsRNA substrate Incubate 30min at 37°C	Preugschat et al. 1996
	1193–1615 (166–589)	423	25mM MOPS (pH 6.8); 5.0mM ATP; 3.0mM MgCl$_2$; 2.0mM DTT 5U RNasin; 100μg/ml BSA; 1–8nM NS34A; 50pM dsDNA substrate Incubate 30min at 37°C	Heilek and Peterson 1997

Table 1. (*Cont.*)

1a	1207–1612 (181–586)	406	25mM MOPS (pH 6.5); 5.0mM ATP; 3.0mM MnCl$_2$; 2.0mM DTT 2.5U RNasin; 100µg/ml BSA; 0.5pmol NS3; 2µl dsRNA substrate Incubate 30min at 37°C	JIN and PETERSON 1995
1a	1209–1608 (182–581)	400	25mM MOPS (pH 6.5); 5.0mM ATP; 3.0mM MnCl$_2$; 2.0mM DTT 2.5U RNasin; 100µg/ml BSA; 1pmol NS3; 0.5pmol dsRNA substrate Incubate 30min at 37°C	KIM et al. 1997a

The NH$_2$-terminal 181 residues of NS3, together with NS4A, consists of the serine protease, while the COOH-terminal 450 amino acids of the NS3 protein contains the NTPase and helicase activities. The table summarizes the strain type, polypeptide length, the number of amino acid, helicase unwinding assay condition, and the corresponding reference for each of these helicase proteins expressed and reported in the literature.

STERN et al. 1997) and a NS3 helicase domain (GWACK et al. 1996; JIN and PETERSON 1995; KIM et al. 1995; PREUGSCHAT et al. 1996; TAI et al. 1996). Similarly, the ATPase activities of the two proteins differ in their sensitivity to polynucleotide stimulation (GALLINARI et al. 1998; MORGENSTERN et al. 1997). Although full-length NS3 has a lower apparent dissociation constant for poly(U) than the helicase domain, both proteins display nearly indistinguishable kinetic parameters for NTP hydrolysis when stimulated with saturating concentration of polynucleotide (KANAI et al. 1995; MORGENSTERN et al. 1997; PREUGSCHAT et al. 1996; Tai et al. 1996). Both enzymes translocate in the 3′ → 5′ direction when unwinding a polynucleotide substrate; full-length NS3 helicase activity does not substantially differ from that of the helicase domain in optimal pH, temperature and divalent cation reaction conditions (Table 1) (GALLINARI et al. 1998). This suggests that neither NS4A nor the NH$_2$-terminal serine protease domain of NS3 play a significant role in the folding, substrate specificity or activity of the NS3 helicase domain.

3 Biochemical Properties of NS3 Helicase

3.1 Basal and Nucleic Acid Stimulated-NTPase Activity

Suzich and coworkers expressed an insoluble protein in *E. coli*, which contained amino acids 166 to 630 of the NS3 protein. The purified protein was refolded in vitro and could hydrolyze all of the basic NTP's and dNTP's. ATP and dATP were the most preferred substrates, and GTP and dGTP were the poorest substrates. The preferred order of substrates is ATP ≅ dATP > CTP > UTP > dCTP ≅ dTTP

> dGTP ≅ GTP (Suzich et al. 1993). This data was confirmed with NS3/4A enzyme purified from COS cells, which hydrolyzed NTPs with the following order of preference: ATP > dATP > CTP > UTP > GTP ≅ dGTP (MORGENSTERN et al. 1997). In contrast, a bacterially expressed, then refolded helicase domain containing amino acids 180–586 of NS3 was reported to have a different order of preference for NTP hydrolysis: ATP > GTP ≫ UTP and CTP was not hydrolyzed (JIN and PETERSON 1995). The reason for the discrepancy between these studies is unclear. Kinetic analyses of NTPase activity associated with the helicase domain confirmed that the enzyme was not selective for the sugar or the base of the NTP substrate, and could even catalyze the release of phosphate from tripolyphosphate (PREUGSCHAT et al. 1996).

NTPase activity is dependent on the addition of Mg^{+2} or Mn^{+2}; it can be stimulated six to 25-fold over basal level by the addition of various single-stranded and double-stranded oligonucleotides (PREUGSCHAT et al. 1996; SUZICH et al. 1993). There is a marked preference for poly(U) and poly(dU) with the NS3 helicase domain alone, which results in a two- and three-fold increase in activity compared to poly(C) and poly(A), respectively (SUZICH et al. 1993). Similar results were obtained with the NS3/4A complex, where the concentration required to reach half-maximal ATPase activity was ~0.18μM as compared to ~27.5μM and ~80μM for poly(A) and poly(C) respectively (MORGENSTERN et al. 1997). For maximum stimulation of ATP hydrolysis by oligo(rU) and oligo(dU), the oligonucleotide must be larger than 12 or 15 nucleobases. Oligonucleotide binding results in a conformational change, which can be measured by quenching of the intrinsic NS3 protein fluorescence. This was used to determine that the affinity of binding is maximal when the oligoribonucleotide is greater than ten nucleobases in length (PREUGSCHAT et al. 1996). Basal ATPase activity is absolutely dependent on Mg^{+2}, but relatively insensitive to increasing concentrations of Mg^{+2} and KCl, up to 10mM and 200mM, respectively (SUZICH et al. 1993). In contrast, at pH6.5 and 0.35mM ATP, poly(U)-stimulated ATPase activity is inhibited by Mg^{+2} above 2.5mM and by increasing concentrations of KCl.

3.2 RNA Binding Activity

KANAI et al. (1995) expressed a protein which contained the COOH-terminal 16 amino acids of NS2 and the NH_2-terminal 581 amino acids of NS3 which had an apparent dissociation constant for poly(U)-Sepharose resin of $2 \times 10^{-7}M$. Deletion of the NH_2-terminal serine protease domain had no effect on RNA binding activity; however, deletions from any other region of the helicase domain led to a significant reduction in RNA binding activity. Recently, it has been hypothesized that the arginine-rich region of motif VI functions as an RNA binding domain (CHO et al. 1998; YAO et al. 1997). However, two constructs containing the arginine-rich region did not bind effectively to poly(U)-Sepharose beads, suggesting that the region might not be responsible for RNA binding activity in the whole NS3 protein (KANAI et al. 1995). Binding of NS3 helicase to poly(U)-Sepharose beads was blocked in a dose-dependent manner by the addition of poly(U), but not poly(A) or poly (C). This suggests that NS3 prefers poly(U). Since the 3′ non-coding region (NCR) of the HCV genome contains a polyuridine stretch (KOLYKHALOV et al. 1996; TANAKA et al. 1995, 1996; YAMADA et al. 1996), (reviewed in (BLIGHT et al. 1998)), HCV NS3 helicase might preferentially bind to its own RNA.

Three other methods which have been used to measure RNA-binding activity are a filter-binding, gel-retardation (GWACK et al. 1996), and fluorescence quench assay (PREUGSCHAT et al. 1996). Using a helicase construct containing residues 166–631 of NS3, GWACK et al. (1996) showed that NS3 RNA binding was insensitive to the absence and presence of Mg^{+2} or 100mM KCl, and that the minimal RNA binding size was between 7 nucleotides (nt) and 20nt. This is consistent with data from PREUGSCHAT et al. (1996) which showed that a similarly sized NS3 helicase domain had a maximal affinity for a 10-base oligoribonucleotide and for a 12-base oligodeoxyribonucleotide. Although there was no selectivity for a particular nucleobase (i.e., adenosine vs deoxyadenosine vs guanosine), the presence of a nucleobase was required for effective binding, and polytriphosphate was a poor substrate.

3.3 Unwinding Activity

Kim and co-workers first described NS3 helicase activity using a His-tagged construct corresponding to the COOH-terminal 466 amino acids of NS3 (165–631 amino acids of NS3) (KIM et al. 1995). HCV RNA helicase activity is coupled to NTPase activity, but the hydrolysis of NTP is not coupled to unwinding of double-stranded RNA (PREUGSCHAT et al. 1996). In common with other helicases, NS3 helicase requires the divalent cation (Mg^{+2} or Mn^{+2})-dependent hydrolysis of NTPs to provide the energy for unwinding (GALLINARI et al. 1998; GWACK et al. 1996; JIN and PETERSON 1995; KIM et al. 1995; PREUGSCHAT et al. 1996; SUZICH et al. 1993; TAI et al. 1996). A similar helicase activity has also been described for the closely related NS3 proteins of BVDV (WARRENER and COLLETT 1995) and HGV (LAXTON et al. 1998).

The full-length NS3/4A enzyme purified from the HSV amplicon expression system was shown to require a 3′ tail in order to unwind RNA/RNA substrates, thus establishing a 3′ to 5′ directionality (HONG et al. 1996). Subsequently, the NS3 helicase domain was reported to exhibit the same 3′ to 5′ directionality on RNA/RNA (TAI et al. 1996), RNA/DNA (GWACK et al. 1996), and DNA/DNA substrates (GWACK et al. 1997).

Although helicases have not been shown to demonstrate any sequence specificity in their nucleic acid substrate requirement, the majority of helicases show a marked preference for RNA or DNA substrates. HCV NS3 helicase is capable of unwinding both RNA and DNA homo- and hetero-duplexes (GWACK et al. 1996; GWACK et al. 1997; HEILEK and PETERSON 1997; TAI et al. 1996). Prior to these reports, SV40 large T antigen (SCHEFFNER et al. 1989), nuclear DNA helicase II

(NDH II) (ZHANG and GROSSE 1994), helicase A (LEE and HURWITZ 1992) and vaccinia virus NPH II (BAYLISS and SMITH 1996) had been the only members of the large family of known helicases described to exibit both DNA and RNA helicase activity.

3.4 Salt Sensitivity

The unwinding activity of the NS3 helicase domain is exquisitely sensitive to salt inhibition; consequently maximum activity is obtained in the absence of potassium ion (GWACK et al. 1996, 1997; TAI et al. 1996). Similarly, strand displacement by the full length NS3 was maximal in the absence of sodium ions (GALLINARI et al. 1998). The NS3 helicase sensitivity to inhibition by salt is unusual; most DNA and RNA helicases are stimulated by salt. For example, the optimal salt concentration for Rep52 DNA helicase is $\sim$50–100mM NaCl (SMITH and KOTIN 1998) and 50–100mM KCl for RNA helicase A (LEE and HURWITZ 1992). Since the physiological ionic strength in the cytoplasm is about 150mM of monovalent cation Na^+ and K^+, it is likely that additional factors are needed to ensure the functionality of HCV helicase, and efficient replication of HCV in cells. Most likely, NS3 helicase is associated with other HCV non-structural proteins or host factors as part of a replication complex in vivo, and is associated with the endoplasmic reticulum membrane through NS4A. Either of these scenarios may exclude salt from the immediate environment of the helicase or change the salt sensitivity of the enzyme.

4 Function of the Conserved Motifs in NS3 Helicase

4.1 Motif I

Mutational and structural analyses of the conserved sequence motifs of HCV helicase (shown in Fig. 1) have identified their separate roles in NTP-binding, NTP-hydrolysis, and nucleic acid binding. Motifs I and II, also known as Walker motifs, or boxes, A and B, respectively (WALKER et al. 1982), are the only sequence elements shared by all the three superfamilies of helicases (GORBALENYA and KOONIN 1993). As shown in Fig. 1, motif I contains the conserved GxGKS/T sequence, which is a common motif in many ATP- and GTP-binding proteins (SARASTE et al. 1990). Motif I forms a phosphate binding loop or P loop, which has been shown to be involved in binding the β phosphate of the nucleotide triphosphate (SUBRAMANYA et al. 1996; WENG et al. 1996). Mutation of HCV helicase Lys-210 in this motif to Glu (KIM et al. 1997b), or to Ala or Gln (HEILEK and PETERSON 1997) inhibited both basal and nucleic acid-stimulated ATPase activities, and completely abrogated RNA helicase activity.

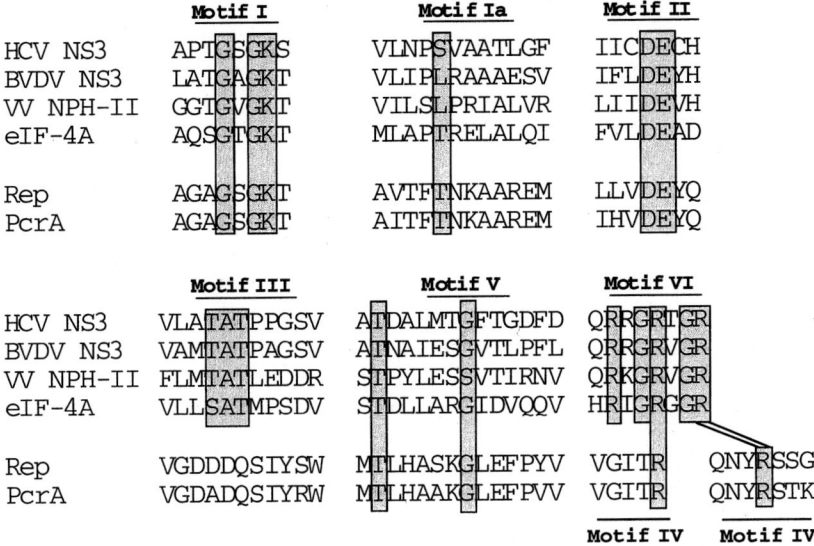

Fig. 1. Alignment of six or seven conserved motifs of the SF1 and SF2 helicases. Sequences of the conserved motifs from four RNA helicases (HCV and BVDV NS3 helicase, vaccinia virus NPH-II helicase, and eucaryotic translation factor eIF4A) of the superfamily 2 and two DNA helicases (*E. coli* Rep helicase and yeast PcrA helicase) of the superfamily 1 are aligned. Residues, which are conserved among four SF2 helicases, are *boxed in shadow*. The *shadowed box* is extended to the SF1 helicases if the conservation of certain residues is observed in Rep and PcrA helicases. It should be noted that Arg-464 and Arg-467 of the HCV helicase are in a homologous position as Arg-610 of motif VI and Arg-287 of motif IV (instead of motif VI) of the PcrA helicase, respectively

4.2 Motif II

Many SF2 helicases were classified as a unique group which contain a DEAD (Asp-Glu-Ala-Asp) box element in motif II. The DEAD box element, which is a variant of the "Walker B" motif, comes in three flavors: DEAD, DEAH, and DEXH. HCV NS3 is a member of the DEXH subfamily and contains the sequence DECH. Motif II is in close proximity to motif I; the first conserved aspartate residue has been shown to bind Mg^{+2} and assist in orienting the Mg^{+2}-ATP substrate for ATP hydrolysis in other kinases (BLACK and HRUBY 1992; PAI et al. 1990; YAN and TSAI 1991). In studies by HEILEK and PETERSON (1997), the mutation of the conserved His-293 residue in the DECH motif to Ala (DECA) in HCV NS3 helicase did not affect ATPase activity, but abolished helicase activity on a M13/DNA substrate. These results are similar to those obtained with vaccinia virus NPH II enzyme in which the RNA helicase activity was abolished by the substitution of alanine for histidine in the DEVH motif (GROSS and SHUMAN 1996). In contrast, in a different study by KIM et al. (1997b), mutation of the same HCV NS3 His-293 residue to Ala decreased poly(U)-stimulated ATPase activity by ~70%, increased poly(U)-independent ATPase activity of approximately tenfold, and had a small effect (~40% reduction) on helicase activity using a RNA:RNA substrate (KIM et al. 1997b). In these experiments the ATPase of the wild type enzyme was stimulated five-fold by the addition of poly(U); in contrast, the basal ATPase activity of the DEVA mutant was inhibited seven-fold by the addition of poly(U) (KIM et al. 1997b).

4.3 Motif III

Motif III contains a conserved $T_{322}ATPP$ sequence which is thought to function as a switch or hinge region to couple conformational changes involved in NTP hydrolysis and unwinding (KADARÉ and HAENNI 1997; SUBRAMANYA et al. 1996; YAO et al. 1997). Mutation of Thr-322 to Ala decreased poly(U)-stimulated ATPase activity approximately fivefold and RNA helicase activity by ~60% (KIM et al. 1997b).

4.4 Motif VI

The role of the $Q_{460}RXGRXGR$ sequence in motif VI in HCV helicase is controversial. Mutagenesis of the corresponding arginine-rich sequences in other helicases has resulted in conflicting results. In the case of plum pox potyvirus (PPV), this region is necessary and sufficient for RNA binding to occur (FERNANDEZ and GARCIA 1996). In contrast, mutations in the corresponding region of eIF-4A resulted in a lost of both RNA binding and ATPase activity (PAUSE et al. 1993), and mutations in vaccinian virus NPH-II protein had no effect on RNA binding, but abolished ATPase and RNA helicase activity (GROSS and SHUMAN 1996). Mutagenesis data to clarify the function of Motif VI can be difficult to interpret because of the requirement of ATP hydrolysis and nucleic acid binding for unwinding activity, and because of the requirement of RNA binding activity for optimal ATP hydrolysis. A significant decrease in ATPase activity was observed when Gln-460 in HCV helicase was mutated to His, Arg-464 to Ala, or Arg-467 to Lys, suggesting that these residues are critical for ATP hydrolysis but not RNA binding (KIM et al. 1997b). These results are consistent with the effect of similar mutations on the corresponding residues in vaccinia virus NPH-II perfomed by Gross and colleagues (GROSS and SHUMAN 1996) who hypothesized that other regions outside the motif VI must bind RNA.

5 Three-Dimensional Structure of NS3 Helicase

The structure of the HCV NS3 helicase in the absence of nucleic acid has been solved to 2.3Å (CHO et al. 1998) and 2.1Å (YAO et al. 1997). In addition, the structure of the helicase domain complexed with a single-stranded DNA oligonucleotide has been solved to 2.2Å resolution (KIM et al. 1998).

As shown in Fig. 2, HCV helicase consists of three structural domains separated by clefts with the oligonucleotide bound in a groove between the first two domains and the third domain. The first two domains have an adenylate kinase-like

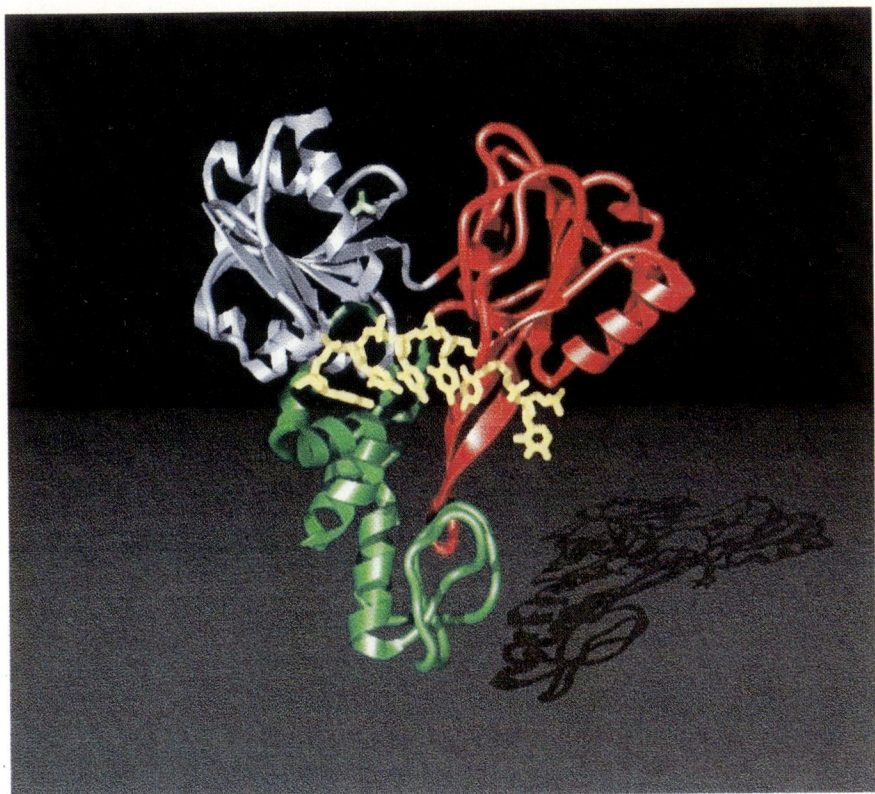

Fig. 2. Fold of HCV NS3 helicase domain complexed with ssRNA. Ribbon diagram illustrating the overall fold of the NS3 helicase with bound dU_8 oligonucleotide. Domain 1 is colored *blue*, domain 2 *red*, and domain 3 *green*. The sulfate ion is shown in *light green* and the DNA oligonucleotide is colored *yellow*

fold, with a phosphate binding loop in the first domain. Domains 1 and 3 share a more extensive interface than either shares with domain 2. The clefts between domains 1 and 2 and domains 2 and 3 are the largest. Domain 2 is flexibly linked to the other two and has been shown to rotate relative to domains 1 and 3 in different crystal forms (YAO et al. 1997).

HCV helicase shows a similar global fold to PcrA DNA helicase from *Bacillus stearothermophilus* (SUBRAMANYA et al. 1996) and *E. coli* Rep DNA helicase of SF1 (KOROLEV et al. 1997) even though there is little shared sequence homology. The structural similarity is found between domains 1A and 2A of Rep DNA helicase and domains 1 and 2 of the HCV helicase, respectively (KOROLEV et al. 1998). It has been suggested that there is preservation of structural scaffold and relationships between helicase motifs across the SF1 and SF2 families of helicase. These homologous domains of PcrA, Rep, and HCV helicases each contain a parallel six-stranded β-sheet flanked by α-helices. In addition, domain 1 of HCV helicase contains a seventh β-strand running anti-parallel to the rest of the sheet. Superposition of domains 1 and 2 of HCV helicase yields an root mean square deviation of 2.0Å for 76 C-α atoms that form the core of each domain. Domain 3 is predominantly α-helical and is associated with domain 2 by a pair of anti-parallel β-strands (Fig. 2).

5.1 Structure of Helicase Conserved Sequence Motifs

NS3 RNA helicase shows high sequence conservation among HCV strains with >80% sequence identity. The most highly conserved segments of these domains correspond to the canonical helicase sequence motifs (GORBALENYA and KOONIN 1993). Mutagenesis studies of individual residues within these motifs in HCV helicase or in other RNA helicases have demonstrated that they are essential for enzyme activity.

As shown in Fig. 3, residues of the conserved motifs correspond to the interdomain cleft between domains 1 and 2. Motifs I and II are both contained in domain 1 which is connected to domain 2 via a flexible linker region corresponding to motif III (CHO et al. 1998; KIM et al. 1998; YAO et al. 1997). Motif Ia extends from the β sheet core of domain 1 to the oligonucleotide binding site. Residues in motif V both contact the oligonucleotide and line the interface between the first two domains. Domain 1 has a similar fold to that of a number of adenosine triphosphate transphosphorylases, such as adenylate and thymidine kinases. In particular, the phosphate binding loop formed by motif I ($GSGK_{210}S$) is virtually identical to the corresponding loop in these kinases where it is involved in binding the β phosphate of ATP. A sulfate ion is found in this exact location in the HCV helicase:dU_8 oligonucleotide binary complex (KIM et al. 1998).

Motif II ($DEXH_{293}$) is proximal to the GXGKS phosphate binding loop and is thought to be involved in binding of the Mg^{+2}-ATP substrate. In adenylate and thymidine kinases, a conserved aspartate binds Mg^{+2}, which helps orient the ATP for nucleophilic attack (BLACK and HRUBY 1992; YAN and TSAI 1991). His-293 is located at the bottom of the interdomain cleft and is essential for the coupling of the ATPase activity to polynucleotide binding.

The conserved arginine residues in the Q_{460}RXGRXGR sequence in motif VI lie on the inner face of domain 2, facing the Mg^{+2}-ATP binding loops in motifs I and II on the inner face of domain 1 (KIM et al. 1998). Studies in several helicases have evaluated the effects of mutations in motif VI (Q_{460}RXGRXGR), however a role for this motif has not been clearly defined. Gln-460 lies at the bottom of the cleft and is thought to interact with His-293 of the $DEXH_{293}$ box found on the opposite side of the cleft on the inner face of domain 1. A similar carbonyl side chain interaction with His is observed in eIF-4A, where the Asp in the DEAD box in motif II is thought to interact with a His in motif VI. Arginines-461, 464 and 467 have been proposed by YAO et al. (1997) to be involved in binding single-stranded RNA in the cleft between domains 1 and 2. However, the structure of HCV helicase complexed with dU_8 in which Arg-461 points away from the cleft and is hydrogen-bonded to Asp-412 and Asp-427 is not consistent with this interpretation (KIM et al. 1998).

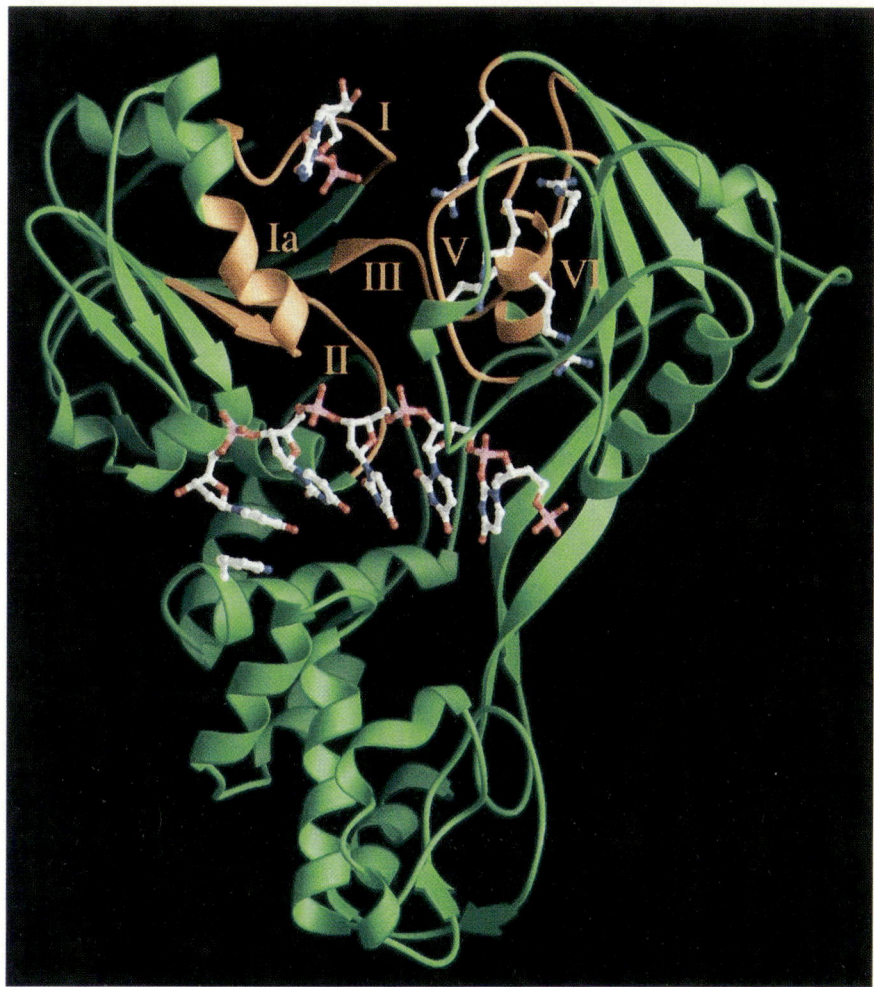

Fig. 3. Location of conserved motifs in the triple complex of HCV NS3 helicase, dU_8 oligonucleotide, and ADP. The orientation of the three domains of the HCV NS3 helicase and DNA oligo is the same as that in Fig. 2. Conserved sequence motifs I through VI, which are located at the interface between domains 1 and 2, are colored *orange* in ribbon and letter. ADP is shown in sticks and binds to the phosphate binding loop or motif I on domain 1. Four arginine residues (Arg-461, Arg-462, Arg-464, and Arg-467) of motif VI (QRRGRTGR) on domain II are shown in sticks. Trp-501 on domain 3, shown in sticks, stacks against the 3′ uridine base of the DNA oligonucleotide

HCV helicase has no region corresponding to motif IV of the SF1 DNA helicases which is responsible for binding the adenosine ring of ATP in Rep and PcrA helicases (KOROLEV et al. 1997; SUBRAMANYA et al. 1996). Mutation of a conserved arginine in this motif in UvrD increases the Km of ATP by 37-fold (HALL and MATSON 1997). In HCV helicase, either another protein segment not in the current structure substitutes for motif IV, or the adenosine ring binds elsewhere. Residues from the putative motif IV in HCV helicase include Ser-370 and Lys-371 which contact the DNA via a water-mediated hydrogen bond and a backbone interaction, respectively. It has been suggested that residues corresponding to the putative HCV helicase motif IV should be designated motif IVa (KIM et al. 1998), which has a different function than motif IV in the SF1 helicases.

5.2 Structure of the ATP Binding Site

The NH_2-terminal region of domain 1 contains a phosphate binding loop that is highly conserved among all helicases and is commonly referred to as the Walker A box or motif I (WALKER et al. 1982). In the structure described by KIM et al. (1998), this loop contained a bound sulfate ion. The phosphate binding loop is structurally similar to those found in a number of other ATPases (SARASTE et al. 1990); the position of the sulfate is similar to that of the β-phosphate of ADP in the crystal structure of the PcrA helicase:ADP complex (SUBRAMANYA et al. 1996). As seen in Fig. 2, the crystal structure of HCV helicase complexed with ADP revealed that the β-phosphate occupies the same position as the bound sulfate. The most conserved residues in the $DECH_{293}$ motif (motif II or Walker box B) are Asp-290 and Glu-291 and their side chains point toward an open area beneath the phosphate binding loop which is thought to be occupied by Mg^{+2} and the γ-phosphate of the bound Mg^{+2}-NTP substrate.

5.3 Structure of the Nucleic Acid Binding Site

As shown in Fig. 2, the single-stranded DNA lies in a channel approximately 16Å in diameter which separates domain 3 from domains 1 and 2. The 5' end of the oligonucleotide resides at the interface of domains 2 and 3 and its 3' end at the interface of domains 1 and 3. This location is consistent with that observed in the 3.0Å resolution structure of Rep DNA helicase complexed with a single-stranded oligonucleotide (KOROLEV et al. 1997), and is roughly perpendicular to the orientation of single-stranded RNA in a model derived from apo HCV helicase structures (CHO et al. 1998; YAO et al. 1997).

In the X-ray structure of the nucleic acid bound HCV helicase (KIM et al. 1998), the oligonucleotide binds in an orthogonal binding site and contacts relatively few conserved residues. The primary interactions between the single-stranded DNA and the HCV helicase are hydrogen bonds directed at the phosphate backbone, but not the base or ribose of the oligonucleotide, as would be expected for a nonspecific protein-nucleic acid complex. The lack of sequence-specific interaction with the oligonucleotide bases is consistent with the independence of HCV helicase from sequences of DNA and RNA homo- and hetero-duplex unwinding substrates.

In the structure of HCV helicase complexed with single-stranded DNA, the oligonucleotide is most tightly bound at the 3' and 5' ends, with few contacts with the central nucleotides. The ends of the oligonucleotide (residues dU_4 and dU_8) are capped by interactions with hydrophobic side chains. Trp-501 stacks with the base of dU_8 while Val-432 interacts with the dU_4 base. These two side-chains act as a pair of bookends of a central binding cavity occupied by five nucleotides. Val-432 and Trp-501 are also completely conserved among all HCV NS3 sequences.

Protein contacts the phosphate backbone primarily through structurally equivalent and symmetrical residues in domains 1 and 2. Thus, the residues in domain 2 (Ser-370 and Thr-411) which interact with the dU_3 and dU_5 backbone phosphates respectively are nearly identical to residues (Ser-231 and Thr-269) in domain 1 which interact with the dU_7 and dU_8 phosphates respectively. Interestingly, Ser-231, Thr-269, Ser-370 and Thr-411 are also absolutely conserved in all the HCV NS3 sequences.

In HCV helicase, an oligonucleotide binds in a channel spanning two protein domains in a manner similar to that seen in replication protein A (RPA) (BOCHKAREV et al. 1997). RPA, single-stranded binding protein (SSB) and aspartyl-tRNA synthetase all contain a nucleic acid binding site termed a L_{45} loop, which consists of anti-parallel strands extending from a protein core (RAGHUNATHAN 1997; RUFF 1991; BOCHKAREV et al. 1997). In RPA, the L_{45} loop binds to the 5' end of the oligonucleotide. As shown in Fig. 2, HCV helicase contains a L_{45} loop related structure in domain 2, where a pair of extended anti-parallel strands bind the 5' end of the oligonucleotide in a manner similar to RPA.

6 Structure-Directed Mutagenesis of Hepatitis C Virus Helicase Motif VI

The role of the arginine rich sequence ($Q_{460}RXGR_{464}XGR_{467}$) in motif VI of domain 2 in HCV helicase has not been clearly defined. Based on independently solved structures of the HCV apo helicase, two groups have postulated that $Q_{460}RXGR_{464}XGR_{467}$ functions as an RNA binding domain (CHO et al. 1998; YAO et al. 1997). In both of these models, single-stranded RNA is predicted to lie in the cleft between domains 1 and 2.

In contrast, based on the crystal structure of the helicase bound with oligonucleotide and the similarity of domains 1 and 2 with adenylate kinase, a third group has proposed that motif VI functions in ATP hydrolysis (KIM et al. 1998). In this model, upon closure of the interdomain 1 and 2 cleft, Arg-464 and Arg-467 would contact the γ and α phosphates of ATP respectively, when ATP is bound in the phosphate binding loop of domain 1. This is similar to the proposed mechanism of ATP hydrolysis for adenylate kinase, which involves contact of the ATP phosphates in domain 1 with conserved basic residues in the second domain. Mutation of the residues corresponding to Arg-464 or Arg-467 to Ala or Gln in vaccinia NPH-II or eIF-4A reduced the ATPase activity to <20% of wild type levels (GROSS and SHUMAN 1996; PAUSE et al. 1993).

In order to further corroborate the helicase: dU_8 structural analysis of the function of motif VI, mutagenesis was performed to determine the contribution of

Table 2. Structure-based mutagenesis of the conserved motif VI of the HCV NS3 helicase

Mutation	Basal ATPase activity	PolyU stimulated ATPase activity[a]	RNA binding	dsRNA/DNA unwinding
Wild-type	100	581	100	100
Q460A	23	32	97	3
R461A	140	193	57	2
R462A	247	337	99	81
R464A	33	21	105	<0.01
R467A	7	14	116	<0.05

Individual mutations were introduced into the HCV helicase using site-directed mutagenesis. The wild-type or mutated HCV NS3 helicase domain proteins were expressed in *E. coli*, purified to homogeneity, and tested for basal ATPase, poly(U)-stimulated ATPase, single-stranded (ss) RNA binding, and RNA/DNA heteroduplex unwinding activities. The activity of each mutant was converted to percentile with the corresponding activity of the wild-type HCV helicase protein as 100%.

[a] Percent of basal ATPase wild type level.

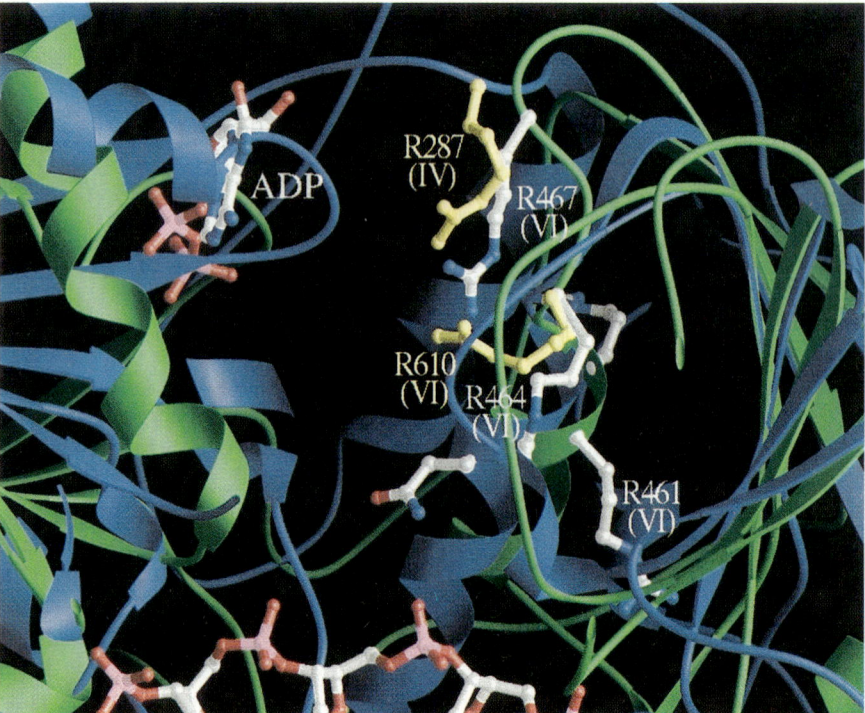

Fig. 4. Superimposition of the interface between domains 1 and 2 of HCV NS3 helicase with the homologous interface of PcrA helicase. The HCV NS3 helicase is colored in *green* and the PcrA helicase is in *blue*. Four residues (Gln-460, Arg-461, Arg-464, and Arg-467) of motif VI (QRRGRTGR) in the HCV NS3 helicase are shown in *white* sticks. Two arginine residues from the PcrA helicase are colored in *yellow*. Arg-464 of the HCV helicase is located in a position homologous to Arg-610 of the PcrA helicase motif VI. Arg-467 of the HCV helicase is aligned very closely with Arg-287 of motif IV (instead of motif VI) of the PcrA helicase

different residues to ATP hydrolysis and nucleic acid binding for unwinding activity, and RNA binding activity for optimal ATP hydrolysis. As shown in Table 2, mutation of Arg-464 or Arg-467 to alanine decreased ATP hydrolysis and completely inhibited unwinding activity, but had no effect on single-stranded RNA binding. This data is consistent with Arg-464 and Arg-467 playing a role in ATP hydrolysis. As shown in Fig. 4, superimposition of the interdomain face of domain 2 of HCV NS3 helicase and PcrA DNA helicase revealed that Arg-610 from motif VI of PcrA is in a comparable position to Arg-464 of HCV helicase. Similarly, Arg-287 of PcrA DNA helicase is analogously juxtaposed as Arg-467 of HCV helicase. However, the Arg-287 residue is located in motif IV and not in motif VI of PcrA DNA helicase. Thus, the arginine residues for ATP hydrolysis are structurally conserved between SF1 and SF2 helicases even though the primary sequence is not.

7 Structure-Based Mutagenesis of the Hepatitis C Virus Helicase Oligonucleotide-Binding Site

Site-directed mutagenesis was used to investigate the role of several residues which were shown to be in close contact with oligonucleotide in the nucleic acid-binding site of the X-ray structure of the NS3 helicase-dU$_8$ oligonucleotide binary complex (KIM et al. 1998). Ser-231, Thr-269, Ser-370, and Thr-411 are strictly conserved among various HCV strains and were predicted to form hydrogen bonds with the phosphate group of the oligonucleotide. Mutation of Ser-231 or Ser-370 to Ala had no effect on basal or poly(U)-stimulated ATPase, unwinding, and single-stranded RNA binding activities of NS3 helicase. In contrast, mutation of Thr-269 or Thr-411 to alanine decreased RNA binding to 20% of wild-type level and completely abolished poly(U) stimulation of ATPase and unwinding activities, although their basal ATPase activity remained intact. Analysis of the interaction of oligonucleotide with helicase observed in the crystal structure suggested that Trp-501 might interact with the uridine rings of dU$_8$. In order to test this hypothesis, the Trp was substituted with Ala, Leu, or Phe. Both Leu and Ala decreased RNA binding and abolished the unwinding and enhancement of ATPase by poly(U), although the basal ATPase level was similar to that of the wild-type helicase. Substitution of Trp-501 with Phe resulted in wild-type levels of basal ATPase, unwinding and single-stranded RNA binding activities. This suggests that the phenyl ring is critical for stacking interactions with the uridine ring.

8 Models of Unwinding

Despite elegant kinetic analysis of helicase II (UvrD) and Rep helicase and the publication of the structure of serveral helicases, a detailed understanding of how

ATP binding and hydrolysis are coupled to unwinding of double-stranded substrate has yet to be determined (LOHMAN and BJORNSON 1996). Although the three-dimensional structures of the NS3 helicase domain in the absence (CHO et al. 1998; YAO et al. 1997) and presence (KIM et al. 1998) of bound oligonucleotide has been determined by X-ray crystallographic analyses, no consensus has been reached with respect to the mechanism of helicase unwinding.

Based on mechanistic and structural studies of Rep DNA helicase, Lohman has classified possible mechanisms for unwinding into "active" or "passive" activity (LOHMAN and BJORNSON 1996). In the "passive" model, helicase binds preferentially to single-stranded nucleic acid and does not actively unwind the duplex. Rather, it binds to single-stranded regions when they become exposed due to "breathing"of the strands. In the "active" model, helicase binds to both single-stranded and double-stranded regions and actively separates the strands in an NTP-dependent reaction. Both models were proposed based on oliogomeric DNA helicases such as SV40 T antigen (MASTRANGELO et al. 1989) and *E. coli* DnaB (REHA-KRANTZ and HURWITZ 1978) which form hexamers; or *E. coli* Rep helicase (WONG et al. 1992; WONG and LOHMAN 1992), HeLa cell DNA helicase (SEO et al. 1991), and herpes simplex virus UL9 (BRUCKNER et al. 1991) which form dimers.

These models do not provide a mechanism for a class of RNA helicases such human p68 (HIRLING et al. 1989), human helicase A (LEE and HURWITZ 1992), vaccinia NPH-II helicase (SHUMAN 1993), or HCV helicase (PREUGSCHAT et al. 1996), for which only a monomeric form is observed in solution. Based on crystal packing interfaces in the crystal structure of the HCV strain 1b helicase domain, Cho and co-workers (CHO et al. 1998) propose that a dimer is formed by interactions between domains 1 and 2 of two HCV helicase molecules and a channel is formed so that a canonical single-stranded RNA can be modeled to pass through a channel formed by the interdomain clefts of both molecules. They predict that ATP hydrolysis leads to a hinge bending motion, which results in a change from the activated form of the enzyme to the resting form and a simultaneous release of the bound RNA. A "descending molecular see-saw" motion is proposed in which the dimer translocates on the single-stranded RNA by rotating toward the 5' end of the bound single-stranded RNA and by cycling between activated (binds RNA) and resting (releases RNA) forms of each member of the dimer. Three sets of data do not support this model of a functional dimer: (1) HCV helicase is a monomer in solution, (2) in the crystal structure of KIM et al. (1998) the oligonucleotide does not bind in the interdomain cleft, (3) Arg-464 and Arg-467 in the Q_{460}RRGRTGR motif are involved in ATP hydrolysis, not RNA binding.

YAO et al. (1997) model the HCV helicase domain as a monomer in complex with RNA. Similar to the functional dimer model described above, the free 3' strand of the single-stranded RNA was fitted into the interdomain cleft between domains 1 and 2 based on putative electrostatic interactions with Arg-461, Arg-464, and Arg-467 in domain 2. Double-stranded RNA was modeled to bind near residues Lys-551, Arg-570, Lys-583, Arg-587, and Lys-589, which cluster in domain 3. Hydrolysis of ATP was predicted to lead to a rotation in domain

2 and to a conformational change within the conserved TATPP sequence of motif III.

Based on a crystal structure of HCV NS3 helicase complexed with an oligonucleotide (KIM et al. 1998), and on mutagenesis studies of motif VI and the RNA binding channel, we have proposed a new model for helicase unwinding. In this model, the 3'-single-stranded tail of a duplex nucleic acid substrate binds to the groove which separates domain 3 from domains 1 and 2. In our model, ATP (or any other NTP) initially binds to the HCV helicase through interaction with motifs I and II on domain 1. As mentioned earlier, the phosphate binding loop or motif I (GXGKS/T) is responsible for binding of the β-phosphate group of ATP, while the Asp-290 of motif II binds to Mg^{+2} and helps orient the ATP-Mg^{+2} complex for nucleotide attack. Binding of ATP leads to closure of the cleft between domains 1 and 2 which is driven by the interaction of Arg-464 and Arg-467 of motif VI in domain 2 with the γ- and α-phosphate groups, respectively, of the bound ATP molecule. Based on multiple crystal forms of the enzyme, domains 1 and 3 of the HCV helicase act as a rigid unit, while domain 2 remains flexible relative to the other two domains. The motion of domain 2 proposed in this model is further supported by observations of BRYANT et al. (1999) who noted a variety of positions of domain 2 in different NS3 helicase crystal forms which represented a 35° rotation about a central hinge point. Both in-plane motions (away from domain 1 and 3) and out-of-plane motions (towards the bound oligo) were observed in different crystal forms. The distance between oligonucleotide phosphate binding moieties (defined by residues G-255 and T-269 in domain 1, and R-393 and T-411 in domain 2) changes by over 8Å as a consequence of movement of this magnitude, suggesting that unwinding of multiple bases in a single NTP-binding event is not impossible. Indeed, none of these crystal forms were produced in the presence of NTP, which would be expected to produce an even bigger change in conformation. This type of NTP-dependent movement of two domains has been observed in other enzymes utilizing NTP, such as adenylate kinase (BILDERBACK et al. 1996; SCHULZ 1992) and mRNA capping enzyme (HAKANSSON et al. 1997). Thus, closure of the ATP-binding cleft is predicted to lead to translocation of domains 1 and 3 in a 3' to 5' direction along the bound polynucleotide strand. The orientation of the other polynucleotide strand, released by unwinding, has not been clearly defined yet, but is unlikely to be bound to the arginine-rich motif VI as proposed by YAO et al. (1997) or CHO et al. (1998) because it would inhibit ATP hydrolysis. The role of two "bookend" amino acids in the RNA-binding channel (Val-432 and especially, Trp-501) is to lock the polynucleotide in place while domain 2 resets, resulting in a unidirectional movement of the helicase in a 3' to 5' direction along the bound polynucleotide.

9 Hepatitis C Virus Helicase Antiviral Drug Development

Hepatitis C Virus infection is a major cause of chronic liver disease, with an estimated worldwide seroprevalence of approximately 1% (PURCELL 1994). Despite urgent medical need, a broadly effective and easily tolerated antiviral therapy for the treatment of infections with HCV has yet to be developed. The NS3 helicase enzyme is an attractive target because of its presumed critical role in virus replication (reviewed in BARTENSCHLAGER 1997; BLIGHT et al. 1998).

9.1 Helicase Targets for Small Molecule Inhibitors

Some of the multiple potential mechanisms of actions by which a molecule could inhibit HCV helicase include: (1) inhibition of ATPase activity by interference with ATP binding, (2) inhibition of ATP hydrolysis or ADP release by blocking the opening or the closing of domain 2, (3) inhibition of RNA binding, (4) inhibition of unwinding by sterically blocking helicase translocation, (5) inhibition of the coupling of ATP hydrolysis to unwinding (reviewed in KIM and CARON 1998; YAO and WEBER 1998). In addition, disruption of the interaction of the NS3 helicase with other proteins in the replication complex may inhibit HCV growth. Elucidation of the crystal structure of HCV helicase, as well as mutagenesis studies, have identified key residues and binding pockets which are essential for enzyme activity. An understanding of the mechanism of action coupled to the knowledge of the three-dimensional structure of the enzyme, provides the basis for structure-based drug design. As described above, currently there is no consensus on the mechanism of unwinding, with three different models proposed by three different groups who have published the crystal structure of HCV helicase.

9.2 High Throughput Helicase Assays

Complementary to structure-based drug design is the use of high throughput primary screening using in vitro assays with purified helicase to identify new anti-HCV inhibitors. ATPase or RNA binding assays can be run to screen for inhibitors of any of the sub-reactions of the unwinding process, or a helicase unwinding assay can be used to identify inhibitors of each of the mechanisms described above. Three types of high throughput helicase assays have been described to date: (1) KYONO et al. (1998) have developed an increasing signal assay using the scintillation proximity assay (SPA) system developed by Amersham. Unwinding of a radiolabelled ^{3}H-DNA:RNA duplex by NS3 helicase is detected by hybridization of the released single-stranded ^{3}H-DNA to a biotinylated complementary oligo; this "capture oligo" is subsequently binds to streptavidin-coated SPA beads, resulting in an increase in scintillation signal. (2) KWONG et al. (1999) developed an increasing signal assay using a ^{33}P-labelled RNA:RNA duplex substrate. Release of the radiolabelled single-stranded ^{33}P-RNA was detected by hybridization to a capture oligo which is adsorbed to the wells of a streptavidin coated FlashPlate PLUS (NEN Life Science Products), resulting in an increase in scintillation counts. (3) HSU et al. (1998) developed a decreasing signal nonradioactive, ELISA-based assay, in which a double-stranded RNA substrate contained one strand which was labeled with biotin, and the other strand was labeled with digoxigenin. Unlike the previous two assays, this assay is performed on a solid phase. The substrate is bound to the wells of a streptavidin-coated 96-well plate. Upon addition of helicase, the digoxigenin strand is released and removed by washing. The remaining duplex substrate in the well is detected by an anti-digoxigenin horseradish peroxidase chromatogenic assay detection system.

9.3 Herpes Simplex Virus Helicase Inhibitors as a Proof of Concept

Recently, proof of concept for helicase inhibitors as antiviral agents was obtained for herpes simplex virus (HSV). Using high throughput screening (JONES 1998), two groups identified the same class of aminothiazole compounds which inhibited the HSV UL5/8/52 helicase/primase complex (FAUCHER et al. 1998; SPECTOR et al. 1998a,b). Optimization of the screening hits resulted in compounds which inhibited HSV growth in cell culture and were orally active in an animal model for HSV (FAUCHER et al. 1998). The mechanism of antiviral action was confirmed when both groups, independently, selected resistant viruses with single point mutations in the UL5 DNA helicase gene, thus demonstrating the possibility of developing selective, potent inhibitors of viral helicases as antiviral agents.

10 Summary

Hepatitis C Virus helicase activity has been mapped to the COOH-terminal 450 residues of the NS3 protein. Due to its complexity and presumed essentiality for viral replication, the helicase is an attractive target for drug discovery. The elucidation of the atomic structure of the HCV NS3 helicase in complex with oligonucleotide and with ADP has helped clarify our understanding of potential sites for inhibitor binding. Molecular details of the mechanism of this enzyme, and in particular, a better understanding of the mechanism by which ATP hydrolysis is coupled to unwinding of double-stranded substrate may facilitate more efficient structure-based drug design.

References

Bartenschlager R (1997) Candidate targets for hepatitis C virus-specific antiviral therapy. Intervirology 40:378–393

Bartenschlager R, Lohmann V, Wilkinson T, Koch JO (1995) Complex formation between the NS3 Serine-type proteinase of the hepatitis C virus and NS4A and its importance for polyprotein maturation. J Virol 69:7519–7528

Bayliss CD, Smith GL (1996) Vaccinia virion protein 18R has both DNA and RNA helicase activities implications for vaccinia virus transcription. J Virol 70:794–800

Bilderback T, Fulmer T, Mantulin WW, Glaser M (1996) Substrate binding causes movement in the ATP binding domain of *Escherichia coli* adenylate kinase. Biochemistry 35:6100–6106

Black ME, Hruby DE (1992) Site-directed mutagenesis of a conserved domain in vaccinia virus thymidine kinase. Evidence for a potential role in magnesium binding. J Biol Chem 267:6801–6806

Blight KJ, Kolyhalov AA, Reed KE, Agapov EV, Rice CM (1998) Molecular virology of hepatitis C virus an update with respect to potential antiviral targets. Antiviral Therapy 3 (Supplement 3) 71–81

Bochkarev A, Pfuetzner RA, Edwards AM, Frappier L (1997) Structure of single-stranded-DNA-binding domain of replication A protein bound to DNA. Nature 385:176–181

Bruckner RC, Crute JJ, Dodson MS, Lehman IR (1991) The herpes simplex virus 1 origin binding protein a DNA helicase. J Biol Chem 266:2669–2674

Bryant GL, Harris MS, Baldwin ET, Tandeske L, Shoemaker KR, Finzel BC (1999) HCV helicase RNA-binding-domain flexibility quantified by comparison of multiple crystal forms. Annual American Crystallographic Association Meeting

Butkiewicz N, Wendel M, Zhang R, Jubin R, Pichardo J, Smith EB, Hart AM, Ingram R, Durkin J, Mui PW, Murray MG, Ramanathan L, Dasmahapatra B (1996) Enhancement of hepatitis C Virus NS3 proteinase activity by association with NS4A-specific synthetic peptides identification of sequence and critical residues of NS4A for the cofactor activity. Virology 225:328–338

Cho H-S, Ha N-C, Kang L-W, Chung KM, Back SH, Jang SK, Oh B-H (1998) Crystal structure of RNA helicase from genotype 1b hepatitis C virus. J Biol Chem 273:15045–15052

Choo Q-L, Kuo G, Weiner AJ, Overby LR, Bradley DW, Houghton M (1989) Isolation of a cDNA clone derived from a blood-born non-A, non-B viral hepatitis genome. Science 244:359–362

Choo Q-L, Richman KH, Han JH, Berger K, Lee C, Dong C, Gallegos C, Coit D, Medina-Selby A, Barr PJ, Weiner AJ, Bradley DW, Kuo G, Houghton M (1991) Genetic organization and diversity of the hepatitis C virus. Proc Natl Acad Sci USA 88:2451–2455

Eagles RM, Balmori-Melián E, Beck DL, Gardner RC, Forster RLS (1994) Characterization of NTPase, RNA-binding and RNA-helicase activities of the cytoplasmic inclusion protein of tamarillo mosaic potyvirus. Eur J Biochem 224:677–684

Failla C, Tomei L, De Francesco R (1994) Both NS3 and NS4A are required for proteolytic processing of hepatitis C virus nonstructural proteins. J Virol 68:3753–3760

Fernandez A, Garcia JA (1996) The RNA helicase CI from plum pox potyvirus has two regions involved in binding to RNA. FEBS Letts 388:206–210

Gallinari P, Brennan D, Nardi C, Brunetti M, Tomei L, Steinkühler C, De Francesco R (1998) Multiple enzymatic activities associated with recombinant NS3 protein of hepatitis C virus. J Virol 72:6758–6769

Gorbalenya AE, Koonin EV (1993) Helicases amino acid sequence comparisons and structure-function relationships. Curr Opin Struc Biol 3:419–429

Gorbalenya AE, Koonin V, Donchenko AP, Blinov VM (1989) Two related superfamilies of putative helicases involved in replication recombination repair and expression of DNA and RNA genomes. Nucleic Acids Res 17:4713–4729

Grakoui A, Wychowski C, Lin C, Feinstone SM, Rice CM (1993) Expression and identification of hepatitis C virus polyprotein cleavage products. J Virol 67:1385–1395

Gross CH, Shuman S (1996) The QRXGRXGRXXXR motif of the vaccinia virus DExH box RNA helicase NPH-II is required for ATP hydrolysis and RNA unwinding but not for RNA binding. J Virol 70:1706–1713

Gwack Y, Kim DW, Han JH, Choe J (1996) Characterization of the RNA binding activity and RNA helicase activity of the hepatitis C virus NS3 protein. Biochem Biophys Res Com 225:654–659

Gwack Y, Kim DW, Han JH, Choe J (1997) DNA helicase activity of the hepatitis C virus nonstructural protein 3. Eur J Bioch 250:47–54

Hakansson K, Doherty AJ, Shuman S, Wigley DB (1997) X-ray crystallography reveals a large conformational change during guanyl transfer by mRNA capping enzymes. Cell 89:545–553

Hall MC, Matson SW (1997) Mutation of a highly conserved arginine in motif VI of Escherichia coli DNA helicase II results in an ATP-binding defect. J Biol Chem 272:18614–18620

Heilek GM, Peterson MG (1997) A point mutation abolishes the helicase but not the nucleoside triphosphatase activity of hepatitis C virus NS3 protein. J Virol 71:6264–6266

Hijikata M, Kato N, Ootsuyama Y, Nakagawa M, Shimotohno K (1991) Gene mapping of the putative structural region of the hepatitis C virus genome by in vitro processing analysis. Proc Natl Acad Sci USA 88:5547–5551

Hirling H, Scheffner M, Restle T, Stahl H (1989) RNA helicase activity associated with human p68 protein. Nature 339:562–564

Hong Z, Ferrari E, Wright-Minogue J, Chase R, Risano C, Seelig G, Lee C-G, Kwong AD (1996) Enzymatic characterization of hepatitis C virus NS3/4A complexes expressed in mammalian cells by using the herpes simplex virus amplicon system. J Virol 70:4261–4268

Houghton M (1996) Hepatitis C viruses, In: Fields BN, Knipe DM, Howley PM (eds) Virology. Raven, New York, pp 1035–1058

Hsu CC, Hwant L-H, Huang Y-W, Chi W-K, Chu Y-D, Chen D-S (1998) An ELISA for RNA helicase activity application as an assay of the NS3 helicase of hepatitis C virus. Biochem Biophys Res Com 253:594–599

Jin L, Peterson DL (1995) Expression isolation and characterization of the hepatitis C virus ATPase/RNA helicase. Arch Biochem Biophys 323:47–53

Jones PS (1998) Strategies for antiviral drug discovery. Antivir Chem and Chemother 9:283–302

Kadaré G, Haenni A-L (1997) Minireview: virus-encoded RNA helicases. J Virol 71:2583–2590

Kaito M, Watanabe S, Tsukiyama-Kohara K, Yamaguchi K, Kobayashi Y, Konishi M, Yokoi M, Ishida S, Suzuki S, Kohara M (1994) Hepatitis C virus particle detected by immunoelectron microscopic study. J Gen Virol 75:1755–1760

Kanai A, Tanabe K, Kohara M (1995) Poly (U) binding activity of hepatitis C virus NS3 protein a putative RNA helicase. FEBS Letters 376:221–224

Kim DW, Gwack Y, Han JH, Choe J (1995) C-terminal domain of the hepatitis C virus NS3 protein contains an RNA helicase activity. Biochem Biophys Res Comm 215:160–166

Kim DW, Gwack Y, Hang JH, Choe J (1997a) Towards defining a minimal functional domain for NTPase and RNA helicase activities of the hepatitis C virus NS3 protein. Virus Res 49:17–25

Kim JL, Caron PR (1998) Crystal structure of hepatitis C virus NS3 RNA helicase reveals a possible enzyme mechanism and suggests multiple potential drug binding sites. International Antiviral News 6:26–28

Kim JL, Morgenstern KA, Griffith JP, Dwyer MD, Thomson JA, Murcko MA, Lin C, Caron PR (1998) Hepatitis C virus NS3 RNA helicase domain with a bound oligonucleotide: the crystal structure provides insights into the mode of unwinding. Structure 6:89–100

Kim JL, Morgenstern KA, Lin C, Fox T, Dwyer MD, Landro JA, Chambers SP, Markland W, Lepre CA, O'Malley ET, Harbeson SL, Rice CM, Murcko MA, Caron PR, Thomson JA (1996) Crystal structure of the hepatitis C virus NS3 protease domain complexed with a synthetic NS4A cofactor peptide. Cell 87:343–355

Kim DW, Kim J, Gwack Y, Han JH, Choe J (1997b) Mutational analysis of the hepatitis C virus RNA helicase. J Virol 71:9400–9409

Kolykhalov A, Feinstone SM, Rice CM (1996) Identification of a highly conserved sequence element at the 3' terminus of hepatitis C virus genome RNA. J Virol 70:3363–3371

Kolykhalov AA, Agapov EV, Blight KJ, Mihalik K, Feinstone S, Rice CM (1997) Transmission of hepatitis C by intrahepatic inoculation with transcribed RNA. Science 277:570–574

Korolev S, Hsieh J, Gauss GH, Lohman TM, Waksman G (1997) Major domain swiveling revealed by the crystal structures of complexes of *E. coli* Rep helicase bound to single-stranded DNA and ADP. Cell 90:635–647

Korolev S, Yao N, Lohman TM, Weber PC, Waksman G (1998) Comparison between the structures of HCV and Rep helicases reveal structural similarities between SF1 and SF2 super-families of helicases. Prot Sci 7:605–610

Kuo M-D, Chin C, Hsu S-L, Shiao J-Y, Wang T-M, Lin J-H (1996) Characterization of the NTPase activity of Japanese encephalitis virus NS3 protein. J Gen Virol 77:2077–2084

Kwong AD, Risano C (1998) Development of a hepatitis C virus RNA helicase high throughput assay, Chapter 9. In: Kinchington D, Schinazi RF (eds) Antiviral Methods and Protocols. Humana Press Inc, Totowa NJ

Kyono K, Miyashiro M, Taguchi I (1998) Detection of hepatitis C virus helicase activity using the scintillation proximity assay system. Anal Biochem 257:120–126

Lain S, Martin MT, Riechmann JL, Garcia JA (1991a) Novel catalytical activity associated with positive-strand RNA virus infection nucleic acid-stimulated ATPase activity of the plum pox potyvirus helicase-like protein. J Virol 65:1–6

Lain S, Riechmann JL, Garcia JA (1991b) RNA helicase a novel activity associated with a protein encoded by a positive-strand RNA virus. Nucleic Acids Res 18:7003–7006

Landro JA, Raybuck SA, Luong Y-C, O'Malley ET, Harbeson SL, Morgenstern KA, Rao G, Livingston DJ (1997) Mechanistic role of an NS4A peptide cofactor with the truncated ns3 protease of hepatitis

C virus. Elucidation of the NS4A stimulatory effect via kinetic analysis and inhibitor mapping. Biochemistry 36:9340–9348

Laxton CD, McMillan D, Sullivan V, Ackrill AM (1998) Expression and characterization of the hepatitis G virus helicase. J Viral Hepatitis 5:21–26

Lee C-G, Hurwitz J (1992) A new RNA helicase isolated from HeLa cells that catalytically translocates in the 3′ to 5′ direction. J Biol Chem 267:4398–4407

Lin C, Lindenbach BD, Pragai BM, McCourt DW, Rice CM (1994) Processing in the hepatitis C virus E2-NS2 region identification of p7 and two distinct E2-specific products with dfferent C termini. J Virol 68:5063–5073

Liuzzi M, Crute J, Grygon C, Hargrave K, Simoneau B, Faucher A, Bolger G, Duan J, Kibler P, Cordingley M (1998) Aminothiazolyl-phenyl-based inhibitors of HSV helicase-primase: a novel class of orally active antiherpetic agents. Antiviral Research 37:A42

Lohman TM, Bjornson KP (1996) Mechanisms of helicase-catalyzed DNA unwinding. Ann Rev Biochem 65:169–214

Love RA, Parge HE, Wickersham JA, Hostomsky Z, Habuka N, Moomaw EW, Adachi T, Hostomska Z (1996) The crystal structure of hepatitis C virus NS3 proteinase reveals a trypsin-like fold and a structural zinc binding site. Cell 87:331–342

Markland W, Petrillo RA, Fitzgibbon M, Fox T, McCarrick R, McQuaid T, Fulghum JR, Chen W, Fleming MA, Thompson JA, Chambers SP (1997) Purification and characterization of the NS3 serine protease domain of hepatitis C virus expressed in Saccharomyces cerevisiae. J Gen Virology 78: 39–43

Martinez R, Shao L, Weller SK (1992) The conserved helicase motifs of the herpes simples virus type-I origin bindng protein UL9 are important for function. J Virol 66:6735–6746

Mastrangelo IA, Hough PVC, Wall JS, Dodson M, Dean FB, Hurwitz J (1989) ATP-dependent assembly double hexamers of SV40 T antigen at the viral origin of DNA replication. Nature 338:658–662

Miller RH, Purcell RH (1990) Hepatitis C virus shares amino acid sequence similarity with pestiviruses and flaviviruses as well as members of two plant virus supergroups. Proc Natl Acad Sci USA 87:2057–2061

Morgenstern KA, Landro JA, Hsiao K, Lin C, Yong G, Su MS-S, Thomson JA (1997) Polynucleotide modulation of the protease nucleoside triphosphatase and helicase activities of a hepatitis C virus NS3-NS4A complex isolated from transfected COS Cells. J Virol 71:3767–3775

Pai EF, Krengel U, Petsko GA, Gody RS, Katsch W, Wittinghofer A (1990) Refined crystal structure of the triphosphate conformation of H-ras p21 at 135Å resoluton implications for the mechanism of GTP hydrolysis. EMBO J 9:2351–2359

Pause A, Methot N, Sonenberg N (1993) The HRIGRXXR region of the DEAD box RNA helicase eukaryotic translation initiation factor 4A is required for RNA binding and ATP hydrolysis. Mol Cell Biol 13:6789–6798

Preugschat F, Averett DR, Clarke BE, Porter DJT (1996) A steady-state and pre-steady-state kinetic analysis of the NTPase activity associated with the hepatitis C virus NS3 helicase domain. J Biol Chem 271:24449–24457

Purcell RH (1994) Hepatitis C virus historical perspective and current concepts. FEMS Microbiol Rev 14:181–192

Raghunathan S, Ricard CS, Lohman TM, Waksman G (1997) Crystal structure of the homo-tetrameric DNA-binding domain of *Escherichia coli* single-stranded DNA-binding protein determined by multiwavelength X-ray diffraction on the selenomethionyl protein at 2.9Å resolution. Proc Natl Acad Sci USA 94:6652–6657

Reha-Krantz LJ, Hurwitz J (1978) The dnaB gene product of *Escheriachia coli*. I. Purification homogeneity and physical properties. J Biol Chem 253:4043–4050

Rice CM (1996) Flaviviridae the viruses and their replication. In: Fields BN, Knipe DM, Howley PM (eds) Virology. Raven, New York, pp 931–960

Ruff M, Moras D (1991) Class II aminoacyl transfer RNA synthetases: crystal structure of yeast aspartyl-tRNA synthetase complexed with tRNA(Asp). Science 252:1682–1689

Saraste M, Sibbald PR, Wittinghofer A (1990) The P-loop – a common motif in ATP- and GTP-binding proteins. Trends Biochem Sci 15:430–434

Scheffner M, Knippers R, Stahl H (1989) RNA unwinding activity of SV40 large T antigen Cell 57: 955–963

Schulz GE (1992) Induced-fit movement in adenylate kinases. Faraday Discuss 93:85–93

Seo YS, Lee SH, Hurwitz J (1991) Isolation of a DNA helicase from HeLa cells requiring the multisubunit human single-stranded DNA-binding protein for activity. J Biol Chem 266:13161–13170

Shuman S (1993) Vaccinia virus RNA helicase directionality and substrate specificity. J Biol Chem 268:11798–11802

Smith DH, Kotin RM (1998) The Rep52 gene product of adeno-associated virus is a DNA helicase with 3′-to-5′ polarity. J Virol 72:4874–4881

Spector FC, Liang L, Giordano H, Sivaraja M, Peerson MG (1998a) T157602 a 2-amino-thiazole inhibits HSV replication by interacting with the UL5 component of the UL5/8/52 helicase primase complex. Antiviral Research 37:A43

Spector FC, Liang L, Giordano H, Sivaraja M, Peterson MG (1998b) Inhibition of herpes simplex virus replication by a 2-amino thiazole via interactions with the helicase component of the UL5-UL8-UL52 complex. J Virol 72:6979–6987

Steinkühler C, Tomei L, De Francesco R (1996) In vitro activity of hepatitis C virus protease NS3 purified from recombinant baculovirus-infected Sf9 cells. J Biol Chem. 271:6367–6373

Subramanya HS, Bird LE, Brannigan JA, Wigley DB (1996) Crystal structure of a DExx box DNA Helicase. Nature 384:379–383

Suzich JA, Tamura JK, Palmer-Hill F, Warrener P, Grakoui A, Rice CM, Feinstone SM, Collett MS (1993) Hepatitis C virus NS3 protein polynucleotide-stimulated nucleoside triphosphatase and comparison with the related pestivirus and flavivirus enzymes. J Virol 67:6152–6158

Tai C-L, Chi W-K, Chen D-S, Hwang L-H (1996) The helicase activity associated with hepatitis C virus nonstructural protein 3 (NS3). J Virol 70:8477–8484

Takamizawa A, Mori C, Fuke I, Manabe S, Murakami S, Fujita J, Onishi E, Andoh T, Yoshida I, Okayama H (1991) Structure and organization of the hepatitis C virus genome isolated from human carrier. J Virol 65:1105–1113

Tanaka T, Kato N, Cho M-J, Shimotohno K (1995) A novel sequence found at the 3′ terminus of hepatitis C virus genome. Biochem Biophys Res Commun 215:744–749

Tanaka T, Kato N, Cho M-J, Sugiyama K, Shimotohno K (1996) Structure of the 3′ terminus of the hepatitis C virus genome. J Virol 70:3307–3312

Tanji Y, Hijikata M, Satoh S, Kaneko T, Shimotohno K (1995) Hepatitits C virus-encoded nonstructural protein NS4A has versatile functions in viral protein processing. J Virol 69:1575–1581

Walker JE MS, Runswick MJ, Gay NJ (1982) Distantly related sequences in the a- and b-subunits of ATP synthase myosin kinases and other ATP-requiring enzymes and a common nucleotide binding fold. EMBO J 1:945–951

Warrener P, Collett M (1995) Pestivirus NS3 (p80) protein possesses RNA helicase activity. J Virol 69:1720–1726

Warrener P, Tamura JK, Collett MS (1993) RNA-stimulated NTPase activity associated with yellow fever virus NS3 protein expressed in bacteria. J Virol 67:989–996

Weng Y, Czaplinski K, Peltz S (1996) Genetic and biochemical characterization of mutations in the ATPase and helicase regions of the Upf1 protein. Mol Cell Biol 16:5477–5490

Wengler G (1991) The carboxyl-terminal part of the NS3 protein of the West Nile flavivirus can be isolated as a soluble protein after proteolytic cleavage and represents an RNA-stimulated NTPase. Virology 184:707–715

Wong I, Chao K, Bujalowski W, Lohman TM (1992) DNA-induced dimerization of the *Escherichia coli* rep helicase. Allosteric effects of single-stranded and duplex DNA. J Biol Chem 267:7596–7610

Wong I, Lohman TM (1992) Allosteric steric effects of nucleotide cofactors on *Escherichia coli* Rep helicase DNA binding. Science 256:350–355

Yamada N, Tanihara NK, Takada A, Yorihuzi T, Tsutsumi M, Shimomura H, Tsuji T, Date T (1996) Genetic organization and diversity of the 3′noncoding region of the hepatitis C virus. Virology 223:255–261

Yan HG, Tsai MD (1991) Mechanism of adenylate kinase. Demonstration of a functional relationship between aspartate 93 and Mg2+ by site-directed mutagenesis and proton phosphorus-31 and magnesium-25 NMR. Biochemistry 30:5539–5546

Yanagi M, Purcell RH, Emerson SU, Bukh J (1997) Transcripts from a single full-length cDNA clone of hepatitis C virus are infectious when directly transfected into the liver of a chimpanzee. Proc Natl Acad Sci USA 94:8738–8743

Yanagi M, St Claire M, Shapiro M, Emerson SU, Purcell RH, Bukh J (1998) Transcripts of a chimeric cDNA clone of hepatitis C virus genotype 1b are infectious in vivo. Virology 244:161–172

Yao N, Hesson T, Cable M, Hong Z, Kwong AD, Le HV, Weber PC (1997) Structure of the hepatitis C virus RNA helicase domain. Nature Structural Biology 4:463–467

Yao N, Weber PC (1998) Helicase a target for novel inhibitors of hepatitis C virus. Antiviral Therapy 3:93–97

Zhang S, Grosse F (1994) Nuclear DNA helicase II unwinds both DNA and RNA. Biochemistry 33:3906–3912

Evading the Interferon Response: Hepatitis C Virus and the Interferon-Induced Protein Kinase, PKR

M.J. Korth[1] and M.G. Katze[2]

1	Introduction	197
2	Hepatitis C Virus and Interferon Therapy	199
2.1	Molecular Biology and Genotypic Variation	199
2.2	Response to Interferon Therapy	200
3	NS5A and the Interferon Sensitivity Determining Region	201
4	The Interferon-Induced Double-Stranded RNA-Activated Protein Kinase, PKR	204
4.1	Physical Characteristics	204
4.2	Cellular Functions	205
4.3	Viral Regulation of PKR	207
4.3.1	Adenovirus and Reovirus: Inhibition of PKR Activation	207
4.3.2	Vaccinia Virus: Inhibition of PKR Catalytic Activity	209
4.3.3	Influenza Virus: Disruption of PKR Dimerization	209
4.3.4	Poliovirus: Directed Degradation of PKR	210
4.3.5	Herpes Simplex Virus: Reversal of PKR-Mediated Phosphorylation	210
4.3.6	Human Immunodeficiency Virus: A Multi-Pronged Approach	211
5	PKR and NS5A	211
5.1	Inhibition of PKR Activity by NS5A	212
5.2	Requirement of the Interferon Sensitivity Determining Region for PKR Inhibition	213
5.3	Implications of PKR Regulation by NS5A	214
6	Future Directions	216
	References	217

1 Introduction

What's a cell to do when faced with the prospect of being invaded by an army of determined viruses? A complex, multi-pronged defense is in order, and luckily, just such a defense is available. The cellular interferon response represents a powerful and multi-faceted approach to dealing with everyday stresses that range from viral infection to keeping cellular growth under control. Mediated by a family of negative growth regulators, collectively referred to as interferons, the interferon response is indeed complex. After binding to specific cell surface receptors,

[1] Regional Primate Research Center, University of Washington, Box 357330, Seattle, WA 98195, USA
[2] Department of Microbiology, University of Washington, Box 357242, Seattle, WA 98195, USA

interferons set in motion a number of signal transduction pathways that lead to the induction of gene expression (reviewed in FOSTER 1997; HAQUE and WILLIAMS 1998; MÜLLER et al. 1994; VILCEK and SEN 1996). Although the majority of interferon-induced gene products have yet to be characterized, several have demonstrated antiviral properties. These include RNase L, 2'-5' oligoadenylate synthetase, the Mx proteins, and the double-stranded (ds)RNA-activated protein kinase, PKR. Together, these proteins are capable of disrupting viral gene expression at multiple levels, including viral mRNA stability, transcription, and translation. Of course, viruses are not without weapons of their own, and many are capable of putting up a good fight, and even winning, in the face of such an arsenal. In this review, we will focus our attention on one particular component of the cellular antiviral response, the protein kinase known as PKR. PKR is one of the better characterized interferon-induced gene products and is a critical component of the antiviral response. As such, PKR is also a primary target for viral inhibition and many viruses have evolved strategies to regulate the activity of PKR. These strategies are surprisingly varied and nearly every aspect of PKR structure and function appears to be fair game when it comes to inhibiting kinase activity.

Given the inclusion of a chapter on PKR in a volume devoted to hepatitis C virus (HCV), the reader is no doubt already aware that HCV has been added to the list of viruses that have found a way to inhibit PKR function. Indeed, we recently reported that the non-structural 5A (NS5A) protein of HCV is capable of binding to PKR and inhibiting kinase activity (GALE et al. 1997). In addition to a review of those findings, we will provide a brief overview of the use of interferon in the treatment of chronic HCV and discuss the series of reports by ENOMOTO et al. (1996; 1995) that initially prompted us to look for an interaction between NS5A and PKR. We will also provide examples of the myriad approaches used by viruses to inhibit PKR activity and show that HCV may utilize yet another novel strategy. Although interferon remains the primary therapeutic agent for the treatment of HCV, its use leads to a clinical remission in only a small percentage of patients. The

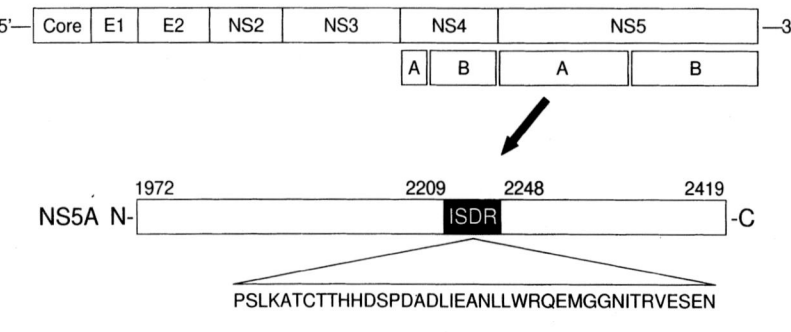

Fig. 1. The hepatitis C virus (HCV) genome, the NS5A protein, and the interferon sensitivity determining region (ISDR). The ISDR sequence shown is that of the prototype HCV-1b isolate, HCV-J (From KATO et al. 1990)

clinical implications of the NS5A-PKR interaction are therefore significant. The ability of NS5A to inhibit PKR suggests a mechanism by which certain HCV isolates may evade the interferon response, thereby contributing to the low response rate of HCV to interferon therapy.

2 Hepatitis C Virus and Interferon Therapy

To set the stage, we will begin by providing a brief background on the molecular biology and classification of HCV, followed by a discussion of the use of interferon as a therapeutic agent against HCV infection, and the host and viral factors that influence therapeutic outcome.

2.1 Molecular Biology and Genotypic Variation

Hepatitis C Virus is an enveloped virus that contains a positive-sense single-stranded RNA genome of approximately 9.5kb. On the basis of its genome organization and virion properties, HCV is classified as a separate genus in the family Flaviviridae, a family that also includes pestiviruses and flaviviruses (OKAMOTO and MISHIRO 1994; SHUKLA et al. 1995). The viral genome consists of a lengthy 5' untranslated region (UTR), a long open reading frame which encodes a polyprotein precursor of approximately 3011 amino acids, and a short 3' UTR. The 5' UTR is the most highly conserved part of the HCV genome and is important for the control of polyprotein translation, a process initiated by a cap-independent mechanism in which ribosomes bind to an internal ribosome entry site (IRES; WANG et al. 1994, 1995).

Although a satisfactory tissue culture system permissive for HCV replication has yet to be developed, viral proteins have been identified by a variety of techniques, including the use of in vitro transcription/translation systems and the transfection of recombinant clones (reviewed in HOUGHTON et al. 1991; NEDDERMANN et al. 1997; VAN DOORN 1994). The HCV polyprotein precursor is cleaved by both host and viral proteases to yield mature viral structural and nonstructural (NS) proteins (Fig. 1). Viral structural proteins include a nucleocapsid core (C) and two envelope glycoproteins, E1 and E2. HCV also encodes two proteases, a zinc-dependent metalloprotease, encoded by the NS2-NS3 region, and a serine protease encoded by the NS3 region. These proteases, together with the NS4A protein (which appears to function as a cofactor of the NS3 protease), are required for cleavage of specific regions of the precursor polyprotein into mature peptides. The COOH-terminal half of the NS5 protein, NS5B, contains the RNA-dependent RNA polymerase. Finally, as we will describe shortly, the NH_2-terminal half of the NS5 protein, NS5A, may provide HCV with a powerful weapon against the host cell interferon response.

HCV isolates have been classified into at least six major genotypes (reviewed in the chapter by D.L. Thomas, this volume). Although this classification is based on nucleotide sequence alignments, different genotypes also exhibit distinctive clinical characteristics and differ greatly in their resistance or sensitivity to interferon. Within the major genotypes, isolates may also be subdivided into more closely related subtypes. Isolates belonging to the same subtype show an average nucleotide sequence homology of greater than 90%, whereas isolates of separate major genotypes share less than 70% homology. Most HCV genotypes exist worldwide, though certain genotypes tend to predominate in specific geographic regions. The most common worldwide variant appears to be genotype 1b, a variant that is particularly prevalent in Japan and Western Europe. Genotype 1b is also common in the United States, although genotype 1a is equally represented (BUKH et al. 1995; MAHANEY et al. 1994).

In addition to genotypic variation, HCV circulates within an infected individual as a population of different but closely related genomes referred to as quasispecies (BUKH et al. 1995; MARTELL et al. 1992). HCV quasispecies may differ throughout the entire genome, but at least two regions exhibit notable sequence variation (ENOMOTO and SATO 1995). One hypervariable region is located in the E2 gene, which encodes the second envelope glycoprotein. This region is thought to encode epitopes for neutralizing antibodies and sequence divergence may be responsible for neutralization-resistant variants (SHIMIZU et al. 1994).

A second hypervariable region is located in the NS5A gene of HCV genotype 1b (ENOMOTO and SATO 1995). During the course of interferon therapy, a proportion of the HCV quasispecies that coexist prior to treatment may become predominant, while other quasispecies disappear. In a retrospective analysis of interferon-resistant and interferon-sensitive HCV quasispecies, ENOMOTO et al. (1994) observed the presence of a hypervariable domain in a region of the HCV genome that encodes the NS5A protein. What made this observation particularly significant was that the sequence of this hypervariable region was found to correlate with resistance or sensitivity to interferon (ENOMOTO et al. 1995, 1996). As will be discussed below, this small region of the NS5A protein has become the focus of considerable attention and controversy in the HCV community.

2.2 Response to Interferon Therapy

The antiviral and antiproliferative properties of the interferons have made them attractive for clinical applications (BARON et al. 1991; BORDEN and PARKINSON 1998; JARAMILLO et al. 1995). Indeed, interferon therapy has proven useful in the treatment of a variety of malignancies, including Kaposi's sarcoma and chronic myelogenous leukemia, as well as viral infections such as hepatitis B. The beneficial effects of interferon in the treatment of chronic HCV were first reported by HOOFNAGLE et al. (1986), and in 1991, interferon-α (a type I interferon) was approved by the Food and Drug Administration for the treatment of chronic HCV. The typical treatment regimen consists of 3 million U of interferon-α administered three times weekly for 6 months, although higher doses and extended therapy regimens have shown some promise and are becoming increasingly common (GRETCH et al. 1996; KOFF 1997; LINDSAY 1997). Unfortunately, the HCV genotype that predominates worldwide, genotype 1, is also the most resistant to the effects of interferon therapy. Only about 50% of patients infected with HCV genotype 1 show an initial response to therapy, with a normalization of serum alanine aminotransferase (ALT) level and a loss of detectable HCV RNA. To make matters worse, 6 months after the cessation of interferon therapy, only 10%–20% of patients achieve a sustained response, as defined by the disappearance of biochemical, histological, and virological indicators of chronic HCV (FRIED and HOOFNAGLE 1995; HOOFNAGLE 1994; IINO et al. 1994; MARTINOT-PEIGNOUX et al. 1995; TSUBOTA et al. 1994).

Given the relatively low efficacy and high cost of interferon treatment, together with potential side effects, efforts have been made to identify pretreatment factors that may be predictive of a long-term response to therapy. Host factors associated with therapeutic outcome include age, duration of infection, and the presence and degree of liver damage, whereas viral factors include pretreatment genotype, viral titer, and the degree of quasispecies genetic diversity (CHAYAMA et al. 1997; DAVIS and LAU 1997; LAM et al. 1994; POLYAK et al. 1997). The infecting viral genotype has proven to be of particular importance in predicting therapeutic outcome and, as indicated above, the majority of patients infected with a type-1 genotype respond poorly to interferon therapy. In contrast, the response rate of patients infected with HCV genotypes 2 or 3 is much higher, with several studies reporting a response rate of greater than 80% (CHEMELLO et al. 1995; TSUBOTA et al. 1994). The high frequency of interferon resistant isolates among HCV genotype 1 suggests that this genotype has evolved a successful mechanism to evade the interferon-mediated antiviral response. Reports that the amino acid sequence of a discrete region of the NS5A protein correlates with response to interferon provided the first hints that HCV resistance to interferon might be mediated, at least in part, by the action of NS5A.

3 NS5A and the Interferon Sensitivity Determining Region

Until recently, little was known about the NS5A protein. NS5A, derived from the NH_2-terminal half of the NS5 protein, is cleaved from the HCV polyprotein by the NS3-encoded serine protease. Although the specific role of NS5A in viral replication is unknown, NS5A can be co-immunoprecipitated with other viral nonstructural proteins and is thought to be integrated into the polymerase complex during replication (HIJIKATA et al. 1993). There are also recent reports that NS5A can act as a transcriptional activator (KATO et al. 1997; TANIMOTO et al. 1997). NS5A is phosphorylated on multiple serine residues and exists in two forms, one exhibiting a basal level of phosphorylation, and a second form that is hyperphosphorylated

(KANEKO et al. 1994; TANJI et al. 1995). The kinase responsible for phosphorylation of NS5A has not been identified, although there is evidence that NS5A is phosphorylated by a cellular serine/threonine kinase (REED et al. 1997). Indeed, NS5A can be phosphorylated in the absence of other viral nonstructural proteins, though hyperphosphorylation is dependent on the NS4A gene product (ASABE et al. 1997). NS4A itself does not, however, appear to have kinase activity. The phosphorylation state of NS5A may play a role in regulating viral replication or influence the interaction of NS5A with specific viral or cellular proteins.

An important clue to a possible function for NS5A came from reports linking the amino acid sequence of a discrete region of NS5A with response to interferon. As defined by ENOMOTO et al. (1995, 1996), this region of NS5A, termed the interferon sensitivity determining region (ISDR), is located in the carboxyl half of NS5A and spans amino acid residues 2209–2248 of the HCV polyprotein (Fig. 1). In a series of analyses, ENOMOTO et al. (1996) observed that patients infected with isolates containing an ISDR sequence identical to that of the prototype genotype 1b strain, HCV-J (KATO et al. 1990), did not respond to interferon therapy. Similarly, 87% of those infected with isolates containing one to three amino acid changes within the ISDR, defined as an intermediate ISDR sequence, also failed to respond to therapy. In marked contrast, patients infected with isolates containing a mutant ISDR, defined as containing four to 11 amino acid changes, showed a complete response to interferon therapy. These studies were the first to suggest that the sequence of the ISDR is a useful predictor of response to interferon therapy and, importantly, that NS5A might somehow mediate HCV resistance to interferon.

Naturally, the reports of an ISDR created considerable excitement and numerous investigators have since attempted to reproduce these findings. Perhaps not surprisingly, the ability of the ISDR sequence to predict therapeutic outcome has not proven to be quite so straightforward, with the results of studies falling along distinct geographic lines. In Japan, subsequent studies appear to confirm the existence on an ISDR. In a study examining 110 patients infected with HCV 1b, the amino acid sequence of the ISDR was again predictive of response to interferon therapy (CHAYAMA et al. 1997). Whereas only 17% of patients infected with isolates containing the prototype ISDR were responsive to treatment, 74% of patients infected with isolates containing a mutant ISDR sequence showed a sustained response. In a similar study analyzing the response of 22 patients to interferon-β (also a type I interferon in common use in Japan for treatment of HCV), 67% of patients infected with HCV 1b that contained a mutant type ISDR had a complete response to therapy (KUROSAKI et al. 1997). None of the patients infected with strains that contained an intermediate or prototype ISDR showed a sustained response. Thus, these studies appear to confirm the correlation between ISDR sequence and response to interferon.

Interestingly, studies outside of Japan have not yielded similar results. In a study of French and Italian patients (48 patients infected with genotype 1b and 18 patients infected with genotype 3a), no correlation was found between ISDR sequence and response to therapy (SQUADRITO et al. 1997). Similar results were found in a study of 43 French patients infected with HCV 1b (KHORSI et al. 1997) and a

German study of 32 patients infected with 1a or 1b (ZEUZEM et al. 1997). Perhaps significantly, these studies also found that multiple mutations within the ISDR are much less common in European HCV 1b isolates. Although a similar analysis of genotype 1b-infected patients has not been conducted in the United States, a study of North American patients infected with genotype 1a found no evidence of a specific amino acid substitution pattern within the ISDR, and no correlation between ISDR sequence and response to interferon (HOFGÄRTNER et al. 1997). Thus, the question of whether the sequence of the ISDR is predictive of therapeutic outcome remains unanswered. Still, in a comprehensive analysis of these studies, HERION and HOOFNAGLE (1997) noted that the response rate of patients infected with HCV containing the prototype ISDR sequence is quite low in both the Japanese and European studies, and that when patients infected with isolates containing intermediate or mutant ISDR sequences are combined, the response rate to interferon therapy is always higher than that in patients infected with HCV strains containing the prototype ISDR. Finally, a recent study from France also reported that, although the prototype ISDR sequence is significantly more frequent in nonresponder quasispecies than in responder quasispecies, no NS5A sequence was found to be intrinsically resistant or sensitive to interferon (PAWLOTSKY et al. 1998). Rather, a sustained response to interferon was more closely associated with a low viral load and a small degree of genetic diversity among circulating quasispecies.

The discrepancies observed in these studies are likely due to several factors, including differences in the treatment regimen employed or the type of interferon administered. For example, patients in the Japanese studies typically received a much higher dose of interferon, which may be required to reveal a correlation between ISDR sequence and therapeutic outcome. This difference in treatment regimen, together with the finding that multiple mutations within the ISDR are much less common in European HCV 1b isolates, might also account for the higher overall response rate to interferon that is observed in Japan than in Western countries (HERION and HOOFNAGLE 1997). It is also possible that amino acid changes outside the ISDR (or even NS5A) may contribute to interferon resistance in HCV isolates found in specific geographic regions. Other factors, such as differences in study parameters and definitions, also make a direct comparison of the results of each of these reports somewhat difficult.

Of course, reports of an ISDR had important implications that extended beyond its use as a predictor of therapeutic outcome. In particular, the ISDR also provided the first glimpse of a molecular mechanism by which HCV might evade the interferon response. Although the mechanism by which this might occur was not examined in these studies, it was suggested that mutations in the ISDR might somehow suppress the replication of HCV and thus increase susceptibility to interferon. Alternatively, NS5A, through an ISDR-dependent mechanism, might interact with one or more antiviral proteins induced by interferon to enable HCV to avoid the effects of interferon therapy. Our laboratory has had a long-standing interest in defining the mechanisms by which certain viruses evade the interferon response, and in particular, their ability to inhibit the interferon-induced protein kinase, PKR. The reports of an ISDR therefore prompted us to investigate whether

NS5A might directly regulate PKR activity. Recently, we demonstrated that NS5A, through an ISDR-dependent mechanism, directly interacts with and inhibits PKR (GALE et al. 1997). Before delving into a discussion of this interaction, however, an overview of PKR is in order.

4 The Interferon-Induced Double-Stranded RNA-Activated Protein Kinase, PKR

One of the most widely studied components of the cellular interferon response is the dsRNA-activated protein kinase, PKR. Best known for its antiviral activities, it is becoming increasingly apparent that PKR also plays an important role in the regulation of normal cell growth and gene expression. In the following sections, we provide an overview of the many cellular functions associated with this kinase and discuss in some detail the various strategies used by viruses to regulate PKR function. Since a number of these strategies target specific aspects of PKR structure, we begin by providing a brief overview of the physical characteristics of PKR (for recent comprehensive reviews on PKR structure, see CLEMENS and ELIA 1997; SAMUEL et al. 1997; WEK 1994).

4.1 Physical Characteristics

PKR can be roughly divided into an NH_2-terminal regulatory domain, and a COOH-terminal catalytic domain (Fig. 2), both of which are targets for viral regulation. The NH_2-terminal region of PKR contains two dsRNA-binding motifs, the sequences of which are conserved among dsRNA-binding proteins (ST. JOHNSTON et al. 1992). The first dsRNA-binding motif is both necessary and sufficient for dsRNA binding, whereas the second motif, although apparently not essential, is required for optimal dsRNA-binding activity (BARBER et al. 1995; MCCORMACK et al. 1994). Binding of dsRNA is necessary for PKR activation, though the exact nature of the activation process is not understood (GREEN et al. 1995; PATEL and SEN 1992; ROMANO et al. 1995). Binding of dsRNA is thought to result in a conformational change, which is followed by autophosphorylation on several serine and threonine residues (TAYLOR et al. 1996). There is also the intriguing suggestion that the dsRNA-binding motifs may also function to target PKR to the ribosome, thereby providing PKR access to its substrate, eIF-2α (ZHU et al. 1997).

Autophosphorylation appears to be the result of an intermolecular interaction between two PKR molecules (THOMIS and SAMUEL 1993, 1995), and there is evidence that dimerization is required for both autophosphorylation and subsequent catalytic activity (PATEL et al. 1995; ROMANO et al. 1995). There is some controversy, however, as to whether dimerization is mediated by dsRNA (acting as a bridge between two PKR molecules) or whether dimerization occurs as a result of

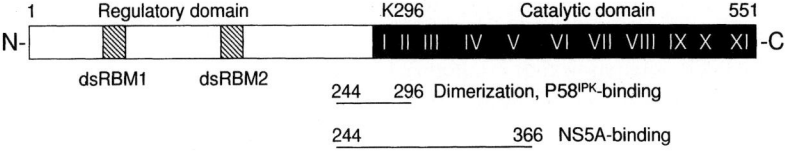

Fig. 2. Structural features of PKR. The position of the two dsRNA-binding motifs (dsRBM1 and dsRBM2) are indicated has *hatched boxes*. The catalytic domain spans amino acids 265–551 and includes the 11 conserved subdomains (indicated by *Roman numerals*) present in all protein kinases (HANKS et al. 1988). Mutation of the conserved lysine in subdomain II (K296) to arginine renders the kinase catalytically inactive. Binding of P58^{IPK} to amino acids 244–296 prevents PKR dimerization, leading to inactivation of the kinase (TAN et al. 1998). The NS5A-binding region of PKR (amino acids 244–366) overlaps this region

direct protein-protein interaction (CARPICK et al. 1997; PATEL et al. 1996; ROMANO et al. 1995; WU and KAUFMAN 1996, 1997). Recently, using the λ phage repressor assay (TAN and KATZE 1998), we identified a region of PKR (amino acids 244–296) that can mediate the dimerization process independently of dsRNA-binding (TAN et al. 1998). As we will describe later, this region is also the target of a number of PKR regulatory proteins, including NS5A.

The protein kinase catalytic domain of PKR, located in the COOH-terminal region of the protein, contains all of the conserved subdomains specific for members of the protein kinase family (Fig. 2). In addition, PKR (as well as other eIF-2α kinases) contains a unique spacer region, referred to as the kinase insert domain, located between kinase subdomains IV and V (MEURS et al. 1990). Deletion of the kinase insert domain abrogates kinase function without affecting substrate recognition, indicating that this unique domain is required for catalytic activity (CRAIG et al. 1996). Activation of PKR is also accompanied by autophosphorylation on multiple threonine residues in the region between catalytic domains VII and VIII, referred to as the activation loop (ROMANO et al. 1998). Phosphorylation of these residues is required for high-level kinase activity. We and others have demonstrated that the region of PKR required for substrate recognition is localized to the COOH-terminal side of the kinase insert domain (GALE et al. 1996; LIU et al. 1997), and at least one viral gene product appears to inhibit PKR function by interfering with substrate recognition.

4.2 Cellular Functions

What are the cellular functions of PKR that make it such an important target for viral regulation? First off, PKR is a primary regulator of mRNA translation initiation rates, mediated through its phosphorylation of serine residue 51 on the α subunit of eukaryotic initiation factor-2 (eIF-2α; CLEMENS 1996; HERSHEY 1991; MERRICK and HERSHEY 1996). In mammalian cells, eIF-2, together with initiator Met-tRNA and guanosine triphosphate (GTP), binds to the 40S ribosomal subunit prior to the binding of mRNA. Upon association of this complex with the 60S ribosomal subunit, eIF-2-bound GTP is hydrolyzed to guanosine diphosphate (GDP). For eIF-2 to promote another round of initiation, the bound GDP must be displaced by GTP, a reaction that is catalyzed by the guanine exchange factor, eIF-2B. Phosphorylation of the α subunit of eIF-2 inhibits the function of eIF-2B, resulting in an inhibition of translation initiation and a decrease in the rate of protein synthesis. Through this mechanism, activation of PKR results in a reduction in viral protein synthesis and replication.

PKR also plays a role in specific signal transduction pathways leading to the transcriptional regulation of gene expression, and in the induction of apoptosis (reviewed in PROUD 1995; WILLIAMS 1995). For example, there is evidence that PKR may activate the transcription factor NF-κB by phosphorylation of its inhibitor, IκB (KUMAR et al. 1994; MARAN et al. 1994). In addition, PKR is required for the antiproliferative activity of interferon regulator factor 1 (IRF-1), a transcription factor that plays an important role in the expression of interferon and interferon-induced genes (KIRCHHOFF et al. 1995). Consistent with these findings, cells derived from *PKR* knockout mice are deficient in the activation of both NF-κB and IRF-1 (DER et al. 1997; KUMAR et al. 1997; YANG et al. 1995). PKR-deficient cells are also resistant to apoptosis in response to dsRNA or tumor necrosis factor-α (TNF-α) (DER et al. 1997). There is also evidence that PKR is activated in response to platelet-derived growth factor (PDGF) and that PKR participates in the PDGF signal transduction pathway, which leads, in turn, to the induction of both c-*fos* and c-*myc* (MUNDSCHAU and FALLER 1995). Finally, calcium depletion from the endoplasmic reticulum also activates PKR (PROSTKO et al. 1995; SRIVASTAVA et al. 1995). The mobilization of calcium plays an important role in cell cycle control (WHITAKER and PATEL 1990) and the activation of PKR in response to depletion of calcium from endoplasmic reticulum stores may represent a unique signaling mechanism. Thus, PKR may participate in the regulation of cellular gene expression through multiple pathways.

Although PKR is best known as a mediator of the antiviral effects of interferon, it is likely that PKR also plays a critical role in the regulation of normal cell growth and gene expression (reviewed in JARAMILLO et al. 1995; KORTH and KATZE 1997; PROUD 1995; SAMUEL et al. 1997). Early evidence for this was observed in yeast, where expression of human PKR results in significant growth suppression due to enhanced levels of eIF-2α phosphorylation. PKR shows significant homology to the yeast protein kinase GCN2 which, when overexpressed in yeast, also mediates a slow-growth phenotype (CHONG et al. 1992; WEK 1994). In mammalian cells, there is evidence that PKR acts as a suppressor of cell proliferation and tumorigenesis. The expression of catalytically inactive forms of human PKR in mouse NIH 3T3 cells results in a transformed phenotype, and the injection of nude mice with PKR mutant-transformed cells results in the rapid development of large tumors (KOROMILAS et al. 1992; MEURS et al. 1993). We have also demonstrated that PKR lacking the dsRNA binding domain I can likewise transform NIH 3T3 cells (BARBER et al. 1995a,b) and that overexpression of a cellular inhibitor of PKR, P58^{IPK}, also leads to malignant transformation (BARBER et al. 1994). Although the mechanism by which mutant PKR transforms cells is not clear, it has been reported that mutant PKR forms inactive heterodimers with wild-type PKR, thereby reducing PKR activity and the level of eIF-2α phosphorylation (COSENTINO et al. 1995; PATEL et al. 1995). Consistent with the idea that PKR mediates cell growth through phosphorylation of eIF-2α, the overexpression of an eIF-2α mutant that cannot be phosphorylated (but not wild-type eIF-2α) also transforms NIH 3T3 cells (DONZÉ et al. 1995). Together, these studies suggest that suppression of PKR, through its expression or activity, may be one pathway leading to malignant cell growth.

4.3 Viral Regulation of PKR

During their replicative cycle, nearly all viruses synthesize dsRNAs, or RNAs with extensive secondary structure, which can serve as potent activators of PKR. To avoid the reduction in viral protein synthesis and replication associated with PKR activation, many viruses have evolved mechanisms to down-regulate PKR function. Viral strategies to down-regulate PKR are surprisingly diverse, and nearly every aspect of PKR structure and function has become a target for viral inhibition. In the following sections, we discuss the strategies used by a number of viruses, which were chosen since they are representative of the different strategies used to down-regulate PKR function (summarized in Fig. 3). We also refer the reader to several additional recent reviews on viral regulation of PKR (GALE and KATZE 1998; KATZE 1995, 1996).

4.3.1 Adenovirus and Reovirus: Inhibition of PKR Activation

The regulatory domain of a protein is an obvious point of attack if one hopes to alter the protein's function, and several viruses have devised strategies to inhibit PKR by interfering with the dsRNA-mediated activation of the kinase. The first reports of such a mechanism came from studies of adenovirus-infected cells, in particular, with the description and characterization of the adenovirus mutant, dl331 (MATHEWS and SHENK 1991). During infection by dl331, which is deficient in the synthesis of a short RNA polymerase III-transcribed RNA termed virus-associated RNA I (VA$_I$), both cellular and viral protein synthesis are inhibited (THIMMAPPAYA et al. 1982). In the absence of VA$_I$, activation of PKR by viral RNAs results in excessive eIF-2α phosphorylation and a depletion of functional eIF-2 (KITAJEWSKI et al. 1986; SIEKIERKA et al. 1985). In contrast, in cells infected by wild-type adenovirus, VA$_I$ RNA accumulates to very high levels within the cytoplasm, where it binds directly the NH$_2$-terminal regulatory domain of PKR (GALABRU et al. 1989; KATZE et al. 1987; MELLITS et al. 1990). In doing so, VA$_I$ RNA appears to inactivate PKR by functioning as a competitive inhibitor of the binding of dsRNA activators (CLARKE and MATHEWS 1995; KATZE et al. 1991). Interestingly, in addition to inactivation of PKR, adenovirus employs a second major strategy to avoid the antiviral effects of interferon, by inhibiting interferon-inducible gene expression through the action of the viral EIA gene products (KELVAKOLANU et al. 1991; REICH et al. 1988).

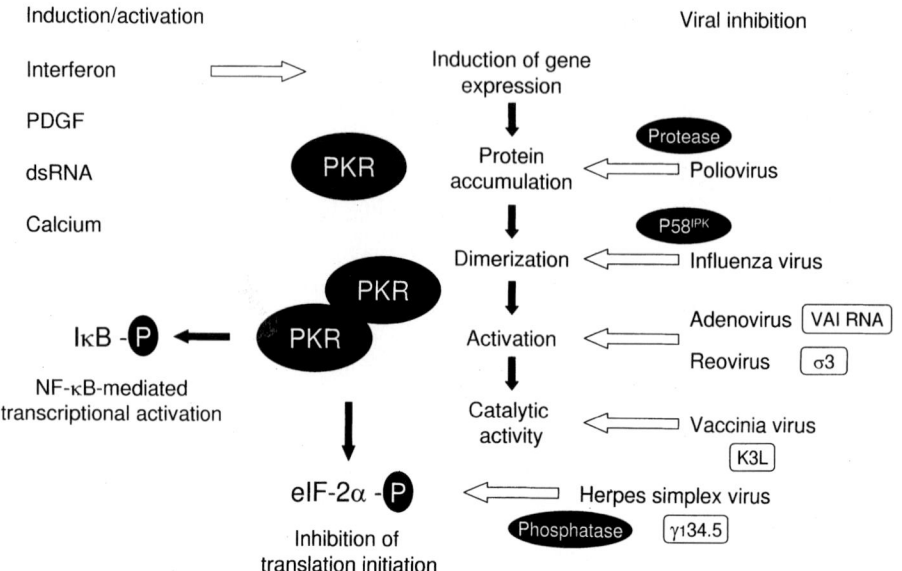

Fig. 3. Cellular functions of PKR and targets of viral inhibition. *Left*, PKR is an interferon-induced dsRNA-activated protein kinase whose substrates include eIF-2α and IκB. PKR-mediated phosphorylation of eIF-2α results in an inhibition of translation initiation and a block to viral protein synthesis and replication. *Right*, Viruses employ a variety of strategies to inhibit PKR activation and function. Adenovirus, reovirus, and vaccinia virus encode viral factors that are responsible for inhibiting the dsRNA-mediated activation of PKR, or for inhibiting PKR's catalytic activity by interfering with substrate binding. Alternatively, viruses such as poliovirus, influenza virus, and HSV recruit cellular proteins to inhibit PKR function. These cellular proteins include a protease to degrade PKR, an inhibitor that interferes with PKR dimerization, and a phosphatase that may reverse PKR-mediated phosphorylation of eIF-2α

Whereas adenovirus encodes an RNA that binds directly to the PKR regulatory domain, thus preventing PKR activation, reovirus uses a variation on this approach to achieve the same end result. Reoviruses are nonenveloped viruses that contain ten segments of dsRNA surrounded by a double shell of capsid proteins. These dsRNAs are potent activators of PKR, making it imperative that reovirus encode a mechanism to down-regulate PKR activity. Somewhat surprisingly, it is a structural component of the outer capsid, the σ3 protein, that appears to serve this purpose. In addition to its role as a structural protein, σ3 is also a dsRNA-binding protein (DENZLER and JACOBS 1994; MILLER and SAMUEL 1992). It appears that σ3, similar to the vaccinia virus *E3L* gene product (discussed below), interferes with PKR activation by sequestering dsRNA activators (IMANI and JACOBS 1988; LLOYD and SHATKIN 1992). Indeed, insertion of the reovirus S4 gene (encoding σ3) into a vaccinia virus mutant lacking *E3L* reverses the interferon-sensitive phenotype of the *E3L* mutant (BEATTIE et al. 1995).

4.3.2 Vaccinia Virus: Inhibition of PKR Catalytic Activity

Vaccinia virus encodes at least two gene products that act by distinct mechanisms to interfere with PKR function. The vaccinia virus *E3L* gene product is a dsRNA-binding protein that inhibits PKR by sequestering dsRNA activators (CHANG et al. 1992; WATSON et al. 1991). Although wild-type vaccinia virus is resistant to the antiviral effects of interferon, viral mutants lacking *E3L* are interferon-sensitive (BEATTIE et al. 1995). In addition to interfering with PKR activation, vaccinia virus has also targeted what may be the second obvious structural feature of PKR, the catalytic domain. The vaccinia virus *K3L* gene product binds directly to PKR and appears to inhibit kinase activity by interfering with the ability of PKR to bind eIF-2α (CARROLL et al. 1993; DAVIES et al. 1992). We and others have demonstrated that K3L binds to PKR between amino acids 366 and 415, deep into the catalytic cleft of the kinase (GALE et al. 1996; LIU et al. 1997). Thus, K3L appears to inhibit PKR by acting as a pseudosubstrate. Deletion of the *K3L* gene from vaccinia virus also results in a loss of PKR inhibitory activity and an increase in sensitivity to the antiviral effects of interferon (BEATTIE et al. 1991).

4.3.3 Influenza Virus: Disruption of PKR Dimerization

Our studies on influenza virus have helped to reveal a novel and intricate PKR regulatory pathway. Influenza virus, rather than encoding a viral gene product to repress PKR, activates a pre-existing cellular stress response pathway to down-regulate PKR function. This influenza virus-activated regulatory pathway is quite complex and likely involves both novel and established members of the stress-response family of proteins. We initially observed that influenza virus infection results in the activation of a cellular 58kDa PKR inhibitor, which we termed P58^{IPK} (for *i*nhibitor of *p*rotein *k*inase) (LEE et al. 1990, 1992). In vitro assays using recombinant proteins indicate that P58^{IPK} directly interacts with PKR, resulting in inhibition of both PKR autophosphorylation and activity (POLYAK et al. 1996). Using the yeast two-hybrid system, we have also demonstrated that P58^{IPK} directly interacts with PKR in vivo and have mapped the P58^{IPK}-interactive region of PKR to amino acids 244–296 (GALE et al. 1996). This region spans the regulatory and catalytic borders of PKR and includes the ATP-binding region of the protein kinase catalytic domain (Fig. 2). We have recently shown that this region of PKR is also required for dimerization, and have demonstrated that binding of P58^{IPK} to this site disrupts the formation of PKR dimers (TAN et al. 1998), thus establishing a novel mechanism by which PKR activity may be regulated. As will be discussed below, we also have evidence that the NS5A protein of HCV may disrupt PKR dimer formation in a similar fashion.

P58^{IPK} is a member of the tetratricopeptide repeat (TPR) family of proteins and possesses nine tandemly arranged TPR motifs. TPR motifs are found in a wide variety of proteins and mediate both homotypic and heterotypic protein-protein interactions (LAMB et al. 1995). Indeed, the TPR6 domain of P58^{IPK} is essential for interaction with PKR and inhibition of kinase function (GALE et al. 1996; TANG et al. 1996). In addition, P58^{IPK} possesses a COOH-terminal DnaJ motif, a motif that likewise has been implicated in mediating protein-protein interactions (SILVER and WAY 1993) and that is also required for inhibition of PKR activity in vivo (TANG et al. 1996).

Although P58^{IPK} is constitutively expressed in mammalian cells (KORTH et al. 1996), the PKR inhibitory activity of P58^{IPK} is kept in check by one or more regulatory proteins (LEE et al. 1994). Influenza virus infection appears to disrupt this regulatory complex, thereby activating the PKR inhibitory activity of P58^{IPK}. We do not yet know whether an influenza virus-encoded protein is responsible for activating P58^{IPK} or whether P58^{IPK} is activated as a result of a more generalized cellular stress response. Using biochemical purification together with in vitro functional assays for P58^{IPK} inhibition, we found that the molecular chaperone, heat shock protein 40 (Hsp40; CRAIG et al. 1993; MORIMOTO et al. 1994), is responsible for keeping P58^{IPK} in an inactive state in uninfected cells (MELVILLE et al. 1997). We have more recently used the yeast two-hybrid system to identify an additional regulator of P58^{IPK}, which we refer to as P52^{rIPK} (for regulator of the inhibitor of protein kinase; GALE et al. 1998). P52^{rIPK} is a novel protein that contains homology to a segment of the molecular chaperone, Hsp90 (HICKEY et al. 1989). Thus, through the regulation of P58^{IPK}, the PKR pathway appears to intersect cellular stress response, growth-regulatory and interferon-regulated pathways.

4.3.4 Poliovirus: Directed Degradation of PKR

Similar to influenza virus, poliovirus appears to recruit cellular factors to deal with PKR. However, rather than modulating PKR function by interfering with kinase activation or activity, poliovirus has chosen to eliminate PKR altogether. We have found that the physical level of PKR is dramatically reduced in poliovirus-infected cells, most likely due to a poliovirus-activated cellular pathway that leads to PKR degradation (BLACK et al. 1989). Although the protease responsible for PKR degradation has not been defined, we believe it is a cellular protease that acts together with viral (or possibly cellular) dsRNA to target PKR for destruction (BLACK et al. 1993).

4.3.5 Herpes Simplex Virus: Reversal of PKR-Mediated Phosphorylation

Instead of directly interfering with PKR function, herpes simplex virus (HSV) may have found a way to maintain normal protein synthetic rates even in the presence of an activated and functional PKR. In cells infected with HSV, PKR is activated, but phosphorylation of eIF-2α is not observed and protein synthesis is unaffected (CHOU et al. 1995). In contrast, in cells infected with a HSV mutant lacking the $\gamma_1$34.5 gene, an increase in eIF-2α phosphorylation is observed, suggesting that $\gamma_1$34.5 is responsible for inhibiting PKR function (HE et al. 1997). The mechanism by which $\gamma_1$34.5 functions to inhibit PKR may, however, be unique. It was recently demonstrated that the $\gamma_1$34.5 gene product interacts with a cellular type 1a phosphatase (HE et al. 1997). It has been proposed that $\gamma_1 34.5$ may interact with and direct the phosphatase to reverse the PKR-mediated phosphorylation of eIF-2α, thereby maintaining normal protein synthetic rates. To make matters even more interesting, there is also evidence that HSV may encode a factor to deliberately activate PKR (MOHR and GLUZMAN 1996). This unusual combination of PKR regulatory mechanisms may enable HSV to selectively prevent phosphorylation of eIF-2α, while allowing PKR to phosphorylate other substrates.

4.3.6 Human Immunodeficiency Virus: A Multi-Pronged Approach

As with almost every aspect of human immunodeficiency virus type-1 (HIV-1), the regulation of PKR in HIV-1-infected cells is quite complex. Rather than relying on a single approach, HIV-1 appears to use multiple strategies to deal with PKR and evade the interferon response. The HIV-1 transactivator responsive region (TAR), present on HIV-1 mRNAs, can form a stable complex with PKR. However, there is some disagreement as to whether this interaction may actually activate, rather than inhibit, the kinase (EDERY et al. 1989; GUNNERY et al. 1990, 1992; MAITRA et al. 1994). It has also been reported that HIV-1 may recruit a cellular TAR RNA-binding protein (TRBP) to inhibit PKR activity (COSENTINO et al. 1995; PARK et al. 1994). At this point, it is not clear whether TRBP inhibits PKR by sequestering activator dsRNAs (such as TAR) or through a direct interaction with the kinase (BENKIRANE et al. 1997). Finally, there is evidence that the HIV-1 transactivator protein, Tat, acts as both a substrate and inhibitor of PKR (BRAND et al. 1997; McMILLAN et al. 1995; ROY et al. 1990). The physiological significance of these interactions in HIV-1-infected cells remains to be determined. Still, it appears that HIV-1 has devoted considerable resources to inhibiting PKR function.

In summary, viruses have evolved diverse strategies to repress PKR function during infection (Fig. 3). In general, viral gene products (or RNAs) are responsible for inhibiting the dsRNA-mediated activation of PKR, or for inhibiting PKR's catalytic activity by interfering with substrate binding. Viruses may also recruit cellular proteins to inhibit PKR function, as exemplified by poliovirus, influenza virus, and HSV. These cellular proteins include a protease to degrade PKR, an inhibitor that interferes with PKR dimerization, and a phosphatase that may reverse PKR-mediated phosphorylation of eIF-2α. As will be discussed in the following section, the strategy used by HCV to repress PKR function is unique, in that a viral gene product is used to regulate PKR function in the same manner as the cellular PKR inhibitor, P58^{IPK}.

5 PKR and NS5A

It is evident that regulation of PKR by a viral gene product is not without precedent. With the publication of reports correlating the sequence of NS5A with sensitivity or resistance to interferon, we became interested in investigating whether NS5A might directly regulate PKR function. These studies led to our recent report that NS5A interacts with and inhibits the function of PKR (GALE et al. 1997) and provide the first evidence for a molecular mechanism underlying HCV resistance to interferon therapy.

5.1 Inhibition of PKR Activity by NS5A

Due to the inability to effectively propagate HCV in tissue culture, we began our analyses using recombinant proteins expressed in yeast. Using the two-hybrid system, we found that NS5A, from two interferon-resistant isolates of HCV-1b and HCV-1a, interacts with the PKR mutant, K296R. This full-length but inactive PKR mutant (containing a Lys → Arg substitution in the catalytic domain) was used for these analyses since expression of the wild-type kinase is growth-suppressive in yeast (ROMANO et al. 1995). We also determined that NS5A interacts with an NH_2-terminal deletion mutant of PKR, which lacks the first dsRNA binding domain and is deficient in its ability to bind dsRNA (BARBER et al. 1995). Thus, NS5A most likely associates with PKR by a direct physical interaction rather than through a dsRNA-dependent mechanism. By employing a series of additional PKR deletion mutants, we also used the two-hybrid system to map the NS5A-interactive region of PKR. We found that NS5A specifically interacts with PKR amino acids 244–366, mapping to a region of PKR that spans the regulatory and catalytic domains (Fig. 2). This region of PKR is predicted to function cooperatively in nucleotide binding and catalysis (BOSSEMEYER 1995; HANKS et al. 1988; TAYLOR et al. 1993) and is required for the formation of active PKR dimers (PATEL et al. 1995; TAN et al. 1998). The implications of this finding with regard to PKR regulation will be discussed shortly.

To determine if the NS5A-PKR interaction resulted in an inhibition of PKR function, we carried out an in vitro analysis of PKR activity in the presence of recombinant NS5A. We found that incubation of purified native PKR with recombinant GST-NS5A results in the inhibition of both PKR autophosphorylation and phosphorylation of an exogenous histone substrate. We confirmed that the loss of PKR activity was not due to degradation of PKR or to NS5A-mediated hydrolysis of ATP. Although it is possible that the inhibition of PKR activation and histone phosphorylation could be due to a GST-NS5A-mediated phosphatase activity, we believe this is unlikely since NS5A does not possess any structural attributes indicative of phosphatase function.

In addition to our in vitro analyses, we examined whether NS5A could regulate PKR function in mammalian cells. Through repression of PKR-mediated eIF-2α phosphorylation, inhibitors of PKR, such as adenovirus VA_I RNA, P58[IPK], and reovirus σ3 protein, can stimulate protein synthesis above basal levels when introduced into mammalian cells (GIANTINI and SHATKIN 1989; KAUFMAN and MURTHA 1987; SELIGER et al. 1992; SVENSSON and AKUSJARVI 1985; TANG et al. 1996). To determine if NS5A expression could likewise alter the level of protein synthesis in mammalian cells, we tested the ability of NS5A to stimulate protein synthesis in COS1 cells using a secreted embryonic alkaline phosphatase (SEAP) reporter assay (TANG et al. 1996). We found that co-expression of SEAP and NS5A in COS1 cells (by transient co-transfection) results in a greater than 300% stimulation of SEAP synthesis over cells co-transfected with SEAP and a vector control. Thus, similar to other PKR inhibitors, NS5A can stimulate protein synthesis in mammalian cells, most likely through a direct inhibitory effect upon PKR. This stimulation of protein synthesis is exactly what infecting viruses need to ensure efficient synthesis of viral proteins.

As an alternative method to more directly examine the effects of NS5A expression on PKR function in vivo, we again employed recombinant proteins expressed in yeast. The growth-suppressive properties of PKR when expressed in yeast provides the basis for a functional assay to directly measure PKR activity (ROMANO et al. 1995). Co-expression of PKR with transdominant inhibitory PKR mutants, or viral-encoded PKR inhibitors such as HIV Tat or vaccinia virus K3L, reverses the PKR-mediated slow-growth phenotype (KAWAGISHI-KOBAYASHI et al. 1997; MCMILLAN et al. 1995). This reversal is due to inhibition of PKR activity, resulting in a decrease in eIF-2α phosphorylation. To assess the effects of NS5A on PKR function in this assay, we expressed NS5A under control of the *GAL1* galactose-inducible promoter in yeast strain RY1-1 (ROMANO et al. 1995). RY1-1 possesses two integrated copies of human *PKR* under control of the *GAL1* promoter and has a PKR-mediated slow growth phenotype when grown on medium containing galactose. We observed that the expression of NS5A in RY1-1 is sufficient to reverse the PKR-mediated growth suppression, demonstrating that NS5A represses PKR function. To confirm that PKR function was repressed in the presence of NS5A, we examined the level of eIF-2α phosphorylation in the RY1-1 transformants. Using isoelectric focusing to resolve phosphorylated and unphosphorylated forms of eIF-2α, we found that expression of NS5A results in an approximately 11-fold increase in the level of unphosphorylated eIF-2α over that observed in control transformants harboring the vector alone. Together, our observations indicate that NS5A, through a direct interaction with PKR, can repress PKR function in vivo resulting in diminished phosphorylation of the PKR substrate, eIF-2α. Thus, by interfering with PKR function, NS5A provides the cell with a mechanism to maintain normal protein synthetic rates and cellular growth. We propose that an analogous situation occurs in HCV-infected liver cells, where NS5A down-regulates PKR function to maintain optimal translation of viral proteins for efficient viral replication.

5.2 Requirement of the Interferon Sensitivity Determining Region for PKR Inhibition

Reports of a correlation between therapeutic outcome and mutations within the ISDR suggest that it is the ISDR that plays a critical role in mediating the interferon-resistant phenotype of HCV. A key question, therefore, was whether the

ISDR contributed to the ability of NS5A to bind to and inhibit PKR. To begin to address this question, we constructed an ISDR deletion mutant (ΔISDR) and examined its ability to interact with PKR in the yeast two-hybrid assay. Significantly, we found that the ΔISDR construct fails to interact with PKR. Deletion of the ISDR from NS5A does not appear to render the mutant protein unstable, since efficient production of the ΔISDR protein can be demonstrated by immunoblot analysis. It is possible, however, that deletion of the ISDR may alter the conformation of NS5A in a fashion that precludes its ability to interact with PKR. Importantly, the ΔISDR protein also fails to reverse PKR-mediated growth suppression in the yeast strain RY1-1. Together, these results indicate that the ISDR is required for both interaction with PKR and inhibition of PKR activity, suggesting a critical role for this region in mediating the function of NS5A.

We have now begun to examine whether specific amino acid substitutions within the ISDR may also compromise the ability of NS5A to repress PKR activity. Preliminary experiments suggest that multiple amino acid substitutions within the ISDR are sufficient to block the ability of NS5A to bind to PKR and inhibit eIF-2α phosphorylation. We are particularly interested in examining the ability of NS5A proteins from interferon-sensitive HCV isolates to interact with and inhibit PKR in order to determine if a correlation exists between the ISDR sequence of these isolates and the ability of NS5A to regulate PKR function. If such a correlation exists, it will provide additional compelling evidence that inhibition of PKR activity is a mechanism used by interferon-resistant HCV isolates to evade the effects of interferon therapy.

5.3 Implications of PKR Regulation by NS5A

NS5A binds to a region of PKR (amino acids 244–366) that includes the region that interacts with the cellular PKR inhibitor, P58^{IPK} (amino acids 244–296; GALE et al. 1996), and which participates in the formation of active PKR dimers (PATEL et al. 1995; TAN et al. 1998). The localization of both the P58^{IPK}-interactive and NS5A-interactive sites to overlapping domains of PKR reflects the importance of this region for kinase function and suggests that these divergent inhibitors may repress PKR by similar mechanisms. Thus, unlike other viral-encoded inhibitors of PKR, which inhibit PKR by functioning as pseudosubstrates or as competitive inhibitors of activator-dsRNA binding, we propose that NS5A may interfere with events leading to PKR activation and catalysis by directly interfering with PKR dimerization (Fig. 4). Although direct experimental evidence is still lacking, NS5A proteins encoded by interferon-sensitive HCV isolates, which contain multiple amino acid substitutions within the ISDR, may lack the ability to interfere with PKR dimerization. Studies to address this issue will be important for an understanding of the significance of the ISDR to PKR regulation and to interferon resistance.

An intriguing possibility is that inhibition of the growth-regulating properties of PKR by NS5A represents a potential mechanism by which HCV infection may lead to tumorigenesis. Although there is a strong association between chronic HCV

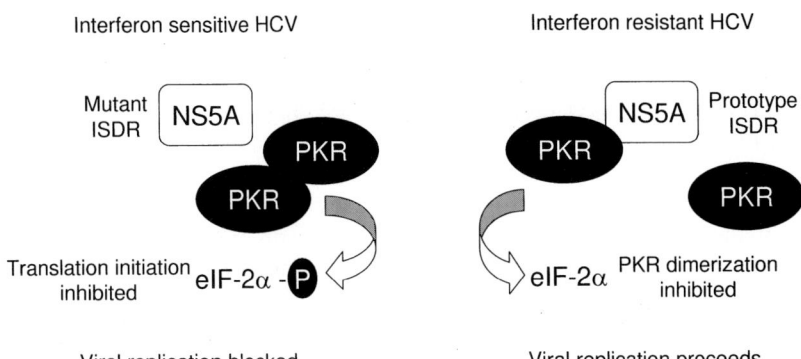

Fig. 4. Model of hepatitis C virus (HCV) resistance to interferon by NS5A-mediated regulation of PKR. *Left*, Interferon sensitive HCV isolates encode an NS5A protein that, due to mutations in the interferon sensitivity determining region (ISDR), is unable to interact with PKR. As a result, PKR phosphorylates eIF-2α, translation initiation is inhibited, and viral replication is blocked. *Right*, Interferon resistant HCV isolates encode an NS5A protein that contains the prototype ISDR sequence and is able to interact with PKR and inhibit kinase activity. We propose that the interaction of NS5A with PKR leads to a disruption of PKR dimer formation. PKR is not active and viral protein synthesis and replication is allowed to proceed

infection and hepatocellular carcinoma (DI BISCEGLIE 1995), it is unclear whether HCV has a direct oncogenic effect on infected hepatocytes or whether continuous cell proliferation, due to chronic liver disease, predisposes infected cells to mutations and malignant transformation (HARUNA et al. 1994). There are indications, however, that HCV proteins may indeed play a direct role in cellular transformation. The HCV NS3 protein transforms NIH 3T3 cells and NS3-transformed cells induce tumor formation in nude mice (SAKAMURO et al. 1995). Similarly, stable expression of the HCV core in Rat-1 cells also results in malignant transformation (CHANG et al. 1998). There are conflicting reports, however, as to whether the HCV core gene, together with H-*ras*, is sufficient to immortalize primary rat embryo fibroblasts (CHANG et al. 1998; RAY et al. 1996). As discussed above, we have shown that overexpression of the cellular PKR inhibitor, P58^{IPK}, results in the malignant transformation of NIH 3T3 cells (BARBER et al. 1994). It is therefore conceivable that inhibition of PKR activity by NS5A may have a similar effect on cell growth and be a contributing factor in the development hepatocellular carcinoma. Studies to examine this possibility are currently being conducted in our laboratory.

Perhaps most importantly, the interaction between NS5A and PKR has significant clinical implications. Given that PKR is one of the primary mediators of the antiviral effects of interferon, the inhibition of PKR activity by NS5A may be a critical factor in the resistance of specific HCV genotypes to interferon therapy. A thorough understanding of the interaction between NS5A and PKR may therefore provide the first step in identifying an agent capable of blocking the interaction. When used in combination with interferon therapy, such an agent may provide a therapeutic benefit by reducing the likelihood of interferon resistance and therefore contribute to more effective treatment regimens against HCV infection.

6 Future Directions

The majority of our studies examining the interaction of NS5A and PKR have been performed in vitro using recombinant proteins, or by the expression of recombinant proteins in yeast. It will be important to carry out analogous studies in mammalian cell culture to more closely approximate the situation in HCV-infected cells. Unfortunately, the task of developing a robust in vitro system to study HCV infection has proven to be a difficult challenge (reviewed in the chapter by N. Kato and K. Shimotohno, this volume). Although there have been reports of HCV replication in tissue culture (DASH et al. 1997; KATO et al. 1995; SHIMIZU et al. 1996; YOO et al. 1995), these systems are currently not optimal for studies examining the interaction of NS5A with PKR or other cellular components. Until such a tissue culture system is developed, the stable expression of NS5A in cell lines provides a useful alternative. We have stably expressed NS5A in NIH 3T3 cells and have also recently developed inducible NS5A-expressing cell lines to examine the effects of NS5A expression on cell growth and PKR function.

In addition to interacting with PKR, NS5A may interact with additional cellular proteins, including other components of the interferon-mediated antiviral response. It is also possible that NS5A may contribute to interferon resistance by interacting with components of an interferon signal transduction pathway, leading to a decrease in the expression of interferon-induced genes. For example, NS5A could potentially affect the Jak-STAT pathway (DARNELL et al. 1994; RANSOHOFF 1998), by acting as an inhibitor of Jak kinases or other pathway components. Such an effect could result in reduced activation of transcription factors, such as ISGF-3 and IRF-1, and a reduction in interferon-induced gene expression. In this regard, it will be important to identify any additional NS5A-interacting proteins and to examine the effects of NS5A expression on the expression of interferon-induced genes.

Finally, it will be important to determine the function of NS5A in HCV replication and pathogenesis to fully understand the role of this protein in mediating the response of HCV to interferon treatment. Additional studies are also needed to further define the ISDR and to determine whether an ISDR exists in HCV genotypes other than 1b. At present, it appears that the sequence of the ISDR is not alone sufficient to predict therapeutic outcome, and the possibility remains that regions outside of the ISDR may also contribute to interferon resistance. For now, the battle between HCV and the host cell continues, with each side claiming its share of victories. However, for the first time, an important viral weapon has been identified. The interaction of NS5A with PKR provides the first evidence of a molecular mechanism underlying HCV resistance to interferon. This interaction may point the way to the development of novel therapeutic strategies to add to the antiviral arsenal and to decrease or eliminate interferon resistance among certain HCV isolates.

Acknowledgements. We thank the past and present members of the Katze laboratory, as well as our many outside collaborators, who contributed to the studies on PKR regulation described in this review. In particular, we acknowledge Michael Gale, Jr., Norina Tang, Seng-Lai Tan, Steven Polyak, David Gretch, and Thomas Dever for their contributions to the NS5A-PKR studies. The work in our laboratory is supported by Public Health Service grants AI 22646, AI 41629, and RR 00166 from the National Institutes of Health, and by grants from the Gustavus and Louise Pfeiffer Research Foundation and RiboGene, Inc. (Hayward, California).

Addendum in proof. Since submission of this chapter, our laboratory has published several additional articles pertaining to the role of the ISDR in mediating the interaction between NS5A and PKR, the antiapoptotic and transforming properties of NS5A, and the ability of NS5A to interact with the cellular adapter protein, Grb2, and to disrupt ERK1/2 phosphorylation in response to growth factor signaling. These articles include:

Gale M, Jr., Blakely CM, Kwieciszewski B, Tan S-L, Dossett M, Korth MJ, Polyak SJ, Gretch DR, Katze MG (1998) Control of PKR protein kinase by hepatitis C virus nonstructural 5A protein: molecular mechanisms of kinase regulation. Mol Cell Biol 18:5208–5218

Gale M, Jr., Kwieciszewski B, Dossett M, Nakao H, Katze MG (1999) Anti-apoptotic and oncogenic potential of hepatitis C virus are linked to interferon resistance by viral repression of the PKR protein kinase. J Virol (in press)

Tan S-L, Nakao H, He Y, Vijaysri V, Neddermann P, Jacobs BL, Mayer BJ, Katze MG (1999) NS5A, a nonstructural protein of hepatitis C virus, binds growth factor receptor-bound protein 2 adaptor protein in a Src homology 3 domain/ligand-dependent manner and perturbs mitogenic signaling. Proc Natl Acad Sci USA 96:5533–5538

References

Asabe SI, Tanji Y, Satoh S, Kaneko T, Kimura K, Shimotohno K (1997) The N-terminal region of hepatitis C virus-encoded NS5A is important for NS4A-dependent phosphorylation. J Virol 71:790–796

Barber GN, Jagus R, Meurs EF, Hovanessian AG, Katze MG (1995a) Molecular mechanisms responsible for malignant transformation by regulatory and catalytic domain variants of the interferon-induced enzyme RNA-dependent protein kinase. J Biol Chem 270:17423–17428

Barber GN, Thompson S, Lee TG, Strom T, Jagus R, Darveau A, Katze MG (1994) The 58-kilodalton inhibitor of the interferon-induced double-stranded RNA-activated protein kinase is a tetratricopeptide repeat protein with oncogenic properties. Proc Natl Acad Sci USA 91:4278–4282

Barber GN, Wambach M, Thompson S, Jagus R, Katze MG (1995b) Mutants of the RNA-dependent protein kinase (PKR) lacking double-stranded RNA binding domain I can act as transdominant inhibitors and induce malignant transformation. Mol Cell Biol 15:3138–3146

Baron S, Tyring SK, Fleischmann WR Jr, Coppenhaver DH, Niesel DW, Klimpel GR, Stanton GJ, Hughes TK (1991) The interferons: mechanisms of action and clinical applications. J Am Med Assoc 266:1375–1383

Beattie E, Denzler KL, Tartaglia J, Perkus ME, Paoletti E, Jacobs BL (1995) Reversal of the interferon-sensitive phenotype of a vaccinia virus lacking E3L by expression of the reovirus S4 gene. J Virol 69:499–505

Beattie E, Tartaglia J, Paoletti E (1991) Vaccinia virus-encoded eIF-2α homolog abrogates the antiviral effect of interferon. Virology 183:419–422

Benkirane M, Neuveut C, Chun RF, Smith SM, Samuel CE, Gatignol A, Jeang K-T (1997) Oncogenic potential of TAR RNA binding protein TRBP and its regulatory interaction with RNA-dependent protein kinase PKR. EMBO J 16:611–624

Black T, Safer B, Hovanessian AG, Katze MG (1989) The cellular 68,000 M_r protein kinase is highly autophosphorylated and activated yet significantly degraded during poliovirus infection: implications for translational regulation. J Virol 63:2244–2252

Black TL, Barber GN, Katze MG (1993) Degradation of the interferon-induced 68,000-M_r protein kinase by poliovirus requires RNA. J Virol 67:791–800

Borden EC, Parkinson D (1998) A perspective on the clinical effectiveness and tolerance of interferon-α. Semin Oncol 25:3–8
Bossemeyer D (1995) Protein kinases – structure and function. FEBS Lett 369:57–61
Brand SR, Kobayashi R, Mathews MB (1997) The Tat protein of human immunodeficiency virus type 1 is a substrate and inhibitor of the interferon-induced, virally activated protein kinase, PKR. J Biol Chem 272:8388–8395
Bukh J, Miller R, Purcell R (1995) Genetic heterogeneity of hepatitis C virus: quasispecies and genotypes. Semin Liver Dis 15:41–63
Carpick BW, Graziano V, Schneider D, Maitra RK, Lee X, Williams BRG (1997) Characterization of the solution structure between the interferon-induced double-stranded RNA-activated protein kinase and HIV-1 transactivating region RNA. J Biol Chem 272:9510–9516
Carroll K, Elroy-Stein O, Moss R, Jagus R (1993) Recombinant vaccinia virus K3L gene product prevents activation of double-stranded RNA-dependent, initiation factor 2 alpha-specific protein kinase. J Biol Chem 268:12837–12842
Chang H-W, Watson JC, Jacobs BL (1992) The E3L gene of vaccinia virus encodes an inhibitor of the interferon-induced, double-stranded RNA-dependent protein kinase. Proc Natl Acad Sci USA 89:4825–4829
Chang J, Yang S-H, Cho Y-G, Hwang SB, Hahn YS, Sung YC (1998) Hepatitis C virus core from two different genotypes has an oncogenic potential but is not sufficient for transforming primary rat embryo fibroblasts in cooperation with the H-ras oncogene. J Virol 72:3060–3065
Chayama K, Tsubota A, Kobayashi M, Okamoto K, Hashimoto M, Miyano Y, Koike H, Koida I, Arase Y, Saitoh S, Suzuki Y, Murashima N, Ikeda K, Kumada H (1997) Pretreatment virus load and multiple amino acid substitutions in the interferon sensitivity-determining region predict the outcome of interferon treatment in patients with chronic genotype 1b hepatitis C virus infection. Hepatology 25:745–749
Chemello L, Bonetti P, Cavalletto L, Talato F, Donadon V, Casarin P, Belussi F, Frezza M, Noventa F, Pontisso P, Benvegnù L, Casarin C, Alberti A, The TriVeneto Viral Hepatitis Group (1995) Randomized trial comparing three different regimens of alpha-2a-interferon in chronic hepatitis C. Hepatology 22:700–706
Chong KL, Feng L, Schappert K, Meurs E, Donahue TF, Friesen JD, Hovanessian AG, Williams BRG (1992) Human P68 kinase exhibits growth suppression in yeast and homology to the translational regulator GCN2. EMBO J 11:1553–1562
Chou J, Chen JJ, Gross M, Roizman B (1995) Association of a M_r 90,000 phosphoprotein with protein kinase PKR in cells exhibiting enhanced phosphorylation of translation initiation factor eIF2α and premature shutoff of protein synthesis after infection with $\gamma 34.5^-$ mutants of herpes simplex virus 1. Proc Natl Acad Sci USA 92:10516–10520
Clarke PA, Mathews MB (1995) Interactions between the double-stranded RNA binding motif and RNA: definition of the binding site for the interferon-induced protein kinase DAI (PKR) on adenovirus VA RNA. RNA 1:7–20
Clemens MJ (1996) Protein kinases that phosphorylate eIF-2 and eIF-2B, and their role in eukaryotic cell translational control. In: Hershey J, Mathews M, Sonenberg N (eds) Translational control. Cold Spring Harbor Laboratory Press, Cold Spring Harbor, pp 139–172
Clemens MJ, Elia A (1997) The double-stranded RNA-dependent protein kinase PKR: structure and function. J Interferon Cytokine Res 17:503–524
Cosentino GP, Venkatesan S, Serluca FC, Green SR, Mathews MB, Sonenberg N (1995) Double-stranded-RNA-dependent protein kinase and TAR RNA-binding protein form homo- and heterodimers in vivo. Proc Natl Acad Sci USA 92:9445–9449
Craig AW, Cosentino GP, Donzé O, Sonenberg N (1996) The kinase insert domain of interferon-induced protein kinase PKR is required for activity but not for interaction with the pseudosubstrate K3L. J Biol Chem 271:24526–24533
Craig EA, Gambill BD, Nelson RJ (1993) Heat shock proteins: molecular chaperones of protein biogenesis. Microbiol Rev 57:402–414
Darnell JE Jr, Kerr IM, Stark GR (1994) Jak-STAT pathways and transcriptional activation in response to IFNs and other extracellular signaling proteins. Science 264:1415–1421
Dash S, Halim A-B, Tsuji H, Hiramatsu N, Gerber MA (1997) Transfection of HepG2 cells with infectious hepatits C virus genome. Am J Pathol 151:363–373
Davies MV, Elroy-Stein O, Jagus R, Moss B, Kaufman RJ (1992) The vaccinia virus K3L gene product potentiates translation by inhibiting double-stranded-RNA-activated protein kinase and phosphorylation of the alpha subunit of eukaryotic initiation factor 2. J Virol 66:1943–1950

Davis GL, Lau JYN (1997) Factors predictive of a beneficial response to therapy of hepatitis C. Hepatology 26:122S–127S

Denzler KL, Jacobs BL (1994) Site-directed mutagenic analysis of reovirus σ3 protein binding to dsRNA. Virology 204:190–199

Der SD, Yang Y-L, Weissman C, Williams BRG (1997) A double-stranded RNA-activated protein kinase-dependent pathway mediating stress-induced apoptosis. Proc Natl Acad Sci USA 94:3279–3283

Di Bisceglie AM (1995) Hepatitis C and hepatocellular carcinoma. Semin Liver Dis 15:64–69

Donzé O, Jagus R, Koromilas AE, Hershey JWB, Sonenberg N (1995) Abrogation of translation initiation factor eIF-2 phosphorylation causes malignant transformation of NIH 3T3 cells. EMBO J 14:3828–3834

Edery I, Petryshyn R, Sonenberg N (1989) Activation of double-stranded RNA-dependent kinase (dsI) by the TAR region of HIV-1 mRNA: a novel translation control mechanism. Cell 56:303–312

Enomoto N, Kurosaki M, Tanaka Y, Marumo F, Sato C (1994) Fluctuation of hepatitis C virus quasispecies in persistent infection and interferon treatment revealed by single-strand conformation polymorphism analysis. J Gen Virol 75:1361–1369

Enomoto N, Sakuma I, Asahina Y, Kurosaki M, Murakami T, Yamamoto C, Ogura Y, Izumi N, Maruno F, Sato C (1996) Mutations in the nonstructural protein 5A gene and response to interferon in patients with chronic hepatitis C virus 1b infection. N Engl J Med 334:77–81

Enomoto N, Sakuma I, Asahina Y, Kurosaki M, Murakami T, Yamamoto C, Izumi N, Marumo F, Sato C (1995a) Comparison of full-length sequences of interferon-sensitive and resistant hepatitis C virus 1b. J Clin Invest 96:224–230

Enomoto N, Sato C (1995b) Hepatitis C virus quasispecies populations during chronic hepatitis C infection. Trends Microbiol 3:445–448

Foster GR (1997) Interferons in host defense. Semin Liver Dis 17:287–295

Fried M, Hoofnagle J (1995) Therapy of hepatitis C. Semin Liver Dis 15:82–91

Galabru J, Katze MG, Robert N, Hovanessian AG (1989) The binding of double-stranded RNA and adenovirus VAI RNA to the interferon-induced protein kinase. Eur J Biochem 178:581–589

Gale M Jr, Katze MG (1998) Molecular mechanisms of interferon resistance mediated by viral-directed inhibition of PKR, the interferon-induced protein kinase. Pharmacol Ther 78:29–46

Gale M Jr, Tan S-L, Wambach M, Katze MG (1996) Interaction of the interferon-induced PKR protein kinase with inhibitory proteins P58^{IPK} and vaccinia virus K3L is mediated by unique domains: implications for kinase regulation. Mol Cell Biol 16:4172–4181

Gale MJ Jr, Blakely CM, Hopkins DA, Melville MW, Wambach M, Romano PR, Katze MG (1998) Regulation of interferon-induced protein kinase PKR: modulation of P58^{IPK} inhibitory function by a novel protein, P52^{rIPK}. Mol Cell Biol 18:859–871

Gale MJ Jr, Korth MJ, Tang NM, Tan SL, Hopkins DA, Dever TE, Polyak SJ, Gretch DR, Katze MG (1997) Evidence that hepatitis C virus resistance to interferon is mediated through repression of the PKR protein kinase by the nonstructural 5 A protein. Virology 230:217–227

Giantini M, Shatkin A (1989) Stimulation of chloramphenicol acetyl transferase mRNA translation by reovirus capsid polypeptide sigma 3 in cotransfected cos cells. J Virol 63:2415–2421

Green SR, Manche L, Mathews MB (1995) Two functionally distinct RNA-binding motifs in the regulatory domain of the protein kinase DAI. Mol Cell Biol 15:358–364

Gretch DR, Polyak SJ, Willson RA, Carithers RL Jr (1996) Treatment of chronic hepatitis C virus infection: a clinical and virological perspective. Adv Exp Med Biol 394:207–224

Gunnery S, Green SR, Mathews MB (1992) Tat-responsive region RNA of human immunodeficiency virus type 1 stimulates protein synthesis in vivo and in vitro: relationship between structure and function. Proc Natl Acad Sci USA 89:11557–11561

Gunnery SA, Rice P, Robertson HD, Mathews MB (1990) Tat-responsive region RNA of human immunodeficiency virus 1 can prevent activation of the double-stranded RNA-activated protein kinase. Proc Natl Acad Sci USA 87:8687–8691

Hanks SK, Quinn AM, Hunter T (1988) The protein kinase family: conserved features and deduced phylogeny of the catalytic domains. Science 241:42–52

Haque SJ, Williams BRG (1998) Signal transduction in the interferon system. Semin Oncol 25:14–22

Haruna Y, Hayashi N, Kamada T, Hytiroglou P, Thung SN, Gerber MA (1994) Expression of hepatitis C virus in hepatocellular carcinoma. Cancer 73:2253–2258

He B, Gross M, Roizman B (1997) The γ34.5 protein of herpes simplex virus 1 complexes with protein phosphatase 1α to dephosphorylate the α subunit of the eukaryotic translation initiation factor 2 and preclude the shutoff of protein synthesis by double-stranded RNA-activated protein kinase. Proc Natl Acad Sci USA 94:843–848

Herion D, Hoofnagle J (1997) The interferon sensitivity determining region: all hepatitis C virus isolates are not the same. Hepatology 25:769–771

Hershey JWB (1991) Translational control in mammalian cells. Annu Rev Biochem 60:717–755

Hickey E, Brandon SE, Smale G, Lloyd D, Weber LA (1989) Sequence and regulation of a gene encoding a human 89 kilodalton heat shock protein. Mol Cell Biol 9:2615–2626

Hijikata M, Mizushima H, Tanji Y, Komoda Y, Hirowatari Y, Akagi T, Kato N, Kimura K, Shimotohno K (1993) Proteolytic processing and membrane association of putative nonstructural proteins of hepatitis C virus. Proc Natl Acad Sci USA 90:10773–10777

Hofgärtner WT, Polyak SJ, Sullivan D, Carithers RL Jr, Gretch DR (1997) Mutations in the NS5A gene of hepatitis C virus in North American patients infected with HCV genotype 1a or 1b. J Med Virol 53:118–126

Hoofnagle JH (1994) Therapy of acute and chronic viral hepatitis. Adv Intern Med 39:241–275

Hoofnagle JH, Mullen KD, Jones DB, Rustgi V, Di Bisceglie A, Peters M, Waggoner JG, Park Y, Jones EA (1986) Treatment of chronic non-A, non-B hepatitis with recombinant human alpha interferon. N Engl J Med 315:1575–1578

Houghton M, Weiner A, Han J, Kuo G, Choo Q-L (1991) Molecular biology of the hepatitis C viruses: implications for diagnosis, development and control of viral disease. Hepatology 14:381–388

Iino S, Hino K, Yasuda K (1994) Current state of interferon therapy for chronic hepatitis C. Intervirology 37:87–100

Imani F, Jacobs BL (1988) Inhibitory activity for the interferon-induced protein kinase is associated with the reovirus serotype 1 sigma 3 protein. Proc Natl Acad Sci USA 85:7887–7891

Jaramillo ML, Abraham N, Bell JC (1995) The interferon system: a review with emphasis on the role of PKR in growth control. Cancer Inv 13:327–338

Kaneko T, Tanji Y, Satoh S, Hijikata M, Asabe S, Kimura K, Shimotohno K (1994) Production of two phosphoproteins from the NS5A region of the hepatitis C viral genome. Biochem Biophys Res Commun 205:320–326

Kato N, Hijikata M, Ootsuyama Y, Nakagawa M, Ohkoshi S, Sugimura T, Shimotohno K (1990) Molecular cloning of the human hepatitis C virus genome from Japanese patients with non-A, non-B hepatitis. Proc Natl Acad Sci USA 87:9524–9528

Kato N, Lan KH, OnoNita SK, Shiratori Y, Omata M (1997) Hepatitis C virus nonstructural region 5 A protein is a potent transcriptional activator. J Virol 71:8856–8859

Kato N, Nakazawa T, Mizutani T, Shimotohno K (1995) Susceptibility of human T-lymphotropic virus type 1 infected cell line MT-2 to hepatitis C virus infection. Biochem Biophys Res Commun 206: 863–869

Katze MG (1995) Regulation of the interferon-induced PKR: can viruses cope? Trends Microbiol 3:75–78

Katze MG (1996) Translational control in cells infected with influenza virus and reovirus. In: Hershey JWB, Mathews MB, Sonenberg N (eds) Translational control. Cold Spring Harbor Laboratory Press, Cold Spring Harbor, pp 607–630

Katze MG, DeCorato D, Safer B, Galabru J, Hovanessian AG (1987) Adenovirus VAI RNA complexes with the 68,000 M_r protein kinase to regulate its autophosphorylation and activity. EMBO J 6: 689–697

Katze MG, Wambach M, Wong M-L, Garfinkel MS, Meurs E, Chong KL, Williams BRG, Hovanessian AG, Barber GN (1991) Functional expression of interferon-induced, dsRNA activated 68,000 Mr protein kinase in a cell-free system. Mol Cell Biol 11:5497–5505

Kaufman RJ, Murtha P (1987) Translational control mediated by eukaryotic initiation factor-2 is restricted to specific mRNAs in transfected cells. Mol Cell Biol 7:1568–1571

Kawagishi-Kobayashi M, Silverman JB, Ung TL, Dever TE (1997) Regulation of the protein kinase PKR by the vaccinia virus pseudosubstrate inhibitor K3L is dependent on residues conserved between the K3L protein and the PKR substrate eIF-2α. Mol Cell Biol 17:4146–4158

Kelvakolanu D, Bandyopohyay S, Harter M, Sen GC (1991) Inhibition of interferon inducible gene expression by adenovirus EIA proteins: block in transcriptional complex formation. Proc Natl Acad Sci USA 88:7459–7463

Khorsi H, Castelain S, Wyseur A, Izopet J, Canva V, Rombout A, Capron D, Capron J-P, Lunel F, Stuyver L, Duverlie G (1997) Mutations of hepatitis C virus 1b NS5A 2209–2248 amino acid sequence do not predict the response to recombinant interferon-alpha therapy in French patients. J Hepatol 27:72–77

Kirchhoff S, Koromilas AE, Schaper F, Grashof M, Sonenberg N, Hauser H (1995) IRF-1 induced cell growth inhibition and interferon induction requires the activity of the protein kinase PKR. Oncogene 11:439–445

Kitajewski J, Schneider RJ, Safer B, Munemitsu SM, Samuel CE, Thimmappaya B, Shenk T (1986) Adenovirus VAI RNA antagonizes the antiviral action of interferon by preventing activation of the interferon-induced eIF-2α kinase. Cell 45:195–200

Koff RS (1997) Therapy in chronic hepatitis C: say goodbye to the 6-month interferon regimen. Am J Gastroenterol 91:2072–2074

Koromilas AE, Roy S, Barber GN, Katze MG, Sonenberg N (1992) Malignant transformation by a mutant of the IFN-inducible dsRNA-dependent protein kinase. Science 257:1685–1689

Korth MJ, Katze MG (1997) mRNA metabolism and cancer. In: Morris D, Harford J (eds) mRNA metabolism and post-transcriptional gene regulation. Wiley-Liss, New York, pp 265–280

Korth MJ, Lyons CN, Wambach M, Katze MG (1996) Cloning, expression, and cellular localization of the oncogenic 58-kDa inhibitor of the RNA-activated human and mouse protein kinase. Gene 170:181–188

Kumar A, Haque J, Lacoste J, Hiscott J, Williams BRG (1994) Double-stranded RNA-dependent protein kinase activates transcription factor NF-κB by phosphorylating IκB. Proc Natl Acad Sci USA 91:6288–6292

Kumar A, Yang Y-L, Flati V, Der S, Kadereit S, Deb A, Haque J, Reis L, Weissmann C, Williams BRG (1997) Deficient cytokine signaling in mouse embryo fibroblasts with a targeted deletion in the PKR gene: role of IRF-1 and NF-κB. EMBO J 16:406–416

Kurosaki M, Enomoto N, Murakami T, Sakuma I, Asahina Y, Yamamoto C, Ikeda T, Tozuka S, Izumi N, Marumo F, Sato C (1997) Analysis of genotypes and amino acid residues 2209 to 2248 of the NS5A region of hepatitis C virus in relation to the response to interferon-β therapy. Hepatology 25:750–753

Lam NP, DeGuzman LJ, Pitrak D, Layden TJ (1994) Clinical and histologic predictors of response to interferon-alpha in patients with chronic hepatitis C viral infection. Dig Dis Sci 39:2660–2664

Lamb JR, Tugendreich S, Hieter P (1995) Tetratrico peptide repeat interactions: to TPR or not to TPR? Trends Biochem Sci 20:257–259

Lee TG, Tang N, Thompson S, Miller J, Katze MG (1994) The 58,000-dalton cellular inhibitor of the interferon-induced double-stranded RNA-activated protein kinase (PKR) is a member of the tetratricopeptide repeat family of proteins. Mol Cell Biol 14:2331–2342

Lee TG, Tomita J, Hovanessian AG, Katze MG (1990) Purification and partial characterization of a cellular inhibitor of the interferon-induced protein kinase of M_r 68,000 from influenza virus-infected cells. Proc Natl Acad Sci USA 87:6208–6212

Lee TG, Tomita J, Hovanessian AG, Katze MG (1992) Characterization and regulation of the 58,000-dalton cellular inhibitor of the interferon-induced, dsRNA-activated protein kinase. J Biol Chem 267:14238–14243

Lindsay KL (1997) Therapy of hepatitis C: overview. Hepatology 26:71S–77S

Liu L-X, Margottin F, Le Gall S, Schwartz O, Selig L, Benarous R, Benichou S (1997) Binding of HIV-1 Nef to a novel thioesterase enzyme correlates with Nef-mediated CD4 down-regulation. J Biol Chem 272:13779–13785

Lloyd RM, Shatkin AJ (1992) Translational stimulation by reovirus polypeptide 3: substitution for VAI RNA and inhibition of phosphorylation of the α subunit of eukaryotic initiation factor 2. J Virol 66:6878–6884

Mahaney K, Tedeschi V, Maertens G, DiBisceglie AM, Vergalla J, Hoofnagle JH, Sallie R (1994) Genotypic analysis of hepatitis C virus in American patients. Hepatology 20:1405–1411

Maitra RK, McMillan N, Desai S, McSwiggen J, Hovanessian AG, Sen G, Williams BRG, Silverman RH (1994) HIV-1 TAR RNA has an intrinsic ability to activate interferon-inducible enzymes. Virology 204:823–827

Maran A, Maitra RK, Kumar A, Dong B, Xiao W, Li G, Williams BRG, Torrence PF, Silverman RH (1994) Blockage of NF-κB signaling by selective ablation of an mRNA target by 2–5A antisense chimeras. Science 265:789–792

Martell M, Esteban JI, Quer J, Genesca J, Weiner A, Esteban R, Guardia J, Gomez J (1992) Hepatitis C virus (HCV) circulates as a population of different but closely related genomes: quasispecies nature of HCV genome distribution. J Virol 66:3225–3229

Martinot-Peignoux M, Marcellin P, Pouteau M, Castelnau C, Boyer N, Poliquin M, Degott C, Descombes I, Le Breton V, Milotova V, Benhamou JP, Erlinger S (1995) Pretreatment serum hepatitis C virus RNA levels and hepatitis C virus genotype are the main and independent prognostic factors of sustained response to interferon alpha therapy in chronic hepatitis C. Hepatology 22:1050–1056

Mathews MB, Shenk T (1991) Adenovirus virus-associated RNA and translation control. J Virol 65:5657–5662

McCormack SJ, Ortega LG, Doohan JP, Samuel CE (1994) Mechanism of interferon action: motif I of the interferon-induced RNA-dependent protein kinase (PKR) is sufficient to mediate RNA-binding activity. Virology 198:92–99

McMillan NAJ, Chun RF, Siderovski DP, Galabru J, Toone WM, Samuel CE, Mak TW, Hovanessian AG, Jeang K-T, Williams BRG (1995) HIV-1 Tat directly interacts with the interferon-induced, double-stranded RNA-dependent kinase, PKR. Virology 213:413–424

Mellits KH, Kostura M, Mathews MB (1990) Interaction of adenovirus VA RNA1 with the protein kinase DAI: nonequivalence of binding and function. Cell 61:843–852

Melville MW, Hansen WJ, Freeman BC, Welch WJ, Katze MG (1997) The molecular chaperone hsp40 regulates the activity of P58^{IPK}, the cellular inhibitor of PKR. Proc Natl Acad Sci USA 94:97–102

Merrick WC, Hershey JWB (1996) The pathway and mechanism of eukaryotic protein synthesis. In: Hershey J, Mathews M, Sonenberg N (eds) Translational control. Cold Spring Harbor Laboratory Press, Cold Spring Harbor, pp 31–70

Meurs E, Chong KL, Galabru J, Thomas N, Kerr I, Williams BRG, Hovanessian AG (1990) Molecular cloning and characterization of the human double-stranded RNA-activated protein kinase induced by interferon. Cell 62:379–390

Meurs EF, Galabru J, Barber GN, Katze MG, Hovanessian AG (1993) Tumor suppressor function of the interferon-induced double-stranded RNA-activated protein kinase. Proc Natl Acad Sci USA 90:232–236

Miller JE, Samuel CE (1992) Proteolytic cleavage of the reovirus sigma 3 protein results in enhanced double-stranded RNA-binding activity: identification of a repeated basic amino acid motif within the C-terminal binding region. J Virol 66:5347–5356

Mohr I, Gluzman Y (1996) A herpesvirus genetic element which affects translation in the absence of the viral GADD34 function. EMBO J 15:4759–4766

Morimoto RI, Tissieres A, Georgopoulos C (1994) Progress and perspectives on the biology of heat shock proteins and molecular chaperones. In: Morimoto RI, Tissieres A, Georgopoulos C (eds) The biology of heat shock proteins and molecular chaperones. Cold Spring Harbor Laboratory Press, Cold Spring Harbor, pp 1–30

Mundschau LJ, Faller DV (1995) Platelet-derived growth factor signal transduction through the interferon-inducible kinase PKR. J Biol Chem 270:3100–3106

Müller U, Steinhoff U, Reis LFL, Hemmi S, Pavlovic J, Zinkernagel RM, Aguet M (1994) Functional role of type I and type II interferons in antiviral defense. Science 264:1918–1921

Neddermann P, Tomei L, Steinkühler C, Gallinari P, Tramontano A, De Francesco R (1997) The nonstructural proteins of the hepatitis C virus: structure and functions. Biol Chem 378:469–476

Okamoto H, Mishiro S (1994) Genetic heterogeneity of hepatitis C virus. Intervirology 37:68–76

Park H, Davies MV, Langland JO, Chang H-W, Nam YS, Tartaglia J, Paoletti E, Jacobs BL, Kaufman RJ, Venkatesan S (1994) TAR RNA-binding protein is an inhibitor of the interferon-induced protein kinase PKR. Proc Natl Acad Sci USA 91:4713–4717

Patel R, Sen GC (1992) Identification of the double stranded RNA-binding domain of the human interferon-inducible protein kinase. J Biol Chem 267:7871–7876

Patel RC, Stanton P, McMillan NMJ, Williams BRG, Sen GC (1995) The interferon-inducible double-stranded RNA-activated protein kinase self-associates in vitro and in vivo. Proc Natl Acad Sci USA 92:8283–8287

Patel RC, Stanton P, Sen GC (1996) Specific mutations near the amino terminus of double-stranded RNA-dependent protein kinase (PKR) differentially affect its double-stranded RNA binding and dimerization properties. J Biol Chem 271:25657–25663

Pawlotsky J-M, Germanidis G, Neumann AU, Pellerin M, Frainais P-O, Dhumeaux D (1998) Interferon resistance of hepatitis C virus genotype 1b: relationship to nonstructural 5 A gene quasispecies mutations. J Virol 72:2795–2805

Polyak SJ, Faulkner G, Carithers RL Jr, Corey L, Gretch DR (1997) Assessment of hepatitis C virus quasispecies heterogeneity by gel shift analysis: correlation with response to interferon therapy. J Infect Dis 175:1101–1107

Polyak SJ, Tang N, Wambach M, Barber GN, Katze MG (1996) The p58 cellular inhibitor complexes with the interferon-induced, double-stranded RNA-dependent protein kinase, PKR, to regulate its autophosphorylation and activity. J Biol Chem 271:1702–1707

Prostko CR, Dholakia JN, Brostrom MA, Brostrom CO (1995) Activation of the double-stranded RNA-regulated protein kinase by depletion of endoplasmic reticular calcium stores. J Biol Chem 270:6211–6215

Proud CG (1995) PKR: a new name and new roles. Trends Biochem Sci 20:241–246

Ransohoff RM (1998) Cellular responses to interferons and other cytokines: the Jak-STAT paradigm. N Engl J Med 338:616–618

Ray RB, Lagging LM, Meyer K, Ray R (1996) Hepatitis C virus core protein cooperates with ras and transforms primary rat embryo fibroblasts to tumorigenic phenotype. J Virol 70:4438–4443

Reed KE, Xu J, Rice CM (1997) Phosphorylation of the hepatitis C virus NS5A protein in vitro and in vivo: properties of the NS5A-associated kinase. J Virol 71:7187–7197

Reich N, Pine R, Levy D, Darnell JE (1988) Transcription of interferon-stimulated genes is induced by adenovirus particles but is suppressed by EIA gene products. J Virol 62:114–119

Romano PR, Garcia-Barrio MT, Zhang X, Wang Q, Taylor DR, Zhang F, Herring C, Mathews MB, Qin J, Hinnebusch AG (1998) Autophosphorylation in the activation loop is required for full kinase activity in vivo of human and yeast eukaryotic initiation factor 2α kinases PKR and GCN2. Mol Cell Biol 18:2282–2297

Romano PR, Green SR, Barber GN, Mathews MB, Hinnebusch AG (1995) Structural requirements for double-stranded RNA binding, dimerization, and activation of the human eIF-2α kinase DAI in *Saccharomyces cerevisiae*. Mol Cell Biol 15:365–378

Roy S, Katze MG, Parkin NT, Edery I, Hovanessian AG, Sonenberg N (1990) Control of the interferon-induced 68-kilodalton protein kinase by the HIV-1 tat gene product. Science 247:1216–1219

Sakamuro D, Furukawa T, Takegami T (1995) Hepatitis C virus nonstructural protein NS3 transforms NIH 3T3 cells. J Virol 69:3893–3896

Samuel CE, Kuhen KL, George CX, Ortega LG, Rende-Fournier R, Tanaka H (1997) The PKR protein kinase: an interferon-inducible regulator of cell growth and differentiation. Int J Hematol 65:227–237

Seliger LS, Giantini M, Shatkin AJ (1992) Translational effects and sequence comparisons of the three serotypes of the reovirus S4 gene. Virology 187:202–210

Shimizu YK, Feinstone SM, Kohara M, Purcell RH, Yoshikura H (1996) Hepatitis C virus: detection of intracellular virus particles by electron microscopy. Hepatology 23:205–209

Shimizu YK, Hijikata M, Iwamoto A, Alter HJ, Purcell RH, Yoshikura H (1994) Neutralizing antibodies against hepatitis C virus and the emergence of neutralization escape mutant viruses. J Virol 68:1494–1500

Shukla DD, Hoyne PA, Ward CW (1995) Evaluation of complete genome sequences and sequences of individual gene products for the classification of hepatitis C viruses. Arch Virol 140:1747–1761

Siekierka J, Mariano TM, Reichel PA, Mathews MB (1985) Translational control by adenovirus: lack of virus-associated RNA1 during adenovirus infection results in phosphorylation of initiation factor eIF-2 and inhibition of protein synthesis. Proc Natl Acad Sci USA 82:1959–1963

Silver PA, Way JC (1993) Eukaryotic DnaJ homologs and the specificity of Hsp70 activity. Cell 74:5–6

Squadrito G, Leone F, Sartori M, Nalpas B, Berthelot P, Raimondo G, Pol S, Bréchot C (1997) Mutations in the nonstructural 5 A region of hepatitis C virus and response of chronic hepatitis C to interferon alpha. Gastroenterology 113:567–572

Srivastava SP, Davies MV, Kaufman RJ (1995) Calcium depletion from the endoplasmic reticulum activates the double-stranded RNA-dependent protein kinase (PKR) to inhibit protein synthesis. J Biol Chem 270:16619–16624

St. Johnston D, Brown NH, Gall JG, Jantsch M (1992) A conserved double-stranded RNA-binding domain. Proc Natl Acad Sci USA 89:10979–10983

Svensson C, Akusjarvi G (1985) Adenovirus VA RNA$_I$ mediates a translational stimulation which is not restricted to viral mRNAs. EMBO J 4:957–964

Tan S-L, Gale MJ Jr, Katze MG (1998) Double-stranded RNA-independent dimerization of interferon-induced protein kinase PKR and inhibition of dimerization by the cellular P58^{IPK} inhibitor. Mol Cell Biol (in press)

Tan S-L, Katze MG (1998) Using genetic means to dissect homo- and heterotypic interactions with PKR, the interferon-induced protein kinase. Methods Companion Methods Enzymol (in press)

Tang NM, Ho CY, Katze MG (1996) The 58-kDa cellular inhibitor of the double stranded RNA-dependent protein kinase requires the tetratricopeptide repeat 6 and DnaJ motifs to stimulate protein synthesis in vivo. J Biol Chem 271:28660–28666

Tanimoto A, Ide Y, Arima N, Sasaguri Y, Padmanabhan R (1997) The amino terminal deletion mutants of hepatitis C virus nonstructural protein NS5A function as transcriptional activators in yeast. Biochem Biophys Res Commun 236:360–364

Tanji Y, Kaneko T, Satoh S, Shimotohno K (1995) Phosphorylation of hepatitis C virus-encoded nonstructural protein NS5A. J Virol 69:3980–3986

Taylor DR, Lee SB, Romano PR, Marshak DR, Hinnebusch AG, Esteban M, Mathews MB (1996) Autophosphorylation sites participate in the activation of the double-stranded-RNA-activated protein kinase PKR. Mol Cell Biol 16:6295–6302

Taylor SS, Knighton DR, Zheng J, Sowadski JM, Gibbs CS, Zoller MJ (1993) A template for the protein kinase family. Trends Biochem Sci 18:84–89

Thimmappaya B, Weinberger C, Schneider RJ, Shenk T (1982) Adenovirus VA1 RNA is required for efficient translation of viral mRNAs at late times after infection. Cell 31:543–551

Thomis DC, Samuel CE (1993) Mechanism of interferon action: evidence for intermolecular autophosphorylation and autoactivation of the interferon-induced, RNA-dependent protein kinase PKR. J Virol 67:7695–7700

Thomis DC, Samuel CE (1995) Mechanism of interferon action: characterization of the intermolecular autophosphorylation of PKR, the interferon-inducible, RNA-dependent protein kinase. J Virol 69:5195–5198

Tsubota A, Chayama K, Ikeda K, Yasuji A, Koida I, Saitoh S, Hashimoto M, Iwasaki S, Kobayashi M, Hiromitsu K (1994) Factors predictive of response to interferon-α therapy in hepatitis C virus infection. Hepatology 19:1088–1094

van Doorn L-J (1994) Molecular biology of the hepatitis C virus. J Med Virol 43:345–356

Vilcek J, Sen GC (1996) Interferons and other cytokines. In: Fields BN, Knipe DM, Howley PM (eds) Fields virology. Lippincott-Raven, Philadelphia, pp 375–399

Wang C, Le S-Y, Ali N, Siddiqui A (1995) An RNA pseudoknot is an essential structural element of the internal ribosome entry site located within the hepatitis C virus 5′ noncoding region. RNA 1:526–537

Wang C, Sarnow P, Siddiqui A (1994) A conserved helical element is essential for internal initiation of translation of hepatitis C virus RNA. J Virol 68:7301–7307

Watson JC, Chang H-W, Jacobs BL (1991) Characterization of a vaccinia virus-encoded double-stranded RNA-binding protein that may be involved in inhibition of the double-stranded RNA-dependent protein kinase. Virology 185:206–216

Wek RC (1994) eIF-2 kinases: regulators of general and gene-specific translation initiation. Trends Biochem Sci 19:491–496

Whitaker M, Patel R (1990) Calcium and cell cycle control. Development 108:525–542

Williams BRG (1995) The role of the dsRNA-activated kinase, PKR, in signal transduction. Semin Virol 6:191–202

Wu S, Kaufman RJ (1996) Double-stranded (ds) RNA binding and not dimerization correlates with the activation of the dsRNA-dependent protein kinase (PKR). J Biol Chem 271:1756–1763

Wu SY, Kaufman RJ (1997) A model for the double-stranded RNA (dsRNA)-dependent dimerization and activation of the dsRNA-activated protein kinase PKR. J Biol Chem 272:1291–1296

Yang Y-L, Reis LFL, Pavlovic J, Aguzzi A, Schäfer R, Kumar A, Williams BRG, Aguet M, Weissmann C (1995) Deficient signaling in mice devoid of double-stranded RNA-dependent protein kinase. EMBO J 14:6095–6106

Yoo BJ, Selby MJ, Choe J, Suh BS, Choi SH, Joh JS, Nuovo GJ, Lee Y-S, Houghton M, Han JH (1995) Transfection of a differentiated human hepatoma cell line (Huh7) with in vitro-transcribed hepatitis C virus (HCV) RNA and establishment of a long-term culture persistently infected with HCV. J Virol 69:32–38

Zeuzem S, Lee J-H, Roth WK (1997) Mutations in the nonstructural 5 A gene of European hepatitis C virus isolates and response to interferon alpha. Hepatology 740–744

Zhu S, Romano PR, Wek RC (1997) Ribosome targeting of PKR is mediated by two double-stranded RNA-binding domains and facilitates in vivo phosphorylation of eukaryotic initiation factor-2. J Biol Chem 272:14434–14441

Hepatitis C Virus RNA-Dependent RNA Polymerase (NS5B Polymerase)

C.H. HAGEDORN, E.H. VAN BEERS, and C. DE STAERCKE

1	Introduction	225
2	Genomic Replication of RNA Viruses	227
3	Structural Features of a RNA-Dependent RNA Polymerase	228
4	Amino Acid Sequence Comparison of Hepatitis C Virus NS5B Regions	231
4.1	Distribution of Conserved Amino Acids	233
4.2	Cysteine Residues	237
4.3	Phosphorylation	237
4.4	The Conservation of Val-284 and Arg-345 Are Unique to Hepatitis C Virus	238
5	Characterization of Recombinant Forms of Hepatitis C Virus NS5B Polymerase	238
5.1	Expression Systems for Recombinant Hepatitis C Virus NS5B Polymerase	239
5.1.1	Insect Cells/Recombinant Baculovirus	240
5.1.2	*Escherichia coli*	241
5.1.3	Mammalian Cells	242
5.2	Enzyme Assays	242
5.3	Biochemical Characteristics of Recombinant NS5B	244
6	Questions and Future Directions	248
6.1	Template Specificity of Hepatitis C Virus NS5B Polymerase	248
6.2	Hepatitis C Virus Replication Complexes	250
6.3	Biomedical Applications	252
References		253

1 Introduction

Approximately 170 million people world-wide are chronically infected with the hepatitis C virus (HCV) (BRADLEY et al. 1983; CHOO et al. 1989; HOUGHTON 1996; ALTER 1997; also see Thomas, this volume). Although chronic infection with HCV frequently produces no symptoms for 5–30 years, it eventually has adverse effects on the host and constitutes a major cause of chronic liver disease and hepatocellular carcinoma (BISSELL 1997; also see Dickens and Fried, this volume). HCV evades host antiviral defenses by mechanisms that remain to be identified and establishes a

---

Division of Digestive Diseases and Genetics – Winship Cancer Center, Rm. 2101 Woodruff Memorial Research Building, 1639 Pierce Drive, Emory University School of Medicine, Atlanta, GA 30322, USA

chronic infection in a majority (70%–90%) of patients. One factor that is likely to play a role in the ability of HCV to avoid host antiviral defenses is the high frequency of mutation that occurs in the HCV genome. This property is typical of RNA viruses and in part reflects the lack of proofreading activity in their polymerases (HOLLAND et al. 1992). Our current knowledge of the mechanisms of HCV genomic RNA replication is rudimentary and encompasses only several biochemical characteristics of recombinant HCV NS5B polymerase. Although the indolent nature of chronic hepatitis C may provide a wide window for therapeutic intervention, current antiviral therapies remain relatively ineffective and have significant side effects. In addition, the lack of an easily established tissue culture system to propagate HCV has impeded progress toward understanding the critical steps in HCV replication and the development of small molecule inhibitors of HCV RNA replication.

In this review our goals are to describe the expression and initial biochemical characterization of recombinant forms of HCV NS5B polymerase, to present computer-assisted comparisons of HCV NS5B sequences that highlight several unanswered questions, and to outline aspects of genomic replication used by other positive strand RNA viruses that suggest future directions for studying HCV NS5B polymerase. Understanding the detailed mechanism of HCV RNA replication may provide a new example of the mechanisms used by positive strand RNA viruses to replicate their genomes. In addition, knowledge of the HCV NS5B polymerase may also have practical applications in developing novel therapeutics for chronic hepatitis C. Although detailed studies of several recombinant HCV nonstructural proteins such as the NS3 protease are in progress (KIM et al. 1996; LANDRO et al. 1997; YAN et al. 1998; URBANI et al. 1998), studies of recombinant NS5B polymerase have just begun (BEHRENS et al. 1996; AL et al. 1997, 1998; LOHMANN et al. 1997; YAMASHITA et al. 1998). Moreover, questions regarding the molecular organization and in vivo regulation of the NS5B polymerase within replication complexes remain unanswered.

RNA-dependent RNA polymerases (RdRp) are a category of viral enzymes that replicate the genomes of RNA viruses. The majority of these viruses have a positive (+) stranded RNA genome that can be translated as an mRNA within the host cell (KOONIN 1991; BUCK 1996). Others have either a double stranded (ds) RNA genome (e.g., reovirus) or a single negative stranded RNA genome (e.g. influenza A virus). The mechanism of genomic replication of positive strand RNA viruses, such as HCV, represents a unique aspect of their biology that distinguishes them from other viruses. Positive strand RNA viruses include a wide variety of bacterial, plant, and animal viruses: arteri-, astro-, carmo-, flavi-, hepaci- (HCV), pesti-, phage-, picorna-, poty-, rubi-, sobemo-, tobamo-, and tymo-viruses (KOONIN 1991; BUCK 1996). The poliovirus RdRp ($3D^{pol}$) is probably the most intensively studied RdRp of this category of RNA viruses (FLANEGAN and BALTIMORE 1977; PATA et al. 1995; XIANG et al. 1995; RICHARDS and EHRENFELD 1998). Most of the studies of poliovirus RdRp have been done with purified recombinant $3D^{pol}$ or viral genetic approaches (NEUFELD et al. 1991; JABLONSKI et al. 1991; DIAMOND and KIRKEGAARD 1994; TETERINA et al. 1995; JABLONSKI and MORROW 1995; RICHARDS et al. 1996; HOPE et al. 1997; RICHARDS and EHRENFELD 1997; BECKMAN and KIRKEGAARD 1998; XIANG et al. 1998). However, some studies of poliovirus RNA replication by 3D^{pol} have been done with the relatively small quantities of native enzyme recovered from extracts of infected cells (EHRENFELD et al. 1970; NEUFELD et al. 1991; BARTON et al. 1996). In contrast, larger quantities of native RdRp from some positive strand RNA viruses of plants have been isolated from infected tissues and have facilitated studies of their enzymatic activity and the identification of host proteins associated with the viral RdRp (QUADT et al. 1993; ADKINS et al. 1998). Nevertheless, isolating intact HCV replication complexes or purifying significant quantities of NS5B polymerase from infected tissue has been problematic. This is in part due to difficulties with HCV cell culture systems and with obtaining infected tissue samples that have not undergone autolysis. As with the example of poliovirus, it seems most likely that recombinant proteins, engineered mammalian cells, and genetic approaches will provide much of the information regarding HCV NS5B polymerase and the HCV replication complex.

2 Genomic Replication of RNA Viruses

The details of RNA replication for positive strand RNA viruses, including the protein and RNA cofactors required for both plus and minus strand synthesis, appear to vary greatly. Unlike some viruses, such as the dsRNA reoviruses or dsDNA hepadnaviruses (e.g. hepatitis B virus), the positive strand RNA viruses do not package their polymerase within the mature virion (SHATKIN and SIPE 1968; SUMMERS et al. 1975; SUMMERS 1988). This makes the positive strand RNA viruses, those coding for one genome-sized mRNA (picorna-, flavi-, and hepaci-) and those coding for one or more subgenomic mRNAs (toga-, corona-, and caliciviruses, for example hepatitis E virus), completely dependent on the host cell protein synthesis machinery to translate genomic viral RNA (ROIZMAN and PALESE 1996). The positive strand RNA viruses that code for one genome-sized mRNA synthesize a large viral polyprotein which is processed by proteases to produce the RdRp required for viral replication. The RNA-dependent RNA polymerases use the viral positive strand genomic RNA as a template to synthesize negative strands. Little is known about the switch from translation of genomic RNA to the use of genomic RNA as a template for RNA replication. In the case of poliovirus, the RdRp is unable to replicate genomic RNA that is being translated into protein due to specific protein-poliovirus RNA interactions (GAMARNIK and ANDINO 1998). Once synthesized, negative strands serve as templates for the synthesis of a quantitatively larger number of positive strands of genomic RNA. These RNA products provide additional templates for viral protein synthesis and genomic RNA for the production of progeny virus. The mechanisms for synthesizing negative strands and genomic RNA, which have different biological functions, appear unique and are regulated differently (BUCK 1996; ANDINO et al. 1993).

Positive strand RNA viruses that have proved amenable to propagation in tissue culture and can also be studied in model reconstituted systems have provided specific examples of how viral RNA replication occurs in vivo. The specifics for each virus appear to vary greatly (Buck 1996). However, a common observation is that relatively few copies of negative strand RNA appear to be synthesized compared to the number of positive strands of genomic RNA (Buck 1996; Strauss and Strauss 1994). The relative quantities of plus and minus strands synthesized within infected cells ranges from 7(+):1(−) to 100(+):1(−) depending on the virus studied (Hayes and Buck 1990; Andino et al. 1990; Novak and Kirkegaard 1991; Strauss and Strauss 1994; Buck 1996). At least for several viruses it is only during the early phase of viral infection when both plus and minus strands of RNA are synthesized (Strauss and Strauss 1994; Buck 1996). This is followed by a phase where only positive strands of genomic RNA are synthesized. The molecular events that mediate this process of template switching remain incompletely understood for even the extensively studied poliovirus. Nevertheless, model systems to study poliovirus RNA replication have provided some insight into this process. In vitro, the catalytic subunit of RNA-dependent RNA polymerases (e.g., poliovirus $3D^{pol}$) can make copies of nonviral RNA templates such as globin mRNA (Hey et al. 1986; Plotch et al. 1989). This raised the question: were additional factors required to provide specificity for viral RNA replication in vivo? Several studies have provided evidence that small proteins derived from the poliovirus polyprotein play a role in the interaction of the poliovirus $3D^{pol}$ with viral RNA templates (Andino et al. 1993; Hope et al. 1997; Richards and Ehrenfeld 1998; Xiang et al. 1998). The observation that recombinant HCV NS5B can effectively replicate globin mRNA templates in vitro suggests that either host factors or virally derived proteins may provide HCV template specificity in vivo (Al et al. 1997, 1998). Moreover, initial studies indicate that recombinant NS5B does not specifically bind to the structural motif at the 3'-end of genomic HCV RNA (Lohmann et al. 1997). These observations suggest a role for additional factors that mediate positive-strand template specificity in vivo.

3 Structural Features of a RNA-Dependent RNA Polymerase

Although little is know about the structure of RNA-dependent RNA polymerases (RdRps), comparisons of predicted secondary structures and the recent crystal structure of poliovirus $3D^{pol}$ provide an outline of some of the basic features that might be expected. The predicted secondary structure of HCV NS5B polymerase is similar to that predicted for $3D^{pol}$ of the poliovirus (Fig. 1). For example, the N-terminal regions that are unique to RdRps and not found in other polymerases are conserved (O'Reilly and Kao 1998). In addition, the arrangement of α-helices and β-strands within the fingers and palm subdomains are similar (Fig. 1).

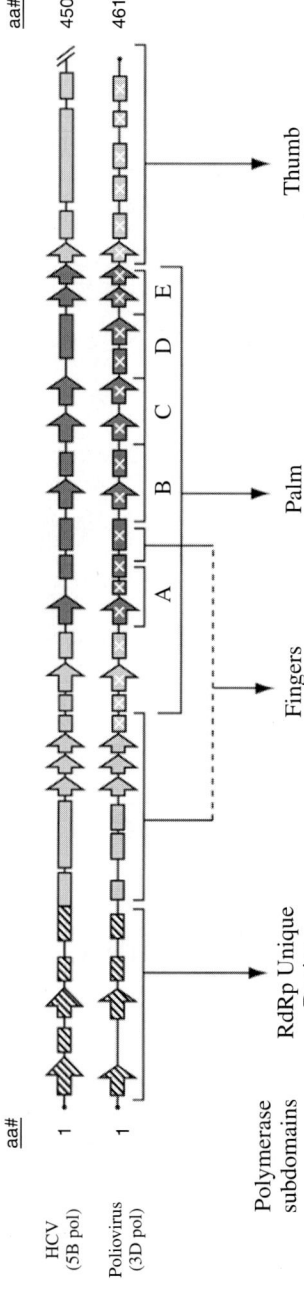

Fig. 1. The predicted secondary structure of HCV NS5B polymerase (5B^{pol}) compared with that predicted 3D^{pol} of poliovirus (adapted from O'Reilly and Kao 1998). *Boxes* indicate α-helices; *arrows*, β-strands. The 3D^{pol} regions that were observed in the crystal structure are indicated with a clear X (Hansen et al. 1997). The NH$_2$-terminal region unique to RNA-dependent RNA polymerases (RdRps) is indicated by the *hatch shading*. This region of HCV NS5B polymerase includes residues 1–97 of the NH$_2$-terminus. The conserved motifs within the core palm structure are designated by the *darkest shading*. This region includes the A-D motif common to all polymerases and the E motif which is found only in RdRps and reverse transcriptases. The highly conserved Gly-Asp-Asp (GDD) sequence is part of the C motif and is located in the loop between β-strand 2 and 3 in 3D^{pol}. Mutagenesis studies of the poliovirus have shown that deletion of Trp-5 or residues 1–6 of 3D^{pol} eliminate polymerase function (Plotch et al. 1989; Hansen et al. 1997). Note: HCV NS5B polymerase has 591 amino acids and the predicted secondary structure shown is truncated at residue 450

Both RNA and DNA polymerases have a general structure resembling a "right hand" with fingers, palm and thumb subdomains (Fig. 2). This structure has been reported for the poliovirus 3D^{pol}, Klenow fragment of *E. coli* DNA polymerase I, and the HIV-I RT (HANSEN et al. 1997; OLLIS et al. 1985; KOHLSTAEDT et al. 1992). The palm subdomain containing the catalytic center of each polymerase is remarkably conserved. It contains four amino acid sequence motifs found in all classes of polymerases, designated A, B, C, and D, plus a fifth motif, E, that is

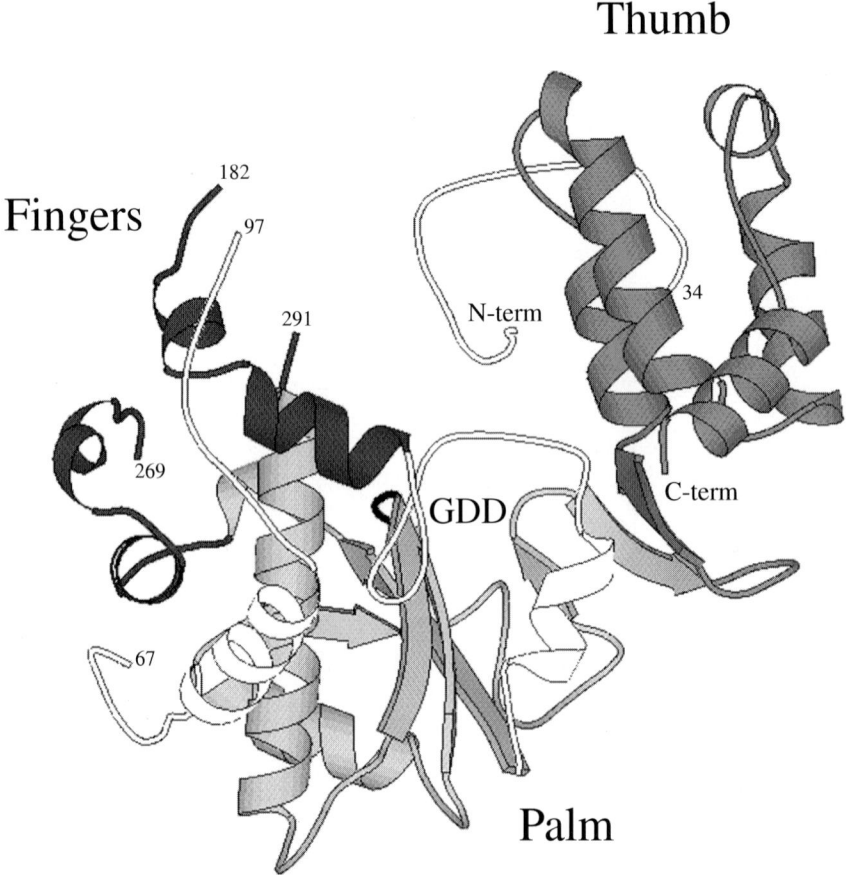

Fig. 2. The Poliovirus 3D^{pol} structure is shown in ribbons (Sybyl; Tripo, Inc.) using the coordinates reported by HANSEN et al. 1997. Motifs A, B, C, D, and E in the palm subdomain are represented by the *lightest shading*. The Gly$_{327}$-Asp$_{328}$-Asp$_{329}$ (GDD) sequence, a Mg^{2+} coordination site, is highly conserved among polymerases and is indicated by the *dark loop region* just off β-strand-2 (left of GDD). The fingers subdomain (*darkest shading*) remains poorly understood in both structure and function. The thumb subdomain (*intermediate shading*) is postulated to form a clamp on the RNA template. Note: residues 1–12, 38–66, 98–181, and 170–290 are not shown because this region was disordered in the crystal. Residues 12–25 of the NH$_2$-terminus are positioned in the active site cleft of an adjacent 3D^{pol} molecule (HANSEN et al. 1997). However, the precise role of this region in RNA polymerization remains unknown. This structure can be viewed in Chime at http://www.bimcore.emory.edu.Research/Hagedorn/hcv.html or e-mail chagedo@bimcore.emory.edu

unique to RdRps and RTs (POCH et al. 1989; O'REILLY and KAO 1998). However, the structure of the fingers and thumb are very different among the known polymerase structures. The NH_2-terminal region of the poliovirus polymerase that is unique to RdRps is believed to play a role in the proposed oligomerization of $3D^{pol}$ (PATA et al. 1995; HANSEN et al. 1997). The sequence motifs in the palm subdomain have the following postulated roles: motif A, Mg^{2+} coordination, sugar selection?; motif B, sugar selection?; motif C, Mg^{2+} coordination at the highly conserved GDD sequence; motif D, completes the palm core structure; and motif E, hydrophobic interaction with the thumb. The unique architecture of the fingers of RdRps has been suggested to determine their specificity for RNA templates. Moreover, the fingers region has been shown to be important for RNA synthesis because mutations in this region of poliovirus $3D^{pol}$, BMV 2a, Qb pII, and Sindbis virus NSP4 polymerase diminish RNA replication (RICHARDS et al. 1996; DIAMOND and KIRKEGAARD 1994; KRONER et al. 1989; MILLS et al. 1988; HAHN et al. 1989; O'REILLY and KAO 1998).

4 Amino Acid Sequence Comparison of Hepatitis C Virus NS5B Regions

A comparison of many RdRp and RdDp (reverse transcriptase) amino acid sequences, from a variety of viruses, revealed four evolutionarily conserved motifs based on secondary structure predictions (KAMER and ARGOS 1984; POCH et al. 1989). In HCV NS5B polymerase they include amino acid positions 220–225, 282–291, 317–319 (GDD) and 342–345 (SHUKLA et al. 1995; CHOO et al. 1991) (Figs. 3, 4). An amino acid sequence alignment comparing the RdRp of 45 RNA viruses including HCV NS5B polymerase has also been described (KOONIN 1991). In this study a total of eight conserved motifs were identified (KOONIN 1991). The functional significance of these motifs, other than the GDD motif, remains to be determined. Interestingly, NS5B nucleotide sequences have been used to classify HCV into specific viral subtypes. In one taxonomic study a limited region of NS5B, the last 1092 nucleotides from the 3'-end of the HCV genome, successfully predicted the subtype of 53 HCV isolates (DE LAMBALLERIE et al. 1997; CHA et al. 1992). Nevertheless, others have pointed out the difficulties in using partial sequences in determining HCV subtypes (BUKH et al. 1995).

The sequence of many HCV isolates have been reported and stored in Gen Bank since the study of KOONIN in 1991. We examined these sequences to determine if they might provide additional information regarding the NS5B polymerase protein. Forty-eight full length NS5B polymerase amino acid sequences (591 residues each) from different HCV isolates were analyzed to identify conserved residues, possible functional domains, and amino acid polymorphisms. This alignment included HCV genotypes 1a, 1b, 1c, 2a, 2b, 2c, 3a, 3b, 4a, 5a, and 6a (Table 1)

```
  1   I   SMSYSWTGAL VTPCAAEEQK LPINALSNSL LRHHNLVYST TSRSACQRQK
      II  SMSYSWTGAL VTPCAAEEQK LPINALSNSL LRHHNLVYST TSRSACQRQK
      III SMSYTWTGAL ITPCAAEESK LPINALSNSL LRHHNMVYAT TSRSASIRQK

 51   I   KVTFDRLQVL DSHYQDVLKE VKAAASKVKA NLLSVEEACS LTPPHSAKSK
      II  KVTFDRLQVL DSHYQDVLKE VKAAASKVKA NLLSVEEACS LTPPHSAKSK
      III KVTFDRLQVL DDHYRDVLKE MKAKASTVKA KLLSVEEACK LTPPHSAKSK

101   I   FGYGAKDVRC HARKAVAHIN SVWKDLLEDS VTPIDTTIMA KNEVFCVQPE
      II  FGYGAKDVRC HARKAVTHIN SVWKDLLEDN VTPIDTTIMA KNEVFCVQPE
      III FGYGAKDVRN LSSKAVNHIR SVWKDLLEDT ETPIDTTIMA KNEVFCVQPE

151   I   KGGRKPARLI VFPDLGVRVC EKMALYDVVS KLPLAVMGSS YGFQYSPGQR
      II  KGGRKPARLI VFPDLGVRVC EKMALYDVVT KLPLAVMGSS YGFQYSPGQR
      III KGGRKPARLI VFPDLGVRVC EKMALYDVVS TLPQAVMGSS YGFQYSPGQR

201   I   VEFLVQAWKS KKTPMGFSYD TRCFDSTVTE SDIRTEEAIY QCCDLDPQAR
      II  VEFLVQAWKS KKTPMGFSYD TRCFDSTVTE SDIRTEEAIY QCCDLDPQAR
      III VEFLVNTWKS KKCPMGFSYD TRCFDSTVTE NDIRVEESIY QCCDLAPEAR
                                               ↓
251   I   VAIKSLTERL YVGGPLTNSR GENCGYRRCR ASGVLTTSCG NTLTCYIKAR
      II  VAIKSLTERL YVGGPLTNSR GENCGYRRCR ASGVLTTSCG NTLTCYIKAR
      III QAIRSLTERL YIGGPLTNSK GQNCGYRRCR ASGVLTTSCG NTLTCYLKAS

                              ■                    ↓
301   I   AACRAAGLQD CTMLVCGDDL VVICESAGVQ EDAASLRAFT EAMTRYSAPP
      II  AACRAAGLQD CTMLVCGDDL VVICESAGVQ EDAASLRAFT EAMTRYSAPP
      III AACRAAKLQD CTMLVCGDDL VVICESAGTQ EDAASLRAFT EAMTRYSAPP

351   I   GDPPQPEYDL ELITSCSSNV SVAHDGAGKR VYYLTRDPTT PLARAAWETA
      II  GDPPQPEYDL ELITSCSSNV SVAHDGAGKR VYYLTRDPTT PLARAAWETA
      III GDPPQPEYDL ELITSCSSNV SVAHDASGKR VYYLTRDPTT PLARAAWETA

401   I   RHTPVNSWLG NIIMFAPTLW ARMILMTHFF SVLIARDQLE QALNCEIYGA
      II  RHTPVNSWLG NIIMFAPTLW ARMILMTHFF SVLIARDQLE QALDCEIYGA
      III RHTPVNSWLG NIIMYAPTLW ARMILMTHFF SILLAQEQLE KALDCQIYGA

451   I   CYSIEPLDLP PIIQRLHGLS AFSLHSYSPG EINRVAACLR KLGVPPLRAW
      II  CYSIEPLDLP PIIQRLHGLS AFSLHSYSPG EINRVAACLR KLGVPPLRAW
      III CYSIEPLDLP QIIQRLHGLS AFSLHSYSPG EINRVASCLR KLGVPPLRVW

501   I   RHRARSVRAR LLSRGGRAAI CGKYLFNWAV RTKLKLTPIA AAGRLDLSGW
      II  RHRARSVRAR LLARGGRAAI CGKYLFNWAV RTKLKLTPIA AAGQLDLSGW
      III RHRARSVRAR LLSQGGRAAT CGKYLFNWAV RTKLKLTPIP AASQLDLSSW

551   I   FTAGYSGGDI YHSVSHARPR WFWFCLLLLA AGVGIYLLPN R
      II  FTAGYSGGDI YHSVSHARPR WIWFCLLLLA AGVGIYLLPN R
      III FVAGYSGGDI YHSISRARPR WFMWCLLLLS VGVGIYLLPN R
```

Fig. 3. Amino acid alignment of three HCV NS5B polymerase sequences using the single letter amino acid code. Sequence I is from a full-length HCV RNA (type 1a) that is capable of initiating an HCV infection in chimpanzees when injected directly into the liver (KOLYKHALOV et al. 1997; GenBank AF009606). Sequence II is from a HCV type 1a clone which encodes an active NS5B polymerase as measured in in vitro assays of recombinant protein (CHOO et al. 1989; GenBank M62321; AL et al. 1997, 1998). Sequence III is from a HCV type 1b clone which encodes an active NS5B polymerase as measured in in vitro assays of recombinant protein (LOHMANN et al. 1997; GenBank Z97730). The numbers between 1 and 591 indicate the relative positions of amino acids in NS5B polymerase encoded by each genome. The two amino acids indicated by an *arrow*, Val-284 and Arg-345, are strictly conserved in HCV NS5B polymerase but not other RNA-dependent RNA polymerases (see Sect. 4). The complete CLUSTAL W alignment can be viewed and downloaded at http://www.bimcore.emory.edu/Research/Hagedorn/hcv.html

(SIMMONDS 1994; SMITH and SIMMONDS 1997). Alignments were computed by the CLUSTAL W algorithm in the MEGALIGN program using DNASTAR software (HIGGINS et al. 1996).

48 HCV sequences were obtained with BLAST-NCBI using the NS5B amino acid region from a HCV genotype 1a (CHOO et al. 1991) as a query and the GenBank database (Table 1). All except one of the 48 sequences represent the entire 591 amino acids encoded by the NS5B genomic region. The one exception is isolate U89019 (HCV type 1b) that had codon deletions for residues 217, 308, and 523 and thus consisted of only 588 amino acids (Table 1) (YEH et al. 1996). At this time it is not known if each of the isolates sequenced and stored in the data bank encode an enzymatically active NS5B polymerase. The lack of proofreading activity in RNA-dependent RNA polymerases results in RNA mutation rates that are relatively high and generally 10^{-4}–10^{-5} per base pair (HOLLAND et al. 1992; WARD and FLANEGAN 1992). This means that some of the HCV isolates sequenced and reported in the database may indeed encode viruses that are defective in some aspect of replication, including NS5B polymerase activity. Moreover, bias in the selection of HCV clones suggests that all isolates may not be truly representative of virus existing in vivo (FORNS et al. 1997). Although many of the isolates represent quasispecies that are presumed to be capable of replication, some may encode defective forms of NS5B polymerase. The NS5B polymerase encoded by a type 1a isolate (CHOO et al. 1991; AL et al. 1998) and several type 1b isolates (BEHRENS et al. 1996; LOHMANN et al. 1997; YAMASHITA et al. 1998; FERRARI et al. 1999) have been expressed and shown to have in vitro RdRp activity.

4.1 Distribution of Conserved Amino Acids

The complete sequence of NS5B from 48 different HCV isolates were analyzed using the LAMA program to determine what residues and motifs were highly conserved (Table 1, Fig. 4) (http://blocks.fhcrc.org/blocks-bin/LAMA_search). Among all the NS5B sequences examined 263 amino acid positions were fully conserved (Fig. 4). A total of 51% of the fully conserved positions consisted of three or more adjacent amino acids (Table 2, Fig. 4). The longest domain of uninterrupted conserved residues was 11 amino acids and corresponds to positions 342–352 (Table 2, Fig. 4). Three other relatively long, fully conserved regions were identified. One consists of nine consecutive amino acids (365–373) and the other seven consecutive amino acids (221–227 and 524–530) (Table 2, Fig. 4). The strict conservation of these residues suggests a functional role for these regions regarding polymerase activity, protein folding or may even reflect a conservation of HCV RNA structure.

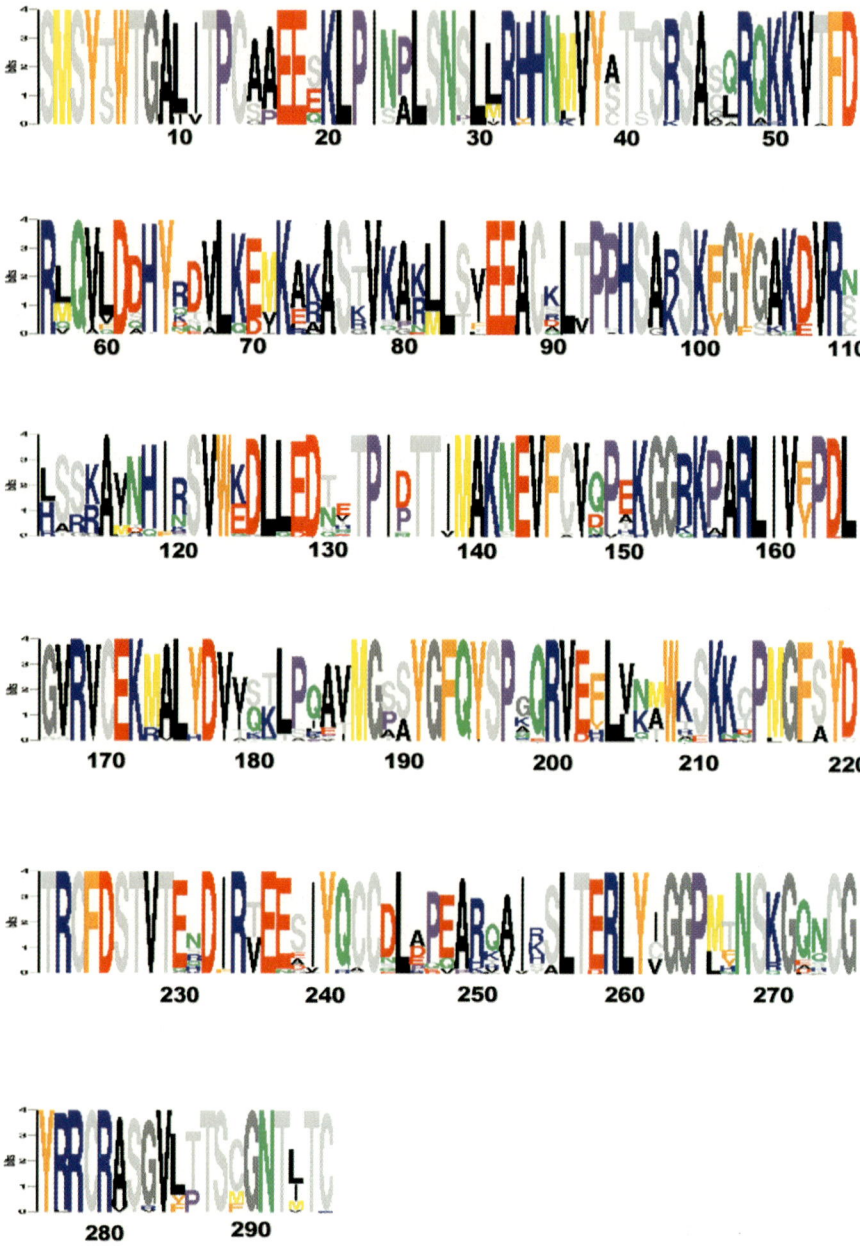

Fig. 4. Conservation of amino acids in NS5B polymerases from 48 different HCV isolates. The *horizontal axis* indicates the amino acid residue in NS5B polymerase (1–591). The *vertical axis* represents the degree of conservation. A letter of maximum height indicates strict conservation among all 48 sequences analyzed. Sequence polymorphisms are indicated by a smaller size of the single letter amino acid code where size is proportional to the frequency of the amino acid at that position. The most frequently occurring amino acid is located at the *top* and the least frequent at the *bottom*. A color coded version of this figure

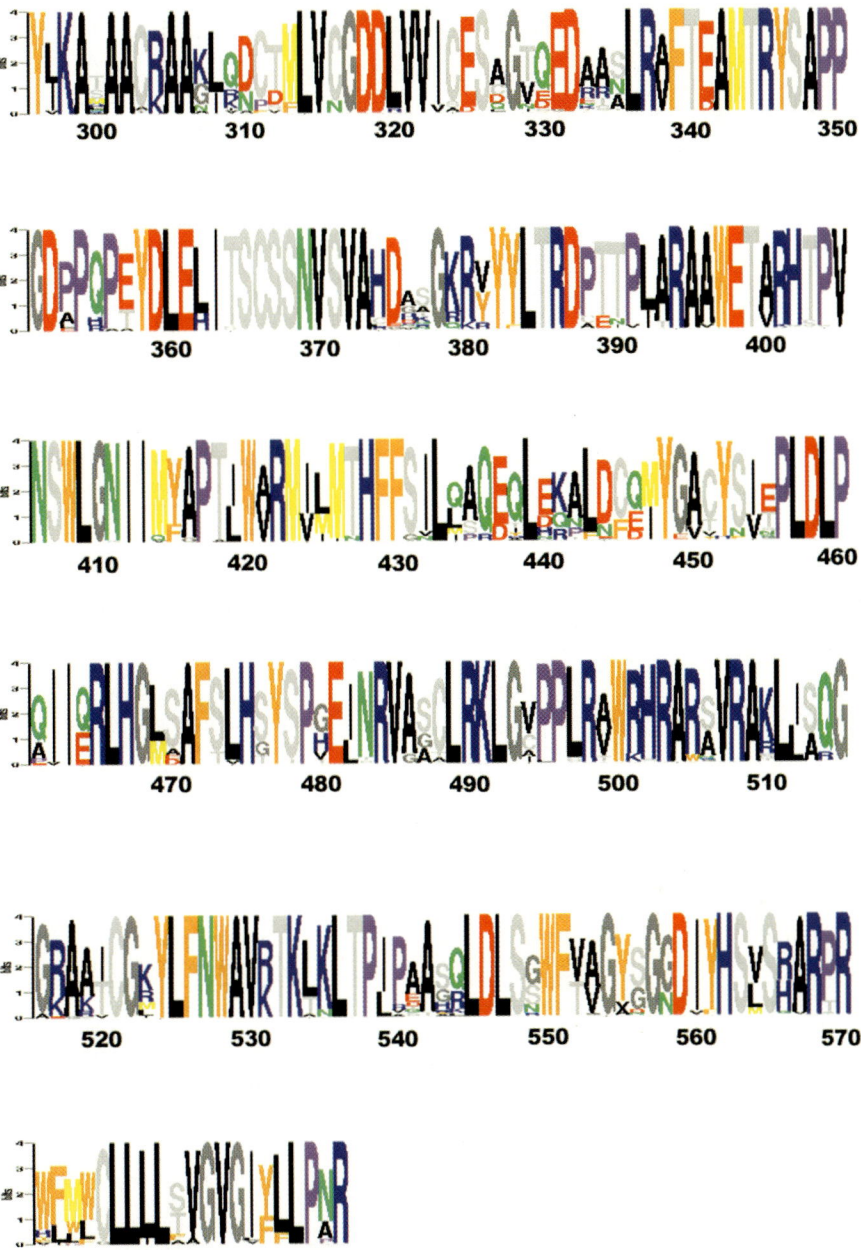

Fig. 4. (*Cont.*) indicating the properties of each R group can be viewed and downloaded at http://www.bimcore.emory.edu/Research/Hagedorn/hcv.html. The color code is: red, acidic amino acids (Glu, Asp); blue, basic (Lys, Arg, His); light grey, polar OH/SH (Ser, Thr, Cys); green, amide (Asn, Gln); yellow, sulfur (Met); black, hydrophobic (Ala, Val, Leu, Ile); orange, aromatic (Tyr, Phe, Trp); purple, proline; and dark grey, glycine

Table 1. Source of hepatitis C virus NS5B polymerase sequences analyzed

Genotype	GenBank accession	Reference
1a-1	M62321	Choo et al. (1991)
1a-H	M67463	Inchauspe et al. (1991)
1a-J1	D10749	Okamoto et al. (1991)
1a-C2	D10934	Wang et al. (1993)
1b-HB	L02836	Bi et al. (unpublished)
1b-J	D90208	Kato. et al. (1990)
1b-JT	D01171	Tanaka et al. (1992)
1b-K1-S2	D50485	Enomoto et al. (1995)
1b-K1-R1	D50480	Enomoto et al. (1995)
1b	Z97730	Lohmann et al. (1997)
1b-H77	AF011753	Yanagi et al.(1997)
1b-K1R2	D50481	Enomoto et al. (1995)
1b	M58335	Takamizawa et al. (1991)
1b-MKC1 A	D45172	Seki and Honda (1995)
1b	D30613	Seki et al. (1995)
1b-A	AJ000009	Trowbridge and Gowans (1998)
1b-L2	U01214	Cho et al. (1995)
1b	U16362	Cho et al. (unpublished)
1b	M96362	Cho et al. (unpublished)
1b	D50483	Enomoto et al. (1995)
1b-HD-1	U45476	Müller et al. (unpublished)
1b-HPCT-2	D16435	Hijikata et al. (1993)
1b-N	D63857	Zheng (1994)
1b-TMORF	D89872	Tanaka (unpublished)
1b-K1-S3	D50484	Enomoto et al. (1995)
1b-HC-J4/83	D13558	Okamoto et al. (1992)
1b-HC-J4/91	D10750	Okamoto et al. (1992)
1b-K1-R3	D50482	Enomoto et al. (1995)
1b-JS	D85516	Sugiyama et al. (1997)
1b-J33	D14484	Cho (unpublished)
1b-JT	D01172	Tanaka et al. (1992)
1b-N	S62220	Hayashi et al. (1993)
1b	U89019	Yeh et al. (1996)
1b-JK1-"full"	X61596	Honda et al. (1992)
1c-HC-G9	D14853	Okamoto et al. (1994)
2a-HC-J6	D00944	Okamoto et al. (1991)
2b-HC-J8	D01221	Okamoto et al. (1992)
2c-BEBE-1	D50409	Nakao et al. (1996)
3a-NZL-1	D17763	Sakamoto et al. (1994)
3a	X76918	Seelig R. et al. (unpublished)
3a-K3a/650	D28917	Yamada et al. (1994)
3b-Tr Kj	D26556	Chayama et al. (1994)
3-JK049	D63821	Tokita et al. (1996)
3b-Tr Kj	D49474	Chayama et al. (1994)
4a-ED43	Y11604	Chamberlain et al. (1997a)
5a	Y13184	Chamberlain et al. (1997b)
6-JK046	D63822	Tokita et al. (1996)
6a-Euhk2	Y12083	Adams et al. (1997)

The source of the 48 HCV NS5B polymerase sequences used in the amino acid comparison and analysis (see Sect. 4) are identified by genotype, GenBank accession number, and a literature reference. These sequences can be found on the same web page cited in Fig. 4.

Table 2. Conserved amino acid positions in NS5B polymerase

Number of consecutive conserved amino acids	1	2	3	4	5	6	7	8	9	10	11
Frequency	75	27	16	8	4	–	2	–	1	–	1
Position	NI	NI	NI	1–4 6–9 20–23 86–89 191–194 384–387 465–468 545–548	157–161 406–410 456–460 489–493		221–227 524–530		365–373		342–352

The number, frequency, and location of conserved amino acids present in HCV NS5B polymerase are shown. NS5B sequences from a total of 48 HCV isolates were analyzed (Table 1). Position one refers to the first residue of the NH$_2$-terminus of NS5B (Ser$_1$-Met-Ser-Tyr...) as determined by polyprotein processing studies (KOLYKHALOV et al. 1996); NI, not indicated.

4.2 Cysteine Residues

All NS5B sequences analyzed had between 16 and 23 cysteine residues. However, no more than ten cysteine residues are conserved across the 48 different HCV isolates studied and were at positions 14, 89, 170, 223, 243, 274, 279, 295, 366, and 521 (Fig. 4). The large number of cysteine residues present in NS5B suggests that the thiol-disulfide status of this protein may be a key issue regarding the folding and production of enzymatically active NS5B polymerase. A sequence comparison that also includes RNA-dependent RNA polymerases from other viruses showed that Cys-366 was fully conserved among a wide variety of viruses; this is not the case for the other nine cysteine residues that are conserved in HCV NS5B polymerase (KOONIN 1991). An alignment of HCV NS5B with the corresponding region of an isolate of HGV (GenBank AB008342), a closely related virus, showed that three of the ten cysteine residues conserved in HCV NS5B (residues 223, 279, and 295) were also found in NS5B of HGV (SAITO et al. 1998; see Simons et al., this volume). At this time no experimental evidence indicates which of the conserved cysteine residues within HCV NS5B might be necessary for protein folding or other essential functions and why Cys-366 is conserved in the RdRp of a wide variety of viruses.

4.3 Phosphorylation

Hepatitis C virus NS5B is phosphorylated when expressed in cultured eukaryotic cells (HWANG et al. 1997). However, the sites of phosphorylation and any physiologic function are not known at this time. The amino acid sequences of HCV NS5B polymerase were analyzed with the ProDom database and the recognition motifs described by KEMP and PEARSON (1990) to identify phosphorylation motifs

(ALTSCHUL et al. 1997) (Table 1). Although multiple potential phosphorylation sites exist within each NS5B polymerase sequence, no site was conserved in all 48 sequences (data not shown). It will be of interest to determine if the in vivo phosphorylation site(s) of NS5B polymerase vary between genotypes or are conserved and not predicted by primary amino acid sequence analysis.

4.4 The Conservation of Val-284 and Arg-345 Are Unique to Hepatitis C Virus

The conservation of Val-284 and Arg-345 appears to be unique to HCV NS5B polymerase as compared to the polymerases of other RNA viruses (POCH et al. 1989; KOONIN 1991). Val-284 is part of an otherwise-well-conserved RdRp sequence motif (residues 282–291 in HCV NS5B) and may have functional significance (POCH et al. 1989). Depending on what family of RNA virus is examined, position 284 of the RNA-dependent RNA polymerase is strictly conserved to encode either Val, Asp, Gln, or Asn (KOONIN 1991). The conserved Arg-345 present in HCV NS5B polymerase does have a functional significance since changing this residue to Lys, found at this location in most other RdRps, increased activity by 52% in an in vitro poly(G) polymerase assay (LOHMANN et al. 1997). It will be of interest to determine if this mutation has a similar effect with HCV templates in mammalian cells.

5 Characterization of Recombinant Forms of Hepatitis C Virus NS5B Polymerase

The recent studies of recombinant NS5B polymerase, both HCV type 1a and 1b, expressed in either insect cells or *E. coli* provide direct evidence that it catalyzes an RNA-dependent RNA polymerase reaction (BEHRENS et al. 1996; AL et al. 1997, 1998; LOHMANN et al. 1997; YAMASHITA et al. 1998) (Fig. 3). In each of these studies, recombinant NS5B was isolated under nondenaturing conditions in which a large percentage of the recombinant protein remained insoluble. One preliminary report describes the isolation of recombinant NS5B polymerase under denaturing conditions and the refolding of active enzyme (YUAN et al. 1997). Although recombinant NS5B utilizes HCV RNA as a template, it also readily replicates other RNA templates in vitro such as poly(A), poly(C), or globin mRNA (model homopolymeric or heteropolymeric RNAs) without additional viral or mammalian cell co-factors (BEHRENS et al. 1996; AL et al. 1997, 1998; LOHMANN et al. 1997). This observation suggests that there are additional factors which regulate NS5B polymerase template specificity that remain to be identified. The information we present regarding recombinant HCV NS5B polymerase covers not only the initial biochemical properties of the enzyme, but also some areas of methodology since this field is so new.

5.1 Expression Systems for Recombinant Hepatitis C Virus NS5B Polymerase

The only source of functional HCV NS5B polymerase at this time is recombinant protein. Efforts to isolate significant quantities of native NS5B polymerase from infected tissues have not resulted in preparations suitable for biochemical studies (unpublished reports). Although expression and isolation of enzymatically active NS5B polymerase using recombinant systems have not been easy, active enzyme has been isolated from insect cells, *E. coli*, and mammalian cells (BEHRENS et al. 1996; AL et al. 1997, 1998; LOHMANN et al. 1997; YAMASHITA et al. 1998; HELLER et al. 1998).

Some of the difficulties experienced in expressing, purifying, and studying recombinant NS5B polymerase may be due to properties of the protein that permit it to function in vivo in what is likely to be a hydrophobic environment where NS5B is anchored to membranes and surrounded by other proteins of the HCV replication complex. NS5B polymerase has a hydrophobic COOH-terminal domain of 21 amino acids that anchors the protein to mammalian cell membranes (YAMASHITA et al. 1998) (Fig. 4). Although expression of the entire NS5B protein can yield sufficient quantities of soluble protein to determine basic biochemical properties of this polymerase, modifications of the current methods may yield recombinant NS5B in larger quantities and with properties that make it more suitable for some studies. Indeed, removal of the COOH-terminal hydrophobic domain of NS5B polymerase makes this protein easier to isolate from *E. coli* expression systems (YAMASHITA et al. 1998; De Staercke, unpublished; FERRARI et al. 1999). This and other modifications of NS5B that increase the yield, aid in purification, or in some other way facilitate the preparation of recombinant protein for biochemical or structural studies have only begun to be tested.

One question has been whether the numerous cysteine residues within NS5B polymerase may increase the difficulty of proper folding of the recombinant protein or require additional efforts to protect essential thiol groups during the purification of NS5B. Further biochemical studies will be needed to determine the function of those cysteine residues. In addition, although at least one covalent modification of NS5B has been reported in mammalian cells, phosphorylation, there is no evidence at this time that it regulates enzymatic activity (HWANG et al. 1997). It therefore seems a reasonable approach to use recombinant proteins that lack such modifications for further studies. Nevertheless, covalent modifications such as phosphorylation may regulate the interaction of NS5B polymerase with other proteins that modify or complement its function in vivo. To address these questions an effective system to study NS5B polymerase activity in intact mammalian cells needs to be developed.

A property of NS5B polymerase that has experimental and possibly biological implications is the polymorphism in the primary amino acid sequence between different HCV isolates (Figs. 3, 4). For example, the sequence of NS5B polymerase from one isolate of HCV type 1a differs from a type 1b isolate by a total of 62 amino acids (Fig. 3). Such differences may affect both the biochemical properties and the ability to purify recombinant NS5B (COLLETT et al. 1998). It will be interesting to see if differences in the primary amino acid sequence of NS5B encoded by the HCV subtypes (e.g., type 1a and type 1b), as well as single amino acid changes present in quasispecies of one genotype, produce measurable changes in the properties of recombinant NS5B polymerase (FERRARI et al. 1999). However, we must consider that some of the NS5B sequences identified may not be representative of virus present in vivo due to errors introduced during RT-PCR and nonrandom selection events that occur while isolating clones of HCV (FORNS et al. 1997).

5.1.1 Insect Cells/Recombinant Baculovirus

Recombinant NS5B polymerase (HCV type 1b) has been expressed, isolated, and characterized using baculovirus vectors (pBlue BacIII, Invitrogen) and a SF9 (*Spodoptera frugiperda* clone 9) insect cell expression system (BEHRENS et al. 1996). To purify untagged recombinant NS5B, one gram of frozen cells was lysed in 10 ml of a nondenaturing solubilization buffer of 20 mM Tris/HCl (pH 7.5), 1 mM EDTA, 10 mM DTT, 1 mM phenylmethylsulfonyl fluoride (PMSF), and 50% glycerol (DEFRANCESCO et al. 1996). The preparation had detergent and salts added to a final concentration of 2% Triton X-100, 500 mM NaCl, and 10 mM $MgCl_2$. DNase I (15 µg/ml) was added to the suspension, it was stirred for 30 min at room temperature and clarified by centrifugation at $100,000 \times g$ for 30 min at 4°C. The supernatant was used in sequential DEAE, heparin, and poly(U)-Sepharose chromatography steps followed by a final Mono-Q FPLC purification step (DEFRANCESCO et al. 1996).

Another report described the rapid purification of recombinant NS5B polymerase (HCV type 1b) with a hexahistidine tag (TSH_6) at the COOH-terminus that was also expressed using recombinant baculovirus vectors (pBac9, Clontech) and insect cells (High5 cells, Clontech) (LOHMANN et al. 1997). The method of cell lysis and extraction of recombinant NS5B differs from the one described above. Cells (4×10^7) containing recombinant NS5B polymerase were washed once with phosphate-buffered saline, resuspended in 1 ml of 10 mM Tris/HCl (pH 7.5), 10 mM NaCl, 1.5 mM $MgCl_2$, 10 mM 2-mercaptoethanol, 1 mM PMSF and leupeptin at 4 µg/ml, and incubated at 4°C for 30 min. Lysates were centrifuged for 10 min at $10,000 \times g$ and the pellet was resuspended by sonication in 1 ml of 20 mM Tris/HCl (pH 7.5), 300 mM NaCl, 10 mM $MgCl_2$, 0.5% Triton X-100, 20% glycerol, and 10 mM 2-mercaptoethanol containing PMSF and leupeptin. This extract was centrifuged for 10 min at $10,000 \times g$ and the pellet resuspended in 1 ml of 20 mM Tris/HCl (pH 7.5), 500 mM NaCl, 10 mM $MgCl_2$, 2% Triton X-100, 50% glycerol, and 10 mM 2-mercaptoethanol with PMSF and leupeptin. Following sonication, the extract was centrifuged for 10 min at $10,000 \times g$ and the supernatant was used to purify NS5B-his_6 by Ni^{2+} affinity chromatography.

5.1.2 *Escherichia coli*

Both untagged and tagged recombinant HCV NS5B polymerase have been produced in *E. coli* and isolated under nondenaturing conditions (AL et al. 1997, 1998; DE STAERCKE et al. 1998; YAMASHITA et al. 1998) Untagged recombinant NS5B (HCV type 1a) expressed in *E. coli* has a strict primer dependent RdRp activity in vitro and no evidence for terminal nucleotidyl transferase activity (AL et al. 1997, 1998). The recombinant NS5B expressed in *E. coli* was solubilized under nondenaturing conditions by mixing cells on ice for 20 min in 50 mM Tris-HCl (pH 7.5), 100 mM KCl, 0.5 mM EDTA, 1 mM DTT, 0.1% NP-40, and 15% glycerol with 30 µg/ml lysozyme. Samples were sonicated in the presence of 1 mM PMSF, 1 µg/ml each of aprotinin, leupeptin, and pepstatin A, and centrifuged at 21,000 × g for 30 min at 4°C. The NH_2-terminus of this recombinant protein is Ala-Ser-Met-Ser-Tyr-Ser-Trp-Thr-Gly as determined by Edman degradation analysis (AL et al. 1998). This differs from the predicted native NS5B, identified as Ser-Met-Ser-Tyr-Ser-Trp-Thr-Gly by polyprotein processing studies, by a single Ala residue at the NH_2-terminal (GRAKOUI et al. 1993; KOLYKHALOV et al. 1994). In addition, NS5B polymerase with a hexahistidine tag at the COOH-terminus and expressed in *E. coli* is enzymatically active (DE STAERCKE et al. 1998; FERRARI et al. 1999).

Based on the experience of others with recombinant poliovirus RdRp ($3D^{pol}$) there has been concern that changes at the NH_2-terminus of recombinant HCV NS5B polymerase might alter enzymatic activity (HANSEN et al. 1997). However, a recent study described the expression of enzymatically active NS5B polymerase (HCV type 1b) fused to GST at the NH_2-terminus (GST-NS5B) (YAMASHITA et al. 1998). The GST domain at the NH_2-terminus facilitated purification and possibly increased yields of properly folded protein. Purified GST-NS5B had RdRp activity that was primer-dependent. Mutation of the Gly-Asp-Asp (GDD) motif of GST-NS5B to Val-Asp-Asp completely inactivated the RdRp activity. On the other hand, the fusion protein lost approximately two thirds of the RdRp activity after being treated with thrombin to produce "free" NS5B polymerase. Based on the plasmid construct described, the NS5B produced after thrombin cleavage is predicted to have 17 amino acids added to the Ser-Met-Ser-Tyr NH_2-terminus of native NS5B (YAMASHITA et al. 1998; KOLYKHALOV et al. 1994). Further studies are needed to determine if the loss of activity using this expression approach is simply due to thrombin cleavage of NS5B at sites other than the intended Leu-Val-Pro-Arg-Gly-Ser site or if another explanation exists.

Precisely what amino acid deletions or additions are tolerated at either the NH_2- or COOH-terminus of HCV NS5B polymerase remains to be determined. This may be a key factor regarding the expression of recombinant forms of NS5B for both biochemical and structural studies. Deletion of either 19 or 40 amino acids from the NH_2-terminus of NS5B produced a complete loss of RdRp activity (LOHMANN et al. 1997; see Sect. 5.3). Moreover, the addition of a hexahistidine tag to the NH_2-terminus of NS5B polymerase decreased the in vitro polymerase activity of this recombinant protein relative to NS5B with a hexahistidine tag at the COOH-terminus (LOHMANN et al. 1997). A more detailed answer to the question of what changes are tolerated at the NH_2-terminus of NS5B will require further studies.

Amino acid changes in the NH_2-terminus of recombinant poliovirus $3D^{pol}$ that seem minor had major effects on enzymatic activity. For example, deletions of Trp-5 or residues 1–6 of $3D^{pol}$ completely eliminated RdRp activity (HANSEN et al. 1997). A possible explanation for the inhibition of recombinant poliovirus $3D^{pol}$ by what appear to be minor modifications at the NH_2-terminus may exist in the recently described crystallographic structure of the poliovirus RdRp. Consistent with studies suggesting that $3D^{pol}$ functions as an oligomer, the structural models predict that the NH_2-terminal polypeptide segments of two polymerase molecules interact (designated interface II) (PATA et al. 1995; HANSEN et al. 1997). Based on both the $3D^{pol}$ structure and modeling studies it has been proposed that the NH_2-terminus region of one polymerase molecule makes a contribution to the active site cleft of an adjacent polymerase molecule (HANSEN et al. 1997). Future studies remain to demonstrate if this hypothesis regarding $3D^{pol}$ can be verified and if it is relevant to understanding the HCV NS5B polymerase.

5.1.3 Mammalian Cells

A detailed report of the RdRp activity of recombinant NS5B polymerase expressed in mammalian cells has not been published at this time. However, preliminary results have provided evidence that expression of enzymatically active NS5B polymerase can be achieved in mammalian cells (HELLER et al. 1998). NS5B expression was driven by the cytomegalovirus (CMV) promoter in pcDNA3 (Invitrogen) which was used to produce stably transfected BKH cells (HELLER et al. 1998). The preliminary results indicated that a large number of stably transfected clones needed to be analyzed before any cells expressing NS5B polymerase could be identified. However, the use of an inducible expression system (pBI vector and Tet-On cells) permitted the co-expression of NS5B polymerase with a reporter gene (luciferase) and facilitated a more rapid identification of Huh-7 cells expressing higher levels of NS5B polymerase (HELLER et al. 1998; BARON et al. 1995). This initial report provides evidence that enzymatically active NS5B polymerase can be expressed in mammalian cells. However, a system that permits recombinant NS5B polymerase to be studied in intact mammalian cells remains to be developed.

5.2 Enzyme Assays

The methods used to assay recombinant HCV NS5B polymerase follow those used to study poliovirus RdRp ($3D^{pol}$) (FLANEGAN and BALTIMORE 1977; HEY et al. 1986; WIMMER et al. 1993). Templates in these assays can be either homopolymeric oligoribonucleotides such as $poly(A)_{400-600}$, nonviral heteropolymeric RNAs such as globin mRNA, or viral RNA synthesized by in vitro transcription.

The reports describing recombinant HCV NS5B polymerase have relied on assays that are similar to each other. The report of untagged recombinant NS5B polymerase isolated from insect cells used assays of 40 µl containing 20 mM Tris-HCl (pH 7.5), 25 mM KCl (later changed to 50 mM NaCl), 5 mM MgCl$_2$, 0.5% Triton X-100, 1 mM DTT, 1 mM EDTA, 20 U RNasin, 50 µg/ml actinomycin D, 5–10 µCi [^{32}P]NTP of one species, and 0.5 mM each of the remaining NTPs (BEHRENS et al. 1996; TOMEI et al. 1998). RNA templates were present at 10 µg/ml (later changed to 10–20 nM) and when indicated oligo(U)$_{12}$ or oligo(dT)$_{12-18}$ primers were present at 25 µg/ml. The heteropolymeric RNA template used in most of the assays was DCoH (399 nucleotide mRNA of dimerization co-factor of hepatocyte nuclear factor-1) because it produced the largest quantity of RNA products for analysis (MENDEL et al. 1991). HCV templates consisting of 618 nucleotides of the 3'-end of the HCV genome, excluding the 3'-end cloverleaf, were prepared using T7 polymerase (DEFRANCESCO et al. 1996; TANAKA et al. 1995; KOLYKHALOV et al. 1996). Subsequent studies used homopolymeric templates of poly(C) at a final concentration of 15 µM with oligo(G) primers at 0.15 µM (TOMEI et al. 1998). Incubations were for 2 h at 22°C. To analyze RNA products the incubations were stopped by the addition of a 2 × proteinase K buffer (300 mM NaCl, 100 mM Tris-HCl/pH 7.5, 1% SDS). Samples (80 µl) were incubated for 30 min at 37°C with 50 µg of proteinase K; RNA products were extracted with phenol/chloroform, and subsequently analyzed by polyacrylamide gels containing 7 M urea (DEFRANCESCO et al. 1996).

Another study of recombinant NS5B polymerase, with a hexahistidine tag located at the COOH-terminal (NS5B^{C-HIS}), also used polymerase assays similar to those described above (LOHMANN et al. 1997). Incubations were 25 µl in volume and contained 20 mM Tris-HCl (pH 7.5), 5 mM MgCl$_2$, 1 mM DTT, 25 mM KCl, 1 mM EDTA, 10–20 U of RNasin, 2–10 µCi of [α-^{32}P]GTP, 500 µM each CTP, ATP, and UTP, and 0.5 µg of HCV RNA or 400 ng of a homopolymeric RNA as template. When full length HCV RNA templates, including the 3'-end cloverleaf, were used no primers were required for RdRp activity (TANAKA et al. 1995). When homopolymeric RNA was used as template, 4 pmol of either oligo(U)$_{12}$ or oligo(G)$_{12}$ primers were included in incubations. Actinomycin D was present at a final concentration of 50 µg/ml and incubations were performed for 2 h at 22°C.

The assays of recombinant NS5B polymerase expressed in *E. coli* were done under slightly different conditions than those reported for NS5B expressed in insect cells. Assays were 50 µl in volume and contained 50 mM each HEPES (pH 8.0), [^{3}H]UTP (20 µCi, specific activity of 30–50 Ci/mmol) at 15 µM, 500 µM each ATP, CTP, and GTP, 4 mM DTT, 10 mM magnesium acetate, 60 µM ZnCl$_2$, 4 µg/ml of actinomycin D, 20 µg/ml poly(A)$_{460-600}$, and 10 µg/ml oligo(U)$_{5-25}$ (AL et al. 1997, 1998). Some assays used 5 µg/ml of globin mRNA as template instead of poly(A). Incubations were for 60 min at 30°C. To analyze the size of RNA products incubations were extracted with phenol/chloroform, precipitated in the presence of carrier tRNA, and analyzed by 1.5% agarose/formaldehyde gel electrophoresis.

The assays used to study GST-NS5B were 40 µl in volume and contained 20 mM Tris-HCl (pH 7.5), 5 mM MgCl$_2$, 1 mM DTT, 1 mM EDTA, 20 U of RNasin, 50 µg/ml actinomycin D, 5 µCi of [α-^{32}P]UTP (800 Ci/ mMol), 500 µM each ATP, CTP, and GTP with 10 µg/ml of HCV RNA template, and were incubated for 2 h at 30°C (YAMASHITA et al. 1998). Poly(U) polymerase assays contained 10 µg/ml of poly(A) template and 1 µg/ml of oligo(U)$_{14}$ primer and were maintained for 2 h at 25°C. In addition, the effect of including different detergents in the GST-NS5B RdRp assays was described. Triton X-100 and Nonidet P-40 were found to have little inhibitory effect at a concentration of 0.1%. Tween-20 may have had a slight stimulatory effect at concentrations of 0.1% and 1%, whereas sarcosyl and SDS were inhibitory at 0.01%.

Although a system to measure NS5B polymerase activity in intact mammalian cells is not yet available, a preliminary report measured RdRp activity in lysates of mammalian cells expressing NS5B (HELLER et al. 1998). Cell lysates were prepared by sonicating 5×10^6 cells in a buffer consisting of 20 mM Tris-HCl (pH 7.5), 500 mM NaCl, 10 mM MgCl$_2$, 1 mM DTT, 0.5% Triton X-100, 0.2% CHAPS, and 30% glycerol with protease inhibitors. Extracts were treated with S7 nuclease to destroy endogenous RNA that might be bound to NS5B and interfere with detection of polymerase activity. The nuclease was quenched with EGTA prior to polymerase assays. Incubations were 60 µl in volume with 20 mM Tris-HCl (pH 7.5), 5 mM MgCl$_2$, 25 mM KCl, 1 mM EDTA, 60 U RNasin, 50 µg/ml actinomycin D, 2.5 µCi of [α-^{32}P]GTP, 500 µM each CTP, ATP, UTP, and 0.4 µg of poly(C) templates with oligo(G) primers.

5.3 Biochemical Characteristics of Recombinant NS5B

Untagged NS5B polymerase (HCV type 1b) expressed in insect cells has been studied to directly demonstrate that this protein has RdRp activity and to determine several biochemical characteristics (BEHRENS et al. 1996). The recombinant NS5B polymerase was Mg^{2+}-dependent and had maximal RdRp activity at pH 7.5. Recombinant NS5B synthesized RNA from templates of HCV (not including the 3'-end cloverleaf) or non-HCV RNA templates (BEHRENS et al. 1996; TANAKA et al. 1995; KOLYKHALOV et al. 1996). However, a significant amount of primer-independent RNA synthesis was observed under most assay conditions with HCV templates suggesting terminal nucleotidyl transferase (TNTase) activity. The presence of this activity raised questions regarding what quantity of the newly synthesized radiolabeled RNA produced was due to de novo RNA synthesis (RdRp activity) or due to the transfer of as few as one radiolabeled nucleotide to the 3'-end of template RNA (TNTase). However, when the 3'-hydroxyl group of template RNA was blocked by oxidation the formation of radiolabeled RNA was strictly dependent on the presence of primers providing further evidence of authentic RdRp activity (BEHRENS et al. 1996). Nevertheless, several experiments in this study also supported the presence of terminal nucleotidyl transferase activity in preparations of recombinant NS5B even after sucrose density gradient fractionation of the purified enzyme. The association of TNTase activity with preparations of RdRp has been reported and suggested to play a role in replicating the RNA of other viruses (GRUN and BRINTON 1986; NEUFELD et al. 1994; ZABEL et al. 1981; SCHIEBEL et al. 1993). In addition, a variety of cellular 3' terminal transferases have been described (ANDREWS and BALTIMORE 1986; LEUNG et al. 1979; BAKALARA et al. 1989; TRIPPE et al. 1998). Although the possibility of NS5B polymerase having intrinsic TNTase activity was raised by these studies, a subsequent report provided evidence that the TNTase activity observed may have been due to minor contaminants derived from insect cells (BEHRENS et al. 1996; LOHMANN et al. 1997). Moreover, TNTase activity has not been detected in preparations of recombinant NS5B polymerase expressed in *E. coli* (AL et al. 1997, 1998; YAMASHITA et al. 1998).

Purified recombinant NS5B polymerase had RdRp activity with heteropolymeric RNA templates up to several hundred nucleotides long (BEHRENS et al. 1996). Analysis of the newly synthesized RNA products using denaturing gel electrophoresis (5% polyacrylamide/7 M urea at room temperature) revealed RNA products of about 200 nucleotides rather than the expected 399 nucleotides of the template RNA. However, increasing the concentration of polyacrylamide and the temperature at which the gels were run (10% polyacrylamide/7 M urea at 70°C) resulted in a different mobility of the RNA products indicating that they were actually twice the length of the RNA templates. This suggested that a stable secondary structure was present when RNA products were separated under less stringent denaturing conditions. This hypothesis was tested by determining the susceptibility of the RNA products to several ribonucleases. RNA products were completely degraded by treatment with RNase A at low ionic strength (50 mM NaCl) while RNA products were resistant to the single strand specific RNase T1 and S1. The size of the RNA products was only affected by increasing concentrations of RNase A (at high ionic strength) and S1 nuclease which eventually resulted in RNA products with an electrophoretic mobility similar to the template RNA. These observations provided evidence that the RNA products were actually double stranded. The authors hypothesized that the 3′ ends of template RNA were used to prime RNA synthesis by a copy-back mechanism to produce a duplex RNA molecule consisting of both the template and a complementary RNA strand that was covalently attached by a hairpin structure (BEHRENS et al. 1996).

The phenomenon of dimer-sized RNA products has also been observed during in vitro studies of the poliovirus RdRp (3D^{pol}) (TUSCHALL et al. 1982; YOUNG et al. 1985; NEUFELD et al. 1991; LUBINSKI et al. 1986; PLOTCH et al. 1989; HEY et al. 1986; CHO et al. 1993). Moreover, dimer-sized genomic RNA products have been isolated from tissue culture cells infected with poliovirus or encephalomyocarditis virus (EMCV), suggesting that such molecules may not be artifacts of in vitro replication assays but may actually be involved in the physiological replication of positive strand RNA viruses (SENKEVICH et al. 1980; YOUNG et al. 1985). The biological role for such dimer-sized RNA products remains unknown at this time.

Additional properties of untagged NS5B polymerase expressed in insect cells using baculovirus have been recently reported (TOMEI et al. 1998). The assays were done in 20 mM Tris-HCl (pH 7.5), 50 mM NaCl, 5 mM MgCl$_2$, 0.5% Triton X-100, 1 mM DTT, and NTPs at concentrations of one- to fivefold their Km. Heteropolymeric RNA templates were present at 10–20 nM. The homopolymeric template used was poly(C) at 15 µM with oligo(G) primers at 0.15 µM. The K_m

values for NTPs were different depending on what RNA template was used. With heteropolymeric templates the apparent K_m for UTP was 2.0 µM, for GTP 0.4 µM, and for ATP 10 µM. When poly(A) templates and oligo(U) primers were used the apparent K_m for UTP was 5.0 µM. However, with poly(C) templates and oligo(G) primers the apparent K_m for GTP was 0.2 µM.

Untagged recombinant HCV NS5B polymerase (HCV type 1a) has also been produced in *E. coli* and isolated under nondenaturing conditions (AL et al. 1997, 1998). This form of recombinant NS5B polymerase exhibited a strict requirement for primers to replicate a heteropolymeric RNA template, globin mRNA, providing evidence that NS5B has no intrinsic TNTase activity. A COOH-terminal deletion mutant provided evidence that the RdRp activity observed was indeed due to the recombinant NS5B protein. The other characteristics of the enzyme expressed in *E. coli* did not differ much from NS5B expressed in insect cells except that the optimal incubation temperature was 30°C for NS5B expressed in *E. coli* instead of 22°C for the NS5B expressed in insect cells (BEHRENS et al. 1996). Recombinant NS5B polymerase from other isolates of HCV type 1a need to be studied regarding their enzymatic activity (FERRARI et al. 1999).

The biochemical properties of histidine-tagged recombinant NS5B polymerase (HCV type 1b) expressed in insect cells have been described (LOHMANN et al. 1997; LOHMANN et al. 1998). The COOH-terminal histidine tag incorporated in recombinant NS5B^{C-HIS} facilitated purification and did not inactivate the enzyme. This recombinant protein exhibited primer-dependent RdRp activity when poly(A)/oligo(U)$_{12}$ were used as template and primer, and was also able to transcribe full-length HCV RNA in vitro.

The RNA templates used in these studies were either full length positive strand HCV RNA or the last 319 nucleotides of the HCV genome which includes the 3' UTR polyuridine tract and the 3'-end cloverleaf element (CHOO et al. 1991; TANAKA et al. 1995; KOLYKHALOV et al. 1996). Polymerase assays were 25 µl in volume and contained 20 mM Tris-HCl (pH 7.5), 5 mM MgCl$_2$, 1 mM DTT, 25 mM KCl, 1 mM EDTA, 10–20 U of RNasin, 2–10 µCi of [α-^{32}P]GTP, 500 µM each (CTP, ATP, UTP), and 0.5 µg of HCV RNA or 400 ng of a homopolymeric RNA as a template (LOHMANN et al. 1997). When full length HCV templates were used the radioactive nucleotide was adjusted to a final concentration of 10 µM and no primers were needed. When homopolymeric RNA, poly(A) or poly(C) was used as template, either oligo(U)$_{12}$ or oligo(G)$_{12}$ primer (4 pmol) was included in incubations.

Recombinant NS5B^{C-HIS} produced RNA products that were approximately twice the size of full length HCV genomic RNA templates which appear to be synthesized by a copy-back priming mechanism based on RNase analysis of the products (LOHMANN et al. 1997; BEHRENS et al. 1996). In addition, the relative activities of recombinant NS5B^{C-HIS} polymerase with different homopolymeric RNA templates were compared. Poly(C) and poly(A) were the only effective templates and poly(C) resulted in approximately three times more incorporation of radiolabeled nucleotide into RNA products than poly(A) (LOHMANN et al. 1997). Studies examining the ability of NS5B^{C-HIS} to bind RNA using filter assays showed that it bound RNA homopolymers with the following specificity poly(U) > poly(G) > poly(A) > poly(C). Interestingly, NS5B^{C-HIS} did not specifically bind HCV RNA that included the authentic 3′-end of genomic RNA as compared to a control RNA (LOHMANN et al. 1996; TANAKA et al. 1995; KOLYKHALOV et al. 1996). In summary, recombinant NS5B^{C-HIS} polymerase was strictly dependent on the presence of primers when poly(A) or poly(C) were used as templates; only poly(A) and poly(C), not poly(G) and poly(U), serve as effective templates. However, when HCV RNA that included the authentic 3′-end of genomic RNA was used as a template, RNA appeared to be synthesized by a copy-back priming mechanism as reported for the DCoH mRNA template (BEHRENS et al. 1996).

A number of site-directed mutants of NS5B^{C-HIS} were tested for both RdRp and TNTase activity. Mutations in conserved domains that inactivated NS5B polymerase in assays with poly(C)/oligo(G) as a substrate included: Asp-220 to Asn, Cys, or Gly; Asp-225 to Asn or Gly; Gly-283 to Arg, Leu or Asp; Thr-287 to Cys or Lys; Asn-291 to Lys; Asp-318 (of the Gly$_{317}$-Asp$_{318}$-Asp$_{319}$ motif) to His, Asn, or Glu; or Asp-319 to His (LOHMANN et al. 1997). Preparations of these mutants of NS5B^{C-HIS} lacked RdRp activity but retained TNTase activity. Alternatively, the substitution of Arg-345 with Lys exhibited a 52% increase in RdRp activity above wild-type NS5B^{C-HIS} polymerase, but no change in TNTase activity. The basis for this stimulation of polymerase activity remains unknown at this time.

Deletion mutants of the NH$_2$-terminal and COOH-terminal regions of NS5B polymerase were also examined regarding in vitro RdRp activity with poly(C)/oligo(G) as a substrate. Preparations of NS5B mutants lacking 19 or 40 NH$_2$-terminal residues had no RdRp activity (LOHMANN et al. 1997). However, as with other preparations of mutants lacking RdRp activity, they still contained TNTase activity. Mutants having 83 or more amino acids deleted from the amino terminus could not be purified using metal (Ni^{2+}) chelating chromatography. Alternatively, COOH-terminal mutants of NS5B (deletions of 25 or 55 residues) with a hexahistidine tag at their NH$_2$-terminal still had as much as 42% of the RdRp activity of full-length NS5B^{C-HIS} (also see FERRARI et al. 1999). Both the mutagenesis results and further analysis of insect cells infected with wild-type baculovirus (no NS5B insert) provided evidence that TNTase activity is not an intrinsic property of HCV NS5B polymerase. It was concluded that the TNTase activity observed in the preparations of NS5B polymerase expressed in insect cells was due to a contaminating cellular enzyme.

At this time relatively little has been done to study HCV NS5B polymerase within mammalian cells. One obstacle in approaching this problem is the lack of a suitable reporter system that can accurately measure RdRp activity in intact mammalian cells. If such a system is developed it may enable genetic approaches to be used in identifying viral or cellular proteins that regulate NS5B polymerase within a physiological context. Nevertheless, a recent report does prove that NS5B polymerase can be expressed and RdRp enzyme activity recovered from lysates of BHK and Huh-7 cells (HELLER et al. 1998). In this study pcDNA3 (Invitrogen) containing NS5B cDNA was used to produce stably transfected BHK cells. In addition, pBI containing NS5B and a reporter cDNA was used to stably transfect

Huh-7 cells. Lysates of stably transfected cells were assayed as described in Section 5.2. Although transfection with pcDNA3/NS5B and selection with neomycin produced many stably transfected cell lines, only a few cell lines expressed detectable levels of NS5B polymerase. However, by using pBI with a luciferase reporter gene and Tet-On cells, the authors were able to isolate a larger number of cell lines expressing HCV NS5B, and the levels of NS5B expression appeared to be greater. NS5B polymerase was measured in these cells by northern blots, immunoprecipitation of NS5B protein, and in vitro RdRp assays of cell lysates. This initial report provides encouragement that stably transfected cell lines may be useful in studying properties of HCV NS5B polymerase.

6 Questions and Future Directions

6.1 Template Specificity of Hepatitis C Virus NS5B Polymerase

Does HCV NS5B polymerase produce complementary copies of both viral and cellular RNA in vivo as it does in vitro? This seems unlikely, especially during the early stages of in vivo HCV replication when copies of viral RNA and molecules of NS5B polymerase are scarce. If NS5B polymerase indeed interacts specifically with HCV RNA templates in vivo, then is this solely due to compartmentalization within replication complexes with membrane-anchored NS5B or do specific NS5B-RNA interactions play a role? Alternatively, do other viral or cellular proteins specifically bind HCV RNA structural elements and mediate template-specific interactions indirectly by protein-protein interactions with NS5B?

The observation that HCV NS5B polymerase expressed in *E. coli* will readily replicate globin mRNA, as does poliovirus $3D^{pol}$, suggests that additional HCV proteins or host factors are required for the specific interaction of polymerase with HCV RNA templates in vivo (AL et al. 1997, 1998; HEY et al. 1986; PLOTCH et al. 1989). In addition, the reported lack of a specific interaction of NS5B polymerase with the 3'-end cloverleaf (also called 3'-X tail) of genomic HCV RNA suggests a mechanism other than a specific NS5B-RNA interaction participating in the initiation of genomic HCV RNA transcription (TANAKA et al. 1995; KOLYKHALOV et al. 1996; LOHMANN et al. 1997). Nevertheless, the possibility that NS5B polymerase binds RNA structural elements within the HCV negative strand or the 5' untranslated region of genomic HCV RNA remains to be tested. The example of poliovirus suggests a possible alternative to a highly specific and singular interaction of NS5B polymerase with HCV RNA. The formation of a RNP complex around RNA structural elements within HCV RNA templates might occur and facilitate template-specific interactions with NS5B polymerase. Multiple lines of evidence support the formation of a RNP complex composed of the poliovirus genomic 5'-terminal cloverleaf, the poliovirus 3AB and $3CD^{pro}$ proteins, and a cellular 38 kDa protein (ANDINO et al. 1990, 1993; ROEHL et al. 1997; MOSIMANN et al. 1997). The formation of this RNP has been proposed to regulate the synthesis of poliovirus genomic RNA from negative strands of RNA (PAUL et al. 1994; XIANG et al. 1995; RICHARDS and EHRENFELD 1998; PARSLEY et al. 1997). Another study has described a nucleolar protein, identified as nucleolin, that specifically binds the 65 nucleotide long 3' untranslated region of the poliovirus genome and is required for efficient synthesis of negative strands of RNA (WAGGONER and SARNOW 1998). This protein-RNA interaction has biological effects because depletion of nucleolin from cell extracts limits an early step of poliovirus replication and decreases the production of infectious virus (WAGGONER and SARNOW 1998). On the other hand, one study has shown that deletions of the 3' untranslated region of poliovirus genomic RNA do not necessarily ablate viral replication (TODD et al. 1997).

It appears that protein-RNA and protein-protein interactions analogous to those regulating poliovirus RNA replication will also be found during HCV RNA replication. Progress has been made in defining the structural basis for protein-RNA interactions and the data suggest that a β-sheet structure with surface loops that confer specificity of RNA binding may be common. This mechanism for protein binding of a specific RNA structure has been observed for ribonucleo-proteins (RNPs) that bind the U series of snRNA and the binding of eIF4E to the 5'-cap structure of eukaryotic mRNA (ALLAIN et al. 1997; KENAN et al. 1991; MATTAJ 1993; DREYFUSS et al. 1993; MARCOTRIGIANO et al. 1997). Recent studies of the effects of the poliovirus 3AB protein on 3D^{pol} activity emphasize the role of 3AB in stabilizing weak interactions between nucleotidyl-protein-primed initiation events during viral RNA replication (RICHARDS and EHRENFELD 1998). It may be that many of the protein-RNA interactions that are physiologically relevant for HCV RNA replication are relatively low affinity interactions that are even weaker than one observed for an essential step in eukaryotic mRNA translation (eIF4E binding to mRNA caps, K_d of 1.2 µM) (CARBERRY et al. 1989; HAGEDORN et al. 1997). If ribonucleoprotein complexes consisting of several proteins are regulating HCV RNA replication then the complete RNA recognition/binding domain may be composed of subdomains occurring on several proteins.

The well-described HCV 5' untranslated region (UTR) and the more recently identified 3'-end cloverleaf (3'-X tail) of HCV genomic RNA seem to be likely targets for protein-RNA interactions (HONDA et al. 1996; REYNOLDS et al. 1996; TANAKA et al. 1995; KOLYKHALOV et al. 1996). Regarding the 5' UTR, one study describes the interaction of a polypyrimidine tract-binding protein (PTB) and the possible function of this protein-RNA complex during internal initiation of HCV RNA translation (ALI and SIDDIQUI 1995; also see Rijnbrand and Lemon, this volume). Recombinant PTB binds several domains of the HCV 5' UTR as measured by UV cross-linking and mutagenesis studies. In addition, PTB isomers from HeLa cells cross-link to the same regions of the HCV 5' UTR (ALI and SIDDIQUI 1995). However, a definitive physiological role of PTB in HCV RNA translation remains to be proven and any role in HCV RNA replication is speculative. If PTB indeed plays a role in regulating HCV RNA translation, then one might expect that another protein required for HCV RNA replication might compete for the PTB RNA binding site since viral RNA engaged with the host cell translational machinery appears unlikely to be a template for RNA replication (GAMARNIK and ANDINO 1998).

The location and predicted structure of the 98-base 3'-end cloverleaf of genomic HCV RNA suggested a biological function (TANAKA et al. 1995, 1996; KOLYKHALOV et al. 1996; BLIGHT and RICE 1997). However, what is the evidence that this RNA structure is required for viral replication? Recent studies showed that HCV RNA lacking the entire 98-base element, the SLII and SLIII region, or the SLI region (extreme 3'-end of the genome) did not produce a detectable infection in chimpanzees following the direct injection of RNA into the liver (KOLYKHALOV et al. 1997; C.M. Rice, personal communication; see Major and Feinstone, this volume). Since the 3'-end cloverleaf of genomic HCV RNA does appear to be essential for infectivity of HCV RNA, then what biological function might it serve? One possibility is that this RNA structural element serves as a molecular beacon and protein-binding motif for the assembly of a RNP complex with host and viral proteins required for HCV RNA replication. Indeed, several cellular proteins have been shown to UV cross-link to the 98-base 3'-end of HCV genomic RNA. These proteins include 38, 57, 87, and 130 kDa proteins (TSUCHIHARA et al. 1997; INOUE et al. 1998). The 57 kDa protein is PTB and is unable to bind the same region of negative strand HCV RNA (TSUCHIHARA et al. 1997). Additional studies are needed to verify these protein-RNA interactions by an alternative experimental approach and to prove that they regulate HCV RNA replication. It is appealing to think that such protein-RNA interactions might provide specificity for selectively replicating either genomic or negative strands of HCV RNA. Such a switch in the use of RNA templates is known to occur in replication cycle of positive strand RNA viruses.

6.2 Hepatitis C Virus Replication Complexes

Other plus strand RNA viruses that code for one genome-sized mRNA and are more easily propagated in tissue culture have provided examples of membrane bound viral replication complexes (BIENZ et al. 1990, 1992). The membranes involved in this process vary among the different viruses. However, HCV NS5B polymerase expressed in COS 7 cells is concentrated in the perinuclear region and appears to be associated with nuclear membranes and the endoplasmic reticulum or Golgi complex at 40 h post-transfection (HWANG et al. 1997). Not surprisingly, NS5B contains a hydrophobic domain at the COOH-terminal that functions as a membrane-anchoring domain in mammalian cells (YAMASHITA et al. 1998). Nevertheless, the subcellular localization of NS5B polymerase and the composition of HCV replication complexes in cells that are producing infectious HCV are not known at this time.

Although studies of HCV replication complexes have not been done, the expression of recombinant HCV proteins in mammalian cells offers clues regarding what viral proteins may be part of the replication complex. A preliminary report provides evidence for the formation of a protein complex consisting of NS5B polymerase, NS3, and NS4A in mammalian cells (ISHIDO et al. 1998). The evidence for an in vivo NS5B polymerase/NS3 interaction, which requires the NH_2-terminus of NS3, includes the immunocytochemical co-localization and co-immunoprecipitation of both proteins following transient expression in HeLa or COS-7 cells. In addition, NS5B polymerase co-localizes and co-immunoprecipitates with NS4A (ISHIDO et al. 1998). Other reports based on the transient expression of recombinant proteins and co-immunoprecipitation assays support the suggestion that a complex forms between the HCV NS5A and NS4A proteins in mammalian cells (LIN et al. 1997). The association of NS4A with NS5A requires amino acids 2135–2139 of NS5A and it appears to be necessary for the phosphorylation of at least some sites of NS5A (ASABE et al. 1997). Interestingly, the phosphorylation of NS5A at different sites at least suggests a possible regulatory role for these modifications in assembling protein-protein complexes within HCV infected cells (REED et al. 1997, 1988). Although these observations raise many questions, it is not known if these protein–protein interactions have a physiologic role in regulating HCV replication or occur within authentic HCV replication complexes.

Addressing questions regarding the HCV replication complex will most likely require a cell culture system capable of propagating the virus at a relatively high titer (see Kato and Shimotohno, this volume). A series of observations during cell culture studies of the hepatitis A virus (HAV), a picornavirus, may be relevant to developing cell culture systems to study HCV replication. Although HAV isolated from infected patients does not replicate in cultured cells, it can adapt to replicate in vitro following serial passage in several cell lines (PROVOST and HILLEMAN 1979). However, mutations within the P2 region (encodes the 2B and 2C proteins) of the HAV genome are required for adaptation to growth in cell culture (EMERSON et al. 1992). Although mutations in the 5′ untranslated region of the HAV genome had no independent effect, they do act cooperatively with mutations in the P2 region to enhance viral replication in cell culture (ZHANG et al. 1995). Mutations in the P2 region also appear to influence the growth and virulence of HAV in different species of primates (RAYCHAUDHURI et al. 1998). Although the biochemical functions of the HAV 2B and 2C proteins remain to be fully characterized, the 2C protein has been shown to have the functions of both binding membranes and RNA (TETERINA et al. 1997; KUSOV et al. 1998). On the other hand, a single amino acid insertion in the HAV 2B coding region, which does not affect polyprotein processing, is lethal to HAV replication in cell culture (HARMON et al. 1995). Perhaps the adaptation of HCV to growth in cell culture will require specific mutations that result in polymorphisms in RNA structural motifs and proteins that are required for RNA replication (see Kato and Shimotohno, this volume).

The NS3 protein of HCV is a distant relative, based on sequence homology, with the poliovirus 2C protein which is derived from the P2 genomic region. The morphological and biochemical characteristics of poliovirus replication complexes have been examined in infected cells (BIENZ et al. 1990, 1992). The replication complex is enclosed in a rosette-like shell of virus-induced vesicles and has a tightly packed second membrane system that encloses nascent positive strands of RNA on the replicative intermediate. Electron microscopy studies using biotinylated RNA probes support the proposal that this structure represents the positive strand RNA-synthesizing machinery of infected cells (BIENZ et al. 1992). Several studies provide evidence that the poliovirus P2 proteins 2BC and 2C localize to intracellular membranes where they induce the formation of new membrane structures and alter the morphology of the cellular protein secretory machinery (BIENZ et al. 1990; CHO et al. 1994). Several studies also support the proposal that the 2C protein attaches the poliovirus RNA to vesicular membranes, that it specifically binds the 3′-terminal cloverleaf of the poliovirus negative strand RNA in vitro, and that it is essential for poliovirus RNA replication in cells (BIENZ et al. 1990; BANERJEE et al. 1997). Studying recombinant HCV NS5B polymerase in vitro will provide insights into the mechanisms of HCV RNA replication. However, the example of poliovirus provides an illustration of just how complicated understanding the viral RNA replication process in a physiological context is likely to be. The possible role of the HCV NS3 protein in binding HCV RNA within replication complexes is one of the questions that remains to be answered.

6.3 Biomedical Applications

The ability to express recombinant HCV NS5B polymerase and to assay this enzyme in vitro opens the possibility of identifying small molecules that inhibit HCV RNA replication in patients with chronic hepatitis C. However, to test the effect of many compounds on NS5B polymerase activity a mammalian cell system will be necessary because these compounds must first be metabolized by a cellular (or viral) enzyme to become pharmacologically active. One example of antiviral compounds requiring such an intracellular enzymatic modification are nucleoside analogues that are phosphorylated to become the active triphosphate anabolites. 3TC, lamivudine, is a nucleoside analogue that is a very effective therapeutic for chronic hepatitis B virus (LEE 1997; MARKOWITZ et al. 1998). However, it remains to be seen if analogous compounds can be developed into therapeutics that target the RdRp of an RNA virus (TISDALE et al. 1995). On the other hand, non-nucleoside analogues or other classes of compounds may prove to be more effective inhibitors of HCV NS5B polymerase.

Another question that remains unanswered is what are the predominant sequences of NS5B polymerase in cells that are actively replicating HCV RNA? Although several specific NS5B sequences that encode proteins with RdRp activity in vitro are known at this time, little is known about what effect changes in the NS5B sequence will have on viral replication in vivo (Figs. 1, 2) (see Major and Feinstone, this volume). How might HCV respond to the selective pressures of a small molecule inhibiting NS5B polymerase? In the example of chronic hepatitis B, caused by a dsDNA virus, mutations of the DNA polymerase that include the YMDD domain occur following treatment with the nucleoside analogue lamivudine and permit viral replication to resume (ALLEN et al. 1998; FU and CHENG 1998). Will "escape mutants" of HCV encoding a NS5B polymerase that is resistant to the effect of a small molecule inhibitor rapidly appear in patients with chronic hepatitis C? Due to the nature of RNA viruses and the lack of proofreading activity in their RdRp it seems likely that such mutants will appear for HCV. This suggests that combination therapies targeting several key molecular events in viral replication will be required to effectively treat chronic hepatitis C. The topic of HCV RNA replication and the application of this knowledge in developing effective treatments for chronic hepatitis C should provide plenty of challenges for investigators that are attracted to difficult problems.

Acknowledgements. We thank other members of the Hagedorn laboratory and colleagues in the Woodruff Memorial Research Building for their comments on this review: Ronald H. Al, Laura J.Taylor, Pi-Chen Hsu, Susan H. Voss, and Richard Whalen. We thank Kim Gernert of the Emory University Biomolecular Computing Resource and Philip A. Hagedorn for their computer analysis and graphics work. CH thanks E. Ehrenfeld for a number of helpful discussions. This work was supported in part by Public Health Service grants AI41424, CA63640, and CA77818.

References

Adams NJ, Chamberlain RW, Taylor LA, Davidson F, Lin CK, Elliott RM, Simmonds P (1997) Complete coding sequence of hepatitis C virus genotype 6a. Biochem Biophys Res Comm 234: 393–396

Adkins S, Stawicki SS, Faurote G, Siegel RW, Kao CC (1998) Mechanistic analysis of RNA synthesis by RNA-dependent RNA polymerase from two promoters reveals similarities to DNA-dependent RNA polymerase. RNA 4:455–470

Al RH, Xie Y, Wang Y, Hagedorn CH (1998) Expression of recombinant hepatitis C virus NS5B in *Escherichia coli.* Virus Res 53:141–149

Al RH, Xie Y, Wang Y, De Staercke C, Van Beers EH, Hagedorn CH (1997) Expression of recombinant hepatitis C virus NS5B. Nucleic Acids Symposium Series No 36:197–199

Ali N, Siddiqui A (1995) Interaction of polypyrimidine tract-binding protein with the 5′ noncoding region of the hepatitis C virus RNA genome and its functional requirement in internal initiation of translation. J Virol 69:6367–6375

Allain FHT, Howe PWA, Neuhaus D, Varani G (1997) Structural basis of RNA-binding specificity of human U1 A protein. EMBO J 16:5764–5774

Allen MI, Deslauriers M, Andrews CW, Tipples GA, Walters KA, Tyrrell DL, Brown N, Condreay LD (1998) Identification and characterization of mutations in hepatitis B virus resistant to lamivudine. Hepatology 27:1670–1677

Alter MJ (1997) Epidemiology of Hepatitis C. In: Bissell DM (ed.) Management of hepatitis C. Hepatology 26 (Suppl. 1), 62S–65S

Altschul SF, Madden TL, Schaffer AA, Zhang J, Zhang Z, Miller W, Lipman DJ (1997) Gapped BLAST and PSI-BLAST: a new generation of protein database search programs. Nucl Acids Res 25: 3389–3402

Andino R, Rieckhof GE, Achacoso PL, Baltimore D (1993) Poliovirus RNA synthesis utilizes an RNP complex formed around the 5′-end of viral RNA. EMBO J 12:3587–3598

Andino R, Rieckhof GE, Baltimore D (1990) A functional ribonucleoprotein complex forms around the 5′ end of poliovirus RNA. Cell 63:369–380

Andrews NC, Baltimore D (1986) Purification of a terminal uridylyltransferase that acts as host factor in the in vitro poliovirus replicase reaction. Proc Natl Acad Sci USA 83:221–225

Asabe SI, Tanji Y, Satoh S, Kaneko T, Kimura K, Shimotohno K (1997) The N-terminal region of hepatitis C virus-encoded NS5 A is important for NS4A-dependent phosphorylation. J Virol 71:790–796.

Bakalara N, Simpson AM, Simpson L (1989) The Leishmania kinetoplast-mitochondrion contains terminal uridylyltransferase and RNA ligase activities. J Biol Chem 264:18679–18686

Banerjee R, Echeverri A, Dasgupta A (1997) Poliovirus-encoded 2 C polypeptide specifically binds to the 3′-terminal sequences of viral negative-strand RNA. J Virol 71:9570–9578

Baron U, Freundlieb S, Gossen M, Bujard H (1995) Co-regulation of two gene activities by tetracycline via a bidirectional promoter. Nucleic Acids Res 17:3605–3606

Barton DJ, Morasco BJ, Flanegan JB (1996) Assays for poliovirus polymerase, 3D (Pol), and authentic RNA replication in HeLa S10 extracts. Methods Enzymol 275:35–57

Beckman MT, Kirkegaard K (1998) Site size of cooperative single-stranded RNA binding by poliovirus RNA-dependent RNA polymerase. J Biol Chem 273:6724–6730

Behrens SE, Tomei L, DeFrancesco R (1996) Identification and properties of the RNA-dependent RNA polymerase of hepatitis C virus. EMBO J 15:12–22

Bienz K, Egger D, Pfister T, Troxler M (1992) Structural and functional characterization of the poliovirus replication complex. J Virol 66:2740–2747

Bienz K, Egger D, Troxler M, Pasamontes L (1990) Structural organization of poliovirus RNA replication is mediated by viral proteins of the P2 genomic region. J Virol 64:1156–1163

Bissell DM (ed.) (1997) Management of hepatitis C. Hepatology 26 (Suppl. 1), 1S–156S.

Blight KJ, Rice CM (1997) Secondary structure determination of the conserved 98-base sequence at the 3′ terminus of hepatitis C virus genome RNA. J Virol 71:7345–7352

Bradley DW, Maynard JE, Popper H, Cook EH, Ebert JW, McCaustland KA, Schable CA, Fields HA (1983) Post-transfusion non-A, non-B hepatitis: Physicochemical properties of two distinct agents. J Infect Dis 148:254–265

Buck KW (1996) Comparison of the replication of positive-stranded RNA viruses of plants and animals. Advances Virus Res 47:159–251

Bukh J, Miller RH, Purcell RH (1995) Genetic heterogeneity of hepatitis C virus: quasispecies and genotypes. Sem Liver Dis 15:41–63

Carberry SE, Rhoads RE, Goss DJ (1989) A spectroscopic study of the binding of m^7GTP and m^7GpppG to human protein synthesis initiation factor 4 E. Biochemistry 28:8078–8083

Cha T-A, Beall E, Irvine B, Kolberg J, Chien D, Kuo G, Urdea MS (1992) At least five related, but distinct, hepatitis C viral genotypes exist. Proc Natl Acad Sci USA 89:7144–7148

Chamberlain RW, Adams N, Saeed AA, Simmonds P, Elliott RM (1997a) Complete nucleotide sequence of a type 4 hepatitis C virus variant, the predominant genotype in the Middle East. J Gen Virol 78:1341–1347.

Chamberlain RW, Adams NJ, Taylor LA, Simmonds P, Elliott RM (1997b) The complete coding sequence of hepatitis C virus genotype 5a, the predominant genotype in South Africa. Biochem Biophys Res Comm 236:44–49

Chayama K, Tsubota A, Koida I, Arase Y, Saitoh S, Ikeda K, Kumada H (1994) Nucleotide sequence of hepatitis C virus (type 3b) isolated from a Japanese patient with chronic hepatitis C. J Gen Virol 75:3623–3628

Cho MW, Richards OC, Dmitrieva TM, Agal V, Ehrenfeld E (1993) RNA duplex unwinding activity of poliovirus RNA-dependent RNA polymerase 3D^{pol}. J Virol 63:3010–3018

Cho MW, Teterina N, Egger D, Bienz K, Ehrenfeld E (1994) Membrane rearrangement and vesicle induction by recombinant poliovirus 2 C and 2BC in human cells. Virology 202:129–145

Cho JM, Park YW, Lee YB, Yang JY, Kim CH, Choo SH, Ryu WS (1995) Molecular cloning of hepatitis C virus genome from a single Korean blood donor. Mol Cells 5:317–324

Choo QL, Richman KH, Han JH, Berger K, Lee C, Dong C, Gallegos C, Coit D, Medina-Selby R, Barr PJ, Weiner AJ, Bradley DW, Kuo G, Houghton M (1991) Genetic organization and diversity of the hepatitis C virus. Proc Natl Acad Sci U S A 88:2451–2455

Choo Q-L, Kuo G, Weiner AJ, Overby LR, Bradley DW, Houghton M (1989) Isolation of a cDNA clone derived from a blood-borne non-A, non-B viral hepatitis genome. Science 244:359–362

Collett MS, Chunduru SK, Young DC, Groarke JM, Benetatos CA, Barone LR, Chang S-C, Garrison LM, Gorczyca WP, Kolykhalov AA, Read C, Rice CM (1998) RNA-dependent RNA polymerase activity of hepacivirus and pestivirus NS5B proteins. 5th International Meeting on Hepatitis C Virus and Related Viruses, Molecular Virology and Pathogenesis, Venice, Italy.

DeFrancesco R, Behrens S-E, Tomei L, Altamura S, Jiricny J (1996) RNA-dependent RNA polymerase of hepatitis C virus. In: Kuo LC, Olsen DB, Carroll SS (eds). Meth Enz 275: viral polymerases and related proteins. Academic Press, San Diego

De Staercke C, Al RH, Xie Y, Hagedorn CH (1998) Expression of hepatitis C virus NS5B polymerase in E. coli and characterization. 5th International Meeting on Hepatitis C Virus and Related Viruses, Molecular Virology and Pathogenesis, Venice, Italy

Diamond SE, Kirkegaard K (1994) Clustered charged-to-alanine mutagenesis of poliovirus RNA-dependent RNA polymerase yields multiple temperature-sensitive mutants defective in RNA synthesis. J Virol 68:863–876

Dreyfuss G, Matunis MJ, Pinol-Roma S, Burd CG (1993) HnRNP proteins and biogenesis of mRNA. Annu Rev Biochem 62:289–321

Ehrenfeld E, Maizel JV, Summers DF (1970) Soluble RNA polymerase complex from poliovirus-infected HeLa Cells. Virology 40:840–846

Emerson SU, Huang YK, McRill C, Lewis M, Purcell RH (1992) Mutations in both the 2B and 2C genes of hepatitis A virus are involved in adaptation to growth in cell culture. J Virol 66:650–654

Enomoto N, Sakuma I, Asahina Y, Kurosaki M, Murakami T, Yamamoto C, Izumi N, Marumo F, Sato C (1995) Comparison of full-length sequences of interferon-sensitive and resistant hepatitis C virus 1b. Sensitivity to interferon is conferred by amino acid substitutions in the NS5 A region. J Clin Invest 96:224–230

Ferrari E, Wright-Minogue J, Fang JWS, Baroudy BM, Lau JYN, Hong Z (1999) Characterization of soluble hepatitis C virus RNA-dependent RNA polymerase expressed in *Escherichia coli*. J Virol 73:1649–1654

Flanegan JB, Baltimore D (1977) Poliovirus-specific primer-dependent RNA polymerase able to copy poly(A). Proc Natl Acad Sci USA 74:3677–3680

Forns X, Bukh J, Purcell RH, Emerson SU (1997) How *Escherichia coli* can bias the results of molecular cloning: preferential selection of defective genomes of hepatitis C virus during the cloning procedure. Proc Natl Acad Sci USA 94:13909–13914

Friedland DE, Shoemaker MT, Xie Y, Wang Y, Hagedorn CH, Goss DJ (1997) Identification of the cap binding domain of human recombinant eukaryotic protein synthesis initiation factor 4 E using a photoaffinity analogue. Protein Sci 6:125–131

Fu L, Cheng YC (1998) Role of additional mutations outside the YMDD motif of hepatitis B virus polymerase in L(−) SddC (3TC) resistance. Biochem Pharmacol 55:1567–1572

Gamarnik AV, Andino R (1998) Switch from translation to RNA replication in a positive-stranded RNA virus. Genes & Dev 12:2293–2304

Grakoui A, Wychowski C, Lin C, Feinstone SM, Rice CM (1993) Expression and identification of hepatitis C virus polyprotein cleavage products. J Virol 67:1385–1395

Grun JB, Brinton MA (1986) Characterization of West Nile virus RNA-dependent RNA polymerase and cellular terminal adenylyl and uridylyl transferases in cell-free extracts. J Virol 60:1113–1124

Hagedorn CH, Spivak-Kroizman T, Friedland DE, Goss DJ, Xie Y (1997) Expression of functional eIF-4Ehuman: Purification, detailed characterization, and its use in isolating eIF-4 E binding proteins. Protein Expression Purif 9:53–60

Hahn YS, Grakoui A, Rice CM, Strauss EG, Strauss JH (1989) Mapping of RNA⁻ temperature-sensitive mutants of Sindbis virus: Complementation group F mutants have lesions in nsP4. J Virol 63:1194–1202

Hansen JL, Long AM, Schultz SC (1997) Structure of the RNA-dependent RNA polymerase of poliovirus. Structure 5:1109–1122

Harmon SA, Emerson SU, Huang YK, Summers DF, Ehrenfeld E (1995) Hepatitis A viruses with deletions in the 2 A gene are infectious in cultured cells and marmosets. J Virol 69:5576–5581

Hayashi N, Higashi H, Kaminaka K, Sugimoto H, Esumi M, Komatsu K, Hayashi K, Sugitani M, Suzuki K, Tadao O (1993) Molecular cloning and heterogeneity of the human hepatitis C virus (HCV) genome. J Hepatol 17 Suppl 3:S94-S107

Hayes RJ, Buck KW (1990) Complete replication of a eukaryotic virus RNA in vitro by a purified RNA-dependent RNA polymerase. Cell 63:363–368

Heller A, Windheim M, Berger N, Pfaff E (1998) Characterization of constitutive and inducible cell lines expressing hepatitis C virus NS5B. 5th International Meeting on Hepatitis C Virus and Related Viruses, Molecular Virology and Pathogenesis, Venice, Italy.

Hey TD, Richards OC, Ehrenfeld E (1986) Synthesis of plus- and minus-strand RNA from poliovirion RNA template in vitro. J Virol 58:790–796

Higgins DG, Thompson JD, Gibson TJ (1996) Using CLUSTAL for multiple sequence alignments. Meth Enzymol 266:383–402

Hijikata M, Mizushima H, Tanji Y, Komoda Y, Hirowatari Y, Akagi T, Kato N, Kimura K, Shimotohno K (1993) Proteolytic processing and membrane association of putative nonstructural proteins of hepatitis C virus. Proc Natl Acad Sci USA 90:10773–10777

Holland JJ, De La Torre JC, Steinhauer DA (1992) RNA virus populations as quasispecies. Curr Topics Micro Immunol 176:1–20

Honda M, Brown EA, Lemon SM (1996) Stability of a stem-loop involving the initiator AUG controls the efficiency of internal initiation of translation on hepatitis C virus RNA. RNA 2:955–968

Honda M, Kaneko S, Unoura M, Kobayashi K, Murakami S (1992) Sequence comparisons for a hepatitis C virus genome RNA isolated from a patient with liver cirrhosis. Gene 120:317–318

Hope DA, Diamond SE, Kirkegaard K (1997) Genetic dissection of interaction between poliovirus 3D polymerase and viral protein 3AB. J Virol 71:9490–9498

Houghton M (1996) Hepatitis C viruses. In: Fields BN, Knipe DM, Howley PM (eds) Virology. Lippincott-Raven, Philadelphia

Hwang SB, Park K-J, Kim Y-S, Sung YC, Lai MM (1997) Hepatitis C virus NS5B protein is a membrane-associated phosphoprotein with a predominantly perinuclear localization. Virology 227: 439–446

Inchauspe G, Zebedee S, Lee DH, Sugitani M, Nasoff M, Prince AM (1991) Genomic structure of the human prototype strain H of hepatitis C virus: comparison with American and Japanese isolates. Proc Natl Acad Sci USA 88:10292–10296

Inoue Y, Miyazaki M, Ohashi R, Tsuji T, Fukaya K, Kouchi H, Uemura T, Mihara K, Namba M (1998) Ubiquitous presence of cellular proteins that specifically bind to the 3' terminal region of hepatitis C virus. Biochem Biophys Res Comm 245:198–203

Ishido S, Fujita T, Hotta H (1998) Complex formation of NS5B with NS3 and NS4 A proteins of hepatitis C virus. Biochem Biophys Res Comm 244:35–40

Jablonski SA, Luo M, Morrow CD (1991) Enzymatic activity of poliovirus RNA polymerase mutants with single amino acid changes in the conserved YGDD amino acid motif. J Virol 65:4565–4572

Jablonski SA, Morrow CD (1995) Mutation of the aspartic acid residues of the GDD sequence motif of poliovirus RNA-dependent RNA polymerase results in enzymes with altered metal ion requirements for activity. J Virol 69:1532–1539

Kamer G, Argos P (1984) Primary structural comparison of RNA-dependent polymerases from plant, animal and bacterial viruses. Nucl Acids Res 12:7269–7282

Kato N, Hijikata M, Ootsuyama Y, Nakagawa M, Ohkoshi S, Sugimura T, Shimotohno K (1990) Molecular cloning of the human hepatitis C virus genome from Japanese patients with non-A, non-B hepatitis. Proc Natl Acad Sci USA 87: 9524–9528

Kemp BE, Pearson RB (1990) Protein kinase recognition sequence motifs. Trends in Biochem Sci 15: 342–346

Kenan DJ, Query CC, Keene JD (1991) RNA recognition: Towards identifying determinants of specificity. Trends Biochem Sci 16:214–220

Kim JL, Morgenstern KA, Lin C, Fox T, Dwyer MD, Landro JA, Chambers SP, Markland W, Lepre CA, O'Malley ET, Harbeson SL, Rice CM, Mureko MA, Caron PR, Thomson JA (1996) Crystal structure of the hepatitis C virus NS3 protease domain complexed with a synthetic NS4 A cofactor peptide. Cell 87:343–355

Kohlstaedt LA, Wang J, Friedman JM, Rice PA, Steltz TA (1992) Crystal structure at 3.5 A resolution of HIV-1 reverse transcriptase complexed with an inhibitor. Science 256:1783–1790

Kolykhalov AA, Agapov EV, Blight KJ, Mihalik K, Feinstone SM, Rice CM (1997) Transmission of hepatitis C by intrahepatic inoculation with transcribed RNA. Science 277:570–574

Kolykhalov AA, Agapov EV, Rice CM (1994) Specificity of the hepatitis C virus NS3 serine protease: effects of substitutions at the 3/4 A, 4 A/4B, 4B/5 A, and 5 A/5B cleavage sites on polyprotein processing. J Virol 68:7525–7533

Kolykhalov AA, Feinstone SM, Rice CM (1996) Identification of a highly conserved sequence element at the 3' terminus of hepatitis C virus genome RNA. J Virol 70:3363–3371

Koonin EV (1991) The phylogeny of RNA-dependent RNA polymerases of positive-strand RNA viruses. J Gen Virol 72:2197–2206

Kroner P, Richards D, Traynor P, Ahlquist P (1989) Defined mutations in a small region of the brome mosaic virus 2a gene cause diverse temperature-sensitive RNA replication phenotypes. J Virol 63:5302–5309

Kusov YY, Probst C, Jecht M, Jost PD, Gauss-Muller V (1998) Membrane association and RNA binding of recombinant hepatitis A virus protein 2 C. Arch Virol 143:931–944

de Lamballerie X, Charrel RN, Attoui H, De Micco P (1997) Classification of hepatitis C virus variants in six major types based on analysis of the envelope 1 and nonstructural 5B genome regions and complete polyprotein sequences. J Gen Virol 78:45–51

Landro JA, Raybuck SA, Luong YPC, O'Malley ET, Harbeson SL, Morgenstern KA, Rao G, Livingston DJ (1997) Mechanistic role of an NS4 A peptide cofactor with the truncated NS3 protease of hepatitis C virus: elucidation of the NS4 A stimulatory effect via kinetic analysis and inhibitor mapping. Biochemistry 36:9340–9348

Lee WM (1997) Medical progress: hepatitis B virus infection. N Engl J Med 337:1733–1745

Leung WC, Leung MF, Rawls WE (1979) Distinctive RNA transcriptase, polyadenylic acid polymerase, and polyuridylic acid polymerase activities associated with Pichinde virus. J Virol 30:98–107

Lin C, Wu J-W, Hsiao K, Su MS-S (1997) The hepatitis C virus NS4 A protein: interactions with the NS4B and NS5 A proteins. J Virol 71:6465–6471

Lohmann V, Roos A, Korner R, Koch J, Bartenschlager R (1998) biochemical and kinetic analyses of NS5B RNA-dependent RNA polymerase of the hepatitis C virus. Virol 249:108–118

Lohmann V, Korner F, Herian U, Bartenschlager R (1997) Biochemical properties of hepatitis C virus NS5B RNA-dependent RNA polymerase and identification of amino acid sequence motifs essential for enzymatic activity. J Virol 71:8416–8428

Lubinski JM, Kaplan G, Raceniello VR, Dasgupta (1986) Mechanism of in vitro synthesis of covalently linked dimeric RNA molecules by poliovirus replicase. J Virol 58:459–467

Marcotrigiano J, Gingras A-C, Sonenberg N, Burley SK (1997) Cocrystal structure of the messenger RNA 5' cap-binding protein (eIF-4 E) bound to 7-methyl-GDP. Cell 89:951–961

Markowitz JS, Martin P, Conrad AJ, Markmann JF, Seu P, Yersiz H, Goss JA, Schmid P, Pakrasi A, Artinian L, Murray NG, Imagawa DK, Holt C, Goldstein LI, Stribling R, Busuttil RW (1998) Hepatology 28:585–589

Mattaj IW (1993) RNA recognition: A family matter? Cell 73:837–840

Mendel DB, Khavari PA, Conley PB, Graves MK, Hansen LP, Admon A, Crabtree GR (1991) Characterization of a cofactor that regulates dimerization of a mammalian homeodomain protein. Science 254:1762–1767

Mills DR, Priano C, DiMauro P, Binderow BD (1988) Qβ replicase: Mapping the functional domains of an RNA-dependent RNA polymerase. J Mol Biol 205:751–764

Mosimann SC, Cherney MM, Sia S, Plotch S, James MN (1997) Refined x-ray crystallographic structure of poliovirus 3 C gene product. J Mol Biol 273:1032–1047

Nakao H, Okamoto H, Tokita H, Inoue T, Iizuka H, Pozzato G, Mishiro S (1996) Full-length genomic sequence of a hepatitis C virus genotype 2c isolate (BEBE1) and the 2c-specific PCR primers. Arch Virol 141:701–704

Neufeld KL, Galarza JM, Richards OC, Summers DF, Ehrenfeld E (1994) Identification of terminal adenylyl transferase activity of the poliovirus polymerase 3Dpol. J Virol 68:5811–5818

Neufeld KL, Richards OC, Ehrenfeld E (1991) Purification, characterization, and comparison of poliovirus RNA polymerase from native and recombinant sources. J Biol Chem 266:24212–24219

Novak JE, Kirkegaard K (1991) Improved method for detecting negative strands used to demonstrate specificity of plus strand encapsidation and the ratio of positive to negative strands in infected cells. J Virol 65:3384–3387

Okamoto H, Kojima M, Sakamoto M, Iizuka H, Hadiwandowo S, Suwignyo S, Miyakawa Y, Mayumi M (1994) The entire nucleotide sequence and classification of a hepatitis C virus isolate of a novel genotype from an Indonesian patient with chronic liver disease. J Gen Virol 75:629–635.

Okamoto H, Kurai K, Okada S, Yamamoto K, Lizuka H, Tanaka T, Fukuda S, Tsuda F, Mishiro S (1992) Full-length sequence of a hepatitis C virus genome having poor homology to reported isolates: comparative study of four distinct genotypes. Virology 188:331–41

Okamoto H, Okada S, Sugiyama Y, Kurai K, Iizuka H, Machida A, Miyakawa Y, Mayumi M (1991) Nucleotide sequence of the genomic RNA of hepatitis C virus isolated from a human carrier: comparison with reported isolates for conserved and divergent regions. J Gen Virol 72:2697–2704

Ollis DL, Brick P, Hamlin R, Xuong NG, Steitz TA (1985) Structure of large fragment of *Escherichia coli* DNA polymerase complexed with dTMP. Nature 313:762–766

O'Reilly EK, Kao CC (1998) Analysis of RNA-dependent RNA polymerase structure and function as guided by known polymerase structures and computer predictions of secondary structure. Virology 252:287–303

Parsley TB, Towner JS, Blyn LB, Ehrenfeld E, Semler BL (1997) Poly (rC) binding protein 2 forms a ternary complex with the 5'-terminal sequences of poliovirus RNA and the viral 3CD proteinase. RNA 3:1124–1134

Pata JD, Schultz SC, Kirkegaard K (1995) Functional oligomerization of poliovirus RNA-dependent RNA polymerase. RNA 1:466–477

Paul AV, Cao X, Harris KS, Lama J, Wimmer E (1994) Studies with poliovirus polymerase 3D^{pol}. Stimulation of poly(U) synthesis in vitro by purified poliovirus protein 3AB. J Biol Chem 269: 29173–29181

Plotch SJ, Palant O, Gluzman Y (1989) Purification and properties of poliovirus RNA polymerase expressed in *E. coli*. J Virol 63:216–225

Poch O, Sauvaget I, Delarue M, Tordo N (1989) Identification of four conserved motifs among the RNA dependent polymerase encoding elements. EMBO J 8:3867–3874

Provost PJ, Hilleman MR (1979) Propagation of human hepatitis A virus in cell culture in vitro. Proc Soc Exp Biol Med 160:213–221

Quadt R, Kao CC, Browning KS, Hershberger RP, Ahlquist P (1993) Characterization of a host protein associated with brome mosaic virus RNA-dependent RNA polymerase. Proc Natl Acad Sci USA 90:1498–1502

Raychaudhuri G, Govindarajan S, Shapiro M, Purcell RH, Emerson SU (1998) Utilization of chimeras between human (HM-175) and simian (AGM-27) strains of hepatitis A virus to study the molecular basis of virulence. J Virol 72:7467–7475

Reed KE, Gorbalenya AE, Rice CM (1998) The NS5 A/NS5 proteins of viruses from three genera of the family flaviviridae are phosphorylated by associated serine/threonine kinases. J Virol 72:6199–6206

Reed KE, Xu J, Rice CM (1997) Phosphorylation of the hepatitis C virus NS5BA protein in vitro and in vivo: properties of the NS5A-associated kinase. J Virol 71:7187–7197

Reynolds JE, Kaminski A, Carroll AR, Clarke BE, Rowlands DJ, Jackson RJ (1996) Internal initiation of translation of hepatitis C virus RNA: the ribosome entry site is at the authentic initiation codon. RNA 2:867–878

Richards OC, Ehrenfeld E (1998) Effects of poliovirus 3AB protein on 3D polymerase-catalyzed reaction. J Biol Chem 273:12832–12840

Richards OC, Ehrenfeld E (1997) One of two NTP binding sites in poliovirus RNA polymerase required for RNA replication. J Biol Chem 272:23261–23264

Richards OC, Baker S, Ehrenfeld E (1996) Mutation of lysine residues in the nucleotide binding segments of the poliovirus RNA-dependent RNA polymerase. J Virol 70:8564–8570

Roehl HH, Parsley TB, Ho TV, Semler BL (1997) Processing of a cellular polypeptide by 3CD proteinase is required for poliovirus ribonucleoprotein complex formation. J Virol 71:578–585

Roizman B, Palese P (1996) Multiplication of viruses: an overview. In: Fields BN, Knipe DM, Howley PM (eds) Virology. Lippincott-Raven, Philadelphia

Saito T, Shiino T, Arakawa Y, Hayashi S, Abe K (1998) Geographical characterization of hepatitis G virus genome: evidence for HGV genotypes based on phylogenetic analysis. Hepatol Res 10:121–130

Sakamoto M, Akahane Y, Tsuda F, Tanaka T, Woodfield DG, Okamoto H (1994) Entire nucleotide sequence and characterization of a hepatitis C virus of genotype V/3a. J Gen Virol 75:1761–1768

Schiebel W, Haas B, Marinkovic S, Klanner A, Sanger HL (1993) RNA-dependent RNA polymerase from tomato leaves. J Biol Chem 163:11858–11867

Seki M, Honda Y, Kondo J, Fukuda K, Ohta K, Sugimoto J, Yamada E. (1995) Effective production of the hepatitis C virus core antigen having high purity in *Escherichia coli*. J Biotechnol 38:229–241

Seki M, Honda Y (1995) Phosphorothioate antisense oligodeoxynucleotides capable of inhibiting hepatitis C virus gene expression: in vitro translation assay. J Biochem (Tokyo) 118:1199–1204

Senkevich TG, Cumakov IM, Lipskaya GY, Agol VI (1980) Palindrome-like dimers of double-stranded RNA of encephalomycarditis virus. Virology 102:339–348

Shatkin AJ, Sipe JD (1968) RNA polymerase activity in purified reoviruses. Proc Natl Acad Sci USA 61:1462–1469

Shukla DD, Hoyne PA, Ward CW (1995) Evaluation of complete genome sequences and sequences of individual gene products for the classification of hepatitis C viruses. Arch Virol 140:1747–1761

Simmonds P (1994) Variability of hepatitis C virus genome. Curr Stud Hematol Blood Transfus 61:12–35

Smith DB, Simmonds P (1997) Characteristics of nucleotide substitution in the hepatitis C virus genome: constraints on sequence change in coding regions at both ends of the genome. J Mol Evol 45: 238–246

Strauss JH, Strauss EG (1994) The alphaviruses: gene expression, replication, and evolution. Microbiol Rev 58:491–562

Sugiyama K, Kato N, Mizutani T, Ikeda M, Tanaka T, Shimotohno K (1997) Genetic analysis of the hepatitis C virus (HCV) genome from HCV-infected human T cells. J Gen Virol 78:329–336

Summers J (1988) The replication cycle of hepatitis B viruses. Cancer 61:1957–1962

Summers J, O'Connell A, Millman I (1975) Genome of hepatitis B virus: restriction enzyme cleavage and structure of the DNA extracted from Dane particles. Proc Natl Acad Sci USA 72:4597–4601

Takamizawa A, Mori C, Fuke I, Manabe S, Murakami S, Fujita J, Onishi E, Andoh T, Yoshida Y, Okayama H (1991) Structure and organization of the hepatitis C virus genome isolated from human carriers. J Virol 65:1105–1113

Tanaka T, Kato N, Cho M-JE, Sugiyama K, Shimotohno K (1996) Structure of the 3' terminus of the hepatitis C virus genome. J Virol 70:3307–3312

Tanaka T, Kato N, Cho MJ, Shimotohno K (1995) A novel sequence found at the 3' terminus of hepatitis C virus genome. Biochem Biophys Res Commun 215:744–749

Tanaka T, Kato N, Nakagawa M, Ootsuyama Y, Cho MJ, Nakazawa T, Hijikata M, Ishimura Y, Shimotohno K (1992) Molecular cloning of hepatitis C virus genome from a single Japanese carrier: sequence variation within the same individual and among infected individuals. Virus Res 23:39–53

Teterina NL, Bienz K, Egger D, Gorbalenya AE, Ehrenfeld E (1997) Induction of intracellular membrane rearrangements by HAV proteins 2 C and 2BC. Virology 237:66–77

Teterina NL, Zhou WD, Cho MW, Ehrenfeld E (1995) Inefficient complementation activity of poliovirus 2 C and 3D proteins for rescue of lethal mutations. J Virol 69:4245–4254

Tisdale M, Ellis M, Klumpp K, Court S, Fort M (1995) Inhibition of influenza virus transcription by 2'-Deoxy-2'-Fluoroguanosine. Antimicrob Agents Chemother 39:2454–2458

Todd S, Towner JS, Brown DM, Semler BL (1997) Replication-competent picornaviruses with complete genomic RNA 3' noncoding region deletions. J Virol 71:8868–8874

Tokita H, Okamoto H, Iizuka H, Kishimoto J, Tsuda F, Lesmana LA, Miyakawa Y, Mayumi M (1996) Hepatitis C virus variants from Jakarta, Indonesia classifiable into novel genotypes in the second (2e and 2f), tenth (10a) and eleventh (11a) genetic groups. J Gen Virol 77:293–301

Tomei L, Vitale L, Serafini S, Altamura S, DeFrancesco R, Vitelli A (1998) In vitro properties of the RNA-dependent RNA polymerase of the hepatitis C virus. 5th International Meeting on Hepatitis C Virus and Related Viruses, Molecular Virology and Pathogenesis, Venice, Italy

Trippe R, Sandrock B, Benecke BJ (1998) A highly specified terminal uridylyl transferase modifies the 3'-end of U6 small nuclear RNA. Nucleic Acids Res 26:3119–3126

Trowbridge R, Gowans EJ (1998) Molecular cloning of an Australian isolate of hepatitis C virus. Arch Virol 143:501–511

Tsuchihara K, Tanaka T, Hijikata M, Kuge S, Toyoda H, Nomoto A, Yamamoto N, Shimotohno K (1997) Specific interaction of polypyrimidine tract-binding protein with the extreme 3'-terminal structure of the hepatitis C virus genome, the 3'X. J Virol 71:6720–6726

Tuschall DM, Hiebert E, Flanegan JB (1982) Poliovirus RNA-dependent RNA polymerase synthesizes full-length copies of poliovirus RNA, cellular mRNA, and several plant virus RNAs in vitro. J Virol 44:209–216

Urbani A, Bazzo R, Nardi MC, Cicero DO, DeFrancesco R, Steinkuhler C, Barbato G (1998) The metal binding site of the hepatitis C virus NS3 protease. A spectroscopic investigation. J Biol Chem 273:18760–18769

Waggoner S, Sarnow P (1998) Viral ribonucleoprotein complex formation and nucleolar-cytoplasmic relocalization of nucleolin in poliovirus-infected cells. J Virol 72:6699–6709

Wang Y, Okamoto H, Tsuda F, Nagayama R, Tao QM, Mishiro S (1993) Prevalence, genotypes, and an isolate (HC-C2) of hepatitis C virus in Chinese patients with liver disease. J Med Virol 40:254–260

Ward CD, Flanegan JB (1992) Determination of the poliovirus RNA polymerase error frequency at eight sites in the viral genome. J Virol 66:3784–3793

Wimmer E, Hellen CU, Cao X (1993) Genetics of poliovirus. Ann Rev Genet 27:353–426

Xiang W, Cuconati A, Hope D, Kirkegaard K, Wimmer E (1998) Complete protein linkage map of poliovirus P3 proteins: interaction of polymerase 3D^{pol} with VPg and with genetic variants of 3AB. J Virol 72:6732–6741

Xiang W, Cuconati A, Paul AV, Cao X, Wimmer E (1995) Molecular dissection of the multifunctional poliovirus RNA-binding protein 3AB. RNA 1:892–904

Yamada N, Tanihara K, Mizokami M, Ohba K, Takada A, Tsutsumi M, Date T (1994) Full-length sequence of the genome of hepatitis C virus type 3a: comparative study with different genotypes. J Gen Virol 75:3279–3284

Yamashita T, Kaneko S, Yukiriro S, Qin W, Nomura T, Kobayashi K, Murakami S (1998) RNA-dependent RNA polymerase activity of the soluble recombinant hepatitis C virus NS5B protein truncated at the C-terminal region. J Biol Chem 273:15479–15486

Yan Y, Li Y, Munshi S, Sardana V, Cole JL, Sardana M, Steinkuehler C, Tomei L, DeFrancesco R, Kuo LC, Chen Z (1998) Complex of NS3 protease and NS4 A peptide of BK strain hepatitis C virus: a 2.2 A resolution structure in a hexagonal crystal form. Protein Science 7:837–847

Yanagi M, Purcell RH, Emerson SU, Bukh J (1997) Transcripts from a single full-length cDNA clone of hepatitis C virus are infectious when directly transfected into the liver of a chimpanzee. Proc Natl Acad Sci USA 94:8738–8743

Yeh CT, Chu CM, Liaw YF (1996) Distinct composition of viral quasispecies between ascites and serum samples from patients with late stage chronic hepatitis C. Biochem Biophys Res Commun 227: 524–529

Young DC, Tuschall DM, Flanegan JB (1985) Poliovirus RNA-dependent RNA polymerase and host cell protein synthesize product RNA twice the size of poliovirion RNA in vitro. J Virol 54:256–264

Yuan ZH, Kumar U, Thomas HC, Wen YM, Monjardino J (1997) Expression, purification, and partial characterization of HCV RNA polymerase. Biochem Biophys Res Comm 232:231–235

Zabel PL, Dossers KW, Kammen A (1981) Terminal uridylyl transferase of Vigna unguiculata: purification and characterization of an enzyme catalyzing the addition of a single UMP residue to the 3'-end of an RNA primer. Nucleic Acids Res 9:2433–2453

Zhang H, Chao SF, Ping LH, Grace K, Clarke B, Lemon SM (1995) An infectious cDNA clone of a cytopathic hepatitis A virus: genomic regions associated with rapid replication and cytopathic effect. Virology 212:686–697

Zheng WZ (1994) Genotype identification of hepatitis C virus (HCV) isolated from a Japanese carrier in Nagasaki prefecture and genome analysis E1 and E2/NS1 envelope glycoprotein regions Jpn J Trop Med Hyg 22:169–177

Systems to Culture Hepatitis C Virus

N. Kato[1] and K. Shimotohno[2]

1 Introduction 261

2 Hepatitis C Virus Culture System in Human Lymphocytes 262
2.1 Hepatitis C Virus Replication in HPB-Ma Cells 262
2.2 Hepatitis C Virus Replication in MT-2 Cells 263
2.3 Other Hepatitis C Virus-Susceptible Lymphocytes 265
2.4 Viral Transmission Mode 265

3 Hepatitis C Virus Culture System in Human Hepatocytes 266
3.1 Hepatitis C Virus Replication in PH5CH Cells 266
3.2 Other Hepatitis C Virus-Susceptible Hepatocytes 267

4 Genetic Analysis of Hepatitis C Virus from Hepatitis C Virus-Infected Cells 268
4.1 Alteration of HVR1 Populations of Hepatitis C Virus During Culture 269
4.2 Complete Structure of Hepatitis C Virus Genome from Hepatitis C Virus-Infected Cells . . . 271
4.3 Cell Tropism of Hepatitis C Virus 272

5 Discussion 274

References 275

1 Introduction

Hepatitis C virus (HCV) frequently causes chronic hepatitis, cirrhosis and hepatocellular carcinoma (Choo et al. 1989; Kuo et al. 1989; Ohkoshi et al. 1990; Saito et al. 1990). HCV is considered to belong to the family Flaviviridae, whose members have a positive-stranded RNA genome of about 9.6kb including a large open reading frame (ORF). The ORF encodes a polyprotein precursor of about 3,000 amino acids (Kato et al. 1990; Tanaka et al. 1995), subsequently cleaved into at least 11 structural and nonstructural viral proteins (Hijikata et al. 1991a, 1993a,b). Although our understanding of the molecular biology of HCV has progressed rapidly (reviewed by Shimotohno 1995; Houghton 1996, Rice 1996, Shimotohno

---

[1] Virology and Glycobiology Division, National Cancer Center Research Institute, 5-1-1 Tsukiji, Chuo-ku, Tokyo 104-0045, Japan
[2] Laboratory of Human Tumor Viruses, Department of Viral Oncology, Institute for Virus Research, Kyoto University, Shogoin, Sakyo-ku, Kyoto 606-8505, Japan

and FEINSTONE 1997; MAJOR and FEINSTONE 1997), all of the data regarding the functions of HCV gene products have been obtained in artificial systems, either in vitro or in vivo, using established mammalian cell expression systems. Therefore, these data need to be confirmed in HCV-infected cells. Unfortunately, to date, no satisfactory tissue culture system has been developed in which HCV can be multiplied to the extent required for such experiments. However, several trials for the establishment of HCV replication and multiplication system using mammalian cultured cells have been performed. In this review, we describe the recent progress regarding the development of systems to culture HCV, and discuss recent results of studies using human lymphocytes and hepatocytes to propagate HCV.

2 Hepatitis C Virus Culture System in Human Lymphocytes

Although it is considered that hepatocytes are considered to be the major natural target cells for HCV infection, replication and multiplication, in early studies, several human lymphocyte cell lines were found to be susceptible to HCV infection and have been used for the establishment of in vitro HCV culture systems. Among-HCV susceptible lymphocyte cell lines, HCV culture systems using the retrovirus-infected T cell lines HPB-Ma and MT-2 have been well characterized.

2.1 Hepatitis C Virus Replication in HPB-Ma Cells

In 1992, SHIMIZU et al. (1992) first reported that murine retrovirus-infected MOLT-4Ma cells, a human T cell line, were susceptible to HCV infection. However, when the infected cells were continuously cultured, intracellular HCV RNA was not constantly detected by RT-nested PCR, suggesting that the efficiency of HCV replication was very low. In this study, it appeared that HCV replication in MOLT-4 cells free of retrovirus was less efficient than in MOLT-4Ma cells, although it is unclear how the retrovirus affects the efficiency of HCV replication. The same group found that another murine retrovirus (an amphotropic murine leukemia virus pseudotype of murine sarcoma virus)-infected human T cell line, HPB-Ma, was able to support replication of HCV (SHIMIZU et al. 1993). Cloning of HPB-Ma cells revealed a clonal variation in susceptibility to HCV infection. Using one of the clones, HPB-Ma clone 10-2, infectivity titers of HCV inocula from different sources were examined. It was found that the in vitro infectivity titers correlated with the reported infectivity titers of the inocula in chimpanzees (SHIMIZU et al. 1993). Using this HCV culture system, interferon-α(IFN-α) and IFN-β were shown to effectively inhibit HCV replication at concentrations of 2×10^2–2×10^3IU/ml. IFN-γ was not as effective as IFN-α and IFN-β, and it appeared that IFNs did not prevent HCV adsorption to cells but did prevent HCV replication in the cells (SHIMIZU and YOSHIKURA 1994). Recently, HCV was successfully cultured for more than 1 year in

HPB-Ma clone 10-2 cells, although intracellular positive-stranded HCV genome titers were not determined, and it was shown that the virion density in sucrose gradient was around 1.12g/ml (NAKAJIMA et al. 1996). This value was in the range of those reported for pestiviruses (RICE 1996). Furthermore, virus-like particles with a diameter of approximately 50nm were detected in cytoplasmic vesicles by immunoperoxidase electron microscopy using antibodies against HCV core and envelope proteins (SHIMIZU et al. 1996).

2.2 Hepatitis C Virus Replication in MT-2 Cells

A cell line other than HPB-Ma that is susceptible to HCV infection, MT-2, a human T cell leukemia virus type I (HTLV-I)-infected cell line, was found by our group in 1995 (KATO et al. 1995). In the initial experiment using inoculum 1B-1, intracellular positive-stranded HCV RNA was detected until 15 days postinoculation (p.i) by RT-nested PCR. Intracellular HCV RNA disappeared within 7 days p.i. in various other human cell lines examined. As further evidence of intracellular replication of HCV, it was found that the limited HCV population became predominant in MT-2 cells, as sequence analysis of hypervariable region 1 (HVR1) revealed that HVR1 species from cells 10 days p.i. had converged to only one species, although HVR1 species from the inoculum 1B-1 showed the typical quasispecies characteristics (KATO et al. 1995). Furthermore, the anti-HVR1 antibody against the HVR1 species which converged at 10 days p.i. was not in the primary inoculum 1B-1, although antibodies against the other HVR1 species were easily detected in inoculum 1B-1. This result is consistent with the observation that serum samples containing HCV which do not form immune complexes with anti-HCV antibodies exhibit good HCV infectivity in vivo (HIJIKATA et al. 1993c), although it is not clear how anti-HCV antibodies, including anti-HVR1 antibody, prevent the replication of HCV. Alterations of HVR1 populations during HCV culture will be described in detail in Sect. 4.1.

To better understand susceptibility to HCV, five MT-2 clones in which HCV replicates more persistently than in parental MT-2 cells were obtained (MIZUTANI et al. 1996b). Positive-stranded HCV RNA could be detected in these clones until at least 21 days p.i., and MT-2C, one of the five clones which supported HCV replication up to 30 days p.i., was used for further characterization. Semi-quantitative analysis of HCV RNA by RT-PCR revealed that the level of RNA synthesis in the infected MT-2C cells increased after inoculation, reached a maximum level at 4 days p.i., and remained at this level until at least 11 days p.i. Negative-stranded HCV RNA, an intermediate in HCV replication, was also detected in the infected cells by two different methods (GUNJI et al. 1994; LANFORD et al. 1994) with strand specificity (MIZUTANI et al. 1996b).

Using the MT-2C cell culture system, HCV replication was shown to be inhibited by an antisense oligonucleotide (4 days culture at 10μM) complementary to the region containing the initiator AUG codon of the HCV core encoding region (MIZUTANI et al. 1995, 1996b). When MT-2C cells were cultured for 7 days after removal of antisense oligonucleotide, HCV RNA was again detectable, suggesting that the antisense oligonucleotide had a specific and reversible inhibitory effect on HCV multiplication (MIZUTANI et al. 1996b).

The inhibitory effect of IFN-α on HCV replication was also demonstrated using HCV-infected MT-2C cells. Semi-quantitative analysis revealed that the HCV RNA level decreased to about 10% 2 days after IFN-α treatment (MIZUTANI et al. 1996b). IFN-α was not detected in the culture medium of non-HCV-infected MT-2C cells by ELISA and was not toxic to MT-2C cells at the inhibitory dose (10^2U/ml). These results indicated that the HCV culture system using MT-2C cells is useful for evaluation of potential antiviral agents.

To investigate whether viral replication is affected by conditions of host cell growth, the effects of several antibiotics on HCV replication were examined. HCV-infected MT-2C cells treated with G418 or hygromycin B sustained HCV RNA for a longer period (maximum 43 days p.i.) than untreated cells. This result indicated that these antibiotics support HCV retention in cells in prolonged culture. Although growth inhibition of MT-2C cells was observed in presence of these antibiotics, it is unlikely that the growth rate the host cells is correlated with HCV replication because other antibiotics, i.e. puromycin and blasticidin S, showed no effect on HCV retention in the cells (MIZUTANI et al. 1996b). There are two possible explanations for this observation: prolongation of virus growth, and prolongation of the cell cycle. However, there was no evidence of cell cycle-dependent HCV replication and multiplication (M. Ikeda et al., unpublished data).

To obtain a more persistent HCV replication system using MT-2C cells, the effects of various culture conditions after virus inoculation were examined. Addition of anti-IFN-α antibody, anti-IFN-β antibody, tumor necrosis factor (TNF)-α, transforming growth factor (TGF)-β, cycloheximide, neuraminidase or low-density lipoproteins (LDL) had no significant effect in extending the period supporting HCV replication. However, a temperature shift from 37°C to 32°C after virus inoculation was effective in supporting HCV replication for a significantly longer period (MIZUTANI et al. 1996a). Under these conditions, intracellular HCV RNA was constantly detected for at least 80 days p.i. and could be detected in the cells up to 198 days p.i., suggesting that cell culture at 32°C provides a better environment for HCV infection and replication than 37°C. Semi-quantitative analysis revealed that the amount of HCV RNA in cells (up to 20 days p.i.) cultured at 32°C, however, was similar to that at 37°C (MIZUTANI et al. 1996a). Prolonged retention of HCV in the cells cultured at 32°C was also observed in parental MT-2 cells, the other MT-2 clones, and the non-neoplastic human hepatocyte cell line PH5CH, which is susceptible to HCV infection (KATO et al. 1996; see Sect. 3.1). The effect of the temperature shift did not depend on the type of inoculum, as similar prolonged retention of HCV in cells was observed when other sera were used as the inoculum (Kato et al., unpublished observation). These findings suggest that cells cultured at 32°C generally become more susceptible to HCV infection or replication, or that the infectivity or stability of HCV produced by cells cultured at 32°C is greater than at 37°C.

As HCV-susceptible human T cell lines HPB-Ma and MT-2 are infected with a mouse retrovirus and HTLV-I, respectively, retrovirus infection may be favorable for HCV infection or replication. However, other HTLV-I infected cell lines, C91/ PL and OCH, did not support HCV replication (KATO et al. 1995), and there were no correlations between the level of HTLV-I expression and the replication of HCV in several MT-2 clones which showed different susceptibilities to HCV infection and replication (MIZUTANI et al. 1996b).

2.3 Other Hepatitis C Virus-Susceptible Lymphocytes

In addition to HPB-Ma clone 10-2 and MT-2C cells, it has been reported that the human B-cell line Daudi supported HCV replication for more than 1 year after virus inoculation, although intracellular HCV RNA titer was not measured throughout the culture period (NAKAJIMA et al. 1996). As Daudi cells became unhealthy after virus inoculation, probably because of IFN activity induced by the infection (SHIMIZU and YOSHIKURA 1995), culture of Daudi cells was maintained by adding fresh cells after removing the old medium at each passage. The density in sucrose gradient of HCV virions produced from Daudi cells was also shown to be around 1.12g/ml, the same as those from HPB-Ma clone 10-2 cells (NAKAJIMA et al. 1996). Furthermore, intracellular HCV particles which were similar to those observed in HPB-Ma clone 10-2 cells were also detected by electron microscopy (SHIMIZU et al. 1996).

Two other HCV culture systems, the human bone marrow-derived lymphoid cell lines CE and TOFE, were reported to be susceptible to HCV infection, and both cell lines supported HCV replication for several months, although HCV RNA titers in these cells are not clear (BERTOLINI et al. 1993; VALLI et al. 1995). H9, a derivative of the human T cell line HT, was also tested for its susceptibility to HCV infection in vitro. The $5'$-NC (noncoding) region of the HCV genome was intermittently detected in the cells and culture medium until 21 days p.i. (NISSEN et al. 1994), although quantitative analysis of the HCV genome was not performed. In vitro HCV infection using peripheral blood mononuclear cells (PBMCs) has also been carried out, and HCV RNA was detected in these cells until 25 days p.i., but the PBMCs of the different donors were not all permissive to HCV (CRIBIER et al. 1995). Quantification of HCV RNA by branched-DNA assay showed that intracellular HCV RNA decreased during the first week of culture and increased during second or third week. However, HCV RNA titer in the culture medium was very low.

2.4 Viral Transmission Mode

As the HCV-infected HPB-Ma clone 10-2 cells produced barely detectable amounts of viral genome in the culture medium, virus transmission was thought to occur by a cell-to-cell mode. This was confirmed by the multicycle transmission of HCV from infected HPB-Ma clone 10-2 cells cocultured with neomycin-resistant and hygromycin B-resistant cells (SHIMIZU and YOSHIKURA 1994).

Cell-free viral transmission was also demonstrated using MT-2 and MT-2C cells (N. Kato et al., unpublished data; MIZUTANI et al. 1996). Successful cell-free viral transmission was also observed using another clone of MT-2 cells, MT-2A, which was shown to support HCV replication as efficiently as MT-2C (MIZUTANI et al. 1996b). Furthermore, it was shown that cell-free virus transmission could be repeated by culture at 32°C. Under these conditions, long-term culture supporting HCV replication was achieved (MIZUTANI et al. 1996a).

Thus, both cell-to-cell and cell-free transmission modes seem necessary for the multiplication of HCV, although cell-to-cell viral transmission has not been confirmed in the MT-2 cell culture system.

3 Hepatitis C Virus Culture System in Human Hepatocytes

Since hepatocytes are thought to be the natural target cells for HCV infection, it is important to establish an efficient HCV replication system using human hepatocytes to understand the life cycle and pathogenesis of HCV. As described above, studies of HCV replication systems using human lymphocytes progressed beyond those using human hepatocytes. Recently, however, we and other groups have found that several human hepatocyte cell lines were susceptible to HCV infection, and we have characterized HCV replication in human hepatocytes.

3.1 Hepatitis C Virus Replication in PH5CH Cells

PH5CH, a recently established non-neoplastic human hepatocyte line immortalized with simian virus 40 large T antigen (NOGUCHI and HIROHASHI 1996), was found to be most susceptible to HCV infection among several human hepatocyte lines examined (KATO et al. 1996). In cells inoculated with sera derived from two HCV-positive blood donors (1B-1 and 1B-3), positive-stranded HCV RNA was detected up to 30 days p.i.. Semi-quantitative analysis of HCV RNA from the cells and culture medium revealed that HCV multiplied during the culture period (KATO et al. 1996). To obtain cells capable of supporting HCV replication more persistently, several PH5CH clones were tested for susceptibility to HCV infection using HCV-positive serum 1B-2. Consequently, three clones (PH5CH1, PH5CH7 and PH5CH8) in which intracellular HCV RNA could be detected up to 35 days p.i. were obtained (IKEDA et al. 1998). Since HCV RNA was detectable in parental PH5CH cells inoculated with serum 1B-2 only until 16 days p.i. (IKEDA et al. 1997), it was suggested that HCV replication is supported efficiently in these three PH5CH clones. Semi-quantitative analysis of HCV RNA indicated that HCV replicated in these cloned PH5CH cells was released into the culture medium. However, HCV

RNA was detected intermittently in the culture medium, whereas intracellular HCV RNA was detected continuously. Similar observations have been reported in MT-2 cells and HPB-Ma cells (IKEDA et al. 1997; SHIMIZU et al. 1993). These results suggest that HCV progeny are intermittently produced from the infected cells. Although the mechanism of intermittent production of HCV is not clear, release of HCV progeny into the culture medium might be associated with, for example, phase of the cell cycle. The establishment of conditions that allow continuous production of HCV to culture medium will markedly improve this HCV culture system. Semi-quantitative analysis of internalized HCV RNA after treatment of cloned PH5CH cells and parental PH5CH cells with proteinase K immediately after virus inoculation revealed that the three highly susceptible PH5CH clones contained about tenfold higher levels of HCV RNA than low susceptible PH5CH clones or parental PH5CH cells, suggesting a correlation between the amount of internalized HCV and the period over which intracellular HCV RNA can be detected after virus inoculation (IKEDA et al. 1998). Using cloned PH5CH cells inoculated with serum 1B-2, it was demonstrated that HCV replication was maintained for 70–100 days when the temperature of cell culture after virus inoculation was reduced from 37°C to 32°C. Since we observed prolonged detection of HCV RNA in MT-2C cells inoculated with serum 1B-1 and cultured at 32°C (MIZUTANI et al. 1996a), these results indicate that the effect of the temperature shift from 37°C to 32°C does not depend on the type of inoculum or the cells.

Recently, it was shown that the HCV-infected cloned PH5CH cell culture system was also useful for evaluating antiviral agents as well as HPB-Ma cells (SHIMIZU and YOSHIKURA 1994) and MT-2C cells (MIZUTANI et al. 1995, 1996b), demonstrating that addition of IFN-α at a final concentration of 10^3U/ml was effective in preventing HCV multiplication in HCV-infected PH5CH8 cells (IKEDA et al. 1998).

3.2 Other Hepatitis C Virus-Susceptible Hepatocytes

In a trial to demonstrate HCV replication in hepatocytes, primary chimpanzee hepatocytes were shown to be susceptible to in vitro HCV infection (LANFORD et al. 1994). In this study, intracellular negative-stranded HCV RNA was detectable 4 days p.i. and throughout the remainder of the experimental period of 25 days by the strand-specific RT-PCR method using tagged primers. Evaluation of this method for the detection of negative-stranded HCV RNA will be discussed in the last section. Although the suppression of HCV multiplication by IFN-α (2×10^3IU/ml) was also shown in this system, further characterization of HCV replication in chimpanzee hepatocytes has not yet been reported. Susceptibility of primary human fetal hepatocytes to HCV infection was also tested, and it was shown that these hepatocytes could support HCV replication and multiplication (IACOVACCI et al. 1993, 1997).

An HCV cultivation system using primary hepatocyte cultures from patients with chronic hepatitis was reported (ITO et al. 1996). Quantitative analysis of HCV genome titers in the cultured cells and culture medium by a competitive RT-PCR method revealed a significant amount of HCV RNA in the cells and culture medium during cultivation.

Trials to find HCV-producing established human hepatocyte cell lines have also been carried out. Tsuboi et al. (1996) reported that HCV RNA was detected from three established human hepatocellular carcinoma cell lines (JHH-1, JHH-4 and JHH-6) and that HCV sequences belonging to genotype 1 and genotype 2b were obtained from JHH-1 cells, and JHH-4 and JHH-6 cells, respectively. However, additional data supporting HCV replication in these hepatocyte cells have not yet been reported.

Tagawa et al. (1995) reported that the human embryonic hepatocyte cell line WRL68 and hepatoblastoma cell line HepG2 sustained HCV RNA until 62 days and 39 days p.i., respectively. However, intracellular HCV RNA levels were very low (about ten copies/mg RNA). Although there have been no additional reports supporting the susceptibility of WRL68 cells to HCV infection, recently Seipp et al. (1997) reported the susceptibility of human hepatoma (HuH7 and HepG2), porcine kidney (PK15) and swine testis (STE) cell lines to HCV infection and replication. In this report, a stabilizing effect on HCV propagation was observed for 50 days in serum-free medium with stimulation of LDL receptor expression by lovastatin, although supplementation with lovastatin during inoculation did not enhance virus replication. Under these culture conditions, long-term persistence of HCV in cells and release of virions into the supernatant were achieved for up to 130 days, although quantification of intracellular HCV RNA revealed that the amount of viral RNA in primary infected HuH7 cells remained at low levels of maximally one copy/40 cells (Seipp et al. 1997). Prolonged retention of HCV in PH5CH cells infected under LDL receptor-stimulated conditions (serum-free medium and lovastatin) was not observed (M. Ikeda et al., unpublished observations), although the level of intracellular HCV RNA in HuH7 cells was similar to that in cloned PH5CH cells. Recently, it has been reported that a non-hepatic cell line, Vero, was also susceptible to HCV infection (Valli et al. 1997).

4 Genetic Analysis of Hepatitis C Virus from Hepatitis C Virus-Infected Cells

It is well known that HCV has a quasi-species nature within individual patients (Kato et al. 1992a,b; Martell et al. 1992). The most characteristic region is HVRl, which is located in the NH_2-terminal region of the second envelope glycoprotein (E2) (Hijikata et al. 1991, 1993; Weiner et al. 1991; Kato et al. 1992a; Kurosaki et al. 1993) . The dynamics of the quasi-species nature of HCV have been reported based on the results of studies using clinical specimens from patients with acute hepatitis and specimens from HCV-infected chimpanzees. In these studies, HVRI was found to be a good molecular marker for distinguishing HCV species (SEKIYA et al. 1994; KATO et al. 1994; DOORN et al. 1995; SAITO et al. 1996). Since most circulating HCV particles are thought to be defective regarding their replicating activity (MARTELL et al. 1992; HIJIKATA et al. 1993c; SHIMIZU et al. 1993), it is important to identify the HCV species which is replicating in the cells. To date, sequence analysis of HVRI has been frequently used to distinguish the replicating HCV species.

4.1 Alteration of HVR1 Populations of Hepatitis C Virus During Culture

Qualitative analysis of HVR1 populations from HCV-infected HPB-Ma cells (HIJIKATA et al. 1995; NAKAJIMA et al. 1996), Daudi cells (NAKAJIMA et al. 1996), MT-2 cells (KATO et al. 1995; IKEDA et al. 1997), MT-2C cells (SUGIYAMA et al. 1997), PH5CH cells (IKEDA et al. 1997) and cloned PH5CH cells (KATO et al. 1998) revealed that HCV species with the limited HVR1 sequence became predominant in the HCV replicating cells cultured at 37°C, despite the presence of a mixed population of HVR1 in the primary inocula.

Alterations in HVR1 populations from MT-2 and cloned MT-2C cells inoculated with serum 1B-1 have been well characterized. Sequence analysis of 100 independent HVR1 cDNA clones showed that HVR1 populations from serum 1B-1 showed a complicated quasi-species nature consisting of four types (I–IV) and 22 distinct species, which were assigned to a phylogenetic tree constructed by the GENETYX-MAC program (unweighted pair-grouping method with arithmetic mean; NEI and GOJOBORI 1986). However, HVR1 populations from the HCV-infected MT-2 cells at 12 days p.i. had converged to only two closely related HVR1 species (II-1 and II-4), belonging to type II, and differing from each other by only one amino acid (IKEDA et al. 1997). A similar convergence of HVR1 populations was observed in cloned MT-2C cells. HVR1 species II-1 became predominant in the MT-2C cells at 12 days p.i. with a frequency of 50%, although its frequency was only 13% in the primary inoculum (SUGIYAMA et al. 1997b). Moreover, HVR1 species II-6 and II-7, which differed from species II-1 by only one amino acid and were not isolated from the inoculum, were also obtained from the MT-2C cells. These results indicated that HCV species belonging to type II were adapted to MT-2 and MT-2C cells after virus inoculation. Although HVR1 populations from the PH5CH cells at 12 days p.i. were not analyzed, because of low infectivity of serum 1B-1 for PH5CH cells, HVR1 species belonging to types I and III became the predominant populations in the PH5CH cells at 8 days p.i. HVR1 species II-1 and II-4, which were converged in the MT-2 cells, were not obtained from the PH5CH cells. In addition, it is noteworthy that 19 (12 distinct HVR1 species) of 60 HVR1 cDNA clones obtained from MT-2 and PH5CH cells inoculated with serum 1B-1 were not included among the 40 cDNA clones isolated from serum 1B-1 (IKEDA et al. 1997).

Serum 1B-2 showed good infectivity for PH5CH and three cloned PH5CH cell lines, and alterations in HVR1 populations from cells inoculated with serum 1B-2 have been well characterized. For the amplification of HVR1 by RT-nested PCR, a new primer set was designed to perfectly match the nucleotide sequences of HCV RNAs from serum 1B-2. HVR1 populations (50 independent HVR1 cDNA clones) from serum 1B-2 also showed a rather complicated quasi-species nature, which could be classified into three types, I (40% frequency), II (52% frequency) and III (8% frequency), and 14 distinct species. Sequence analysis of HVR1 populations revealed that a limited HVR1 species (II-1) belonging to type II became predominant (80%–90% frequency) in HCV-infected PH5CH cells at 8 and 16 days p.i. (IKEDA et al. 1997). The convergence to HVR1 species II-1 was also observed in cloned PH5CH cells (KATO et al. 1998). It was demonstrated that such convergence of HVR1 populations occurred in PH5CH and cloned PH5CH cells cultured at 32°C after inoculation with serum 1B-2 (IKEDA et al. 1997a; KATO et al. 1998). Sequence analysis of HVR1 cDNA clones from the HCV-infected cells revealed that only type II was present in PH5CH, cloned PH5CH1, PH5CH7 and PH5CH8 cells after 40 days p.i., and that HVR1 species II-1 became predominant (more than 70%) in these cells, although the frequency of this HVR1 species in primary inoculum, serum 1B-2, was 38%.

Following long-term culture of HCV-infected HPB-Ma clone 10-2 cells, HVR1 populations gradually converged to the limited populations in cells inoculated with serum H77, which contained 10^7 HCV-1a genomes/ml and a chimpanzee infectious dose of $10^{6.5}$/ml (NAKAJIMA et al. 1996). Although 13 different HVR1 species were obtained from primary inoculum H77 and seven different HVR1 species were recovered immediately after adsorption, only three HVR1 species and one HVR1 species (H1-2) were recovered from the cells at 67 and 221 days p.i., respectively. A similar phenomenon was observed in another independent experiment, although the converged HVR1 species (H1-1) differed by only one nucleotide from the H1-2 species in the former experiment (NAKAJIMA et al. 1996). These results revealed that the HVR1 species converged in HPB-Ma clone 10-2 cells was the same or very close to the major HVR1 species recovered from the sera of chimpanzees at 10 days after infection with serum H77. Using the same inoculum H77, alterations in HVR1 populations in Daudi cells have been similarly examined. In Daudi cells, it was also observed that convergence to only one HVR1 species (H1-2), which was the same as that in HPB-Ma clone 10-2 cells, occurred after 25 days p.i. (NAKAJIMA et al. 1996). These results indicated that HCV with the same particular sequence of HVR1 was able to replicate well in both HPB-Ma clone 10-2 and Daudi cells. Interestingly, it has been shown that the HVR1 species recovered from the PBMCs of a chimpanzee at 10 days p.i. was H1-2 although serum and liver specimens from the same chimpanzee contained HCV with a variety of other HVR1 species (SHIMIZU et al. 1997). These results suggest that HCV with the HVR1 sequence of H1-2 may have a replication advantage in lymphocytes, both in vitro and in vivo. The convergence to limited HCV populations in both HPB-Ma clone 10-2 and Daudi cells after virus infection was confirmed in both E2 and NS5 regions (NAKAJIMA et al. 1996). A similar phenomenon was also reported by the same group using HPB-Ma clone 10-2 cells and Daudi cells inoculated with plasma 6, which was obtained from a healthy carrier of HCV, and which contained a chimpanzee infectious dose of 10^5/ml, although the observation period was only from 6 to 10 days p.i. (HIJIKATA et al. 1995).

As observed in the HCV-infected HPB-Ma clone 10-2, Daudi, MT-2, MT-2C, PH5CH and cloned PH5CH cells, convergence to a particular HVR1 species occurs in a time-dependent manner, suggesting that some limited HCV populations gradually acquire a growth advantage within the cells. An alternative possibility is that the convergence of HVR1 populations may reflect the minor infectious HCV populations in the inocula, as it has been suggested that only a proportion of circulating HCV virions are infectious (MARTELL et al. 1992; HIJIKATA et al. 1993c; SHIMIZU et al. 1993).

However, there has been one contradictory report in the HCV culture system using primary hepatocytes from patients with chronic hepatitis (ITO et al. 1996). In this system, the ratio of three major types in HVR1 populations was almost the same in hepatocytes prior to and following culture (day 56), while genetic alterations of HVR1 populations in the serum of the patient occurred even within a period of 1 month. Although the reasons for this discrepancy between the results obtained from this system and those of the other HCV culture systems described above is not apparent, the differences in growth rate or clonal characteristics of cultured cells or might be determining factors.

Sequence analysis of the 5'-NC region of HCV RNA before and after long-term in vitro infection using human bone marrow derived lymphoid cells, TOFE cells, revealed one base substitution in the 5'-NC region (VALLI et al. 1995). This base substitution in the regulatory element of the 5'-NC region may be related to the ability of the virus to grow in cell culture or to the cell tropism of the virus. Furthermore, SHIMIZU et al. (1996) also reported similar results regarding the 5'-NC region using HPB-Ma cells. However, since no base substitutions in the 5'-NC region have been observed in HCV culture systems using MT-2 and PH5CH cells (KATO et al. 1998), further analysis is needed for definitive conclusions regarding genetic alterations in the 5'-NC region.

4.2 Complete Structure of Hepatitis C Virus Genome from Hepatitis C Virus-Infected Cells

As described above, it has been shown using several HCV replication systems that an HCV population with a limited HVR1 sequence becomes predominant in cultured cells, despite the complicated quasi-species nature of HVR1 in the primary inoculum. To determine whether the same phenomenon occurs throughout the HCV genome, we carried out RT-nested PCR on the whole viral genome from the MT-2C cells at 12 days after inoculation with serum 1B-1. The nucleotide sequences of these PCR products were determined and compared with those of clones isolated from a cDNA library derived from serum 1B-1 used as an inoculum. Molecular evolutionary analysis comparing the sequences of the HCV clones obtained from both sources revealed that the HCV populations became homogeneous in more than half of the 16 regions compared (SUGIYAMA et al. 1997b). This finding suggests that limited HCV populations are able to replicate in MT-2C cells. In addition, several cDNA clones containing a 3'X-tail sequence, which was recently identified as the bona fide 3' terminus of the HCV genome (TANAKA et al. 1995), were isolated from HCV-infected MT-2C cells. It was confirmed that the nucleotide sequence of the 3'X-tail was highly conserved, suggesting its involvement in HCV replication (SUGIYAMA et al. 1997b). The poly(U) stretch, which is located just upstream of the 3'X-tail, was also amplified and characterized. As observed in previous studies (TANAKA et al. 1995, 1996), the poly(U) stretch from serum 1B-1 also showed marked heterogeneity in length, varying from 47 to 71 nucleotides (average of 60). However, interestingly, the poly(U) stretch from the HCV-infected MT-2C cells was shorter than that from serum 1B-1, and the range of lengths was more restricted, from 29 to 32 nucleotides (average of 30), suggesting that a shorter poly(U) stretch might affect the replication of HCV in MT-2C cells. Within the poly(U) region, as reported previously (TANAKA et al. 1995, 1996), a variable distribution of C, CC or CCC residues was also observed (SUGIYAMA et al. 1997b). On the basis of the sequences of HCV cDNA clones obtained from HCV-infected MT-2C cells, the entire nucleotide sequence of the HCV genome (HCV-JS) containing the 3'X-tail was determined as a candidate for an infectious HCV molecular clone (SUGIYAMA et al. 1997b). To date, there have been no reports regarding complete HCV genomes from cells than MT-2C cells.

4.3 Cell Tropism of Hepatitis C Virus

Although hepatocytes are the natural target cells for HCV infection, it has been reported that PBMCs may also be targets for HCV infection (GUNJI et al. 1994; LERAT et al. 1996; MULLER et al. 1993; SAITO et al. 1996). We recently found relatively high HCV genome titers in the lymph nodes, but not the sera, of patients with gynecological cancers, suggesting that the lymph nodes play an important role in the carrier state and the persistence of HCV infection (SUGIYAMA et al. 1997a). In addition, clinical studies have shown a significant association between HCV infection and lymphoproliferative disorders, mixed cryoglobulinemia (FERRI et al. 1993; GABRIELLI et al. 1994), and non-Hodgkin's lymphoma (LUPPI et al. 1996; SILVESTRI et al. 1996), and it has been proposed that chronic HCV infection should be considered as a multifaceted clinical syndrome rather than a simple liver disease (FERRI et al. 1997). These studies suggested that the cell tropism of HCV is a predominant factor in the etiology of lymphoproliferative disorders.

As described briefly in Sect. 4.1, it has been suggested that HCV possesses cell tropism, as an HCV population with a limited HVR1 sequence became predominant in cultured cells despite the complicated quasi-species nature of the HVR1 in the primary inoculum. Analysis of the infectivity of sera from HCV-positive blood donors for MT-2 and PH5CH cells also suggested the cell tropism of HCV, since some of the HVR1 populations from MT-2 (T-cell) and PH5CH (hepatocyte) cells were different (IKEDA et al. 1997). To further characterize the cell tropism of HCV, the dynamics of HCV populations over 30 days of culture after virus inoculation were examined using three MT-2 clones (MT-2A, MT-2B and MT-2C) and three PH5CH clones (PH5CH1, PH5CH7 and PH5CH8), in which HCV infection and replication occurred more efficiently than in the respective parental cells, and using serum 1B-2, an inoculum with good infectivity for these cloned cells. Sequence analysis of HVR1, which is frequently used as a molecular marker of HCV species, in these cloned cells and *Hpa*II digestion analysis, which can distinguish three major HVR1 types (I, II and III) derived from inoculum 1B-2, indicated that HVR1 type I became the predominant HCV species in MT-2A, MT-2B and MT-2C cells, and HVR1 type II became predominant in PH5CH1, PH5CH7 and PH5CH8 cells during culture after virus inoculation, while both types I and II were almost equally present in inoculum 1B-2 or in the cells immediately after virus inoculation. These results suggested that inoculum 1B-2 contains both lymphotropic and hepatotropic HCVs, which can be distinguished by HVR1 type. To search for cell type-specific sequences in regions other than HVR1, three HCV cDNA clones (3.4kb of 5'-noncoding region to nonstructural (NS) 2) containing HVR1 type I, obtained from HCV-infected MT-2C cells, and three HCV cDNA clones containing HVR1 type II, obtained from HCV-infected PH5CH7 cells, were sequenced. The nucleotide sequences of the 5'-NC region of all cDNA clones showed close identity, with no apparent differences between the MT-2C and PH5CH7 cell-derived clones. However, comparison of the sequences revealed that 37 of 1008 amino acids commonly differed between cDNA clones containing HVR1 types I and II, and that 24 of these 37 amino acids were localized in the E2-encoding region, although 14 of 24 amino acids were in the HVR1. From comparison with the sequence of HCV-JS genome (SUGIYAMA et al. 1997b), which was obtained from MT-2C cells inoculated with the other serum 1B-1, it was found that 11 amino acids were commonly obtained from HCV-infected MT-2C cells, but not from the PH5CH7 cells as follows: amino acid positions 70 (core), 399 (HVR1), 454, 464 (E2), 765 (p7), 824, 837, 853, 887, 951 and 968 (NS2) (KATO et al. 1998). These specific amino acid residues may be important determinants of the cell tropism of HCV. However, further investigations are required to clarify this point using MT-2 and PH5CH cloned cells inoculated with other HCV-positive sera.

To directly determine which of these amino acid positions are involved in the cell tropism of HCV, new experimental systems, such as the HCV proliferation system using infectious HCV cDNA clones, which have been recently used in chimpanzee (KOLYKHALOV et al. 1997; YANAGI et al. 1997) and in tissue culture (YOO et al. 1995; DASH et al. 1997), will be required. In addition, since relatively high HCV genome titers were recently found in the lymph nodes (SUGIYAMA et al. 1997a), comparison of HCV genomes from lymph nodes and liver tissues might also help in clarification of the cell tropism of HCV. Furthermore, sequence analysis of the NS3 region through to the 3'-terminus of HCV genomes from infected cells will be also needed to fully identify all major determinants of cell tropism of HCV.

5 Discussion

A number of cultured cell lines derived from lymphocytes and hepatocytes have been found to be susceptible to HCV replication. Viral replication and multiplication have also been demonstrated in several human cultured cell lines. As HCV presumably replicates via a negative-stranded RNA intermediate which is synthesized from positive-stranded genomic HCV RNA, specific detection of negative-stranded HCV RNA is important to confirm active replication in the infected cells. We and other groups (GUNJI et al. 1994; LANFORD et al. 1994, 1995; LERAT et al. 1996) have developed strand-specific RT-PCR methods for the detection of negative-stranded HCV RNA that prevent false priming of the incorrect strand (LANFORD et al. 1994, 1995; LERAT et al. 1996), self priming due to the complex secondary structure (GUNJI et al. 1994), or random priming by cellular oligonucleotides (LANFORD et al. 1995). Using these strand-specific RT-PCR methods, negative-stranded HCV RNA was detected in HCV-infected cultured lymphocytes (MIZUTANI et al. 1996b) and hepatocytes (ITO et al. 1996; SEIPP et al. 1997). However, we and others showed that false signals for negative-stranded HCV RNA could still be observed even using the tagged system if sufficient quantities of positive-stranded HCV RNA were present (MIZUTANI et al. 1998; LERAT et al. 1996). Therefore, the amount of positive-stranded HCV RNA present in the sample must be quantified to ensure specific detection of negative-stranded HCV RNA. Recently, we developed a novel strand-specific RT-PCR method, by which as few as 100 copies of negative-stranded HCV RNA could be specifically detected even with the coexistence of a 100-fold excess of positive-stranded HCV RNA (MIZUTANI et al. 1998). This new method might be useful for the specific detection of negative-stranded HCV RNA from in vitro HCV-infected cells to confirm replication of HCV in the cells.

During the course of development of HCV culture systems, it was found that both MT-2C and PH5CH cells were susceptible to recently identified hepatitis G virus (HGV) (SIMONS et al. 1995; LINNEN et al. 1996), which, like HCV, belongs to the *Flaviviridae* (IKEDA et al. 1997b). When these cells were inoculated with serum 1B-3 containing both HCV and HGV obtained from a blood donor, 5'-NC region of HCV RNA and HGV RNA were detected in both cell lines more than 30 days p.i. This finding suggests that common cellular factors are involved in the replication and multiplication of both HCV and HGV, and that factors which suppress against HGV also suppress HCV replication. This HGV-infected culture system will be useful for various biological and virological studies of HGV, although it is not well accepted that HGV itself causes human hepatitis (MIYAKAWA and MAYUMI 1997).

Although the culture systems developed to replicate HCV are useful for studying the mechanism of HCV replication, developing antiviral agents and vaccines, etc., the degree of HCV multiplication in these culture systems is not yet satisfactory because only RT-PCR was used to detect HCV RNA in all reports describing such systems. Cell lines or culture conditions allowing the detection of

HCV RNA by northern blot analysis or HCV proteins by western blot analysis have not yet been established. Further improvements are, therefore, required. However, several current HCV culture systems using human cultured cells such as HPB-Ma, MT-2, PH5CH, etc., may allow further experiments on antiviral strategies.

References

Bertolini L, Lacovacci S, Ponzetto A, Gorini G, Battaglia M, Carloni G (1993) The human bone-marrow-derived B-cell line CE, susceptible to hepatitis C virus infection. Res Virol 144:281–285

Choo QL, Kuo G, Weiner AJ, Overby LR, Bradley DW, Houghton M (1989) Isolation of a cDNA clone derived from a blood-born non-A, non-B viral hepatitis genome. Science 244:359–362

Cribier B, Schmitt C, Bingen A, Kirn A, Keller F (1995) In vitro infection of peripheral blood mononuclear cells by hepatitis C virus. J Gen Virol 76:2485–2491

Dash S, Halim AB, Tsuji H, Hiramatsu N, Gerber MA (1997) Transfection of HepG2 cells with infectious hepatitis C virus genome. Am J Pathol 151:363–373

Doorn LE, Capriles I, Maertens G, Deleys R, Murray K, Kos T, Schellekens H, Quint W (1995) Sequence evolution of the hypervariable region in the putative envelope region E2/NS1 of hepatitis C virus is correlated with specific humoral immune responses. J Virol 69:773–778

Ferri C, Monti M, Civita LL, Longombardo G, Greco F, Pasero G, Gentilini P, Bombardieri S, Zignego AL (1993) Infection of peripheral blood mononuclear cells by hepatitis C virus in mixed cryoglobulinemia. Blood 82:3701–3704

Ferri C, Civita LL, Zignego AL, Pasero G (1997) Hepatitis-C-virus infection and cancer. Int J Cancer 71:1113–1115

Gabrielli A, Manzin A, Candela M, Caniglia ML, Paolucci S, Danieli MG, Clementi M (1994) Active hepatitis C virus infection in bone marrow and peripheral blood mononuclear cells from patients with mixed cryoglobulinaemia. Clin Exp Immunol 97:87–93

Gunji T, Kato N, Hijikata M, Hayashi K, Saitoh S, Shimotohno K (1994) Specific detection of positive and negative stranded hepatitis C viral RNA using chemical RNA modification. Arch Virol 134: 293–302

Higashi Y, Kakumu S, Yoshioka K, Wakita T, Mizokami K, Ohba Y, Ito T, Ishikawa T, Takayanagi M, Nagai Y (1993) Dynamics of genome change in the E2/NS1 region of hepatitis C virus in vivo Virology 197:659–668

Hijikata M, Kato N, Ootsuyama Y, Nakagawa M, Ohkoshi S, Shimotohno K (1991) Hypervariable regions in the putative glycoprotein of hepatitis C virus. Biochem Biophys Res Commun 175:220–228

Hijikata M, Kato N, Ootsuyama Y, Nakagawa M, Shimotohno K (1991) Gene mapping of the putative structural region of the hepatitis C virus genome by in vitro processing analysis. Proc Natl Acad Sci USA 88:5547–5551

Hijikata M, Mizuno K, Rikihisa T, Shimizu YK, Iwamoto A, Nakajima N, Yoshikura H (1995) Selective transmission of hepatitis C virus in vivo and in vitro. Arch Virol 140:1623–1628

Hijikata M, Mizushima H, Akagi T, Mori S, Kakiuchi N, Kato N, Tanaka T, Kimura K, Shimotohno K (1993a) Two distinct proteinase activities required for the processing of a putative nonstructural precursor protein of hepatitis C virus. J Virol 67:4665–4675

Hijikata M, Mizushima H, Tanji Y, Komoda Y, Hirowatari Y, Akagi T, Kato N, Kimura K, Shimotohno K (1993b) Proteolytic processing and membrane association of putative nonstructural proteins of hepatitis C virus. Proc Natl Acad Sci USA 90:10773–10777

Hijikata M, Shimizu YK, Kato H, Iwamoto A, Shih JW, Alter HJ, Purcell RH, Yoshikura H (1993c) Equilibrium centrifugation studies of hepatitis C virus: evidence for circulating immune complexes. J Virol 67:1953–1958

Houghton M(1996) Hepatitis C virus. In: Fields BN, Knipe DM, Howley PM et al. (eds) Fields virology. Lippincott-Raven, Philadelphia, pp 1035–1058

Iacovacci S, Sargiacomo M, Parolini I, Ponzetto A, Peschle C, Carloni G (1993) Replication and multiplication of hepatitis C virus genome in human foetal liver cells. Res Virol 144:275–279

Iacovacci S, Manzin A, Barca S, Sargiacomo M, Serafino A, Valli MB, Macioce G, Hassan HJ, Ponzetto A, Clementi M, Peschle C, Carloni G (1997) Molecular characterization and dynamics of hepatitis C virus replication in human fetal hepatocytes infected in vitro. Hepatology 26:1328–1337

Ikeda M, Kato N, Mizutani T, Sugiyama K, Tanaka K, Shimotohno K (1997a) Analysis of the cell tropism of HCV by using in vitro HCV-infected human lymphocytes and hepatocytes. J Hepatol 27:445–454

Ikeda M, Sugiyama K, Mizutani T, Tanaka T, Tanaka K, Shimotohno K, Kato N (1997b) Hepatitis G virus replication in human cultured cells displaying susceptibility to hepatitis C virus infection. Biochem Biophys Res Commun 235:505–508

Ikeda M, Sugiyama K, Mizutani T, Tanaka T, Tanaka K, Sekihara H, Shimotohno K, Koto N (1998) Human hepatocyte clonal cell lines that support persistent replication of hepatitis C virus. Virus Res 56:157–167

Ito T, Mukaigawa J, Hirabayashi Y, Mitamura K, Yasui K (1996) Cultivation of hepatitis C virus in primary hepatocyte culture from patients with chronic hepatitis C results in release of high titer infectious virus. J Gen Virol 77:1043–1054

Kato N, Hijikata M, Ootsuyama Y, Nakagawa M, Ohkoshi S, Sugimura T, Shimotohno K (1990) Molecular cloning of the human hepatitis C virus genome from Japanese patients with non-A, non-B hepatitis. Proc Natl Acad Sci USA 87:9524–9528

Kato N, Ikeda M, Mizutani T, Sugiyama K, Noguchi M, Hirohashi S, Shimotohno K (1996) Replication of hepatitis C virus in cultured non-neoplastic human hepatocytes. Jpn J Cancer Res 87:787–792

Kato N, Ikeda M, Sugiyama K, Mizutani T, Tanaka T, Shimotohno K (1998) Hepatitis C virus population dynamics in human lymphocytes and hepatocytes infected in vitro. J Gen Virol 79:1859–1869

Kato N, Nakazawa T, Mizutani T, Shimotohno K (1995) Susceptibility of human T-lymphotropic virus type I infected cell line MT-2 to hepatitis C virus infection. Biochem Biophys Res Commun 206: 863–869

Kato N, Ootsuyama Y, Ohkoshi S, Nakazawa T, Sekiya H, Hijikata M, Shimotohno K (1992a) Characterization of hypervariable regions in the putative envelope protein of hepatitis C virus. Biochem Biophys Res Commun 189:119–127

Kato N, Ootsuyama Y, Sekiya H, Ohkoshi S, Nakazawa T, Hijikata M, Shimotohno K (1994) Genetic drift in hypervariable region 1 of the viral genome in persistent hepatitis C virus infection. J Virol 68:4776–4784

Kato N, Ootsuyama Y, Tanaka T, Nakagawa M, Nakazawa T, Muraiso K, Ohkoshi S, Hijikata M, Shimotohno K (1992b) Marked sequence diversity in the putative envelope protein of hepatitis C viruses. Virus Res 22:107–123

Kolykhalov AA, Agapov EV, Blight KJ, Mihalik K, Feinstone SM, Rice CM (1997) Transmission of hepatitis C by intrahepatic inoculation with transcribed RNA. Science 277:570–574

Kuo G, Choo QL, Alter HJ, Gitnick GL, Redeker AG, Purcell RH, Miyamura T, Dienstag JL, Alter MJ, Stevens CE, Tegtmeier GE, Bonino F, Colombo WS, Lee WS, Kuo C, Berger K, Shuster JR, Overby LR, Bradley DW, Houghton M (1989) An assay for circulating antibodies to a major etiologic virus of human non- A, non-B hepatitis. Science 244:362–364

Kurosaki M, Enomoto N, Marumo F, Sato C (1993) Rapid sequence variation of the hypervariable region of hepatitis C virus during the course of chronic infection. Hepatology 18:1293–1299

Lanford RE, Sureau C, Jacob JR, White R, Fuerst TR (1994) Demonstration of in vitro infection of chimpanzee hepatocytes with hepatitis C virus using strand-specific RT/PCR. Virology 202: 606–614

Lanford RE, Chavez D, Chisari FV, Sureau C (1995) Lack of detection of negative-strand hepatitis C virus RNA in peripheral blood mononuclear cells and other extrahepatic tissues by the highly strand-specific rTth reverse transcriptase PCR. J Virol 69:8079–8083

Lerat H, Berby F, Trabaud MA, Vidalin O, Major M, Trepo C, Inchauspe G (1996) Specific detection of hepatitis C virus minus strand RNA in hematopoietic cells. J Clin Invest 97:845–851

Linnen J, Wages J, Zhang-Keck ZY, Fry KE, Krawczynski KZ, Alter H, Koonin E, Gallagher M, Alter M, Hadziyannis S, Karayiannis P, Fung K, Nakatsuji Y, Shih JWK, Young L, Piatak M Jr, Hoover C, Fernandez J, Chen S, Zoul JC, Morris T, Hyams KC, Ismay S, Lifson JD, Hess G, Foung SKH, Thomas H, Bradley D, Margolis H, JP (1996) Molecular cloning and disease association of hepatitis G virus: a transfusion-transmissible agent. Science 271:505–508

Luppi M, Ferrari MG, BonaccorsiO G, Longo G, Narni F, Barozzi P, Marasca R, Mussini C, Torelli G (1996) Hepatitis C virus infection in subsets of neoplastic lymphoproliferations not associated with cryoglobulinemia. Leukemia 10:351–355

Major ME, Feinstone SM (1997) The molecular virology of hepatitis C. Hepatology 25:1527–1538

Martell M, Esteban JI, Quer J, Genesca J, Weiner A, Esteban R, Guardia J, Gomez J (1992) Hepatitis C virus (HCV) circulates as a population of different but closely related genomes quasispecies nature of HCV genome distribution. J Virol 66:3225–3229

Miyakawa Y, Mayumi M (1997) Hepatitis G virus – a true hepatitis virus or an accidental tourist? N Engl J Med 336:795–796

Mizutani T, Ikeda M, Saito S, Sugiyama K, Shimotohno K, Kato N (1998) Detection of negative-stranded hepatitis C virus RNA using a novel strand-specific reverse transcription-polymerase chain reaction. Virus Res 53:209–214

Mizutani T, Kato N, Hirota M, Sugiyama K, Murakami A, Shimotohno K (1995) Inhibition of hepatitis C virus replication by antisense oligonucleotide in culture cells. Biochem Biophys Res Commun 212:906–911

Mizutani T, Kato N, Ikeda M, Sugiyama K, Shimotohno K (1996a) Long-term human T cell culture system supporting hepatitis C virus replication. Biochem Biophys Res Commun 227:822–826

Mizutani T, Kato N, Saito S, Ikeda M, Sugiyama K, Shimotohno K (1996b) Characterization of hepatitis C virus replication in cloned cells obtained from a human T cell leukemia virus type 1-infected cell line, MT-2. J Virol 70:7219–7223

Muller HM, Pfaff E, Goeser T, Kallinowski B, Solbach C, Theilmann L (1993) Peripheral blood leukocytes serve as a possible extrahepatic site for hepatitis C virus replication. J Gen Virol 74: 669–676

Nakajima N, Hijikata M, Yoshikura H, Shimizu YK (1996) Characterization of long-term cultures of hepatitis C virus. J Virol 70:3325–3329

Nei M, Gojobori T (1986) Simple methods for estimating the numbers of synonymous and nonsynonymous nucleotide substitutions. Mol Biol Evol 3:418–426

Nissen E, Hohne M, Schreier E (1994) in vitro replication of hepatitis C virus in a human lymphoid cell line (H9) J. Hepatol. 20:437

Noguchi M, Hirohashi S (1996) Cell Lines from non-neoplatic liver and hepatocellular carcinoma tissue from a single patient. In Vitro Cell Dev Biol Anim 32:135–137

Ohkoshi S, Kojima H, Tawaraya H, Miyajima T, Kamimura T, Asakura H, Satoh A, Hirose S, Hijikata M, Kato N, Shimotohno K (1990) Prevalence of antibody against non-A, non-B hepatitis virus in Japanese patients with hepatocellular carcinoma. Jpn J Cancer Res 81:550–553

Rice CM (1996) Flaviridae: the viruses and their replication. In: Fields BN, Knipe DM, Howley PM (eds) Virology, 3rd edn. Lippincott-Raven, Philadelphia, pp 935–942

Saito S, Kato N, Hijikata M, Gunji T, Itabashi M, Kondo M, Tanaka K, Shimotohno K (1996) Comparison of hypervariable regions (HVR1 and HVR2) in positive- and negative- stranded hepatitis C virus RNA in cancerous and noncancerous liver tissue, peripheral blood mononuclear cells, and serum from a patient with hepatocellular carcinoma. Int J Cancer 67:199–203

Saito I, Miyamura T, Ohbayashi A, Harada H, Katayama T, Kikuchi S, Watanabe Y, Koi S, Onji M, Ohta Y, Choo QL, Houghton M, Kuo G (1990) Hepatitis C virus infection is associated with the development of hepatocellular carcinoma. Proc Natl Acad Sci USA 87:6547–6549

Seipp S, Mueller HM, Pfaff E, Stremmel W, Theilmann L, Goeser T (1997) Establishment of persistent hepatitis C virus infection and replication in vitro. J Gen Virol 78:2467–2476

Sekiya H, Kato N, Ootsuyama Y, Nakazawa T, Yamauchi K, Shimotohno K (1994) Genetic alterations of the putative envelope proteins encoding region of the hepatitis C virus in the progression to relapsed phase from acute hepatitis: humoral immune response to hypervariable region 1. Int J Cancer 57:664–670

Shimizu YK, Iwamoto A, Hijikata M, Purcell RH, Yoshikura H (1992) Evidence for in vitro replication of hepatitis C virus genome in a human T cell line. Proc Natl Acad Sci USA 89:5477–5481

Shimizu YK, Purcell RH, Yoshikura H (1993) Correlation between the infectivity of hepatitis C virus in vivo and its infectivity in vitro. Proc Natl Acad Sci USA 90:6037–6041

Shimizu YK, Yoshikura H (1994) Multicycle infection of hepatitis C virus in cell culture and inhibition by alpha and beta interferons. J Virol 68:8406–8408

Shimizu YK, Yoshikura H (1995) In-vitro systems for the detection of hepatitis C virus infection. Viral Hepatitis Rev. 1:59–65

Shimizu YK, Feinstone SM, Kohara M, Purcell RH, Yoshikura H (1996) Hepatitis C virus: detection of intracellular virus particles by electron microscopy. Hepatology 23:205–209

Shimizu YK, Igarashi H, Kanematu T, Fujiwara K, Wong DC, Purcell RH, Yoshikura H (1997) Sequence analysis of the hepatitis C virus genome recovered from serum, liver, and peripheral blood mononuclear cells of infected chimpanzees. J Virol 71:5769–5773

Shimotohno K (1995) Hepatitis C virus as a causative agent of hepatocellular carcinoma. Intervirology 38:162–169
Shimotohno K, Feinstone SM.(1997) Hepatitis C virus and Hepatitis G virus. In: Richman DD, Whitley RJ, Hayden FG (eds) Clinical virology. Churchill Livingstone, Edinburg, pp 1187–1216
Simons JN, Leary TP, Dawson GJ, Pilot-Matias TJ, Muerhoff AS, Desai SM, associated with human hepatitis. Nat Med 1:564–569
Silvestri F, Pipan C, Barillari G, Zaja F, Fanin R, Infanti L, Russo D, Falasca E, Botta GA, Baccarani M (1996) Prevalence of hepatitis C virus infection in patients with lymphoproliferative disorders. Blood 87:4296–4301
Sugiyama K, Kato N, Ikeda M, Mizutani T, Shiomotohno K, Kato T, Sugiyama Y, Hasumi K (1997a) Hepatitis C virus in pelvic lymph nodes and female reproductive organs. Jpn J Cancer Res 88:925–927
Sugiyama K, Kato N, Mizutani T, Ikeda M, Tanaka T, Shimotohno K (1997b) Genetic Analysis of the hepatitis C Virus (HCV) genome from HCV-infected human T Cells. J Gen Virol 78:329–336
Tanaka T, Kato N, Cho MJ, Shimotohno K (1995) A novel sequence found at the 3' terminus of hepatitis C virus genome. Biochem Biophys Res Commun 215:744–749
Tanaka T, Kato N, Cho MJ, Sugiyama K, Shimotohno K (1996) Structure of the 3' terminus of the hepatitis C virus genome. J Virol 70:3307–3312
Tagawa M, Kato N, Yokosuka O, Ishikawa T, Ohto M, Omata M (1995) Infection of human hepatocyte cell lines with hepatitis C virus in vitro. J Gastroenterol Hepatol 10:523–527
Tsuboi S, Nagamori S, Miyazaki M, Mihara K, Fukaya K, Teruya K, Kosaka T, Tsuji T, Namba M (1996) Persistence of hepatitis C virus RNA in established human hepatocellular carcinoma cell lines. J Med Virol 48:133–140
Valli MB, Bertolini L, Iacovacci S, Ponzetto A, Carloni G (1995) Detection of a 5' UTR variation in the HCV genome after a long-term in vitro infection. Res Virol 146:285–288
Valli MB, Carloni G, Manzin A, Nasorri F, Ponzetto A, Clementi M (1997) Hepatitis C virus infection of a Vero cell clone displaying efficient virus-cell binding. Res Virol 148:181–186
Weiner AJ, Brauer MJ, Rosenblatt J, Richman KH, Tung J, Crawford K, Bonino F, Saracco G, Choo QL, Houghton M, Han J (1991) Variable and hypervariable domains are found in the regions of HCV corresponding to the flavivirus envelope and NS1 proteins and the pestivirus envelope glycoproteins. Virology 180:842–848
Yanagi M, Purcell RH, Emerson SU, Bukh J (1997) Transcripts from a single full-length cDNA clone of hepatitis C virus are infectious when directly transfected into the liver of a chimpanzee. Proc Natl Acad Sci USA 94:8738–8743
Yoo BJ, Selby MJ, Choe J, Suh BS, Choi SH, Joh JS, Nuovo GJ, Lee HS, Houghton M, Han JH (1995) Transfection of a differentiated human hepatoma cell line (Huh7) with in vitro-transcribed hepatitis virus (HCV) RNA and establishment of a long-term culture persistently infected with HCV. J Virol 69:32–38

Characterization of Hepatitis C Virus Infectious Clones in Chimpanzees: Long-Term Studies

M.E. Major and S.M. Feinstone

1 Introduction	279
2 The Chimpanzee and Viral Persistence	280
3 Early Attempts to Generate a Hepatitis C Virus Infectious Clone	281
4 Determination of the Authentic 3′ End of Hepatitis C Virus	282
5 Development of a Consensus Clone	284
6 Disease Progression Following Inoculation with the Hepatitis C Virus Consensus Clone	286
7 Sequence Evolution and Persistence	291
8 Conclusion	295
References	296

1 Introduction

The chimpanzee (*Pan troglodyte*) is genetically more than 98.5% identical to man. As such a close relative, chimpanzees have been a valuable model for a variety of human diseases and have been especially useful for infectious disease research. Human hepatitis viruses have a generally very restricted host range. However, all of the six established human hepatitis viruses, hepatitis A, B, C, D (delta), E and G will infect chimpanzees (Maynard et al. 1975; Barker et al. 1975; Alter et al. 1978; Rizzetto et al. 1980; Tsarev et al. 1993; Bukh et al. 1998), though hepatitis E virus and hepatitis G virus have not been shown to induce disease (Bukh et al. 1998). Some of these viruses will infect other primates and the delta agent can be introduced into several non-primate species that harbor their own host specific hepadna viruses. Of the human hepatitis viruses, only hepatitis A will infect and replicate reliably in cell culture. Therefore, animal models have been extremely important for viral hepatitis research and have been key in the development of effective vaccines for both hepatitis B and A. The hepatitis C virus itself was first identified by molecular cloning of the viral genetic material extracted from the

Laboratory of Hepatitis Viruses, Division of Viral Products, CBER/FDA, Bldg. 29A Rm1D14, HFM448, 8800 Rockville Pike, Bethesda, MD 20892, USA

plasma of a chimpanzee that was chronically infected by a human non-A, non-B hepatitis virus later identified as hepatitis C virus (HCV) (Choo et al. 1989). The chimpanzee studies leading up to the initial identification of HCV have been reviewed by Bradley in a separate chapter of this book. Since the initial identification of the genome of HCV in the chimpanzee plasma, this animal model has remained a valuable tool for research whenever an infectivity readout is required. Despite extensive efforts in a number of laboratories, the *in vitro* propagation of HCV in any cell culture system remains at a primitive level and has not been useful for most types of infectivity analyses required.

2 The Chimpanzee and Viral Persistence

The chimpanzee model of HCV infection has been particularly useful for understanding the mechanisms of chronicity, immune response to HCV, recovery and protection. HCV infections become persistent in at least 75% of infected humans. HCV is an RNA virus with no DNA intermediate and no obvious potential for genomic integration, therefore, the cause of these persistent infections has been difficult to understand. Chimpanzee studies have shown that, while not as frequent as in humans, chronic infections occur in 30%–50% of infected animals. Details of the viral persistence in chimpanzees have been studied using experimental infections with characterized inocula coupled with sequence analysis of the hypervariable region of the HCV E2 glycoprotein. Much of this work has focused on the variation in the hypervariable region that is contained within the 30 NH_2-terminal amino acids of the E2 envelope glycoprotein, termed HVR1 (Weiner et al. 1991; Kato et al. 1992; Hijikata et al. 1991a). The genome of HCV has a high degree of sequence variability based on a generally high error rate of the RNA dependent RNA polymerase. The HVR1 however, seems to change at an even faster rate than the remainder of the genome (Ogata et al. 1991; Kurosaki et al. 1993) which likely indicates a high degree of tolerance for mutation at this site. Generally, following inoculation with HCV-positive human serum a virus population with a single HVR1 sequence quickly becomes dominant in the chimpanzee. Over time, the original virus may be replaced with a genetic variant with a different HVR1 sequence. Analysis of the immune response in both humans and chimpanzees has demonstrated that the new HVR1 sequence may not react with antibody that arose in response to the HVR of the initial dominant virus (Sekiya et al. 1994; Weiner et al. 1992; Kato et al. 1993; Taniguchi et al. 1993). Therefore, the new dominant virus might represent an immune escape variant. Repeated rounds of antibody induction followed by selection of a virus with a non-cross-reacting HVR1 which then becomes the dominant species is believed to account for the viral persistence. Other studies in chimpanzees have shown that cytotoxic T lymphocyte (CTL) epitopes may also vary during the course of an infection under immune pressure (Weiner et al. 1995; Chang et al. 1997; Kaneko et al. 1997). However, broadly spread sequence changes outside of the HVR1 that might also contribute to the maintenance of persistence have not been carefully studied.

Other chimpanzee experiments have shown that, following infection, seroconversion and viral clearance, an animal could be reinfected with a second infectious inoculum (FARCI et al. 1992; PRINCE et al. 1992). This reinfection was not dependent on genotype heterogeneity. That is, even if both inocula contained virus of the same genotype, the chimpanzee could be reinfected. Even more surprising, a recovered chimpanzee that had developed antibody, including antibody to the envelope glycoproteins and the HVR1, could be reinfected using exactly the same inoculum (FARCI et al. 1992). Sequence analysis of the viral genomes from these chimpanzees revealed that the predominant virus in the reinfected animal had a different HVR1 sequence compared to that of the predominant virus in the serum of the chimpanzee following the initial inoculation (FARCI et al. 1992). The obvious explanation is that the virus exists in the host as a population with genomic, and subsequently, amino acid sequence variability. The HVR1, which seems to tolerate a very high level of variability, may represent an important epitope towards which neutralizing antibodies may be directed. If antibody to the predominant HVR1 arises in the serum, that virus becomes suppressed and a virus with a non-cross-reacting HVR1 is selected which then becomes the predominant virus present in that host. This continuous antibody selection resulting in new dominant viruses is an attractive hypothesis to explain viral persistence. Nevertheless, while this type of selection has been observed, it has not been proved that this is the primary mechanism responsible for viral persistence *in vivo*.

3 Early Attempts to Generate a Hepatitis C Virus Infectious Clone

Without a useful cell culture system, it has not been possible to virologically clone HCV by, for example, repeated plaque purification in order to produce a virus with a relatively uniform sequence. Therefore, it has been difficult to study directly the role of the HVR1 in either neutralization of the virus or the maintenance of viral persistence, as any inoculum contains a population of viruses differing at the nucleotide and amino acid levels. One way to produce a clonally identical population of virus is to initiate the infection with RNA transcribed from a cDNA clone representing the entire viral genome. We began an effort to produce such a clone soon after the identification of the HCV genome was published by CHOO et al. (1989). Our initial clones were derived from RNA extracted from a liver biopsy obtained from a chimpanzee acutely infected with the H strain of HCV (genotype 1a) (FEINSTONE et al. 1981). This material was used because it was considered a renewable resource of high titer HCV RNA. Fragments of 2000–5000 nucleotides in length were amplified by reverse transcriptase polymerase chain reaction (RT-PCR), cloned into a plasmid vector behind the T7 polymerase promoter and spliced together at appropriate restriction enzyme sites (Wychowski et al. unpublished). Attempts to extend the sequence at either end by various PCR based technologies did not result in cloning additional bases beyond that already published (CHOO et al. 1989). This original putative full-length clone was then modified in Rice's laboratory to eliminate all extra bases at the ends except the initial 5′ G, which is necessary for efficient transcription by the T7 polymerase. This clone was transcribed into RNA. The RNA was mixed with cationic lipid (Lipofectin) and injected directly into the liver of a chimpanzee. The liver had been exposed by a surgical incision and the RNA was inoculated under direct visualization into multiple sites. A negative control RNA was also prepared that was identical to the experimental RNA except that it had a 20 amino acid in-frame deletion in the NS5B coding region encompassing the GDD sequence believed to be necessary for the RNA dependent RNA polymerase activity (KOLYKHALOV et al. 1997). We included this negative control to safeguard against the possibility that inoculation of a large quantity of RNA could result in some false positive measurements leading us to believe that the RNA was indeed replicating and producing infectious virus. For example, the RNA could be translated and function as a nucleic acid vaccine inducing antibody responses similar to those produced by infection or small quantities of RNA might escape the liver cells and somehow resist degradation causing positive results in highly sensitive nested RT-PCR reactions. The trauma of surgery, the injections, or possibly toxicity associated with the inoculum might have resulted in elevations of liver function tests, especially the alanine amino transferase (ALT) assay that is relied upon as a sensitive measure of hepatitis, inflammation or necrosis in the liver. The two chimpanzees were followed for 6 months beginning 3 days after the inoculation with weekly or bi-weekly serum samples and liver biopsies. During the entire period of observation, no HCV RNA was detected in the serum, no HCV-specific antibody arose and no ALT elevations were detected.

4 Determination of the Authentic 3′ End of Hepatitis C Virus

This initial clone was made from liver-derived RNA and not from RNA extracted from virus circulating in the blood. It was possible that much of this RNA was defective and would not be infectious. For that reason, a new set of cDNA clones was prepared by Kolykhalov in Rice's lab. These clones were made directly from RNA extracted from the serum of the original patient H. This serum had been titered in chimpanzees and was known to contain approximately $10^{6.5}$ chimpanzee infectious doses per ml. (FEINSTONE et al. 1981). In addition, a large effort was made to extend the ends of the clone beyond the known published sequence. No additional bases could be found at the 5′ end of the genome. However, using a technique in which oligonucleotides with known sequences were ligated to the 3′ ends of the viral RNA using T4 RNA ligase and the RNA amplified by RT-PCR using a primer complementary to the ligated oligonucleotide, a new sequence, believed to be the terminus, was shown to exist at the 3' end beyond the poly(U) stretch. It was shown that the 3' noncoding region of the HCV genome consisted of four elements; a short highly variable sequence, a homopolymeric poly(U) tract, a polypyrimidine stretch of primarily Us with interspersed C residues and a novel highly conserved sequence of 98 bases which made up the ultimate 3' end (KOLYKHALOV et al. 1996). This 98 nucleotide sequence is predicted to have a high degree of secondary structure (Fig. 1) which, together with the fact that it is nearly identical in every HCV isolate so far studied representing all genotypes, implies that this structure is required for RNA replication. Extensive studies gave confidence that indeed the ultimate 3' end had been determined. In addition, at about the same time, Shimotohno and colleagues in Japan, using a different methodology, identified a nearly identical 3' end sequence in a genotype 1b virus (TANAKA et al. 1995, 1996). More recently YANAGI et al. (1999) demonstrated the importance of the poly (U/UC) tract and the 98 base pair tail for *in vivo* infection using the chimpanzee model and RNA transcribed from cDNA clones. Deletions in these regions of the clones rendered RNA transcripts non infectious. Mutation of the variable region, however, was tolerated and led to viremia in the chimpanzee within 1 week postinoculation.

A combinatorial library of about 10^5 full-length clones was produced from RNA extracted from the H plasma by Kolykhalov using high-fidelity RT-PCR and subcloning into a recipient plasmid vector that contained the 5'- and 3'-terminal HCV sequences and a T7 promoter. Over 200 clones from this library

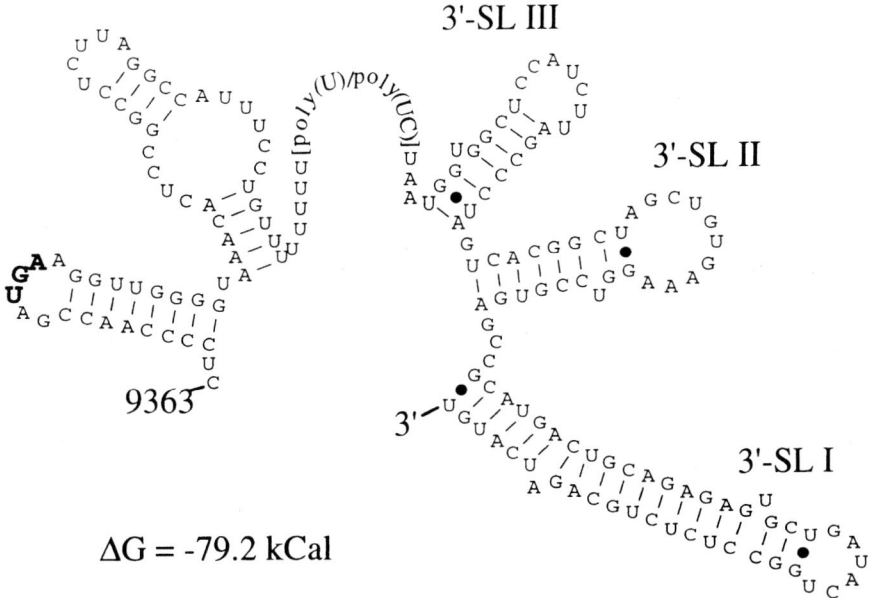

Fig. 1. Computer-predicted secondary structure of the hepatitis C virus (HCV)-H 3' nontranslated region (NTR) RNA and limited upstream sequence. Stem loop (*SL*) structures in the 3'-terminal 98-base RNA element are labeled (3'-SL I to III). The open reading frame UGA termination codon located at nucleotide 9375 of the HCV-H genome is shown in *boldface type*

were screened by restriction digests, and analysis of the polyprotein processing and completeness of translation to the COOH-terminal of the polyprotein. Previous studies in Rice's and other labs had demonstrated the pattern of the polyprotein cleavage products on SDS PAGE (HIJIKATA et al. 1991b; GRAKOUI et al. 1993). Each band had been specifically identified by radioimmunoprecipitation techniques using specific antibody. A group of 17 clones that had the expected translation/cleavage pattern were identified. In addition, a group of 17 clones that appeared to have complete translation products but that had alternative cleavage patterns were also selected with the idea that the originally described cleavage pattern may have been an aberrant one. RNA was transcribed from each of these clones and two chimpanzees were inoculated at multiple sites in the liver under direct visualization. The 17 clones with the "classic" cleavage pattern were inoculated into one animal and the clones with the alternate cleavage patterns were inoculated into a second. Finally, a third animal received a similar total quantity of control RNA as in the first experiment. The NS5B-deleted control clone as well as the 34 experimental clones all included the newly described 98 nucleotide sequence at the 3' end. The animals were again bled and biopsied at weekly or bi-weekly intervals beginning 2 days after inoculation. As in the first experiment, there was no evidence of viral replication, no hepatitis and no antibody after 6 months of follow-up.

5 Development of a Consensus Clone

It had long been suspected that, as all our clones had been made by initially amplifying the RNA by RT-PCR, even though we used the most faithful polymerases available and took other steps to reduce the error rate, multiple PCR mistakes could have been included in any given clone that would have been lethal for viral replication. It was also possible that many of the RNA molecules extracted from the serum were already defective in some way that would have rendered them non-replicating. It had been hoped that, as the genome of this virus had been noted to vary greatly and only clones with full-length translation products were used, the virus would be "tolerant" of these mistakes. However, the failure of the last inoculation, which included 17 distinct clones all of which had the proper translations/cleavage patterns and the extended 3' end, gave reason to investigate the likelihood of lethal PCR errors. Kolykhalov selected six clones from the 17 and determined the sequence for each in its entirety. Once the sequence of these six clones was determined, it became clear that each clone contained numerous mistakes that could easily have rendered it non-infectious. From those six complete sequences, a single consensus sequence was derived. He then pieced together a single clone that represented the consensus (Fig. 2). Sequencing of these multiple clones also showed that the length of the poly(U/UC) stretch was variable. It could not be determined if this variability was represented in viral genomes or was an

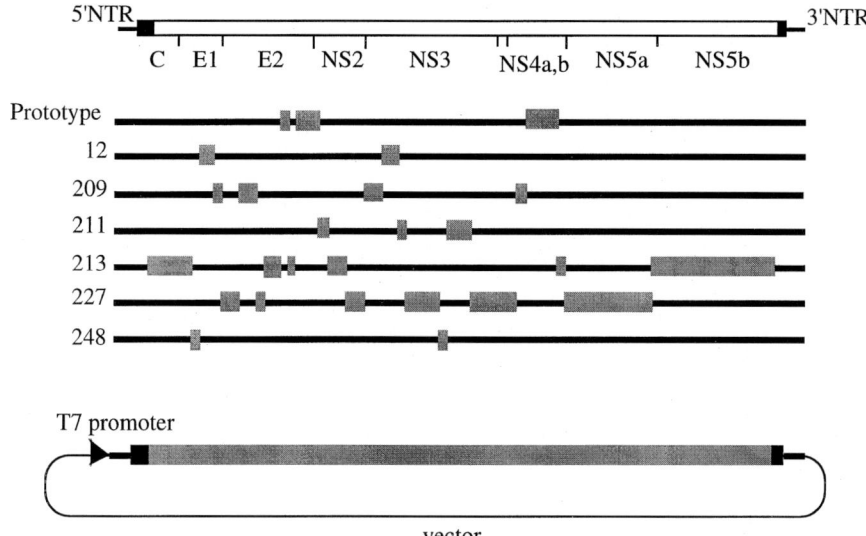

Fig. 2. Assembly of HCV-H77 consensus clone. *Top*, hepatitis C virus (HCV)-H77 RNA with 5' and 3' nontranslated region (NTRs) indicated as *solid black lines and boxes*. Locations of the full-length polyprotein cleavage products are shown. Below, fragments (*shaded blocks*) assembled from individual full-length clones to generate a consensus sequence clone in a vector containing the T7 promoter

artifact of the PCR amplification across this long stretch of pyrimidines. For completeness, clones were prepared with poly(U/UC) of either short (75 nucleotides) or long (133 nucleotides) stretches. In addition, as there had been reports of a single additional base on the 5' end of the genome, clones were prepared with either the single extra G to initiate T7 polymerase transcription or the G plus either an A, C, G or T. Thus there were clones with five different 5' ends each of which had either the long or the short 3' poly(U/UC) for a total of ten clones (Fig. 3). The remaining sequence of the ten clones was identical to the consensus sequence that had been determined.

False positive results had never appeared in the previous negative experiments, therefore it was decided in this experiment to eliminate the negative control. However, to ensure that our method for transfection of the RNA into chimpanzee liver worked we included a positive control. Sue Emerson of NIAID supplied us with an infectious cDNA clone of hepatitis A virus that would infect and produce disease in chimpanzees (EMERSON et al. 1992). In this experiment, we inoculated two animals with HCV-specific RNA. The first animal, 1535, was inoculated at two sites in the liver with the complete unpurified transcription mix containing the template DNA and approximately 3000µg of transcribed RNA diluted to 500µl in PBS. For the second chimpanzee, 1536, we prepared the RNA by the technique that we had previously used. The DNA template used for the RNA transcription reaction was digested with DNase and the RNA was precipitated in alcohol. About

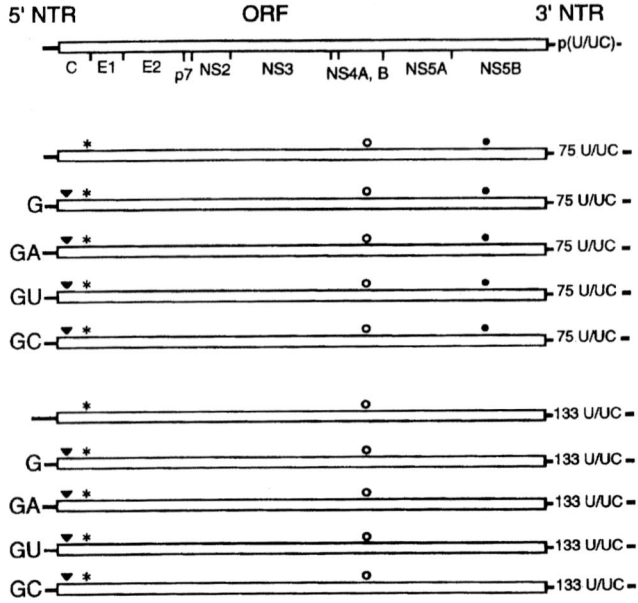

Fig. 3. Features of the 10 full-length consensus clone derivatives tested in chimpanzees Ch1535 and Ch1536. Top, hepatitis C virus (HCV)-H77 RNA. 5' and 3' nontranslated regions (NTRs) are represented as *solid lines*, locations of the polyprotein cleavage products are shown. Below, the 10 RNA transcripts used for inoculation of Ch1535 and Ch1536. Additional 5' nucleotides and 75-base vs 133-base poly(U/UC) tracts are indicated. Clones with additional 5' bases contained a silent mutation inactivating the *Xho*I site at position 514 (*solid triangles*). Clones with 75-base vs 133-base poly(U/UC) tracts were distinguished by A (*solid circle*) versus G at position 8054, respectively. All clones included two silent nucleotide substitutions: position 899 (C instead of U, asterisk); position 5936 (C instead of A, *open circles*). The substitution at position 5936 inactivated an internal *Bsm*I site in the H77 cDNA so that an engineered *Bsm*I site could be used for production of runoff RNA transcripts with the exact 3' terminus of the HCV genome RNA. Nucleotide positions are according to the published HCV H sequence. (From INCHAUSPÉ et al. 1991)

22μg of RNA mixed with about 9μg of lipofectin in 150μl of PBS was inoculated into three sites. The positive control animal was inoculated with hepatitis A virus RNA by the method used to inoculate chimpanzee 1535.

6 Disease Progression Following Inoculation with the Hepatitis C Virus Consensus Clone

The animal that received the hepatitis A RNA developed typical mild hepatitis A as expected in about 4 weeks. Both of the animals that received the consensus sequence HCV RNA developed increasing quantities of HCV RNA in their serum, determined by RT-PCR, beginning at 1 (1535) or 2 (1536) weeks after inoculation with titers at between 10^4 and 10^5 copies/ml. This RNA appeared to be encapsidated, as it was shown to be RNase resistant. In the previous, negative, experiments, no viral RNA was ever detected in the serum even as early as 2 days following either of the inoculations. Analysis of ALT levels, liver histology, serum RNA copy number, and HCV-specific antibody showed that during the first 26 weeks following inoculation both animals developed similar and classical responses to HCV infection (Fig. 4). Chimpanzee 1535 developed evidence of acute hepatitis just 2 weeks after RNA inoculation. The ALT level more than doubled from the preinoculation levels and there were histologic changes in the liver biopsy consistent with early acute hepatitis (Fig. 4A). This early rise in ALT has been observed in many chimpanzees inoculated intravenously with HCV infectious serum. The early ALT elevation gradually declined over the next 6 weeks but again increased at about week 9 post-inoculation (p.i.) reaching a peak at week 17 p.i. Chimpanzee 1536 did not exhibit the early ALT elevation but did have a rapid rise in ALT beginning at week 10 p.i. which peaked at week 12 p.i. (Fig. 4D). ALT levels returned to normal by week 21 for both chimpanzees, liver enzymes were monitored throughout the study but no further elevations were observed.

Inflammatory changes observed in the liver biopsies were assessed semi-quantitatively on a scale of 0–4 (none, minimal, mild, moderate, marked) and generally followed the pattern of liver enzymes in terms of peak severity and endurance. Early biopsies (at 2–3 weeks p.i.) showed focal apoptosis of liver cells and mitotic activity without evidence of inflammation. Mild piecemeal necrosis appeared first in Ch1536 at week 6, and in Ch1535 at week 10 with maximum inflammatory responses between weeks 11–12 (Ch1536) and 15–17 (Ch1535). Inflammation in Ch1535 never advanced beyond the mild level initially seen at week 10 while that in Ch1536 was characterized by numerous foci of lobular inflammation, frequent apoptotic hepatocytes and mild piecemeal necrosis in many portal areas (Fig. 5A,B). Similar to the ALT levels, inflammation in both animals had decreased by week 22 and was scored as minimal in subsequent biopsies.

HCV RNA levels in the serum reached peaks of approximately 10^6 copies/ml at 14 (1535) and 11 (1536) weeks p.i. (Fig. 4B,E). In both animals the ALT and HCV-specific antibody levels increased above baseline directly following the peak in serum RNA. Antibody, measured by commercial EIA, first appeared in the serum of 1535 at week 13 and in 1536 at week 10. HCV envelope antibody levels were measured by ELISA using recombinant E1E2p7 antigen which had an amino acid sequence corresponding to that encoded by the RNA transcripts used for the inoculation of 1535 and 1536. Initially, anti-E1E2 antibody levels were comparable in both animals with the response in Ch1536 preceding that in Ch1535 by approximately 5 weeks (Fig. 4C,F). At week 28 Ch1536 began to show increasing levels of anti-E1E2 antibody. This was in contrast to Ch1535 in which anti-E1E2 antibody levels remained at a P/N of 3–4 throughout the study. In spite of this elevation in serum antibody, 1536 remained positive for HCV RNA, though at a log lower than that of 1535. The cause of this antibody increase is unclear, it did not appear to follow any increase in viral replication, as determined by serum RNA or ALT levels.

Despite a sharp decrease in serum HCV RNA levels following the immune responses, neither Ch1535 nor Ch1536 completely cleared the infection but both

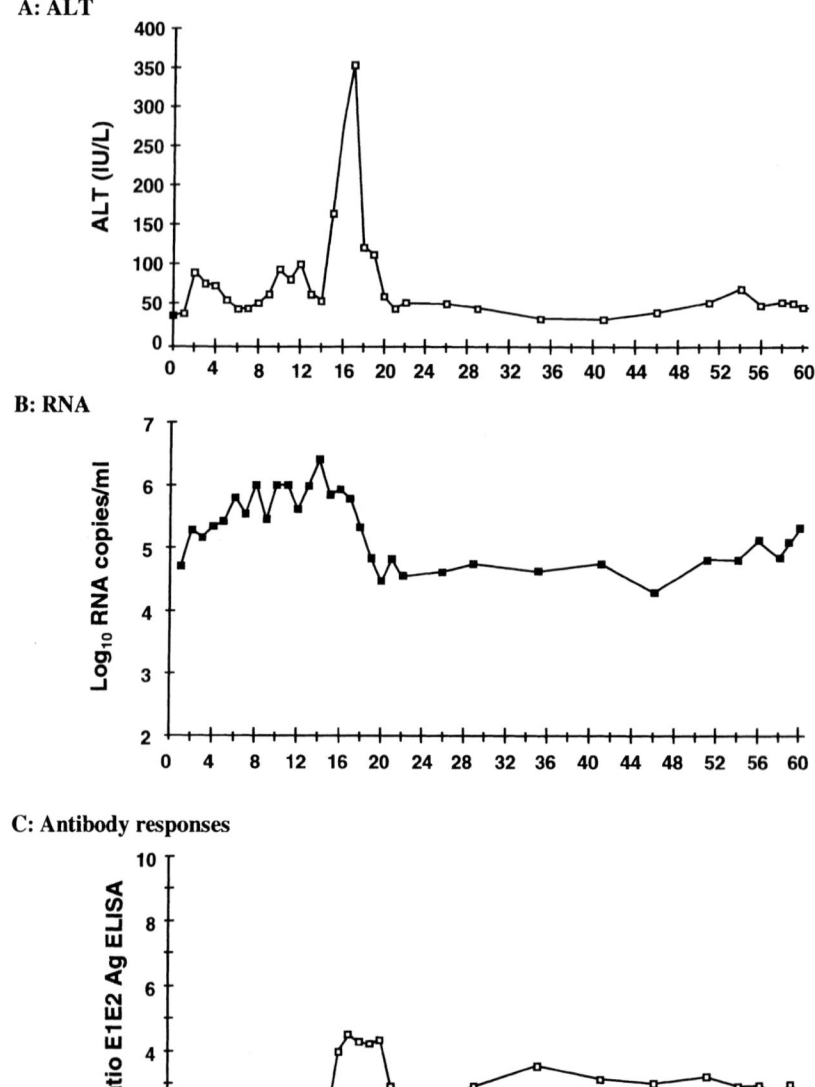

Fig. 4A–F. Clinical, virologic, and serologic responses in Ch1535 (**A–C**) and Ch1536 (**D–F**) to inoculation with HCV consensus clone. **A, D**, Serum ALT levels (IU/L); **B, E** Serum RNA copy number ($\log_{10}$ RNA copies/ml); **C, F** Anti-E1E2 antibody responses measured by ELISA using recombinant E1E2 antigen. *Dotted line* represents cutoff (P/N = 2) (P/N ratio represents the OD_{405} for the test sample divided by that for a pre serum sample from the same chimpanzee)

Ch1536

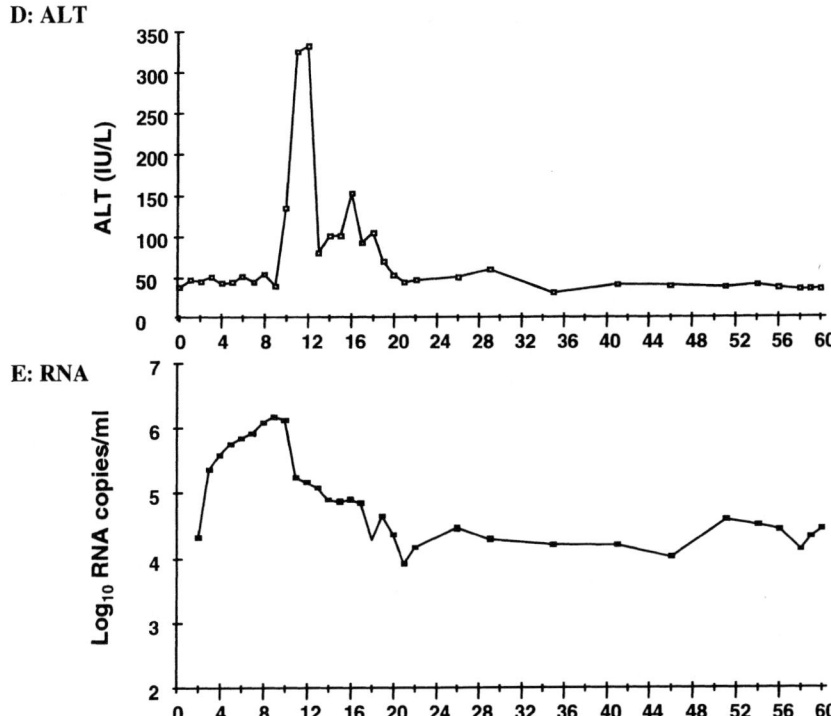

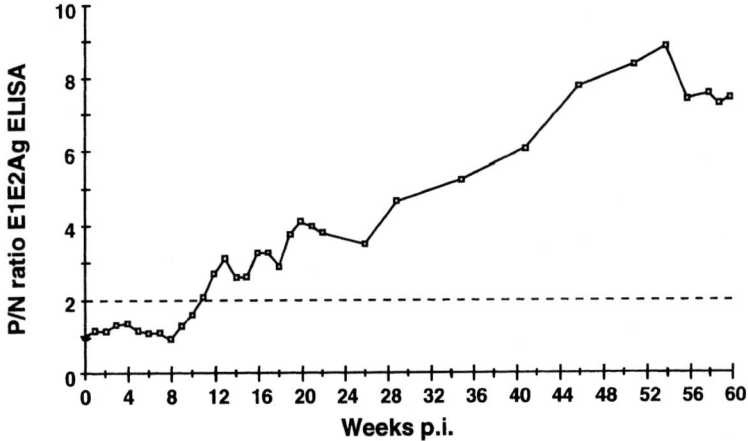

Fig. 4 (*Cont.*)

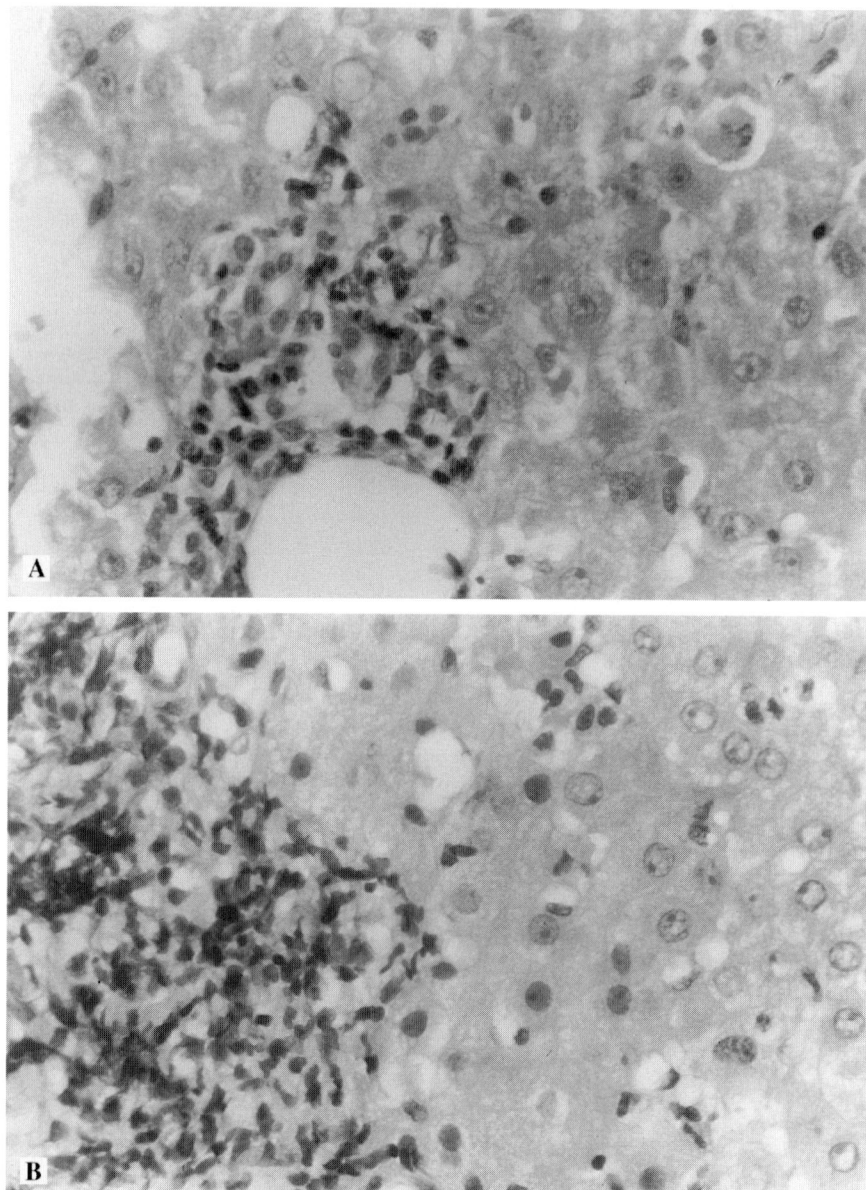

Fig. 5A,B. Histological analysis of liver biopsies from Ch1535 and Ch1536. **A** Portal area inflammation from Ch1535 at week 15 p.i. **B** Piecemeal necrosis from Ch1536 at week 11 p.i. These dates represent peak liver inflammatory responses in both animals

became chronically infected, with RNA titers remaining at approximately 10^5 (Ch1535) and 10^4 (Ch1536) copies/ml. This was in the presence of detectable humoral immune responses to several HCV antigens (Fig. 4A–F).

Thus, it appears that the full-length consensus sequence clones indeed represented closely the genomic sequence of HCV strain H. The ultimate proof that the virus being produced in these animals was indeed infectious would have been accomplished by passaging the serum from one of them to a naive chimpanzee. As chimpanzees are a valuable and very expensive resource, it was decided not to perform this experiment as there could be no other explanation for the data than complete virus replication. In the two previous, negative experiments, the second of which included clones that were nearly identical to the consensus clone, no evidence of viral replication was observed. We never detected RNA in the serum or in liver biopsies. There was no biochemical or histological evidence of hepatitis and no seroconversion. On the other hand, inoculation of the consensus sequence clones resulted in the appearance of HCV RNA in the serum in both animals in just 1–2 weeks, hepatitis in one animal at 2 weeks and in the other at 9 weeks and seroconversion at 13 and 10 weeks. In addition, the RNase resistant RNA levels increased over time to levels that could not be explained as input RNA, and restriction enzyme digestion and sequence analysis of cDNA amplicons of the recovered viral RNA revealed the presence of engineered markers, proving the infections stemmed from the inoculated RNA transcripts. Silent markers identifying the 5′ terminal sequences and the length of the poly(U/UC) tract indicated that RNA transcripts with and without additional 5′ nucleotides were infectious as were transcripts containing either the 75-base or the 133-base poly(U/UC) tracts.

Following the initial report by KOLYKHALOV et al. (1997), infectious clones were produced by YANAGI and colleagues of genotypes 1a (H strain) and 1b using the construction of a consensus sequence clone (YANAGI et al. 1997, 1998).

7 Sequence Evolution and Persistence

In some ways it was surprising that the disease produced by RNA transfection was so typical of HCV infections in chimpanzees observed after intravenous inoculation of infectious serum or plasma. The RNA transcripts represented a single coding region sequence with only minor modifications at the 5′ and 3′ termini, thereby eliminating the quasispecies population that is thought to contribute significantly to the persistence of this virus and possibly to the clinical picture, including interferon sensitivity. Explanations for viral persistence in both 1535 and 1536 were sought through sequence analysis of the HCV genome isolated from serum-derived RNA.

As stated above, it is thought that the HCV HVR1 may contain an antibody neutralization epitope suggesting that when subject to immune pressure there is the potential for escape mutants (SEKIYA et al. 1994; WEINER et al. 1992; KATO et al. 1993; TANIGUCHI et al. 1993; SHIMIZU et al. 1994). Observations have been made in humans and chimpanzees that this region changes during the course of a chronic infection with new dominant viral species arising that encode HVR1 sequences distinct from previous isolates (WEINER et al. 1992; KATO et al. 1993; TANIGUCHI et al. 1993).

One of the major unanswered questions is whether these variants were already present in the original inoculum, though at a low level, or whether they are due to *de novo* mutations which occur as a result of the non-proof reading ability of the HCV RNA-dependent RNA polymerase (RdRp) and which prove both tolerable and opportunistic for the virus. Such questions have been impossible to address without a pure single sequence virus as source inoculum. Using RNA transcribed from a consensus clone to inoculate chimpanzees provided a means to analyze the natural evolution of HCV and determine its true potential for incorporating mutations in its genome and the role of these in the maintenance of a persistent infection.

The evolution of the virus was examined primarily through sequence analysis of HVR1 which, based on published sequences, is generally believed to be the region most prone to mutation (OGATA et al. 1991; KATO et al. 1993, 1994; VAN DOORN et al. 1994). Figure 6 shows amino acid sequence alignments of clones generated from serum RNA and representing the HVR1 of Ch1535 and Ch1536 at 60 weeks p.i. There was no change in the majority HVR1 sequence at this time point in either chimpanzee, although two substitutions were detected in RNA from 1535 upstream of HVR1, in the E1 signal sequence, which were contained within all clones. Analysis of clones obtained from serum RNA throughout the study showed that these arose at weeks 29 ($V_{371}A$) and 41 ($V_{373}L$) p.i. At all time points the degree of variability was very low, consisting of point mutations most of which were not

A: Ch1535

```
aa367                                                           aa426          No.
NWAKVLVVLL LFAGVDAETH VTGGSAGRTT AGLVGLLTPG AKQNIQLINT NGSWHINSTA
---------- ---------- ---------- ---------- ---------- ----------    0
----A-L--- ---------- ---------- ---------- ---------- ----------   35
--V-A-L--- -------K-- ---------- ---------- ---------- ----------    1
----A-L--- ---------- ---------- ---------- ----T----- ----------    1
----A-L--- ---------- -----T---- ---------- ---------- ----------    1
----A-L--- ---------- ---------- ------P--- ---------- ----------    1
```

B: Ch1536

```
aa367                                                           aa426          No.
NWAKVLVVLL LFAGVDAETH VTGGSAGRTT AGLVGLLTPG AKQNIQLINT NGSWHINSTA
---------- ---------- ---------- ---------- ---------- ----------   25
---------- ---------- ---------- ---A------ ---------- ----------    4
---------- ---------- ---------- ---------- --------A- ----------    1
---------- ---------- ---------- ---------- ----T----- ----------    1
---------- --------A- ---------- ---------- ---------- ----------    1
---------- ---------- --E------- ---------- ---------- ----------    1
```

Fig. 6A,B. Sequence analysis of clones representing HVR1 of HCV isolated from RT-PCR amplified serum RNA. Amino acid sequence alignment of the original HCV-H77 HVR1 encoded by the consensus clone (sequence shown) and those of clones representing quasispecies present at 60 weeks p.i. in Ch1535 (**A**) and Ch1536 (**B**). HVR1 is *underlined*. Amino acids homologous to consensus clone are indicated with *dashes*. Numbers of clones corresponding to each amino acid sequence are indicated on the *right hand side*

observed in subsequent samples suggesting these may represent PCR-induced mutations. Clearly in the case of these two animals, mutations in HVR1 were not responsible for persistence of HCV.

Therefore, the question of why these animals did not clear the infection still remained. A CD4 T cell response to a conserved epitope contained within the NS3 region has been detected in a majority of patients with varying haplotypes that have resolved acute infections (DIEPOLDER et al. 1995, 1997). The epitope was shown to have high binding affinity for several common HLA-DR alleles and may play an important role in viral clearance. It has also been suggested that low or late anti-envelope antibody responses may correlate with chronicity, while earlier responses, within 3–6 months of infection, are associated with resolution of acute infections (ALLANDER et al. 1997; ZIBERT et al. 1997a,b). This could be also be true of cellular responses. Ch1535 and Ch1536 developed detectable antibodies to several viral antigens within 9–13 weeks p.i. (Fig. 4 C,F); however, quantitative analyses of antibodies to the E1E2 region and HVR1 indicated that the responses were low during this time and therefore may not have been adequate to clear the virus even though serum RNA titers were reduced significantly (Fig. 4A,D). Similarly, although the elevated ALT levels and inflammatory responses are suggestive of a cellular response this may also have been too weak to completely clear the virus.

Further sequence analysis of RNA isolated from serum at week 60 revealed several nucleotide and corresponding amino acid mutations in portions of the genome other than HVR1 (Table 1); none of these were shared between the two animals. From this data the overall mutation rate for HCV was calculated as 1.57×10^{-3} and 1.48×10^{-3} nucleotide substitutions per site per year for Ch1535 and Ch1536, respectively. This is similar to that previously reported for samples taken 20 years apart from a chronically infected patient ($\sim 1.9 \times 10^{-3}$ base substitutions per genome site per year), although this study found a significantly high rate of change in HVR1, 28.2% variation at the nucleotide level (OGATA et al. 1991). For Ch1535 and Ch1536, substitutions were clustered within the envelope, NS2, NS3 and NS5. The function of the p7 protein is unidentified and it is yet to be established whether or not it forms part of the viral envelope. However, despite the short length of this region, serum RNA from both chimpanzees at week 60 carried mutations that resulted in amino acid substitutions.

The identification of other antibody epitopes in the E1E2 region of HCV has been difficult due to the conformational nature of these proteins. Linear B cell epitopes other than those in HVR1 have been identified in E1 and E2 (CHING et al. 1992; RAY et al. 1994; MINK et al. 1994) although none of the mutations observed in Ch1535 and Ch1536 were located within these epitopes. Numerous reports have been published describing MHC class I CTL recognizing antigens of HCV (KOZIEL et al. 1992, 1993, 1997; KANEKO et al. 1996; BATTEGAY et al. 1995; CERNY et al. 1995; NELSON et al. 1997; KUROKOHCHI et al. 1996; FERRARI et al. 1994; ERICKSON et al. 1993; REHERMANN et al. 1996), taken together these studies indicate that CTLs have been detected to all viral proteins. There is some evidence for the existence of CTL escape variants in natural infections involving chimpanzees (WEINER et al. 1995)

Table 1. Sequence changes in HCV genome at week 60 post-infection

HCV region (nt length)	Ch1535				Ch1536			
	nt position	nt change	AA position	AA change	Nt position	Nt change	AA position	AA change
5'NCR (341)	–	–	–	–	–	–	–	–
Core (573)	362	T-C	–	–	518[a]	A-T	–	–
	518[a]	A-T	–	–				
E1 (576)	1453	T-C	371	V-A	1134	C-A	265	L-I
	1458	G-C	373	V-L				
E2 (1089)	1613	C-T	–	–	1728	A-G	463	T-A
	1773	A-G	478	S-G	2468	C-T	–	–
	2499	G-A	720	V-I				
p7 (189)	2603	C-T	–	–	2718	A-G	793	M-V
	2638	T-C	766	V-A				
NS2 (651)	2916	G-A	859	V-M	3047	T-C	–	–
	3173	T-C	–	–				
NS3 (1893)	3632	C-G	–	–	3883	G-A	1181	R-K
	4938	G-C	1533	A-P	5042	T-C		
	5244	G-A	1635	V-I				
NS4A (164)	–	–	–	–	–	–	–	–
NS4B (781)	6104	C-T	–	–	–	–	–	–
NS5A (1343)	6947	G-A	–	–	7382	T-C	–	–
					7386	A-G	2349	T-A
					7575	G-A	2412	A-T
NS5B (1774)	7796	G-T	2485	Q-H	7698	C-A	2453	H-N
	8624	G-A	–	–	7707	C-A	2456	L-M
					8004	G-A	2555	D-N
					8054[b]	G-A	–	–
					9135	C-T	–	–
					9376	G-A	TGA	TAA
Total[c]	17/9376		9/3011		15/9363		9/3011	

AA, amino acid; nt, nucleotide.
[a] Silent marker mutation for RNA transcripts from cDNA clones containing additional 5' bases.
[b] Silent marker mutation for RNA transcripts from cDNA clones with 75-base (A_{8054}) vs 133-base (G_{8054}) poly(U/UC) tracts in 3'UTR (Kolykhalov et al. 1997).
[c] Totals indicate the number of changes observed for the number of nucleotides sequenced and do not include silent mutations.

and humans (Chang et al. 1997) and an association has been observed between the presence of peptide variants and CTL responses suggesting some form of selection may take place although early in infection (Chang et al. 1997).

Of the mutations detected in the viral RNA isolated at week 60, only further vaccine or neutralization experiments with single sequence viruses will determine directly which, if any, represent escape mutations and therefore which may have contributed towards persistent infection in these animals. However, if these mutations arose early in infection, shortly after the immune responses occurred, there is a strong possibility that they may represent escape mutants whereas if they arose late they are probably more a consequence of persistence rather than the cause.

Analysis of serum RNA from weeks 26 (1535) and 22 (1536) p.i. revealed that 55.6% and 44.4% of the mutations detected at week 60 were already present as the majority sequence at these times in Ch1535 (aa 371, 478, 766, 1533, and 1635) and Ch1536 (aa 265, 463, 1181, and 2453), respectively. These are regions that could be targets for antibody or CTL responses. None of these mutations, nor any of those detected at week 60 in either chimpanzee, were located in previously identified immune epitopes; however, they could still represent mutations that enabled the virus to evade the host response enough to remain in circulation. Therefore, although the persistence of HCV in these animals is clearly not due to escape mutations in HVR1, it may be caused by mutations in other regions of the genome that represent immune epitopes.

8 Conclusion

The results that we have obtained in the experiments with the full-length clone have allowed us to understand for the first time the evolution of this virus. The very rapid incorporation of mutations that have been reported in HCV infections and the large sequence diversity observed between isolates appears not to be due to an exceptionally high mutation rate, but to a high tolerance for sequence diversity which accumulates over long periods of time and must be passed from one host to another as the entire population of quasispecies. In addition, data collected from the two chimpanzees that became persistently infected following transfection with transcribed RNA suggest that the viral persistence cannot be explained solely by selection of viruses with alternate HVR1 sequences that do not cross-react immunologically with the original HVR1 sequence. These chimpanzees were both infected with virus of a uniform sequence; therefore, there was no diversity in the population to be selected and we have found no new mutations within HVR1 that escaped the antibody response to the original. While it remains possible that selection of HVR1 variants is part of the mechanism of persistence, in the case of these chimpanzees, there must have been some other means used by the virus to maintain itself in the host for more than 1 year. The mutations that did arise and persist in regions outside of HVR1 must have conferred some selective advantage over wild-type virus. It is possible that some of these mutations resulted in a virus or viruses that replicated better. It is also possible that these mutations also code for antigenic epitopes that are recognized by T cells and may constitute escape mutants. This possibility is being investigated.

While the chimpanzee may not be a perfect model for human hepatitis C, it resembles the infection in humans sufficiently to be useful for studies of pathogenesis, the immune response and chronicity. The program to develop a full-length, functional cDNA clone of the HCV genome would not have been possible without the availability of chimpanzees, and chimpanzees will undoubtedly play an important role in vaccine development.

References

Allander T, Beyene A, Jacobson SH, Grillner L, Persson MA (1997) Patients infected with the same hepatitis C virus strain display different kinetics of the isolate-specific antibody response. J Infect Dis 175:26–31

Alter HJ, Purcell RH, Holland PV, Popper H (1978) Transmissible agent in non-A, non-B hepatitis. Lancet i, 459–463

Barker L.F, Maynard JE, Purcell RH, Hoofnagle JH, Berquist KR, London WT (1975) Viral hepatitis, type B, in experimental animals. Am J Med Sci 270:189–195

Battegay M, Fikes J, Di BA, Wentworth PA, Sette A, Celis E, Ching WM, Grakoui A, Rice CM, Kurokohchi K, Berzofsky JA, Hoofnagle JH, Feinstone SM, Akatsuka T (1995) Patients with chronic hepatitis C have circulating cytotoxic T cells which recognize hepatitis C virus-encoded peptides binding to HLA-A2.1 molecules. J Virol 69:2462–2470

Bukh J, Kim JP, Govindarajan S, Apgar CL, Foung SK, Wages JJ, Yun AJ, Shapiro M, Emerson SU, Purcell RH (1998) Experimental infection of chimpanzees with hepatitis G virus and genetic analysis of the virus. J Infect Dis 177:855–862

Cerny A, McHutchison, JG, Pasquinelli C, Brown ME, Brothers MA, Grabscheid B, Fowler P, Houghton M, Chisari FV (1995) Cytotoxic T lymphocyte response to hepatitis C virus-derived peptides containing the HLA A2.1 binding motif. J Clin Invest 95:521–530

Chang KM, Rehermann B, McHutchison JG, Pasquinelli C, Southwood S, Sette A, Chisari FV (1997) Immunological significance of cytotoxic T lymphocyte epitope variants in patients chronically infected by the hepatitis C virus. J Clin Invest 100:2376–2385

Ching WM, Wychowski C, Beach MJ, Wang H, Davies CL, Carl M, Bradley DW, Alter HJ, Feinstone SM, Shih JW (1992) Interaction of immune sera with synthetic peptides corresponding to the structural protein region of hepatitis C virus. Proc Natl Acad Sci USA 89:3190–3194

Choo QL, Kuo G, Weiner AJ, Overby LR, Bradley DW, Houghton M (1989) Isolation of a cDNA clone derived from a blood-born non-A, non-B viral hepatitis genome. Science 244:59–362

Diepolder HM, Gerlach JT, Zachoval R, Hoffmann RM, Jung MC, Wierenga EA, Scholtz S, Santantonio T, Houghton M, Southwood S, Sette A, Pape GR (1997) Immunodominant CD4+ T-cell epitope within nonstructural protein 3 in acute hepatitis C virus infection. J Virol 71:6011–6019

Diepolder HM, Zachoval R, Hoffmann RM, Wierenga EA, Santantonio T, Jung MC, Eichenlaub D, Pape GR (1995) Possible mechanism involving T-lymphocyte response to non-structural protein 3 in viral clearance in acute hepatitis C virus infection. Lancet 346:1006–1007

Emerson SU, Lewis M, Govindarajan S, Shapiro M, Moskal T, Purcell RH (1992) cDNA clone of hepatitis A virus encoding a virulent virus: induction of viral hepatitis by direct nucleic acid transfection of marmosets. J Virol 66:6649–6654

Erickson AL, Houghton M, Choo QL, Weiner AJ, Ralston R, Muchmore E, Walker CM (1993) Hepatitis C virus-specific CTL responses in the liver of chimpanzees with acute and chronic hepatitis C. J Immunol 151:4189–4199

Farci P, Alter HJ, Govindarajan S, Wong DC, Engle R, Lesniewski R, Mushahwar IK, Desai SM, Miller RH, Ogata N, Purcell RH (1992) Lack of protective immunity against reinfection with hepatitis C virus. Science 258:140

Feinstone SM, Alter HJ, Dienes HP, Shimizu Y, Popper H, Blackmore D, Sly D, London WT, Purcell RH (1981) Non-A, non-B hepatitis in chimpanzees and marmosets. J Infect Dis 144:588–598

Ferrari C, Valli A, Galati L, Penna A, Scaccaglia P, Giuberti T, Schianchi C, Missale G, Marin MG, Fiaccadori F (1994) T-cell response to structural and nonstructural hepatitis C virus antigens in persistent and self-limited hepatitis C virus infections. Hepatology 19:286–295

Grakoui A, Wychowski C, Lin C, Feinstone SM, Rice CM (1993) Expression and identification of hepatitis C virus polyprotein cleavage products. J Virol 67:1385–1395

Hijikata M, Kato N, Ootsuyama Y, Nakagawa M, Ohkoshi S, Shimotohno K (1991a) Hypervariable regions in the putative glycoprotein of hepatitis C virus. Biochem Biophys Res Commun 175:220–228

Hijikata M, Kato N, Ootsuyama Y, Nakagawa M, Shimotohno K (1991b) Gene mapping of the putative structural region of the hepatitis C virus genome by in vitro processing analysis. Proc Natl Acad Sci USA 88:5547–5551

Inchauspé G, Zebedee, S, Lee DH, Sugitani M, Nasoff M, Prince AM (1991) Genomic structure of the human prototype strain H of hepatitis C virus: comparison with American and Japanese isolates. Proc Natl Acad Sci USA 88:10292–10296

Kaneko T, Moriyama T, Udaka K, Hiroishi K, Kita H, Okamoto H, Yagita H, Okumura K, Imawari M (1997) Impaired induction of cytotoxic T lymphocytes by antagonism of a weak agonist borne by a variant hepatitis C virus epitope. Eur J Immunol 27:1782–1787

Kaneko T, Nakamura I, Kita H, Hiroishi K, Moriyama T, Imawari M (1996) Three new cytotoxic T cell epitopes identified within the hepatitis C virus nucleoprotein. J Gen Virol 77:1305–1309

Kato N, Ootsuyama Y, Ohkoshi S, Nakazawa T, Sekiya H, Hijikata M, Shimotohno K (1992) Characterization of hypervariable regions in the putative envelope protein of hepatitis C virus. Biochem Biophys Res Commun 189, 119–127

Kato N, Ootsuyama Y, Sekiya H, Ohkoshi S, Nakazawa T, Hijikata M, Shimotohno K (1994) Genetic drift in hypervariable region 1 of the viral genome in persistent hepatitis C virus infection. J Virol 68:4776–4784

Kato N, Sekiya H, Ootsuyama Y, Nakazawa T, Hijikata M, Ohkoshi S, Shimotohno K (1993) Humoral immune response to hypervariable region 1 of the putative envelope glycoprotein (gp70) of hepatitis C virus. J Virol 67, 3923–3930

Kolykhalov AA, Agapov EV, Blight K, Mihalik K, Feinstone SM, Rice CM (1997) Transmission of hepatitis C by intrahepatic inoculation with transcribed RNA. Science 277:570–574

Kolykhalov AA, Feinstone SM, Rice CM (1996) Identification of a highly conserved sequence element at the 3' terminus of hepatitis C virus genome RNA. J Virol 70:3363–3371

Koziel MJ, Dudley D, Afdhal N, Choo QL, Houghton M, Ralston R, Walker BD (1993) Hepatitis C virus (HCV)-specific cytotoxic T lymphocytes recognize epitopes in the core and envelope proteins of HCV. J Virol 67:7522–7532

Koziel MJ, Dudley D, Wong JT, Dienstag J, Houghton M, Ralston R, Walker BD (1992) Intrahepatic cytotoxic T lymphocytes specific for hepatitis C virus in persons with chronic hepatitis. J Immunol 149:3339–3344

Koziel MJ, Wong DK, Dudley D, Houghton M, Walker BD (1997) Hepatitis C virus-specific cytolytic T lymphocyte and T helper cell responses in seronegative persons. J Infect Dis 176:–866

Kurokohchi K, Akatsuka T, Pendleton CD, Takamizawa, A, Nishioka M, Battegay M, Feinstone SM, Berzofsky JA (1996) Use of recombinant protein to identify a motif-negative human cytotoxic T-cell epitope presented by HLA-A2 in the hepatitis C virus NS3 region. J Virol 70:232–240

Kurosaki M, Enomoto N, Marumo F, Sato C (1993) Rapid sequence variation of the hypervariable region of hepatitis C virus during the course of chronic infection. Hepatology 18:1293–1299

Maynard JE, Lorenz D, Bradley DW, Feinstone SM, Krushak DH, Barker LF, Purcell RH (1975) Review of infectivity studies in nonhuman primates with virus-like particles associated with MS-1 hepatitis. Am J Med Sci 270:81–85

Mink MA, Benichou S, Madaule P, Tiollais P, Prince AM, Inchauspe G (1994) Characterization and mapping of a B-cell immunogenic domain in hepatitis C virus E2 glycoprotein using a yeast peptide library. Virology 200:246–255

Nelson DR, Marousis CG, Davis GL, Rice CM, Wong J, Houghton M, Lau JY (1997) The role of hepatitis C virus-specific cytotoxic T lymphocytes in chronic hepatitis C. J Immunol 158:1473–1481

Ogata, N, Alter HJ, Miller, RH, Purcell RH (1991) Nucleotide sequence and mutation rate of the H strain of hepatitis C virus. Proc Natl Acad Sci USA 88, 3392–3396

Prince AM, Brotman B, Huima T, Pascual D, Jaffery M, Inchauspé G (1992) Immunity in hepatitis C infection. J Infect Dis 165, 438–443

Ray R, Khanna A, Lagging LM, Meyer K, Choo QL, Ralston R, Houghton M, Becherer PR (1994) Peptide immunogen mimicry of putative E1 glycoprotein-specific epitopes in hepatitis C virus [published erratum appears in J Virol 1994 Sep;68(9):6136]. J Virol 68:4420–4426

Rehermann B, Chang KM, McHutchison JG, Kokka R, Houghton M, Chisari FV (1996) Quantitative analysis of the peripheral blood cytotoxic T lymphocyte response in patients with chronic hepatitis C virus infection. J Clin Invest 98:1432–1440

Rizzetto M, Canese MG, Gerin JL, London WT, Sly DL, Purcell RH (1980) Transmission of the hepatitis B virus-associated delta antigen to chimpanzees. J Infect Dis 141:590–602

Sekiya H, Kato N, Ootsuyama Y, Nakazawa T, Yamauchi K, Shimotohno K (1994) Genetic alterations of the putative envelope proteins encoding region of the hepatitis C virus in the progression to relapsed phase from acute hepatitis: humoral immune response to hypervariable region 1. Int J Cancer 57:664–670

Shimizu YK, Hijikata M, Iwamoto A, Alter HJ, Purcell RH, Yoshikura H (1994) Neutralizing antibodies against hepatitis C virus and the emergence of neutralization escape mutant viruses. J Virol 68, 1494–1500

Tanaka T, Kato N, Cho MJ, Shimotohno K (1995) A novel sequence found at the 3' terminus of hepatitis C virus genome. Biochem Biophys Res Commun 215:744–749

Tanaka T, Kato N, Cho MJ, Sugiyama K, Shimotohno K (1996) Structure of the 3' terminus of the hepatitis C virus genome. J Virol 70:3307–3312

Taniguchi S, Okamoto H, Sakamoto M, Kojima M, Tsuda F, Tanaka T, Munekata E, Muchmore EE, Peterson, DA, Mishiro S (1993) A structurally flexible and antigenically variable N-terminal domain of the hepatitis C virus E2/NS1 protein: implication for an escape from antibody. Virology 195: 297–301

Tsarev SA, Tsareva TS, Emerson SU, Kapikian AZ, Ticehurst J, London W, Purcell RH (1993) ELISA for antibody to hepatitis E virus (HEV) based on complete open-reading frame-2 protein expressed in insect cells: identification of HEV infection in primates. J Infect Dis 168:369–378

van Doorn LJ, Quint W, Tsiquaye K, Voermans J, Paelinck D, Kos T, Maertens G, Schellekens H, Murray K (1994) Longitudinal analysis of hepatitis C virus infection and genetic drift of the hypervariable region. J Infect Dis 169:1226–1235

Weiner, A, Erickson, AL, Kansopon, J, Crawford, K, Muchmore, E, Hughes, AL, Houghton M, Walker, C.M (1995) Persistent hepatitis C virus infection in a chimpanzee is associated with emergence of a cytotoxic T lymphocyte escape variant. Proc Natl Acad Sci USA 92:2755–2759

Weiner AJ, Brauer MJ, Rosenblatt J, Richman KH, Tung J, Crawford K, Bonino F, Saracco G, Choo QL, Houghton M (1991) Variable and hypervariable domains are found in the regions of HCV corresponding to the flavivirus envelope and NS1 proteins and the pestivirus envelope glycoproteins. Virology 180, 842–848

Weiner AJ, Geysen HM, Christopherson C, Hall JE, Mason TJ, Saracco G, Bonino F, Crawford K, Marion CD, Crawford KA, Brunetto M, Barr PJ, Miyamura T, McHutchinson J, Houghton M (1992) Evidence for immune selection of hepatitis C virus (HCV) putative envelope glycoprotein variants: Potential role in chronic HCV infections. Proc Natl Acad Sci USA 89:3468–3472

Yanagi M, Purcell RH, Emerson SU, Bukh J (1997) Transcripts from a single full-length cDNA clone of hepatitis C virus are infectious when directly transfected into the liver of a chimpanzee. Proc Natl Acad Sci USA 94:8738–8743

Yanagi M, Shapiro M, Emerson SU, Purcell RH, Bukh J (1998) Transcripts of a chimeric cDNA clone of hepatitis C virus genotype 1b are infectious in vivo. Virology 244:161–172

Yanagi M, St Claire M, Emerson SU, Purcell RH, Bukh J (1999) In vivo analysis of the 3' untranslated region of the hepatitis C virus after in vitro mutagenesis of an infectious cDNA clone. Proc Natl Acad Sci USA 96:2291–2295

Zibert A, Meisel H, Kraas W, Schulz A, Jung G, Roggendorf M (1997a) Early antibody response against hypervariable region 1 is associated with acute self-limiting infections of hepatitis C virus. Hepatology 25:1245–1249

Zibert A, Roggendorf M, Schreier E, Dudziak P (1997b) Characterization of antibody response to hepatitis C virus protein E2 and significance of hypervariable region 1-specific antibodies in viral neutralization. Arch Virol 142:523–534

Cell Mediated Immune Response to the Hepatitis C Virus

B. REHERMANN[1] and F.V. CHISARI[2]

1	Introduction	299
2	Components of the Antiviral Immune Response	300
2.1	CD4-Positive T Helper Cells	300
2.2	CD8-Positive Cytotoxic T Cells	302
3	Role of the Cellular Immune Response in Hepatitis C Virus Infection	302
3.1	Evidence for a Protective Role of the Cellular Immune Response: Contribution to Hepatitis C Virus Clearance	303
3.1.1	The Hepatitis C Virus-Specific T Helper Cell Response	303
3.1.2	The Hepatitis C Virus-Specific Cytotoxic T Cell Response	305
3.2	Evidence for a Pathogenic Role of the Cellular Immune Response: Contribution to Liver Cell Injury	309
3.2.1	Intrahepatic HCV Quasispecies and Antigen Expression	309
3.2.2	The Intrahepatic Inflammatory Infiltrate	309
3.2.3	Functional Capacities of Intrahepatic T Cells	310
4	Potential Mechanisms of Hepatitis C Virus Persistence	311
5	Outlook: Immunotherapy of Hepatitis C	315
References		317

1 Introduction

The hepatitis C virus (HCV) is a small, positive stranded RNA virus that causes an inflammatory disease of the liver. Parenteral transmission is common; however, because de novo hepatitis C virus (HCV) infection is usually clinically inapparent, in almost 40% of the patients neither the time of onset nor the route of infection are known. Although the immune response probably contributes to viral clearance in a minority of patients, the factors and mechanisms that allow the virus to circumvent the host's immune response and to persist in most patients are not completely understood.

---

[1] Liver Diseases Section, DDB, NIDDK, National Institutes of Health, Building 10, Room 9B16, 10 Center Drive MSC 1800, Bethesda, MD 20892-1800, USA
[2] Department of Molecular and Experimental Medicine, The Scripps Research Institute, 10550 North Torrey Pines Road, La Jolla, CA 92037, USA
This is manuscript number 11404-MEM from the Scripps Research Institute

Persistent HCV infection, which occurs in more than 70% of patients, is associated with the symptoms of chronic hepatitis and its complications: liver cirrhosis and hepatocellular carcinoma (CHOO et al. 1989; ALTER et al. 1992; SANSONO and DAMMACCO 1992). Due to the large number of chronically infected patients, extensive studies on the immunopathology of chronic hepatitis C have been performed. Collectively, these data imply that the immune response mediates chronic liver cell injury if it is not able to clear the infection.

Identification of the immunological correlates of viral clearance is, therefore, pivotal to developing vaccines to prevent infection and efficient therapies to cure patients with chronic hepatitis C.

2 Components of the Antiviral Immune Response

In the earliest phase of any viral infection, viral factors such as the size, route and genetic composition of the inoculum and the rate of viral replication determine the kinetics with which host cells become infected and an opposing immune response is induced.

The first line of defense is antigen non-specific and mediated by natural killer (NK) cells, neutrophils and macrophages (MORETTA et al. 1994). Simultaneously, or soon thereafter, virus-specific immunity is induced by professional antigen presenting cells that process and present viral antigens to T and B lymphocytes in the regional lymph nodes. Specifically, the interaction of virus specific T helper (Th) cells with B cells and cytotoxic T cells (CTL) forms the acquired antiviral immune response. This review will focus on the activation and function of HCV-specific helper and cytotoxic T cells.

2.1 CD4-Positive T Helper Cells

CD4-positive T helper cells are stimulated when they recognize viral peptides in the HLA class II binding groove of antigen presenting cells. Peptides presented on HLA class II molecules, ranging from about 10 to up to 25 amino acids long, are generated by proteolytic cleavage from larger viral proteins. Activated, virus-specific, CD4-positive T cells can stimulate antigen-nonspecific inflammatory cells (MOSMANN et al. 1986), but also regulate the antigen-specific immune response (VITETTA et al. 1989) by providing help to CD8-positive, cytotoxic T and B cells (DOHERTY et al. 1992). These effects are mediated by distinct cytokines (KIM et al. 1985; MOSMANN et al. 1986).

T helper 1 (Th1) cytokines, e.g. interleukin (IL)-2, interferon (IFN)-γ, and tumor necrosis factor (TNF)-α, stimulate a number of antiviral host defense mechanisms (CARDING et al. 1993; RAMSAY et al. 1993). They enhance immune recognition of viral antigens by induction of HLA-molecules on infected cells (FARRAR and SCHREIBER 1993) and by activation of CD8-positive T cells and NK cells (ZINKERNAGEL et al. 1993). In addition, they also exert antiviral effects by stimulation of nitric oxide (NO) production in macrophages (CROEN 1993; KARUPIAH et al. 1993; HARRIS et al. 1995; SHARARA et al. 1997) and hepatocytes (NUSSLER et al. 1992), a mechanism that can induce resistance against viral infection in neighboring cells (FARRAR and SCHREIBER 1993). In hepatitis B virus (HBV) infection, Th1 cytokines have also been described to inhibit viral replication and gene expression (GUIDOTTI et al. 1994a, 1996). In HIV infection, Th1 cytokines favor resolution of infection, while cytokines secreted by Th2 cells (IL-4, IL-5, IL-6, IL-9, IL-10 and IL-13) promote progression of disease. Th2 cytokines support the humoral immune response and induce activation and differentiation of B cells. They are therefore useful for the induction of protective antibodies against extracellular pathogens (URBAN et al. 1991), and may actually down-regulate the cytotoxic T cell response by inhibition of Th1 cytokine secretion, MHC expression and antigen presenting capacities of monocytes (MOSMANN and SUBASH 1996; ABBAS et al. 1996; NAGLER et al. 1988; MARTINEZ et al. 1990).

Identifying the factors that determine Th 1/2 differentiation in response to a viral infection is a major aspect of current research: Analysis of murine T cell responses has demonstrated that the T helper cell profile is determined by cytokines produced in the very early phase of infection (LE GROS et al. 1990; SWAIN et al. 1990; BETZ and FOX 1990; CHATELAIN et al. 1992; NAKAJIMA et al. 1992; SCOTT et al. 1989; GAZZINELLI et al. 1993; ABBAS et al. 1996; SYPEK et al. 1993) as well as specific HLA alleles (TITE et al. 1987; MUKRAY et al. 1989), costimulatory molecules (CAI and SPRENT 1996), and quantity and quality of antigen (THOMPSON 1995; LIEW 1990; EL GHAZALI 1993; PAUL and SEDER 1994). Peptide/MHC class II complexes that interact strongly with the T cell receptor (TCR) favor generation of Th1-like cells while those that bind weakly induce Th2-like cells (PFEIFFER et al. 1995). These studies also suggest that priming of the T cell response in lymph nodes by professional antigen presenting cells with a high density of MHC class II/peptide complexes favors Th1-like responses, while low ligand densities favor Th2 like responses. While the mechanisms underlying these observations are not completely understood, it is possible that different signal thresholds are needed for distinct cytokine gene expression. The strength of TCR activation can be determined by the time of (TCR) occupancy and by differential participation of costimulatory signals, since sustained signaling by serial engagement of many TCRs with subsequent TCR down-modulation is a critical aspect for complete T cell activation (VALITUTTI et al. 1995; LE GROS et al. 1990).

Which cell population provides the earliest and largest source of cytokines and thereby determines the differentiation of T helper cells and the quality of the immune response is not yet defined. A good candidate is an unusual CD4-positive T cell population that also expresses several NK cell markers and is restricted by nonclassical class I MHC molecules like CD1 (BENDELAC et al. 1995). These cells can very rapidly produce IL-2 and IL-4 (YOSHIMOTO and PAUL 1994; HAYAKAWA et al. 1992) that prime effector functions of CD8-positive T cells. NK1.1-positive cells have been shown to play a role as nonclassical Th cells in resistance against intracellular pathogens such as *Toxoplasma gondii* (DENKERS and SHER 1997), but their role in persistent viral infections such as hepatitis C has not been analyzed yet.

Finally, CD4-positive T cells have been reported to exert antiviral as well as cytotoxic functions (MINUTELLO et al. 1993; JACOBSON et al. 1984; FLEISCHER et al. 1985). Specifically, HBV envelope-specific Th1 cells can suppress HBV replication noncytolytically in vivo upon injection into HBV-transgenic mice (FRANCO et al. 1997a). They also kill hepatic nonparenchymal cells that phagocytose, process and present soluble HBs antigen and thereby contribute to the pathogenesis of inflammatory liver disease (FRANCO et al. 1997a). In vitro, HBV envelope-specific Th cells have been shown to kill antigen presenting CD8-positive T cells (FRANCO et al. 1992). This effect has not been demonstrated in vivo, yet might represent an important regulatory loop to down-regulate the immune response after the acute phase of disease.

2.2 CD8-Positive Cytotoxic T Cells

The immune response of CD8-positive T cells represents the main effector limb during viral infection. These cells recognize endogenously synthesized peptides that are generated in the cytosol of infected cells, transported across the membrane of the endoplasmatic reticulum by the heteromeric transporter complex TAP 1 and 2 and bound to newly assembled MHC class I molecules (GERMAIN, 1994). They are then presented in the antigen binding groove of HLA class I molecules on the surface of virus-infected cells. HLA class I-peptide interactions are allele-specific in that the peptide must be between eight and 11 amino acids long and contain an HLA allele-specific binding motif. Peptide-activated CD8-positive T cells then kill virus-infected cells by induction of apoptosis, mediated via the perforin/granzyme- and Fas-activated death pathways (BERKE 1995). In addition, CD8-positive T cells can also clear HBV (GUIDOTTI et al. 1994a,b; GILLES et al. 1992), cytomegalovirus (CMV; PAVIC et al. 1993) and rotavirus (FRANCO et al. 1997b) without killing the infected cells. These curative antiviral effects are mediated by T cell-derived cytokines, especially by IFN-γ and TNF-α. Other cytokines, such as Rantes and MIP-1α, MIP-1β, have been described as human immunodeficiency virus (HIV)-suppressive factors produced by CD8-positive T cells (COCCHI et al. 1995).

3 Role of the Cellular Immune Response in Hepatitis C Virus Infection

Similar to the inability of the cellular immune response to control HIV, the apparent resistance of HCV to immunological control is an area of great interest and importance. Research has been hampered by the fact that the acute stage of HCV infection is usually clinically inapparent and also because efficient in vitro models of infection do not exist. Of necessity, therefore, most human studies have concentrated on patients with chronic HCV infection. To date, these reports suggest that the HCV-specific immune response exerts some control over virus replication but is unable to resolve or to terminate persistent infection and chronic hepatitis in most cases.

3.1 Evidence for a Protective Role of the Cellular Immune Response: Contribution to Hepatitis C Virus Clearance

3.1.1 The Hepatitis C Virus-Specific T Helper Cell Response

Clinical observations suggest that viral and immunological events occurring during the initial few weeks after infection may determine the outcome of hepatitis C. For example, it has been reported that, when patients recover from acute HCV infection, they usually clear the virus within three months of clinical onset. It has also been shown that chronic evolution of infection is not a frequent outcome in patients who develop clinically evident, acute hepatitis (GIUBERTI et al. 1994).

Accordingly, analysis of the protective and/or the curative effects of the HCV-specific immune response should be most effectively pursued in patients with acute self-limited HCV infection or patients who have recovered from chronic HCV infection. Indeed, circulating, HCV-specific Th cells that proliferate after in vitro stimulation with recombinant HCV core and NS3 and NS4 antigens have been detected during the first few weeks of acute hepatitis C. As shown in Table 1, multiple peptides that are recognized by Th cells and include immunodominant epitopes have already been identified in HCV. The NS3 protein is of special importance since a CD4-positive T cell response against its protease and helicase domain is much stronger and more frequently found in patients who resolve acute hepatitis C than in patients who develop persistent infection (DIEPOLDER et al. 1995). Furthermore, a strong NS3-specific T cell response seems to be necessary for viral clearance (DIEPOLDER et al. 1995). NS3-specific Th cells recognize a short immunodominant region at amino acid position 1248–1261 with the putative minimal epitope at amino acid position 1251–1259. This epitope promiscuously binds to ten common HLA class II alleles with a high binding affinity and is recognized in the context of five different HLA alleles (DIEPOLDER et al. 1997). Moreover, its sequence is completely conserved within HCV 1a, 1b, 1c, 2a and 2b genotypes.

After recovery from acute hepatitis C, the NS-3 specific Th cell response can be maintained in the peripheral blood for years, suggesting the existence of a strong immunological memory. Whether NS3-specific T cells eradicate the HCV completely, or whether they control and restrict its replication to very low levels at immunoprivileged sites, is not known. The latter mechanism has been reported to occur after clearance of the HBV (REHERMANN et al. 1996c) and allows periodic de novo induction of virus-specific T cells to keep the virus in persistent subclinical latency.

Table 1. Immunodominant hepatitis C virus (HCV) sequences recognized by T helper cells

HCV Protein	Amino Acid Position	Immunogenic HCV Sequences	T Cell Compartment	MHC Restriction	Reference
Core	1–24		PBMC	n.k*	Woitas et al. 1997
Core	20–44		PBMC	n.k	Lechmann et al. 1996; Woitas et al. 1997
Core	23–42		PBMC	n.k	Hoffmann et al. 1995
Core	39–63		PBMC	DR4	Lechmann et al. 1996
Core	66–85		PBMC	n.k	Hoffmann et al. 1995
Core	73–92	GRTWAQPGYPWPLYGNEGCG	PBMC	n.k	Leroux-Roels et al. 1996
Core	79–103		PBMC	n.k	Lechmann et al. 1996; Woitas et al. 1997
Core	111–130	DPRRRSRNLGKVIDTFTCGL	PBMC	DRB1*08032	Kaneko et al. 1997
Core	128–152		PBMC	n.k	Lechmann et al. 1996; Woitas et al. 1997
Core	121–140	KVIDTLTCGFADLMGYIPLV	PBMC	n.k	Leroux-Roels et al. 1996
Core	131–150		PBMC	n.k	Hoffmann et al. 1995
Core	145–164	GGAARALAHGVRVLEDGVNY	PBMC	n.k	Leroux-Roels et al. 1996
Core	148–172		PBMC	DR11	Lechmann et al. 1996; Woitas et al. 1997
Core	157–176	VLEDGVNYATGNLPGCSFSI	PBMC	n.k	Leroux-Roels et al. 1996
Core	161–180	SYNYATGNLPGCSFSIFLLA	PBMC	DRB1*08032	Kaneko et al. 1996
Core	168–192		PBMC	n.k	Lechmann et al. 1996; Woitas et al. 1997
NS3	1251–1259	VLVLNPSVA	PBMC	DR4, DR 11, DR12, DR13, DR16	Diepolder et al. 1997
NS3	1388–1407		PBMC	DRB*1501, DRB5*0101	Diepolder et al. 1997
NS3	1450–1469		PBMC	DRB*1302	Diepolder et al. 1997

PBMC, peripheral blood mononuclear cell; n.k, not known.

Th cells mediating the NS3-specific immune response frequently display a Th1or Th0 cytokine profile while the activation of Th2 responses seems to be involved in the development of chronic hepatitis C. TSAI et al. (1997) reported recently that peripheral blood mononuclear cells (PBMCs) of 17 patients with self-limited acute hepatitis C displayed a Th1 phenotype upon stimulation with HCV antigens, while PBMC of 11 patients who developed chronic hepatitis were characterized by Th2 cytokine profiles and significantly lower proliferative T cells responses. In another study, the levels of circulating IL-2 IL-4, IL-10 and IFN-γ were significantly higher in patients with chronic hepatitis C than in uninfected controls (CACCIARELLI et al. 1996).

In chronic HCV infection, a concomitant immune defect has been associated with faster progression of hepatitis C (EYSTER et al. 1993; LIM et al. 1994; BJORO et al. 1994), implying a cytopathogenic role of HCV or lack of protection by the virus-specific immune response (MARTIN et al. 1989). For example, immunodeficiency worsened the outcome of HCV infection to the extent of subsequent liver failure (ROSSI et al. 1997), and several cases of fulminant hepatitis C have been observed on suspension of immunosuppressive therapy (FAN et al. 1991; GRUBER et al. 1993) or withdrawal of chemotherapy (VENTO et al. 1996).

Even in chronic hepatitis C, a strong Th cell response with Th1 phenotype seems to be associated with a more favorable, less inflammatory course of disease and thus seems to control viral load. Patients without any clinical or histological disease activity mount a Th1 cytokine response upon in vitro stimulation of PBMCs with HCV core antigen that is stronger than that of patients with more active disease (WOITAS et al. 1997). Accordingly, titers of non conformational antibodies are much lower and CD30 expression and IL-10 production by their PBMCs is found less frequently in patients without clinical or histological evidence of disease activity than in patients with chronic active disease (LECHMANN et al. 1996). These observations are in accordance with the notion that Th1 cytokine responses favor cellular and Th2 responses humoral immune reactions and that Th1 responses may be regarded as the more protective immune responses in HCV infection.

3.1.2 The Hepatitis C Virus-Specific Cytotoxic T Cell Response

Due to the small number of patients who present with acute hepatitis C and due to technical and practical limitations inherent in CTL analysis, the CTL response to HCV during acute hepatitis has not yet been characterized. Instead, the immune response of HCV-seronegative, healthy persons with frequent exposure to HCV was analyzed in order to identify potentially protective CTL responses. These persons were repetitively exposed to HCV either occupationally (KOZIEL et al. 1997) or via HCV-infected spouses (BRONOWICKI et al. 1997). These persons mounted HCV-specific helper and cytotoxic T cell responses that were found in the peripheral blood in the absence of either a humoral immune response or viremia. Interestingly, the Th cell response was predominantly targeted against the NS3 protein, as has also been shown in acute, self-limited hepatitis C (BRONOWICKI et al.

1997). A similar cellular immune response in the absence of a humoral response has also been reported in HIV-exposed but uninfected persons (CLERICI et al. 1992; ROWLAND-JONES et al. 1995), compatible with the hypothesis that T cell memory may be induced by periodic reintroduction of viral antigen by recurring, subclinical viral exposure (SCHUPPER et al. 1993).

Studies performed in humans (BATTEGAY et al. 1995; CERNY et al. 1995; KOZIEL et al. 1992, 1993, 1995) and chimpanzees (ERICKSON et al. 1993) chronically infected with HCV identified CTL epitopes in all viral proteins. A list of all epitopes known to date is given in Table 2. Most of these epitopes have been confirmed by independent studies, even across species barriers (BRUNA-ROMERO et al. 1997) and in HLA-A2.1 transgenic mice (SHIRAI et al. 1995; WENTWORTH et al. 1996). Supertype binding motifs seem to ensure that given epitopes can be presented by several rather than individual HLA haplotypes. In addition, many epitopes have been shown to be endogenously processed and immunodominant (KOZIEL et al. 1992; BRUNA-ROMERO et al. 1997).

Since repetitive in vitro stimulation with HCV-derived peptides is required to expand CTLs from the peripheral blood, the frequency of progenitor CTLs in the circulation is assumed to be rather low. Indeed, the frequency of CTL precursors targeted against individual epitopes has been determined to be in the range of 1 in 10^6 to 1 in 10^5 PBMCs (CERNY et al. 1995), which is lower than the frequency of CTLs targeting against HIV (5 in 10^4 PBMCs, CARMICHAEL et al. 1993) or HCMV (1 in 10^4 PBMCs to 5 in 10^5 PMBCs, BORYSIEWICZ et al. 1988) in patients infected with those viruses and against the influenza matrix epitope as a recall antigen (REHERMANN et al. 1996b). In contrast, HCV-specific CTLs must be present at a higher frequency in the liver, because antigen-nonspecific stimulation is sufficient to expand HCV-specific CTLs from liver biopsies (KOZIEL et al. 1992, 1993, 1995; NELSON et al. 1997). However, even the intrahepatic frequency of HCV-specific CTL precursors is much lower than the intrahepatic frequency of lymphocytic choriomeningitis virus (LCMV)-specific murine CTLs during self-limited LCMV infection (BYRNE and OLDSTONE 1984). On the other hand, the relatively low and weak HCV-specific CTL response is not due to generalized immunosuppression since the CTL response against influenza virus (REHERMANN et al. 1996b) and Epstein Barr virus (HIROISHI et al. 1997) is normal in HCV-infected patients.

At best, therefore, the HCV-specific CTL response may exert some control over viral load. Indeed, a stronger polyclonal CTL response in the peripheral blood (REHERMANN et al. 1996a) and the liver (NELSON et al. 1997) was associated with lower levels of HCV viremia. Similar observations have been made in long-term survivors of HIV infection (BORROW et al. 1994; KOUP et al. 1994; SCHRAGER et al. 1994), and HIV disease progression is associated with reduction of the virus-specific CTL response (CARMICHAEL et al. 1993) and increase of viral load (Ho et al. 1989). The HCV-specific CTL response may therefore be too weak to achieve viral clearance but strong enough to control viral load to some extent. It appears, however, that this occurs at the expense of an inflammatory liver disease which after many years can lead to irreversible liver cirrhosis and hepatocellular carcinoma.

Table 2. MHC class I restricted hepatitis C virus (HCV) epitopes

HCV protein	Amino Acid position	Immunogenic peptide/epitope	T Cell compartment	MHC restriction	Reference
Core	2–10	STNPKPQQKK	Liver	A11	KOZIEL et al. 1992
Core	28–36	GQIVGGVYL	PBMC	B60	KANEKO et al. 1996
Core	35–44	YLLPRRGPRL	PBMC	A2	BATTEGAY et al. 1995; CERNY et al. 1995; BÀR et al. 1997
Core	37–50	LPRRGPRLGVRATR	PBMC	H-2d	BRUNA-ROMERO et al. 1997
Core	41–49	GPRLGVRAT	Liver	B7	KOZIEL et al. 1995
Core	41–49	GPRLGVRAT	PBMC	B7	KOZIEL et al. 1997
Core	42–50	PRLGVRATR	PBMC	n.k	KOZIEL et al. 1997
Core	85–98	LYGNEGCGWAGWLL	PBMC	A2	BÀR et al. 1997
Core	88–97	NEGCGWAGWL	PBMC	B44	KITA et al. 1993
Core	127–140	TCGFADLMGYIPLV	PBMC	A2	BÀR et al. 1997
Core	131–140	ADLMGYIPLV	PBMC	A2	CERNY et al. 1995
Core	132–140	DLMGYIPLV	PBMC	A2	SHIRAI et al. 1994; WENTWORTH et al. 1996
Core	132–145	DLMGYIPLVGAPLG	PBMC	H-2d	BRUNA-ROMERO et al. 1997
Core	151–164	LAHGVRVLEDGVNY	PBMC	H-2d	BRUNA-ROMERO et al. 1997
Core	165–178	ATGNLPGCSFSIFL	PBMC	H-2d	BRUNA-ROMERO et al. 1997
Core	167–176	GNLPGCSFSI	PBMC	A2	BÀR et al. 1997
Core	273–286	SFSIFLLALLSCLT	PBMC	H-2d	BRUNA-ROMERO et al. 1997
Core	178–187	LLALLSCLTV	PBMC	A2	BATTEGAY et al. 1995; SHIRAI et al. 1995
Core	178–187	LLALLSCLTI	PBMC	A2	WENTWORTH et al. 1996
Core	181–190	LLSCLTVPAS	PBMC	A2	BÀR et al. 1997
Core	220–227	ILHTPGCV	PBMC	A2	SHIRAI et al. 1995
E1	233–241	GNASRCWVA	Liver	Patr-B16	KOWALSKI et al. 1996
E1	233–242	GNASRCWVAM	Liver	B35	KOZIEL et al. 1992
E1	234–242	NASRCWVAM	PBMC	B35	KOZIEL et al. 1997
E1	273–286	GHRMAWDM	PBMC	H-2d	BRUNA-ROMERO et al. 1997
E1	363–371	SMVGNWAKV	PBMC	A2	SHIRAI et al. 1995
E2	401–411	SLLAPGAKQNV	PBMC	A2	SHIRAI et al. 1995
E2	460–469	RPLTDFDQGW	Liver	B53	KOZIEL et al. 1995
E2	489–496	YPPKPCGI	Liver	B51	KOZIEL et al. 1992
E2	489–496	YPPKPCGI	PBMC	B51	KOZIEL et al. 1997
E2/NS1	542–550	TRPPLGNWF	Liver	Patr-B13	KOWALSKI et al. 1996
E2/NS1	569–578	CVIGGAGNNT	Liver	B50	KOZIEL et al. 1991

Table 2. (Cont.)

Region	Position	Sequence	Source	HLA	Reference
E2	588–596	KHPDATYSR	Liver	Patr-A04	Kowalski et al. 1996
E2	621–628	TINYTIFK	Liver	A11	Koziel et al. 1995
E2	621–628	TINYTIFK	PBMC	A11	Koziel et al. 1997
E2	711–718	SWAIKWEY	Liver	Patr-A11	Kowalski et al. 1996
E2	725–733	FLLLADARV	PBMC	A2	Wentworth et al. 1996
NS2	826–838	LMALTLSPYYKRY	Liver	A29	Koziel et al. 1992
NS2	838–845	YISWCMWW	Liver	A23	Koziel et al. 1995
NS3	1073–1081	CINGVCWTV	PBMC	A2	Cerny et al. 1995; Koziel et al. 1997
NS3	1073–1081	CINGVCWTV	Liver	A2	Koziel et al. 1995
NS3	1169–1177	LLCPAGHAV	PBMC	A2	Cerny et al. 1995
NS3	1287–1296	TGAPVTYSTY	PBMC	A2	Kurokohchi et al. 1996
NS3	1357–1365	VPHPNIEEV	Liver	Patr-B13	Kowalski et al. 1996
NS3	1395–1403	HSKKKCDEL	Liver	B8	Koziel et al. 1995
NS3	1395–1403	HSKKKCDEL	PBMC	B8	Koziel et al. 1997
NS3	1406–1415	KLVALGINAV	PBMC	A2	Cerny et al. 1995
NS3	1444–1452	YTGDFDSVI	Liver	Patr-B01	Kowalski et al. 1996
NS3	1446–1454	GDFDSVIDC	Liver	Patr-B16	Erickson et al. 1993; Kowalski et al. 1996
NS4A	1585–1593	YLVAYQATV	PBMC	A2	Wentworth et al. 1996
NS4A	1666–1675	VLAALAAYCL	PBMC	A2	Wentworth et al. 1996
NS4B	1769–1777	HMWNFISGI	PMBC	A2	Wentworth et al. 1996
NS4B	1789–1797	SLMAFTAAV	PBMC	A2	Cerny et al. 1995
NS4B	1807–1816	LLFNILGGWV	PBMC	A2	Battegay et al. 1995; Cerny et al. 1995; Wentworth et al. 1996
NS4B	1851–1859	ILAGYGAGV	PMBC	A2	Wentworth et al. 1996
NS5A	2252–2260	ILDSFDPLV	PBMC	A2	Cerny et al. 1995
NS5B	2588–2596	RVCEKMALYDV	Liver	A3	Koziel et al. 1995
NS5B	2727–2735	GLQDCTMLV	PBMC	A2	Battegay et al. 1995

3.2 Evidence for a Pathogenic Role of the Cellular Immune Response: Contribution to Liver Cell Injury

3.2.1 Intrahepatic HCV Quasispecies and Antigen Expression

It is interesting that viral quasispecies as well as T cell subpopulations differ profoundly in the liver and the peripheral blood. In the livers of patients with chronic hepatitis C, approximately 5% of the hepatocytes (range 0%–35%) harbor HCV RNA in their cytoplasm, as detected by in-situ RT-PCR (LAU et al. 1996). Even fewer hepatocytes (1%–10%) express HCV core, envelope or NS3 proteins, as detected by immunohistochemical analysis, and they are sparsely distributed throughout the liver (HIRAMATSU et al. 1994).

While there is a quantitative correlation between HCV RNA content in the liver and the blood (LAU et al. 1996), sequence analysis reveals qualitative differences between HCV quasispecies isolated from these two compartments (CABOT et al. 1997; MAGGI et al. 1997; FUJII et al. 1996; SAITO et al. 1996). Liver-derived HCV quasispecies display more synonymous sequence variations, while HCV quasispecies circulating in the peripheral blood are characterized by a narrower sequence spectrum with randomly distributed silent mutations. Hence, a more ancient origin has been ascribed to intrahepatic viral populations, a more recent origin to those from the blood (CABOT et al. 1997).

Despite these qualitative differences, it has been proposed that extrahepatic replication contributes little to the total virus pool and that the liver seems to be the primary site of HCV replication. Indeed, when techniques such as tagged PCR are employed to avoid mispriming, negative strand RNA is only detected in the liver, not in PBMCs (LANFORD et al. 1995). In the periphery, quantitative analysis of serum HCV RNA levels after liver transplantation suggests that the life span of HCV in the circulation may be as short as 4h (ZEUZEM et al. 1996). In the liver, the majority of circulating viruses are produced through continuous rounds of de novo infection and rapid turnover of cells, not by release from chronically infected cells (FUKUMOTO et al. 1996). While some hepatocytes may replicate HCV quasispecies at low levels, others may be recently infected by highly replicative quasispecies, attracting more CTLs and leading to rapid hepatocyte turnover. Obviously, differences in HCV quasispecies lead to differences in the quality and display of HCV antigens. This may be immunologically relevant, since most T cells are induced at extrahepatic sites, i.e., in the lymph nodes, and may therefore not have the appropriate TCR specificity to recognize and eliminate intrahepatic HCV populations efficiently.

3.2.2 The Intrahepatic Inflammatory Infiltrate

Little is known about antigen-nonspecific effector mechanisms that might control HCV infection. Expression of nitric oxide synthase (NOS), which is inducible by T cell-derived cytokines (NATHAN and XIE 1994), is increased in the liver of many HCV-infected patients, correlating with serum alanine aminotransferase levels

(KANE et al. 1997). In HCV-infected livers, the cellular infiltrate is principally localized in areas of focal or piecemeal necrosis in the periportal areas (ONJI et al. 1992; GONZALEZ-PERALTA et al. 1994). MHC class I and ICAM-1 expression is enhanced in hepatocytes and bile duct cells (BALLARDINI et al. 1995b; BARBATIS et al. 1981), correlating with the severity of liver disease (MOSNIER et al. 1994). While CD68-positive macrophages/monocytes are increased in more aggressive forms of hepatitis C, T cells still represent the predominate cell type in HCV-infected livers. Lymphocytes in the portal areas belong to the CD4-positive Th cell population (WEINER et al. 1992; WEYSTAL et al. 1992), whereas lymphocytes in hepatic lobules are mainly CD8-positive suppressor or cytotoxic T cells (KOZIEL et al. 1992; YUK et al. 1986). Their number correlates with serum alanine aminotransferase levels and the histologic grade of the liver, suggesting that these cells promote liver injury.

CD8-positive T cells belong to the CD45RO-positive memory subset (IMADA et al. 1997) and express Fas-ligand (HIRAMATSU et al. 1994; OHISHI et al. 1995). Since hepatocyte expression of Fas, a mediator of apoptosis (HIRAMATSU et al. 1994) is up-regulated, especially near liver-infiltrating lymphocytes at the advancing edges of piecemeal necrosis (HIRAMATSU et al. 1994), and since inflammation in portal and periportal areas is more severe in Fas antigen-positive than in Fas antigen-negative samples (HIRAMATSU et al. 1994), apoptotic cell death probably occurs. HCV-infected hepatocytes are killed by HCV-specific CTL clones via Fas ligand, TNF-α and perforin-based mechanisms. Uninfected, neighboring hepatocytes (bystander cells) can also be killed by antigen-specific CTLs via Fas-Fas ligand interaction, while perforin-mediated bystander lysis is less efficient (ANDO et al. 1997). Finally, expression of membrane bound (KINKHABWALA et al. 1990) and soluble TNF-α (CUTURI et al. 1987; SUNG et al. 1988; STEFFEN et al. 1988) may lead to the death of HCV-infected hepatocytes with altered RNA or protein synthesis: Interestingly, the latter mechanism does not require close cell-cell contact and may, therefore, play a role in the restriction of virus spread.

CD4-positive T cells of the CD45RA-positive, naive subset are found close to follicular centers that consist of CD20-positive B cells and may function as lymphoid follicles in which presentation of HCV antigens to naive B lymphocytes occurs (IMADA et al. 1997). Intrahepatic T helper cells mainly secrete Th1 cytokines. Accordingly, IFN-γ and IL-2 levels are significantly increased in chronic hepatitis C as compared to uninfected controls, correlating with the severity of both the inflammatory and fibrotic components of hepatitis. In contrast, expression of IL-10, a cytokine that down-regulates the Th1 response, is decreased (NAPOLI et al. 1996).

3.2.3 Functional Capacities of Intrahepatic T Cells

Most of the intrahepatic T cells are not HCV-specific (CARDING et al. 1993; BERTOLETTI et al. 1997), because antigen-nonspecific cloning of activated, CD69- or HLA class II-expressing memory T cells yields cytokine producing, but not HCV-specific, T cell clones (BERTOLETTI et al. 1997). Nevertheless, intrahepatic compartmentalization of T cells specific for certain HCV epitopes seems to occur.

MINUTELLO et al. (1993) reported selective TCR expression in the liver that was not found in the peripheral blood compartment. Though the overall TCR-Vβ expression of intrahepatic T cells was polyclonal with more than 13 different Vβ regions in the reported patient, the intrahepatic T cells were predominantly Vβ5.1-positive (KASHII et al. 1997). Moreover, the percentage of intrahepatic T cells with restricted TCR Vβ5.1 usage was higher than that in peripheral blood lymphocytes in chronic hepatitis C and higher than in intrahepatic lymphocytes in chronic hepatitis B (KASHII et al. 1997), suggesting that their accumulation was virus- as well as organ-specific.

These Vβ5.1-positive T cells appear to be relevant for the maintenance of a local inflammatory process in response to a specific HCV-derived antigen. First, they accumulated in the portal areas where inflammatory reactions occur. Second, CD8 and CD45RO were predominantly expressed on these cells. Third, a distinct CDR3 sequence encoded by the V-(D)-J junctional region of the β chain of these T cells was identified. While the germline-encoded CDR1 and CDR2 loops bind the α-helices of the MHC molecule, the CDR3 loop binds the presented peptide (DAVIS and BJORKMAN 1988; CHOTHIA et al. 1988; BJORKMAN et al. 1987; JORGENSEN et al. 1992) and its sequence determines the antigen specificity of the TCR (DAVIS and BJORKMAN 1988). Therefore, these T cells may all have been specific for a common immunodominant HCV antigen.

Similar observations have been reported for HCV-specific Th cells: the specificity of intrahepatic Th cells was focused on the HCV-NS4 protein and the clonotypic TCR of these cells was not found in the PBMC fraction (MINUTELLO et al. 1993). Functionally, these intrahepatic T cells were ten times more efficient in providing help for polyclonal immunoglobulin (Ig)A production by B cells than PBMC-derived T cell clones of the same NS4 specificity (MINUTELLO et al. 1993). This compartmentalization at the site of disease may explain why continuous antibody production can be observed in the absence of detectable HCV-specific Th cell responses in the peripheral blood. It also emphasizes that virological and immunological studies performed solely on peripheral blood samples should be interpreted with caution, since there is compelling evidence that viral quasispecies as well as T cell subpopulation are different between peripheral blood and liver.

4 Potential Mechanisms of Hepatitis C Virus Persistence

The mechanisms whereby HCV circumvents the immune response, persists and causes chronic inflammatory liver disease are currently undefined.

One hypothesis is that the HCV-specific immune response is quantitatively and/or qualitatively inadequate to eliminate the large number of infected cells in an organ as large as the liver. The intensity of the primary cellular immune response depends on antigenic load, costimulatory signals, the type of the antigen presenting cell and on differentiation and cytokine profile of Th cells, and each of these factors may not be optimal in acute HCV infection. The load of viral antigen depends on HCV replication, viremia and consequently, HCV protein expression. Indeed, display of HCV antigens is relatively low, in contrast to HBV which is usually cleared following acute infection of adults. With small doses of antigen, T cell activation and proliferation are generally suboptimal (CAI and SPRENT 1996). In the case of HCV, this may be reflected by the relatively low number of HCV-specific cytotoxic T cells in the peripheral blood of chronically infected patients. Importantly, there is a hierarchy of T cell effector functions with respect to the antigen concentration needed to elicit the corresponding responses. Low peptide concentrations are sufficient to stimulate the killing activity of CTLs. This, however, does not necessarily reflect full T cell activation (VALITUTTI et al. 1996). Higher peptide concentrations resulting in the triggering of at least 20%–50% of the TCRs are required to elicit calcium influx, IFN-γ production and IL-2 responsiveness of the T cells. Finally, only the highest peptide concentration can induce the proliferation needed to amplify virus-specific T cells.

Small doses of viral antigens may therefore preclude strong HCV-specific proliferative T cell responses due to limited IL-2 production. Although several mechanisms exist that may potentially compensate for low antigen doses and enhance antiviral immune responses, they may fail in infections with hepatotropic viruses such as HCV. For example, costimulatory signals, such as CD28-B7 interaction or engagement of CD4, CD8 coreceptors, promote the contact between TCRs and peptide-MHC complexes. They also enhance signaling by increasing TCR contact with tyrosine kinases such as p56lck (MICELI and PARNES 1993). Costimulatory molecules, however, are predominantly expressed by professional antigen presenting cells in the lymph nodes and not by hepatocytes, the preferred cell type for HCV replication. In addition, limited IL-2 production due to insufficient T cell stimulation can be overcome by Th cell-derived cytokines (CAI and SPRENT 1994). Indeed, all studies to date suggest that early Th cell activation does play an important role and may be the crucial determinant for the outcome of acute HCV infection. Whether the appropriate cytokine profile is present at the site of inflammation, however, is not known since the onset of infection is most often clinically inapparent, and for that reason, the intrahepatic Th cell response has not been studied in acute HCV infection. As has been shown in a TCR-transgenic mouse model, not only the number but also the nature of peptide-MHC complexes affects the outcome of the Th subset response. Specifically, low-affinity peptides induce primarily Th1 responses while high-affinity peptides induce Th 2 responses as well as a higher degree of peripheral programmed cell death (PEARSON et al. 1997). The sum of all immunogenic high-and low-affinity peptides generated in HCV infection may therefore influence the quality and strength of the T cell response.

Another hypothetical explanation for HCV persistence, sequence variation due to the quasispecies nature and the high mutation rate of HCV, has often been discussed. Amino acid changes in immunodominant epitopes may permit HCV to escape from the antiviral immune response. Indeed, evidence for immune selection pressure has been obtained from several clinical studies. In immunocompromised patients the viral nucleotide sequence variability in the 27 amino acid hypervariable region 1 (HVR1) of HCV was markedly lower than in immunocompetent HCV-positive patients suggesting that immune selection, primarily by HVR1-specific antibodies, influences sequence variation in this region (ODEBERG et al. 1997). The cellular immune response may also select HVR1 variants, since in the same study amino acid sequence changes were also found in agammaglobulinemic patients without HCV-specific antibody response (ODEBERG et al. 1997). The most convincing evidence of this phenomenon is the lack of immune protection and the infectibility of chimpanzees rechallenged with the same HCV inoculum (FARCI et al. 1992; PRINCE et al. 1992). In addition to lack of protection by the humoral immune response, there is also evidence that the cellular immune response may be subverted during HCV infection, since in subsequent experiments WEINER et al. (1995) have described the emergence of an HCV mutant that was able to escape the HCV-specific CTL response in an infected chimpanzee.

Far more potent than viral variants that escape CTL recognition are viral variants that antagonize the induction of CTLs. In another antigen system, simultaneous presentation of both the antagonist and the agonist peptide on the same antigen presenting cell has been shown to block TCR internalization, thereby preventing the multiple TCR engagements necessary to reach the threshold to activate wild-type specific T cells (LANZAVECCHIA 1997). Such variants would suppress the immune response against both the original and the variant epitope. Only T cells that are not cross-reactive with the variant and not susceptible to antagonism would be stimulated to proliferate. T cells that are cross-reactive with the variant, however, would not be able to reach the activation threshold for proliferation. This explains why T cells cannot expand in the presence of minute amounts of a peptide antagonist with a high TCR binding affinity. The remaining immune response to the wild-type epitope would be significantly weakened and focused on cells infected exclusively with the original virus.

Indeed, CTL escape variants have been demonstrated in many viral infections such as HBV, HTLV and HIV (BERTOLETTI et al. 1994; KLENERMAN et al. 1994; NIEWIESK et al. 1995; BORROW et al. 1997; GOULDER et al. 1997). This phenomenon has recently also been described in patients with chronic hepatitis C (CHANG et al. 1997; KANEKO et al. 1997). In the largest study published on this topic, HCV sequence analysis was performed for ten CTL epitopes in 13 patients (CHANG et al. 1997). Sequence variations in CTL epitopes were significantly more frequently found in the presence than in the absence of a CTL response. In most cases, sequence variations resulted in nonimmunogenic, non-cross-reactive variant peptides of low binding affinity, compatible with non-recognition of the given peptide in the variant virus. In one case, amino acid changes in an immunodominant NS3 epitope yielded an antagonist peptide that was able to inhibit recognition of both wild-type and variant virus. Since the CTL response was multispecifically targeted against several other epitopes in this patient, the clinical course did not change in the study period of 10 months during which this variant emerged.

Apart from a complete abrogation of immune recognition of individual epitopes, viral variants may also alter the immune response in a more subtle way. Since most studies of the HCV-specific immune response rely on in vitro proliferation and cytotoxicity assays as detection system, partial signaling patterns that may be induced by viral partial agonists/antagonists are frequently not analyzed. Partial agonists can induce T cell responses that consist of cytokine production (EVAVOLD and ALLEN 1991), increase in cell size and cell surface expression of cytokine receptors (SLOAN-LANCASTER et al. 1993), or down-modulation of TCR expression (VALITUTTI et al. 1995) in the absence of a proliferative or cytotoxic response. Partial agonists may also induce anergic T cells (SLOAN-LANCASTER et al. 1993) that fail to induce IL-2 mRNA transcription and IL-2 secretion after TCR stimulation and, thus, fail to proliferate. This mechanism may be relevant for the persistence of HCV not only because of the large spectrum of quasispecies and variation of T cell epitopes, but also because anergy can be induced by insufficient costimulation through B7-CD28 (MÜLLER and JENKINS 1995), as would be the case with nonprofessional antigen presenting cells, such as hepatocytes (SLOAN-LANCASTER et al. 1993, 1994). As outlined previously, these mechanisms may weaken the immune response against individual, possibly immunodominat, epitopes sufficiently to prevent complete eradication of HCV. Simultaneously, however, chronic inflammatory liver injury may still progress since even anergic T cells can restore certain effector functions if exogenous IL-2 is provided by T cells of different specificity. Certainly, this can be the case with an immune response as multispecific as in HCV infection.

Nevertheless, it is not known whether viral escape plays a role during acute infection. Some authors argue that the emergence of quasispecies and even viral antagonist or partial agonist epitopes may be the consequence rather than the cause of a persistent infection, since only a prolonged time of viral persistence allows generation and selection of viral variants (CHISARI and FERRARI 1995; GUIDOTTI and CHISARI 1996; OLDSTONE 1996). As far as HCV is concerned, there is only a single report demonstrating that CTL escape variants develop during acute HCV infection of a chimpanzee and may contribute to a chronic outcome of infection (WEINER et al. 1995). In order to evaluate the significance of this mechanism for the high percentage of chronicity after acute HCV infection in humans, prospective analysis of viral nucleotide sequences in the context of the immune response will be necessary during the early phases of infection.

Theoretically, HCV could also interfere with antigen processing or presentation by the hepatocyte. It may diminish its visibility to the immune system sufficiently to permit chronic low grade inflammation, but not enough for the immune response to kill all of the infected cells. Other viruses have been shown to interfere with antigen processing. For example, proteins encoded by the HCMV genes US11 and US12 can dislocate newly synthesized class I heavy chains from the endoplasmic reticulum to the cytosol (MACHOLD et al. 1997). The effect of coexpression of HCV proteins on the presentation of an independent antigen to the corresponding CTL will be needed to test this hypothesis.

Finally, it is possible that HCV, unlike other viruses such as HBV (GUIDOTTI et al. 1996), CMV (PAVIC et al. 1993), HIV (COCCHI et al. 1995) or rotaviruses (FRANCO et al. 1997b) may not be susceptible to control by CTL-derived cytokines.

In HBV infection, these cytokines have been shown to inhibit viral gene expression and replication very efficiently (GUIDOTTI et al. 1994a,b, 1996) and to clear hepatocytes of the infecting virus without causing liver disease. It is possible that HCV is less susceptible to these cytokines to permit viral clearance, but that the CTLs are still capable of contributing to or even causing liver disease. If the virus is naturally resistant to the antiviral effects of T cell-derived cytokines (or if resistance is acquired by selection), viral clearance would depend entirely on the destruction of infected hepatocytes by the immune response. Because direct CTL lysis is an inherently inefficient process, requiring direct cell-cell contact, if this is the only mechanism available to eradicate an overwhelming number of infected cells in a large organ such as the liver, one would expect the process to be incomplete, leading to the destruction of some but not all of the infected hepatocytes, i.e., chronic hepatitis.

5 Outlook: Immunotherapy of Hepatitis C

Understanding the correlates of protective immunity in HCV infection is a necessary prerequisite to the development of vaccines or immunotherapeutic strategies to induce and enhance the HCV-specific T cell response.

To date, IFN-α therapy, recently supplemented by new antiviral agents, is still the only available therapy for patients with chronic hepatitis C. While as many as 50% of treated patients respond with an initial decrease of serum alanine aminotransferase activity as well as HCV RNA levels, less than 10% enjoy a sustained serological, virological and clinical response. Evidence that IFN effects may partly be based on an activation of the immune response is given by the observation that in vitro proliferation of T cells stimulated with HCV core antigen increased. Simultaneously, an enhanced IFN-γ production (IWATA et al. 1995) and decrease in type 2 cytokine responses (CACCIARELLI et al. 1996) was observed. Virus-specific CTL responses have been demonstrated in the blood of IFN-responders that cleared the HBV virus (REHERMANN et al. 1996d). With respect to HCV infection, the analysis is more difficult since peripheral blood CTL responses seem to decrease, rather than to be enhanced by IFN therapy, and accordingly most patients fail to produce a complete serological and virological response (REHERMANN et al. 1996a).

Several groups have since tried to selectively expand T cells with defined antigen specificity in the blood of patients with viral hepatitis. The earliest and to date, most comprehensive studies have been performed in hepatitis B. Based on these studies, and on the observation that peptides, administered in adjuvants, liposomes or via direct attachment to lipids, may be powerful immunogens, a lipopeptide-based vaccine has been developed that contains an HLA-A2-restricted, HBV CTL epitope linked to a universal Th epitope derived from tetanus toxoid and two molecules of palmitic acid. Booster-vaccination of HLA-A2-positive normal volunteers in a phase I clinical trial induced CTL specific for the HBV-derived, HLA-A2-restricted epitope (VITIELLO et al. 1995). These CTL were not only able to recognize and kill peptide-loaded target cells, but also target cells that produced the corresponding HBV protein endogenously, indicating that they may be functional and relevant during in vivo HBV infection (VITIELLO et al. 1995). CTL responses were of similar magnitude as observed in patients who successfully cleared HBV after acute infection, comparable to the strength of memory responses against influenza virus and were detectable in the peripheral blood for at least 9 months (LIVINGSTON et al. 1997). Based on these studies, compound peptides consisting of virus-derived CTL epitopes linked to several immunodominant Th epitopes have also been used to induce HIV-specific CTL responses (SHIRAI et al. 1994). The combination of several immunodominant epitopes is a crucial factor to be considered, since the adoptive transfer of a CTL clone of only one antigenic specificity to an HIV-infected patient resulted in the selection of mutant HIV variants and fatal disease progression (KOENIG et al. 1995).

Endogenous protein processing and expression of all relevant, immunodominant T cell epitopes in a given viral protein can also be induced by intramuscular immunization with DNA expression vectors. Those vectors encode the sequence of the desired immunogenic protein under the control of an appropriate promoter, that leads to the production of recombinant proteins in the muscle which are delivered to bone marrow-derived, antigen presenting cells (DOE et al. 1996) and induce a strong immune response at the T and B cell level in mice. HCV plasmid constructs expressing the HCV core and NS4 protein have been used to generate strong CTL activity in vivo as well as in vitro (TOKUSHIGE et al. 1996). Plasmids coexpressing HCV antigens with cytokines have been demonstrated to further modulate the HCV-specific immune response. For example, HCV core and IL-2/granulocyte-macrophage colony-stimulating factor (GM-CSF)-encoding plasmids have already been shown to increase B and T cell activation in the mouse, while IL-4-producing constructs induced differentiation of Th cells toward a Th0 subtype and suppressed HCV core-specific CTL activity (GEISSLER et al. 1997). If this vaccination strategy could be applied to humans, it could replace many of the current protein vaccines used against a large variety of infectious agents, since DNA vaccines would be more powerful, less expensive and easier to produce and to handle.

If the weakness of the antiviral immune response is indeed the reason for HCV persistence, specific enhancement of the immune response to HCV could lead to viral clearance, hopefully without causing severe liver immunopathology. Clearly, more research is required to dissect the various virus-host mechanisms that contribute to HCV persistence and chronic liver disease. Thus far, these studies have been hampered by the absence of appropriate systems to study HCV replication in vitro. Similarly, animal models to study infection with defined HCV inoculates of a homogenous, known sequence are also not available. The recent identification of the complete HCV sequence and the production of stable infectious molecular clones, coupled with transgenic mouse technology, may now provide the necessary tools to perform these analyses (KOLYKHALOV et al. 1996, 1997; YANAGI et al. 1997).

References

Abbas AK, Murphy KM, Sher A (1996) Functional diversity of helper T lymphocytes Nature 383:787–793
Alter MJ, Margolis HS, Krawczynski K, Judson FN, Mares A, Alexander WJ, Hu PY, Miller JK, Gerber MA, Sampliner RE, Meeks EL, Beach MJ (1992) The natural history of community-acquired hepatitis C in the United States. N Engl J Med 327:1899–1905
Ando K, Hiroishi K, Kaneko T, Moriyama T, Muto Y, Kayagaki N, Yagita H, Okumura K, Imawari M (1997) Perforin, fas/fas ligand, and TNF-alpha pathways as specific and bystander killing mechanisms of hepatitis C virus-specific human CTL. J Immunol 158:5283–5291
Bär S, Langhans B, Lechmann M, Ihlenfeldt HG, Jung G, Sauerbruch T, Spengler U (1997) Limited repertoire of HCV core CTL epitopes in the caucasoid population? J Hepatol 26 [Suppl 1]:111
Bachmaier K, Pummerer C, Shahinian A, Ionescu J, Neu N, Mak TW, Penninger JM (1996) Induction of autoimmunity in absence of CD28 costimulation. J Immunol 157:1752–1757
Ballardini G, Groff P, Pontisso P, Giostra F, Francesconi R, Lenzi M, Zauli D, Albeti A, Bianchi FB (1995a) Hepatitis C virus (HCV) genotype, tissue HCV antigens, hepatocellular expression of HLA-A, B, C, and intercellular adhesion-1 molecules. J Clin Invest 95:2967–2975
Ballardini G, Groff P, Pontisso P, Gistora F, Francesconi R, Lenzi M, Zauli D, Albeti A, Bianchi FB (1995b) Hepatitis C virus (HCV) genotype, tissue HCV antigens, hepatocellular expression of HLA-A, B, C and intercellular adhesion-1 molecules. J Clin Invest 95:2067–2075
Barbatis C, Woods J, Morton JA, Fleming KA, McMichael A, McGee JO (1981) Immunohistchemical analysis of HLA (A, B, C) antigens in liver disease using a monoclonal antibody. Gut 22:985–991
Battegay M, Fikes J, DiBisceglie AM, Wentworth PA, Sette A, Celis E, Ching W-M, Grakoui A, Rice CM, Kurokochi K, Berzofsky JA, Hoofnagle JH, Feinstone SM, Akatsuka T (1995) Patients with chronic hepatitis C have circulating cytotoxic T cells which recognize hepatitis C virus-encoded peptides binding to HLA-A2.1 molecules. J Virol 49:2462–2470
Bendelac A, Lantz O, Quimby ME, Yewdell JW, Bennink JR, Brutkiewicz RR (1995) CD1 recognition by mouse NK1+ T lymphocytes. Science 268:863–865
Berke G (1995) The CTL's kiss of death. Cell 81:9–12
Bertoletti A, D'Elios MM, Boni C, De Carli M, Zignego AL, Durazzo M, Missale G, Penna A, Fiaccadori F, Del Prete G, Ferrari C (1997) Different cytokine profiles of intrahepatic T cells in chronic hepatitis B and hepatitis C virus infections. Gastroenterology 112:193–199
Bertoletti A, Sette A, Chisari FV, Penna A, Levrero M, DeCarli M, Fiaccadori F, Ferrari C (1994) Natural variants of cytotoxic epitopes are T cell receptor antagonists for antiviral cytotoxic T cells. Nature 369:407–410
Betz M, Fox BS (1990) Regulation and development of cytochrome c-specific IL-4 producing T cells. J Immunol 145:1046–1052
Bjorkman PJ, Saper MA, Samraoui B, Bennett WS, Strominger JL, Wiley DC (1987) The foreign antigen binding-site and T cell recognition regions of class-I histocompatibility antigens. Nature 329: 512–518
Bjoro K, Froland SS, Yun Z, Samdal HH, Haaland T (1994) Hepatitis C in primary hypogammaglobulinemia. J Hepatol [Suppl] 1:21-S25
Borrow P, Lewicki H, Hahn BH, Shaw GM, Oldstone MB (1994) Virus-specific CD8+ cytotoxic T-lymphocyte activity associated with control of viremia in primary human immunodeficiency virus type 1 infection. J Virol 68:6103–6110
Borrow P, Lewicki H, Wie X, Horwity MS, Pfeffer N, Meyers H, Nelson JA, Gairin JE, Hahn BH, Oldstone MBA, Shaw GM (1997) Antiviral presssure exerted by HIV-1-specific cytotoxic T lymphocytes (CTLs) during primary infection demonstrated by rapid selection of CTL escape virus. Nat Med 3:212–217
Borysiewicz LK, Graham S, Hickling JK, Mason PD, Sissons (1988) Human cytomegalovirus-specific cytotoxic T lymphocytes: their precursor frequency and stage specificity. Eur J Immunol 18:269–275
Botarelli P, Brunetto MR, Minutello MA, Calvo P, Unutmaz D, Weiner AJ, Choo Q-L, Shuster JR, Kuo G, Bonino F, Houghton M, Abrignani S (1993) T-lymphocyte response to hepatitis C virus in different clinical courses of infection. Gastroenterology 104:580–587
Bronowicki J-P, Vetter D, Uhl G, Hudziak H, Uhrlacher A, Vetter J-M, Doffoel M (1997) Lymphocyte reactivity to hepatitis C virus (HCV) antigens shows evidence for exposure in HCV-seronegative spouses of HCV-infected patients. J Infect Dis 176:518–522

Bruna-Romero O, Lasarte JJ, Wilkinson G, Grace K, Clarke B, Borras-Cuesta F, Prieto J (1997) Induction of cytotoxic T-cell response against hepatitis C virus structural antigens using a defective recombinant adenovirus. Hepatology 25:470–477

Byrne JA, Oldstone MBA (1984) Biology of cloned cytotoxic T lymphocytes specific for lymphocytic choriomeningitis virus: clearance of virus in vivo. J Virol 51:682–686

Cabot B, Esteban JI, Martell M, Genesca J, Vargas V, Esteban R, Guardia J, Gomez J (1997) Structure of replicating hepatitis C virus (HCV) quasispecies in the liver may not be reflected by analysis of circulating HCV virions. J Virol 71:1732–1734

Cacciarelli TV, Martinez OM, Gish RG, Villanueva JC, Krams SM (1996) Immunoregulatory cytokines in chronic hepatitis C virus infection: pre- and posttreatment with interferon alpha. Hepatology 24:6–9

Cai Z, Sprent J (1994) Resting and activated T cells display different requirements for CD8 molecules. J Exp Med 179:2005–2015

Cai Z, Sprent J (1996) Influence of antigen dose and costimulation on the primary response of CD8+ T cells in vitro. J Exp Med 183:2247–2257

Carding SR, Allan W, McMickle A, Doherty PC (1993) Activation of cytokine genes in T cells during primary and secondary murine influenza pneumonia. J Exp Med 177:475–482

Carmichael A, Jin X, Sissons P, Borysiewicz L (1993) Quantitative analysis of the human immunodeficiency viruss type 1 (HIV-1)-specific cytotoxic T lymphocyte (CTL) response at different stages of HIV-1 infection: differential CTL responses to HIV-1 and Epstein-Barr virus in late disease. J Exp Med 177:249–256

Cerino A, Mondelli MU (1991) Identification of an immunodominant B cell epitope on the hepatitis C virus nonstructural region defined by human monoclonal antibodies. J Immunol 147:2692–2696

Cerny A, McHutchison JG, Pasquinelli C, Brown ME, Brothers MA, Grabscheid B, Fowler P, Houghton M, Chisari FV (1995) Cytotoxic T lymphocyte response to hepatitis C virus – derived peptides containing the HLA A2.1 binding motif J Clin Invest 95:521–530

Chang KM, Rehermann B, McHutchison JG, Pasquinelli C, Southwood S, Sette A, Chisari FV (1997) Immunological significance of cytotoxic T lymphocyte epitope variants in patients chronically infected by the hepatitis C virus. J Clin Invest 100:2376–2385

Chatelain R, Varkila K, Coffman RL (1992) IL-4 induces a Th2 response in Leishmania major-infected mice. J Immunol 148:1182–1187

Chisari FV, Ferrari C (1995) Hepatitis B virus immunopathogenesis. In: Paul WE, (ed) Annual review of immunology, vol 13. Annual Reviews, Palo Alto, pp 13:29–60

Choo Q-L, Kuo G, Weiner AJ, Overby LR, Bradley DW, Houghton M (1989) Isolation of a cDNA clone derived from a blood-borne non-A, non-B viral hepatitis genome. Science 244:359–362

Chothia C, Boswell DR, Lesk AM (1988) The outline structure of T-cell alpha beta receptor. EMBO J 7:3745–3755

Clerici M, Giorgi JV, Chou CC, Gudeman VK, Zack JA, Gupta P, Ho HN, Nishanian PG, Berzofsky JA, Shearer GM (1992) Cell-mediated immune response to human immunodeficiency virus (HIV) type 1 in seronegative homosexual men with recent sexual exposure to HIV-1. J Infect Dis 165: 1012–1019

Cocchi F, deVico AL, Garzino-Demo A, Arya SK, Gallo RC, Lusso P (1995) Identification of Rantes, MIP-1alpha and MIP-1beta as the major HIV-suppressive factors produced by CD8+ T cells. Science 270:1811–1815

Croen KD (1993) Evidence for an antiviral effect of nitric oxide. J Clin Invest 91:2446–2452

Cuturi MC, Murphy M, Costa-Giomi MP, Weinmann R, Perussia B, Trinchieri G (1987) Independent regulation of tumor necrosis factor and lymphotoxin production by human peripheral blood lymphocytes. J Exp Med 165:1581–1594

Davis MM, Bjorkman PJ (1988) T cell antigen receptor genes and T cell recognition. Nature 334: 395–402

Denkers EY, Sher A (1997) Role of natural killer and NK1+ T-cells in regulating cell-mediated immunity during Toxoplasma gondii infection. Biochem Soc Trans 25:699–703

Diepolder HM, Zachoval R, Hoffmann RM, Wierenga A, Santantonio T, Jung M-C, Eichenlaub D, Pape GR (1995) Possible mechanism involving T lymphocyte response to non-structural protein 3 in viral clearance in acute hepatitis C virus infection. Lancet 346:1006–1007

Diepolder HM, Gerlach J-T, Zachoval R, Hoffmann RM, Jung MC, Wirrenga EA, Scholz S, Santantonio T, Houghton M, Southwood S, Sette A, Pape GR (1997) Immunodominant CD4+ T-cell epitope within nonstructural protein 3 in acute hepatitis C virus infection. J Virol 71:6011–6019

Doe B, Selby M, Barnett S, Baenziger J, Walker CM (1996) Induction of cytotoxic T lymphocytes by intramuscular immunization with plasmid DN is facilitated by bone-marrow derived cells. Proc Natl Acad Sci USA 93:8578–8583

Doherty PC, Allan W, Eichelberger M (1992) Roles of ab and gd T cell subsets in viral immunity. Annu Rev Immunol 10:123–151

El Ghazali (1993) Number of interleukin-4 and interferon gamma-secreting T cells reactive with tetanus toxoid and the mycobaterial antigen PPD or phytohemagglutinin: distinct response profiles depending on the type of antigen and for activation. Eur J Immunol 23:2740–2745

Erickson AL, Houghton M, Choo Q-L, Weiner AJ, Ralston R, Muchmore E, Walker CM (1993) Hepatitis C virus-specific CTL responses in the liver of chimpanzees with acute and chronic hepatitis C. J Immmunol 151:4189–4199

Evavold BD, Allen PM (1991) Separation of IL-4 production from Th cell proliferation by an altered T cell receptor ligand. Science 252:1308–1310

Eyster ME, Diamondstone LS, Lien JM, Ehmann WC, Quan S, Goedert JJ, for the Multicenter Hemophilia Cohort Study (1993) Natural history of hepatitis C virus infection in multitransfused hemophiliacs: effect of coinfection with human immunodefiency virus. J Aquir Immunodef Synd 6:602–610

Fan FS, Tzeng CH, Hsiao KI, Hu ST, Lin WT, Chen PM (1991) Withdrawal of immunosuppressive therapy in allogeneic bone marrow transplantation reactivates chronic viral hepatitis C. Bone Marrow Transplant 8:417–420

Farci P, Alter HJ, Govindarajan S, Wong DC, Engle R, Lesniewski RR, Mushawhar IK et al (1992) Lack of protective immunity against reinfection with hepatitis C. virus Science 258:135–140

Farci P, Smedile A, Lavarini C, Piantino P, Crivelli O, Caporaso N, Toti M, Bonino F, Rizzetto M (1983) Delta hepatitis in inapparent carriers of hepatitis B surface antigen. A disease simulating acute hepatitis B progressive to chronicity. Gastroenterology 85:669–673

Farrar MA, Schreiber RD (1993) The molecular cell biology of interferon gamma and its receptor. Annu Rev Immunol 11:571–611

Ferrari C, Valli A, Galati L, Penna A, Scaccaglia P, Giuberti T, Schianchi C, Missale G, Marin MG, Fiaccadori F (1994) T-cell response to structural and nonstructural hepatitis C virus antigens in persistent and self-limited hepatitis C virus infections. Hepatology 19:286–295

Fleischer B, Brecht H, Rott R (1985) Recognition of viral antigens by human influenza A virus-specific T lymphocyte clones. J Immunol 135:2800–2805

Fleischer B, Kreth HW (1983) Clonal expansion and functional analysis of virus-specific T lymphocytes from cerebrospinal fluid in measles encephalitis. Hum Immunol 7:239–248

Franco A, Paroli M, Testa U, Benvenuto R, Peschle C, Balsano F, Barnaba V (1992) Transferrin receptor mediates uptake and presentation of hepatitis B envelope antigen by T lymphocytes. J Exp Med 175:1195–1205

Franco A, Guidotti LG, Hobbs MV, Pasquetto V, Chisari FV (1997a) Pathogenetic effector function of CD4-positive T helper 1 cells in hepatitis B virus transgenic mice. J Immunol 2001–2008

Franco MA, Tin C, Rott LS, van Cotte JL, McGhee JR, Greenberg HB (1997b) Evidence for CD8+ T-cell immunity to murine rotavirus in the absence of perforin, fas and gamma interferon. J Virol 71:479–486

Fujii K, Hino K, Okazaki M, Okuda M, Kondoh S, Okita K (1996) Differences in hypervariable region 1 quasispecies of hepatitis C virus between human serum and peripheral blood mononuclear cells. Biochem Biophys Res Commun 225:771–776

Fukumoto T, Berg T, Ku Y, Bechstein WO, Knoop M, Lemmens H-P, Lobeck H, Hopf U, Neuhaus P (1996) Viral dynamics of hepatitis C early after orthotopic liver transplantation: evidence for rapid turnover of serum virions. Hepatology 24:1351–1354

Gazzinelli RT, Hieny S, Wynn TA, Wolf S, Sher A (1993) IL-12 is required for the T-lymphocyte-independent induction of interferon gamma by an intracellular parasite and induces resistance in T cell deficient hosts. Proc Natl Acad Sci USA 90:6115–6119

Geissler M, Gesien A, Tokushige K, Wands JR (1997) Enhancement of cellular and humoral immune responses to hepatitis C virus core protein using DNA-based vaccines augmented with cytokine-expressing plasmids. J Immunol 158:1231–1237

Germain RN (1994) MHC-dependent antigen processing and peptide presentation: providing ligands for T lymphocyte activation. Cell 76:287–299

Gilles PN, Fey G, Chisari FV (1992) Tumor necrosis factor-alpha negatively regulates hepatitis B virus gene expression in transgenic mice. J Virol 66:3955–3960

Giuberti T, Marin MG, Ferrari C, Marchelli S, Schianchi C, Antoni AMD, Pizzocolo G, Fiaccadori F (1994) Hepatitis C virus viremia following clinical resolution of acute hepatitis C. J Hepatol 20:666–671
Gonzalez-Peralta RP, Fang JWS, Davis GL, Gish RG, Wuk PC, Mizokami M, Lau JYN (1994) Immunopathobiology of chronic hepatitis C virus infection. Hepatology 20:232A
Goulder PJR, Phillips RE, Colbert RA, McAdam S, Ogg G, Nowak MA, Giangrande P, Luzzi G, Morgan B, Edwards A, McMichael AJ, Rowland-Jones S (1997) Late escape from an immunodominant cytotoxic T-lymphocyte response associated with progression to AIDS. Nat Med 3:312–217
Grob PJ, Binswanger U, Zaruba K, Joller JHI, Schmid M, Hacki W, Blumberg A, Abplanalp A, Herwig W, Iselin H, Descoeudres C (1983) Immunogenicity of a hepatitis B subunit vaccine in hemodialysis and in renal transplant recipients. Antiviral Res 3:43–52
Gruber A, Lundberg LG, Bjoerkholm M (1993) Reactivation of chronic hepatitis C after withdrawal of immunosuppressive therapy. J Intern Med 234:223–225
Guidotti LG, Ando K, Hobbs MV, Ishikawa T, Runkel RD, Schreiber RD, Chisari FV (1994a) Cytotoxic T lymphocytes inhibit hepatitis B virus gene expression by a noncytolytic mechanism in transgenic mice. Proc Natl Acad Sci USA 91:2764–3768
Guidotti LG, Guilhot S, Chisari FV (1994b) Interleukin 2 and interferon alpha/beta downregulate hepatitis B virus gene expression in vivo by tumor necrosis factor dependent and independent pathways. J Virol 68:1265–1270
Guidotti LG, Ishikawa T, Hobbs MV, Matzke B, Schreiber R, Chisari FV (1996) Intracellular inactivation of the hepatitis B virus by cytotoxic T lymphocytes. Immunity 4:35–36
Guidotti LG, Chisari FV (1996b) To kill or to cure: Options in host defense against viral infection. In: Zinkernagel R, Bloom B (eds) Current Opinion in immunology, vol 8. Current biology, London, pp 478–483
Harris N, Buller RML, Karupiah G (1995) Interferon-gamma induced, nitric oxide mediated inhibition of vaccinia viral replication. J Virol 69:910–915
Hayakawa K, Lin BT, Hardy RR (1992) Murine thymic CD4+ subsets: a subset (Thy0) that secretes diverse cytokines and overexpresses the V beta 8 T cell receptor gene family. J Exp Med 176:269–274
Heath SL, Tew JG, Szakal AK, Burton GF (1995) Follicular dendritic cells and human immunodeficiency virus infectivity. Nature 377:740–744
Hiramatsu N, Hayashi N, Katayama K, Mochizuki K, Kawanishi Z, Kasahara A, Fusamoto H, Kamada T (1994) Immunohistochemical detection of fas antigen in liver tissue of patients with chronic hepatitis C. Hepatology 19:1354–1358
Hiroishi K, Kita H, Kojima M, Okamoto H, Moriyama T, Kaneko T, Ishikawa T, Ohnishi S, Aikawa T, Tanaka N, Yazaki Y, Mitamura K, Imawari M (1997) Cytotoxic T lymphocyte response and viral load in hepatitis C virus infection. Hepatology 25:705–712
Ho DD, Moudgil T, Alam M (1989) Quantitation of human immunodeficiency virus type 1 in the blood of infected persons. N Engl J Med 321:1621–1625
Hoffmann RM, Diepolder HM, Zachoval R, Zwiebel F-M, Jung M-C, Scholz S, Nitschko H, Riethmüller G, Pape GR (1995) Mapping of immunodominant CD4+ T lymphocyte epitopes of hepatitis C virus antigens and their relevance during the course of chronic infection. Hepatology 21:632–638
Hsieh CS, Heimberger A, Gold J, O'Gara A, Murphy K (1992) Differential regulation of T helper phenotype development by IL-4 and IL-10 in alpha, beta T-cell receptor transgenic system. Proc Natl Acad Sci USA 89:6065–6069
Hsieh CS, Macatonia SE, Tripp CS, Wolf SF, O'Gara A, Murphy KM (1993) Development of Th1 CD4+ T cells through IL-12 produced by Listeria-induced macrophages Science 260:547–549
Imada K, Fukuda Y, Koyama Y, Nakano I, Yamada M, Katano Y, Hayakawa T (1997) Naive and memory T cell infiltrates in chronic hepatitis C: phenotypic changes with interferon treatment. Clin Exp Immunol 109:59–66
Iwata K, Wakita T, Okumura A, Yoshioka K, Takayanagi M, Wands JR, Kakumu S (1995) Interferon gamma production by peripheral blood lymphocytes to hepatitis C virus core protein in chronic hepatitis C infection. Hepatology 22:1057–1064
Jacobson S, Richert JR, Biddison WE, Satinsky A, Harztmann RJ, McFarland HF (1984) Measles virus specific T4+ human cytotoxic T cell clones are restricted by class II HLA antigens. J Immunol 133:754–757
Jorgensen JL, Esser U, Fazekas de S, Groth B, Reay PA, Davis MM (1992) Mapping T-cell receptor-peptide contacts by variant peptide immunization of single-chain transgenics. Nature 355:224–230

Kahn RA, Johnson G, Aach RD, Hines A, Ellis FR, Miller WV (1982) The distribution of serum alanine aminotransferase levels in a blood donor population. Am J Epidemiol 115:929–940

Kane JM III, Shears LL Jr, Hierholzer C, Ambs S, Billiar TR, Posner MC (1997) Chronic hepatitis C virus infection in humans: induction of hepatic nitric oxide synthase and proposed mechanisms for carcinogenesis. J Surg Res 69:321–324

Kaneko T, Moriyama T, Udaka K, Hiroishi K, Kita H, Okamoto H, Yagita H, Okumura K, Imawari M (1997) Impaired induction of cytotoxic T lymphocytes by antagonism of a weak agonist borne by a variant hepatitis C virus epitope. Eur J Immunol 27:1782–1787

Kaneko T, Nakamura I, Kita H, Hiroishi K, Moriyama T, Imawari M (1997) Three new cytotoxic T cell epitopes identified within the hepatitis C virus nucleoprotein. J Gen Virol 77:1305–1309

Karupiah G, Xie Q-W, Buller RML, Nathan C, Durate C, Mac Miching JD (1993) Inhibition of viral replication by interferon gamma induced nitric oxide synthased. Science 261:1445–1448

Kashii Y, Shimizu Y, Nambu S, Minemura M, Okada K, Higuchi K, Watanabe A (1997) Analysis of T-cell receptor V beta repertoire in liver-infiltrating lymphocytes in chronic hepatitis C. J Hepatol 26:462–470

Kawamura T, Furusaka A, Koziel MJ, Chung RT, Wang TC, Schmidt EV, Liang TJ (1997) Transgenic expression of hepatitis C virus structural proteins in the mouse. Hepatology 25:1014–1021

Kim J, Woods A, Becker-Dunn E, Bottomly K (1985) Distinct functional phenotypes of cloned Ia-restricted helper T cells. J Exp Med 162:188–201

Kinkhabwala M, Sehajpal P, Skolnik E, Smith D, Sharma VK, Vlassara H, Cerami A, Suthanthiran M (1990) A novel addition to the T cell repertory: cell surface expression of tumor necrosis factor/cachectin by activated normal human T cells. J Exp Med 171:941–946

Kita H, Moriyama T, Kaneko T, Harase I, Nomura M, Miura H, Nakamura I, Yazaki Y, Imawari M (1993) HLA B44-restricted cytotoxic T lymphocytes recognizing an epitope on hepatitis C virus nucleocapsid protein. Hepatology 18:1039–1044

Klenerman P, Rowland-Jones S, McAdams S, Edwards J, Daenke S, Lalloo D, Koppe B, Rosenberg W, Boyd D, Edwards A et al (1994) Cytotoxic T-cell acitivty antatgoniyed by naturally occurring HIV-1 Gag variants. Nature 369:403–407

Koenig S, Conley AJ, Bewah YA, Jones GM, Leath S, Boots LJ, Davey V, Pantaleo G, Demarest JF, Carter C et al (1995) Transfer of HIV-1-specific cytotoxic T lymphocytes to an AIDS patient leads to selection for mutant HIV variants and subsequent disease progression. Nat Med 1:330–336

Kolykhalov AA, Agapov EV, Blight KJ, Mihalik K, Feinstone SM, Rice CM (1997) Transmission of hepatitis C by intrahepatic inoculation with transcribed RNA. Science 277:570–574

Kolykhalov AA, Feinstone SM, Rice CM (1996) Identification of a highly conserved sequence element at the 3' terminus of hepatitis C virus genome RNA. J Virol 70:3363–3371

Koup RA, Safrit JT, Cao Y, Andrews CA, McLeod G, Borkowsky W, Farthing C, Ho DD (1994) Temporal association of cellular immune responses with the initial control of viremia in primary human immunodeficiency virus type 1 syndrome. J Virol 68:4650–4655

Kowalski H, Erickson AL, Cooper S, Domena JD, Parham P, Walker CM (1996) Patr-A and B, the orthologues of HLA-A and B, present hepatitis C virus epitopes to CD8+ cytotoxic T cells from two chronically infected chimpanzees. J Exp Med 183:1761–1775

Koziel JM, Dudley D, Afdhal N, Choo QL, Houghton M, Ralston R, Walker BD (1993) Hepatitis C virus (HCV)-specific cytotoxic T lymphocytes recognize epitopes in the core and envelope proteins of HCV. J Virol 67:7522–7532

Koziel MJ, Dudley D, Wong JT, Dienstag J, Houghton M, Ralston R, Walker BD (1992) Intrahepatic cytotoxic T lymphocytes specific for hepatitis C virus in persons with chronic hepatitis. J Immunol 149:3339–3344

Koziel MJ, Dudley D, Afdhal N, Grakoul A, Rice CM, Choo Q-L, Houghton M, Walker BD (1995) HLA class I-restricted cytotoxic T lymphocytes specific for hepatitis C virus. Identification of multiple epitopes and characterization of patterns of cytokine release. J Clin Invest 96:2311–1221

Koziel MJ, Wong DKH, Dudley D, Houghton M, Walker BD (1997) Hepatitis C Virus-specific cytolyotic T lymphocyte and T helper cell responses in seronegative persons. J Infect Dis 176:859–866

Kündig TM, Bachmann MF, Oehen S, Hoffmann UW, Simaard JJL, Kalberer CP, Pircher H, Ohashi PS, Hengartner H, Zinkernagel RM (1996) On the role of antigen in maintaining cytotoxic T-cell memory. Proc Natl Acad Sci USA 93:9716–9723

Kurokohchi K, Akatsuka T, Pendleton CD, Takamizawa A, Nishioka M, Battegay M, Feinstone SM, Berzofsky JA (1996) Use of recombinant protein to identify a motif-negative human cytotoxic T-cell epitope presented by HLA-A2 in the hepatitis C virus NS3 region. J Virol 70:232–240

La Face DM, Couture C, Anderson K, Shih G, Alexander J, Sette A, Mustelin T, Altman T, Altman A, Grey HG (1997) Differential T cell signaling induced by antagonist peptide-MHC complexes and the associated phenotypic responses. J Immunol 158:2057–2064

Lanford RE, Chavez D, Chisari FV, Sureau C (1995) Lack of detection of negative-strand hepatitis C virus RNA in peripheral blood mononuclear cells and other extrahepatic tissues by the highly strand-specific rTth reverse transcriptase PCR. J Virol 69:8079–8083

Lanzavecchia A (1997) Understanding the mechanisms of sustained signaling and T cell activation. J Exp Med 185:1717–1719

Lau GKK, Davis GL, Wu SPC, Gish RG, Balart LA, Lau JYN (1996) Hepatic expression of hepatitis C virus RNA in chronic hepatitis C: a study by in situ reverse-transcription polymerase chain reaction. Hepatology 23:1318–1323

Le Gros G, Ben-Sasson SZ, Seder R, Finkelman FD, Paul WE (1990) Generation of interleukin 4 (IL-4)-producing cells in vivo and in vitro: IL-2 and IL-4 are required for in vitro generation of IL-4 producing cells. J Exp Med 172:921–929

Lechmann M, Ihlenfeldt HG, Braunschweiger I, Giers G, Jung G, Matz B, Kaiser R, Sauerbruch T, Spengler U (1996) T and B cell responses to different hepatitis C virus antigens in patients with chronic hepatitis C infection and in healthy anti-HCV blood donors without viremia. Hepatology 24:790–795

Leeroux-Roels G, Esquivel CA, DeLeys R, Stuyver L, Elewaut A, Philippé J, Desombere I, Paradijs J, Maertens G (1996) Lymphoproliferative responses to hepatitis C virus core, E1, E2, and NS3 in patients with chronic hepatitis C infection treated with interferon alpha. Hepatology 23:8–16

Liew FY (1990) Regulation of cell-mediated immunity in leishmaniasis. In: Kaufmann SH (ed) T-cell paradigms in parasitic and bacterial infections. Springer, Berlin Heidelberg New York, pp 53–64

Lim HL, Lau GKK, Davis GL, Dolson DJ, Lau JYN (1994) Cholestatic hepatitis leading to hepatic failure in a patient with organ-transmitted hepatitis C virus infection. Gastroenterology 106:248–251

Livingston BD, Crimi C, Grey H, Ishioka G, Chisari FV, Fikes J, Chesnut RW, Sette A (1997) The hepatitis B virus-specific CTL responses induced in humans by lipopeptide vaccination are comparable to those elicited by acute viral infection. J Immunol 159:1383–1392

Machold RP, Wiertz EJHJ, Jones TR, Ploegh HL (1997) The HCMV gene products US11 and US2 differ in their ability to attack allelic forms of murine major histocompatibility complex (MHC) class I heavy chains. J Exp Med 185:363–366

Maggi F, Fornai C, Vatteroni ML, Giorgi M, Morrica A, Pistello M, Cammarota G, Marchi S, Ciccorossi P, Bionda A, Bendinelli M (1997) Differences in hepatitis C virus quasispecies composition between liver, peripheral blood mononuclear cells and plasma. J Gen Virol 78:1521–1525

Makimura M, Miyake S, Akino N, Takamori K, Matsuura Y, Miyamura T, Saito Y (1996) Induction of antibodies against structural proteins of hepatitis C virus in mice using recombinant adenovirus. Vaccine 14:28–34

Martin P, Di Bisceglie AM, Kassianides C, Lisker-Melman M, Hoofnagle JH (1989) Rapid progressive non-A, non-B hepatitis in patients with human immunodeficiency virus infection. Gastroenterology 97:1559–1561

Martinez OM, Gibbons RS, Garovoy MR, Aronson FR (1990) IL-4 inhibits IL-2 receptor expression and IL-2 dependent prolifeation of human T cells. J Immunol 144:2211–2215

Miceli MC, Parnes JR (1993) Role of CD4 and CD8 in T cell ativation and differentiation. Adv Immunol 53:59–122

Minutello MA, Pileri P, Unutmaz D, Censini S, Kuo G, Houghton M, Brunetto MR, Bonino F, Abrignani S (1993) Compartmentalization of T-lymphocyte to the site of disease: Intrahepatic CD4$^+$ T-cells specific for the protein NS4 of hepatitis C virus in patient with chronic hepatitis. J Exp Med 178:17–26

Moretta L, Ciccone E, Mingari MC, Biassoni R, Moretta A (1994) Human natural killer cells: origin, clonality, specificity, receptors. Adv Immunol 55:341–358

Moriyama T, Guilhot S, Klopchin K, Moss B, Pinkert CA, Palmiter RD, Brinster RL, Kanagawa O, Chisari FV (1990) Immunobiology and pathogenesis of hepatocellular injury in hepatitis B virus transgenic mice. Science 248:361–364

Mosmann TR, Cherwinski H, Bond MW, Giedlin MA, Coffman RL (1986) Two types of murine helper T cell clones. I. Definition according to profiles of lymphokine activities and secreted proteins. J Immunol 136:2348–2357

Mosmann TR, Sad S (1996) The expanding universe of T-cell subsets: Th1, Th2 and more. Immunol Today 17:138–146

Mosnier JF, Scoaze JY, Marcellin P, Degott C, Benahmou JP, Feldmann G (1994) Expression of cytokine-dependent immune adhesion molecules by hepatocytes. Gastroenterology 107:1457–1468

Müller D, Jenkins M (1995) Molecular mechanisms underlying functional T-cell unresponsiveness. Curr Opin Immunol 3:375–381

Murray JS, Madri J, Tite J, Carding SR, Bottomly K (1989) MHC control of CD4+ T cell subset activation. J Exp Med 170:2135–2140

Nagler A, Lanier LL, Phillips JH (1988) The effects of IL-4 on human natural killer cells: a potent regulator of IL-2 activation and proliferation. J Immunol 141:2349–2351

Nakajima H, Iwamoto I, Tomoe S, Matsumara R, Tomioka H, Takatsu K, Yoshida S (1992) CD4+ T-lymphocytes and interleukin-5 mediate antigen-induced eosinophil infiltration into mouse trachea. Ann Rev Respir Dis 146:374–377

Napoli J, Bishop GA, McGuiness PH, Painter DM, McCaughan GW (1996) Progressive liver injury in chronic hepatitis C infection correlates with increased intrahepatic expression of T h 1 – associated cytokines. Hepatology 24:759–765

Nasoff MS, Zebedee SL, Inchauspé G, Prince AM (1991) Identification of an immunodominant epitope with in the capsid protein of hepatitis C virus. Proc Natl Acad Sci USA 88:5462–5466

Nathan C, Xie Q-W (1994) Regulation of biosynthesis of nitric oxide. J Biol Chem 269:13725–13728

Nelson DR, Marousis CG, Davis GL, Rice CM, Wong J, Houghton M, Lau JYN (1997) The role of hepatitis C virus-specific cytotoxic T lymphocytes in chronic hepatitis C. J Immunol 158:1473–1481

Niewiesk S, Daenke S, Parker CE, Taylor G, Weber J, Nightingale S, Bangham CR (1995) Naturally occurring variants of human T-cell leukemia virus type I Tax protein impair its recognition by cytotoxic T lymphocytes and the transactivation function of Tax. J Virol 69:2649–2653

Nussler AK, Di Silvio M, Billiar TR, Hoffman RA, Geller DA, Selby R, Madariaga J, Simmons RL (1992) Stimulation of the nitric oxide synthase pathway in human hepatocytes by cytokines and endotoxin. J Exp Med 176:261–264

O'Sullivan D, Arrhenius T, Sidney J, Del Guercio MF, Albertson M, Wall M, Oseroff C, Southwood S, Colon SM, Gaeta FCA, Sette A (1991) On the interaction of promiscuous antigenic peptides with different DR alleles. Identification of common structural motifs. J Immunol 147:2663–2669

Odeberg J, Yun Z, Sönnerborg A, Bjoro K, Uhlen M, Lundeberg J (1997) Variation of hepatitis C virus hypervariable region 1 in immunocompromised patients. J Infect Dis 175:938–943

Ohishi M, Sakisaka S, Koji T, Nakane PK, Tanikawa K (1995) The localization of HCV and the expression of Fas antigen in the liver of HCV-related chronic liver disease. Acta Histochem Cytochem 28:341

Oldstone MBA (1996) Principles of viral pathogenesis. Cell 87:799–801

Onji M, Kikuchi T, Kumon I, Masumoto T, Nadano S, Kajino K, Horiike N, Ohta Y (1992) Intrahepatic lymphocyte subpopulations and HLA class I antigen expression by hepatocytes in chronic hepatitis C. Hepatogastroenterology 39:340–343

Pavic I, Polic B, Crnkovic I, Lucin P, Jonjic S, Koszinowki UH (1993) Participation of endogenous tumor necrosis factor alpha in host resistance to cytomegalovirus infection. J Gen Virol 74:2215–2223

Paul WE, Seder RA (1994) Lymphocyte responses and cytokines. Cell 76:241–251

Pearson CI, van Ewijik W, McDevitt HO (1997) Induction of apoptosis and T helper 2 (Th2) responses correlates with peptide affinitiy for the major histocompatibility complex in self-reactive T cell receptor transgenic mice. J Exp Med 185:583–599

Pfeiffer C, Stein J, Southwood S, Ketelaar H, Sette A, Bottomly K (1995) Altered peptide ligands can control CD4 T lymphocyte differentiation in vivo. J Exp Med 181:1569–1574

Prince AM, Brotman B, Huima T, Pascual D, Jaffery M, Inchaupe G (1992) Immunity in hepatitis C infection. J Infect Dis 165:438–443

Ramsay AJ, Ruby J, Ramshaw IA (1993) A case for cytokines as effector molecules in the resolution of virus infection. Immunol Today 14:155–157

Rehermann B, Chang KM, McHutchison J, Kokka R, Houghton M, Rice CM, Chisari FV (1996a) Differential cytotoxic T lymphocyte responsiveness to the hepatitis B and C viruses in chronically infected patients. J Virol 70:7092–7102

Rehermann B, Chang KM, McHutchison JG, Kokka R, Houghton M, Chisari FV (1996b) Quantitative analysis of the peripheral blood cytotoxic T lymphocyte response, disease activity and viral load in patients with chronic hepatitis C virus infection. J Clin Invest 98:1432–1440

Rehermann B, Ferrari C, Pasquinelli C, Chisari FV (1996c) The Hepatitis B Virus persists for decades after patients' recovery from acute viral hepatitis despite active maintenance of a cytotoxic T-lymphocyte response. Nat Med 2:1104–1108

Rehermann B, Lau D, Hoofnagle JH, Chisari FV (1996d) Cytotoxic T lymphocyte responsiveness after resolution of chronic hepatitis B virus infection. J Clin Invest 97:1655–1665

Rossi G, Tucci A, Cariani E, Ravaggi A, Rossini A, Radaeli E (1997) Outbreak of hepatitis C virus infection in patients with hematologic disorders treated with intravenous immunoglobulins: different prognosis according to the immune status. Blood 90:1309–1314

Rowland-Jones S, Sutton J, Ariyoshi K, Dong T, Gotch F, McAdams S, Whitley D, Sabally S, Gallinore A, Corrah T et al (1995) HIV-specific cytotoxic T-cells in HIV-exposed but uninfected Gambian women. Nat Med 1:59–64

Saito S, Kato N, Hijikata M, Gunji T, Itabashi M, Kondo M, Tanaka K, Shimotohno K (1996) Comparison of hypervariable regions (HVR1 and HVR2) in positive- and negative-stranded hepatitis C virus RNA in cancerous and non-cancerous liver tissue, peripheral blood mononuclear cells and serum from a patient with hepatocellular carcinoma. Int J of Cancer 67:199–203

Sansono D, Dammacco F (1992) Hepatitis C virus related chronic liver disease of sporadic type: clinical, serological and histological features. Digestion 51:115–120

Sata M, Ide T, Noguchi S, Hashimoto O, Ryukichi K, Suzuki H, Tanikawa K (1997) Timing of IFN therapy initiation for acute hepatitis C after accidental needlestick. J Hepatol 27:425–426

Schrager LK, Young JM, Fowler MG, Mathieson BS, Vermund SH (1994) Long-term survivors of HIV-1 infection: definitions and research challenges. AIDS 8:S95-S108

Schupper H, Hayashi P, Scheffel J, Aceituno S, Paglieroni T, Holland PV, Zeldis JB (1993) Peripheral blood mononuclear cell responses to recombinant hepatitis C virus antigens in patients with chronic hepatitis C. Hepatology 18:1055–1060

Scott P, Pearce E, Cheever A, Coffman R, Sher A (1989) Role of cytokines and CD4+ T cell subsets in the regulation of parasitic immunity and disease. Immunol Rev 112:161–182

Sharara AI, Perkins DJ, Misukonis MA, Chan SU, Dominitz JA, Weinberg JB (1997) Interferon (IFN)-alpha activation of human blood mononuclear cells in vitro and in vivo for nitric oxide synthase (NOS) type 2 mRNA and protein expression: possible relationship of induced NOS2 to the anti-hepatitis C effects of IFN-alpha in vivo. J Exp Med 186:1495–1502

Shirai M, Pendleton CD, Ahlers J, Takeshita T, Newman M, Berzofsky JA (1994) Helper-Cytotoxic T lymphocyte (CTL) determinant linkage required for priming of anti-HIV CD8+ CTL in vivo with peptide vaccine constructs. J Immunol 452:549–556

Shirai M, Arichi T, Nishioka M, Nomura T, Ikeda K, Kawanishi K, Engelhard VH, Feinstone SM, Berzofsky JA (1995) CTL responses of HLA-A2.1-transgenic mice specific for hepatitis C viral peptides predict epitopes for CTL of humans carrying HLA-A2.1. J Immunol 154:2733–2742

Sloan-Lancaster J, Evavold B, Allen P (1993) Induction of T-cell anergy by altered T-cell-receptor ligand Nature 363:156–159

Sloan-Lancaster J, Evavold B, Allen P (1994) Th2 cell clonal anergy as a consequence of partial activation. J Exp Med 180:1195–1205

Steffen M, Ottmann O, Moore M (1988) Simultaneous production of tumor necrois factor-alpha and lymphotoxin by normal T cells after induction with IL-2 and anti-T3. J Immunol 140:2621–2640

Sung S-S, Bjordahl JM, Wng CY, Kao HT, Fu SM (1988) Production of tumor necrosis factor/cachectin by human T cell lines and peripheral blood T lymphocytes stimulated by phorbolmyristate acetate and anti-CD3 antibody. J Exp Med 168:1539–1551

Swain SL, Weinberg AD, English M, Huston G (1990) IL-4 directs the development of TH2 like helper effectors. J Immunol 145:3796–3806

Sypek JP, Chung CL, Mayor SEH, Subramanyam LJM, Goldman SJ, Sieburth DS, Wolf SF, Schaub RG (1993) Resolution of cutaneous leishmaniasis: interleukin 12 initiates a protective T helper type 1 immune response. J Exp Med 177:1797–1802

Tedeschi V, Akatsuka T, Shih JW-K, Battegay M, Feinstone SM (1997) A specific antibody response to HCV E2 elicited in mice by intramuscular inoculation of plasmid DNA containing coding sequences for E2. Hepatology 25:459–462

Thompson CB (1995) Distinct roles for the costimulatory ligands B7-1 and B7-2 in T helper cell differentiation? Cell 81:979–982

Tite JP, Foellmer HG, Mardi JP, Janeway CA Jr (1987) Inverse Ir gene control of the antibody and T cell proliferative response to human basement membrane collagen. J Immunol 139:2892–2898

Tokushige K, Wakita T, Pachuk C, Moradpour D, Weiner DB, Zurawski VR Jr, Wands JR (1996) Expression and immune response to hepatitis C virus core DNA-based vaccine constructs. Hepatology 24:14–20

Tsai S-L, Liaw Y-L, Chen M-H, Huang C-Y, Kuo GC (1997) Detection of type 2-like T-helper cells in hepatitis C virus infection: implications for hepatitis C virus chronicity. Hepatology 25:449–458

Unutmaz D, Pileri P, Abrignani S (1994) Antigen-independent activatin of naive and memory resting T cells by a cytokine combination. J Exp Med 180:1159–1164

Urban JF Jr, Katona IM, Paul WE, Finkelman FD (1991) Interleukin-4 is important in protective immunity to a gastrointestinal nematode infection in mice. Proc Natl Acad Sci USA 88:5513–5517

Valitutti S, Müller S, Cella M, Padovan E, Lanzavecchia A (1995) Serial triggering of many T-cell receptors by a few peptide-MHC complexes. Nature 375:148–151

Valitutti S, Müller S, Dessing M, Lanzavecchia A (1996) Different responses are elicited in cytotoxic T lymphocytes by different levels of T cell receptor occupancy. J Exp Med 183:1917–1921

Vento S, Cainelli F, Mirandola F, Cosco L, Di Perri G, Solbiati M, Ferraro T, Concia E (1996) Fulminant hepatitis on withdrawal of chemotherapy in carriers of hepatitis C virus. Lancet 347:92–93

Vitetta ES, Fernandez-botran R, Myers CD, Sanders VM (1989) Cellular interactions in the humoral immune response. Adv Immunol 45:1–105

Vitiello A, Ishioka G, Grey HM, Rose R, Farness P, LaFond R, Yuan L, Chisari FV, Furze J, Bartholomeuz R, Chesnut RW (1995) Development of a lipopeptide-based therapeutic vaccine to treat chronic HBV infection. J Clin Invest 95:341–349

Weiner A, Erickson AL, Kansopon J, Crawford K, Muchmore E, Hughes AL, Houghton M, Walker CM (1995) Persistent hepatitis C virus infection in a chimpanzee is associated with emergence of a cytotoxic T lymphocyte escape variant. Proc Natl Acad Sci USA 92:2755–2759

Weiner AJ, Geysen HM, Christopherson C, Hall JE, Mason TJ, Saracco G, Bonino F, Crawford K, Marion CD, Crawford KA et al. (1992) Evidence for immune selection of hepatitis C virus (HCV) putative envelope glycoprotein variants: potential role in chronic HCV infection. Proc Natl Acad Sci USA 89:3468–3472

Wentworth PA, Vitiello A, Sidney J, Keogh E, Chesnut RW, Grey H, Sette A (1996) Differences and similarities in the A2.1 restricted cytotoxic T cell repertoire in humans and HLA-transgenic mice. Eur J Immunol 26:97–101

Wentworth PA, Sette A, Celis E, Sidney J, Southwood S, Crimi C, Stitely, S, Keogh E, Wong, NC, Livingston B, Alazard D, Vitiello A, Grey HM, Chisari FV, Chesnut RW, Fikes J (1995) Identification of A2-restricted HCV-specific CTL epitopes from highly conserved regions of the viral genome. Int Immunol 8:651–659

Wejstal R, Norkrans R, Weiland O, Fuchs D, Wachter H, Fryden A, Glaumann H (1992) Lymphocyte subsets and B2 microglobulin expression in chronic hepatitis C/nonA-nonB: effect of interferon-alpha treatment. Clin Exp Immunol 87:340–345

Woitas RP, Lechmann M, Jung G, Kaiser R, Sauerbruch T, Spengler U (1997) CD30 induction and cytokine profiles in hepatitis C virus core-specific peripheral blood T lymphocytes. J Immunol 150:1012–1018

Yanagi M, Purcell RH, Emerson SU, Bukh J (1997) Transcripts from a single full-length cDNA clone of hepatitis C virus are infectious when directly transfected into the liver of a chimpanzee. Proc Natl Acad Sci USA 94:8738–8743

Yoshimoto Y, Paul WE (1994) NK1.1pos T cells promptly produce interleukin 4 in response to in vivo challenge with anti-CD3. J Exp Med 179:1285–1295

Yuh K, Shimizu M, Aoyama S, et al (1986) Analysis of lymphocyte subsets in liver biopsy specimens with lymphoid follicle like structures. Acta Hepat Jpn 27:720–725

Zeuzem S, Schmidt JM, Lee JH, Ruster B, Roth WK (1996) Effect of interferon alpha on the dynamics of hepatitis C virus turnover in vivo. Hepatology 23:366–371

Zinkernagel RM, Moskophidis D, Kundig T, Oehen S, Pircher HP, Hengartner H (1993) Effector T-cell induction and T-cell memory versus peripheral deletion of T cells. Immunol Rev 131:198–223

Strategies and Prospects for Vaccination Against the Hepatitis C Viruses

M. Houghton

1 Introduction	327
2 Correlates of Immunity to Hepatitis C Virus: Status	329
3 Recombinant Subunit Proteins as Vaccine Candidates	331
4 Hepatitis C Virus Nucleic Acids as Vaccine Candidates	333
5 Concluding Remarks	335
References	335

1 Introduction

In the USA, it has been shown that much of the hepatitis C virus (HCV)-associated disease burden arose from transmission through recreational drug use in which contaminated needles or syringes (or possibly cocaine straws; Conry-Cantilena et al. 1996) are shared (Alter 1993). Currently, roughly 30,000 newly acquired acute HCV infections occur annually in the USA, most of which have intravenous drug use as the risk factor (Alter 1993). However, in the remaining cases, multiple sexual partners remains an identified risk factor as does any activity subjecting the individual to exposure to contaminated blood (e.g. health-care workers, dialysis patients, organ recipients etc.; Alter 1993). Therefore, while a prophylactic HCV vaccine would probably be used in high-risk groups initially, there would be a case for implementing universal vaccination of children (as in the case for the HBV vaccine) in order to prevent infection associated with adult activities. While there are similarities in the epidemiology of HCV between the USA and other developed countries, it appears that many developing countries suffer from a huge incidence of HCV infection due to past use of nondisposable injection needles/syringes in medical practices as well as cultural practices involving blood transfer. In these situations, public health measures and education programs are of paramount importance.

Chiron Corporation, 4560 Horton Street, Office 4.3138, Emeryville, CA 94608, USA

Certain properties of HCV imply that the development of an HCV vaccine may be difficult. Firstly, there is the ability of this agent to cause persistent infection in the majority of cases (ALTER 1993), this persistence occurring despite the induction of HCV-specific B and T cell responses (CHIEN et al. 1993; WONG et al. 1998). Once chronic infection has developed, the vast majority of individuals remain infected apparently for life with varying degrees of clinical sequelae but often with only minimal liver disease (ALTER 1989). The mechanism(s) responsible for the persistence of this flavi-type RNA virus remains a major question for future research as does the elucidation of those factors responsible for the development of severe HCV-associated liver disease. Various hypotheses for viral persistence have been forwarded such as immune-masking of virus by molecules such as β-lipoprotein (THOMSSEN et al. 1992), the generation of immune-escape mutants (WEINER et al. 1992), and the existence of immune-privileged sites of viral replication. Secondly, since the identification of HCV in 1989, it has become apparent that this positive stranded RNA virus exhibits a great deal of global heterogeneity. Currently, at least six basic genotypes have been distinguished phylogenetically with well over 30 subtypes (BUKH et al. 1997; SIMMONDS et al. 1997). Such heterogeneity is responsible for amino acid divergences of up to 50% in the two envelope glycoproteins for example (BUKH et al. 1993). While there is currently no efficient in vitro propagation system for HCV and therefore a lack of an in vitro virus neutralizing antibody assay, it seems likely that the observed genetic heterogeneity corresponds with the existence of numerous serotypes. As such, the development of a global HCV vaccine may be complex, especially if it relies on anti-envelope humoral immunity for efficacy. Apart from this high level of global heterogeneity, the virus can change rapidly within the infected individual (OGATA et al. 1991) due to the inherent quasi-species nature of this RNA virus as well as selection for mutant fitness (MARTELL et al. 1992) that likely includes immune selection. Mutations within the NH_2-terminal hypervariable region of glycoprotein E2 (gpE2 HVR) have been shown to emerge rapidly within the infected individual which are then non-reactive with pre-existing antibody responses to this region (WEINER et al. 1992; KATO et al. 1994; SHIMIZU et al. 1994). Using the chimpanzee infection model as an assay for virus neutralizing antibody, antibodies to this region have been shown to neutralize infectious virus (FARCI et al. 1996). There is also preliminary evidence in this animal model for the emergence of mutants within epitopes recognized by MHC class-1 restricted $CD8^+$ cytotoxic lymphocytes (CTLs; WEINER et al. 1995). Thirdly, at present there is no compelling evidence for enduring natural immunity in those individuals that resolve and recover from acute HCV infection. To the contrary, experimental rechallenges of convalescent chimpanzees have indicated that re-infections and the development of chronic infection readily take place (FARCI et al. 1992; PRINCE et al. 1992), although the immune response to HCV in chimpanzees may be weaker than in humans (CHOO et al. 1994). However, limited human studies have also documented the re-infection of convalescent children leading to chronic hepatitis (LAI et al. 1994).

On a more encouraging note, there are now growing indications for the role of HCV-specific immune responses in the resolution or amelioration of infection and disease (to be discussed below). There can therefore be some reasonable confidence at recapitulating such immune responses in vaccinees using the large array of vaccine methods currently available.

2 Correlates of Immunity to Hepatitis C Virus: Status

Hepatitis C virus-specific T cell responses have been measured during acute human infections that either led to viral clearance or to the development of chronic infection. In one study, an early T helper response to nonstructural (NS) protein 3 (NS3) was most closely associated with resolution of acute infection (DIEPOLDER et al. 1995), involving a conserved and highly promiscuous (with respect to MHC class II restriction) $CD4^+$ T-cell epitope (DIEPOLDER et al. 1997). The cytokine profile associated with this protective T cell response was consistent with a Th0/Th1 phenotype. Other studies also found that $CD4^+$ T helper responses correlated directly with recovery from acute infection although in these cases, Th0/Th1 immune responses were directed to most of the viral encoded proteins (MISSALE et al. 1996; TSAI et al. 1997). One study has concluded that those individuals displaying predominant Th0/Th1 $CD4^+$ T-helper responses to HCV resolved their infections whereas weaker Th2-type helper responses were observed in some patients progressing to chronicity (TSAI et al. 1997). These investigations present a unified view of HCV-specific T helper responses being involved in the protective immune response, although only peripheral T cell responses were studied and not intrahepatic responses due to difficulties inherent in obtaining liver biopsy samples from such individuals. This is potentially important since compartmentalization of specific T cells within the infected organ has been documented (MINUTELLO et al. 1993; NELSON et al. 1997). Analyses of peripheral $CD4^+$ T cell responses in patients with various levels of HCV-associated disease have shown that clinically asymptomatic chronic infection was also associated with strong $CD4^+$ T cell responses to HCV proteins, in particular to the nucleocapsid (C) protein (BOTARELLI et al. 1993; LECHMANN et al. 1996).

MHC class I-restricted $CD8^+$ cytotoxic lymphocytes (CTLs) have been identified in chronically infected patients in both the periphery (KITA et al. 1993; BATTEGAY et al. 1995; CERNY et al. 1995; REHERMANN et al. 1996a,b) and within the liver tissue (WONG et al. 1998; NELSON et al. 1997). Accordingly, a large number of different epitopes have been mapped within multiple HCV proteins with various MHC class I restrictions. In some chronic patients, multiple CTLs specific to different HCV epitopes have been identified within the liver (WONG et al. 1998; NELSON et al. 1997). The mechanism(s) responsible for the failure of this CTL response in clearing infected hepatocytes remains to be identified. One study investigating HCV-specific CTL responses in the peripheral blood of chronically infected patients has suggested that the HCV CTL precursor frequency is much lower than that directed towards a recall influenza antigen, suggesting that the

HCV-specific CTL response may be weak and limiting in chronic patients (REHERMANN et al. 1996a). In a chimpanzee study, the development of chronic infection was associated with the emergence of a viral mutant within NS3 that could no longer be recognized by pre-existing CTLs (WEINER et al. 1995). While complete evasion of a broad multi-specific CTL response may be difficult, the mutation of one or more immunodominant CTL epitopes within the viral population could conceivably play a role in the development and/or maintenance of chronicity.

Recently, a detailed study has been carried out in the chimpanzee model which has allowed sampling of liver tissue throughout the course of acute infections that led to either resolution and clearance of virus or chronicity. Dramatically, resolution of acute infection was seen in animals that, while failing to make a significant antibody response to HCV, produced an early and broad CTL response to multiple HCV epitopes (COOPER et al. 1999). This preliminary result clearly suggests an important role for an early and broad MHC class I-restricted $CD8^+$ CTL response in preventing the development of chronic HCV infection. In this regard, resolution of acute HCV infections has also been observed in hypogammaglobulinemic patients failing to make humoral immune responses to HCV (BJORO et al. 1994; CHRISTIE et al. 1997; ADAMS et al. 1997).

While the above argues persuasively for an important role of HCV-specific T cell responses in controlling HCV infection, there are several publications indicating a potentially important role also for HCV-specific antibodies. Using a recombinant envelope subunit vaccine, chimpanzees have been completely protected against both acute and chronic infection following challenge with a low dose of homologous virus (CHOO et al. 1994). The study indicated that sufficiently high levels of HCV anti-envelope antibodies could provide sterilizing immunity, at least against homologous virus (CHOO et al. 1994). Chimpanzees have also been passively immunized with human Ig preparations containing anti-HCV after experimental challenge. A significant delay in the onset of acute hepatitis was observed clearly suggesting that anti-HCV antibodies can influence the course of infection (KRAWCZYNSKI et al. 1993). This possibility is also supported by three other investigations: Firstly, the resolution of acute HCV infections in humans has been associated with the early development of anti-gpE2 antibodies (KOBAYASHI et al. 1997). Two other groups have concluded that the early induction of antibodies to gpE2 HVR associates with recovery from acute infection and viral clearance (ALLANDER et al. 1997; LECHNER et al. 1998). While in these cases, HCV-specific T cell responses were not investigated, they do suggest a potential role of antibody in influencing the course of infection and disease. This possibility is further supported by older studies (KNODELL et al. 1976, 1977; SANCHEZ-QUIJANO et al. 1988; KIKUGHI and TATEDA 1980) showing that administration of human Ig preparations prior to blood transfusion reduced the incidence of non-A, non-B hepatitis (now mainly attributable to HCV). More recent studies have also indicated the protective efficacy of human Ig (PIAZZA et al. 1997; FÉRAY et al. 1998). Finally, while spontaneous resolution of chronic HCV infections is a very rare event, it has been suggested that such events are associated with an increase in anti-gpE2 antibody titers, (ISHII et al. 1998), as assayed using a format measuring the ability of patient serum to block the binding of recombinant gpE2 to a human T cell-line (the "NOB" assay; ROSA et al. 1996).

In conclusion, it is encouraging to vaccine development that correlates of immunity to both acute and chronic HCV infection are beginning to emerge. While findings are generally preliminary at this stage and require confirmation, they clearly provide a direction for HCV vaccine development strategies. It is important to note however that other components of the immune response and their potential role in prevention of chronic infection deserve attention such as interferon levels and the role of natural killer (NK) cells. A review of the two main areas of vaccine development to-date now follows.

3 Recombinant Subunit Proteins as Vaccine Candidates

Hepatitis C virus is thought to encode two envelope glycoproteins (gpE1 and gpE2) which are processed from the polyprotein precursor by signal peptidase located within the lumen of the endoplasmic reticulum (ER) (CHOO et al. 1991; HIJIKATA et al. 1991). In mammalian cells transfected with HCV cDNA, gpE1 and gpE2 have been found anchored to the lumen of the ER via COOH-terminal transmembrane anchor sequences (SPAETE et al. 1992; RALSTON et al. 1993). When expressed from their complete gene templates, gpE1 and gpE2 exist as high mannose-containing immature glycoproteins in the form of a noncovalently bonded heterodimer (gpE1/gpE2) that is thought to reflect at least in part, the structure of these glycoproteins in the native virion (RALSTON et al. 1993; DUBUISSON et al. 1994; DELEERSNYDER et al. 1997). This heterodimer has therefore been the subject of vaccine studies. Initially, gpE1/gpE2 was purified from HeLa cells infected with a recombinant vaccinia virus expressing a C-gpE1-gpE2-p7-NS2' gene cassette from the HCV-1 isolate. This was then combined with an oil/water microemulsified adjuvant for vaccination of chimpanzees. In most cases, animals received a three-dose regimen at months 0, 1 and 7, after which 10 chimpanzee infectious doses ($10CID_{50}$) of homologous HCV-1 virus were administered i.v. 2–3 weeks following the third immunization. While this inoculum reproducibly infected control animals, five of the seven chimpanzee vaccinees failed to show any signs of viral infection even using very sensitive RT-PCR assays to monitor viremia in the circulation and within liver biopsy samples (CHOO et al. 1994). The five completely protected animals exhibited the highest anti-gpE1/gpE2 (ELISA) antibody titers as well as the highest anti-gpE2 antibody titers as measured in the NOB assay (CHOO et al. 1994; HOUGHTON et al. 1997). There is some evidence suggesting that neutralisation of HCV is mediated through anti-gpE2 HVR antibodies and as such may be isolate-specific (WEINER et al. 1992; KATO et al. 1994; SHIMIZU et al. 1994; FARCI et al. 1996). However, protection of the vaccinated chimpanzees did not correlate with anti-gpE2 HVR levels (HOUGHTON et al. 1997).

Further vaccine studies have also been conducted using gpE1/gpE2 heterodimer derived from a constitutively-expressing Chinese hamster ovary cell-line (SPAETE et al. 1992; HOUGHTON et al. 1997). Although exhibiting lower immunogenicity than the HeLa-derived vaccine thus leading to infection of vaccinees following experimental challenge with homologous virus, resolution of acute infection occurred at a higher rate than in the unvaccinated control group (HOUGHTON et al. 1997). The two lowest-responders to the HeLa gpE1/gpE2 vaccine also resolved their acute infections (CHOO et al. 1994). Reboosting of three chimpanzee vaccinees (that were previously completely protected against homologous viral challenge) followed by experimental rechallenge with the heterologous HCV-H isolate led to infection in all cases, although infection was delayed and inhibited in two of the three vaccinees (HOUGHTON et al. 1997). It is unclear at present if this result indicates an absence or reduced levels of cross-neutralizing antibody generated by the vaccine or whether the ensuing HCV-H infections were as a result of the higher dose employed as compared with the homologous viral challenges. In this regard, guinea pigs immunized with the HeLa gpE1/gpE2 vaccine have been shown to generate antibodies that can block the binding of recombinant gpE2 derived from either subtype 1a or 1b to Molt-4 T cells in the NOB assay (HOUGHTON et al. 1997). While this suggests that this vaccine may be capable of generating cross-neutralizing antibodies, further chimpanzee challenge work is required to confirm this key point. However, recent data from additional chimpanzee vaccinees challenged with HCV-H indicate that as in the case of homologous viral challenges, prior vaccination results in resolution of the acute infections in most animals (M. Houghton; unpublished).

Viral-like particles (VLPs) have been reported to be produced in insect cell systems expressing a C-gpE1-gpE2-p7-NS2' gene cassette. These 40 nm intracellular VLPs appear to comprise a lipid envelope containing gpE1 and gpE2 encapsidating C (BAUMERT et al. 1998). If confirmed and able to be produced at high levels, these VLPs may offer an attractive formulation for a recombinant protein vaccine. It is also possible that recombinant gpE2 alone may be a protective antigen since protection of chimpanzees vaccinated with the gpE1/gpE2 heterodimer was associated with anti-gpE2 antibody titers (HOUGHTON et al. 1997). While gpE1 requires gpE2 for correct folding, evidence has been presented that gpE2 folds independently of gpE1 (MICHALAK et al. 1997). Various COOH-terminal truncated forms of gpE2 have been investigated with respect to folding and secretion. Surprisingly, gpE2 terminating at amino acid 661 appears to assume a more native structure than a longer version terminating at amino acid 715 (MICHALAK et al. 1997). The secreted version of gpE2(661) may also represent a vaccine candidate (SPAETE et al. 1992; MICHALAK et al. 1997). Partly because gpE1 requires the presence of gpE2 for correct folding, the role of anti-gpE1 in protective immunity is currently unknown.

Clearly, important questions remain concerning the development of a recombinant subunit protein vaccine for HCV. These include the question of cross-neutralizing antibody responses, the duration of effective immunity, the effectiveness of the vaccine against higher viral doses, and the efficacy (and mechanism) of vaccinating with envelope glycoproteins on resolving subsequent infection. However, the preliminary data reported so far encourages further work in this area.

4 Hepatitis C Virus Nucleic Acids as Vaccine Candidates

The original demonstration in 1990 of protein expression following i.m. injection of naked DNA (WOLFF et al. 1990) has led to extensive application of this technology to vaccine development (TANG et al. 1992). Many groups studying a variety of different pathogens have shown the successful induction of humoral and cellular immune responses which can confer protective immunity in a variety of species, including primates (WILLIAMS et al. 1993; ULMER et al. 1993; XIANG et al. 1994; WANG et al. 1994; YOKOYAMA et al. 1995; MANICKAN et al. 1995; MARTINS et al. 1995; DONELLY et al. 1995). One of the major advantages of nucleic acid vaccines is in their ability to prime MHC class I-restricted $CD8^+$ CTL responses as a result of stimulating endogenous protein synthesis in vivo. Since CTLs may be important in HCV clearance, an HCV vaccine that primes such responses could be effective. Other potential advantages of an HCV nucleic acid vaccine include the relative ease of manufacture as compared with recombinant proteins and the ability to deliver either a single nucleic acid with multiple HCV gene cassettes or multiple nucleic acids in order to protect against the wide spectrum of HCVs. There is also the option of using a vaccine regimen in which nucleic acids are delivered first, in order to prime the immune response, followed by boosting with recombinant protein. Such a regimen may strengthen as well as broaden the immune response.

Several groups have investigated immune responses in mice following immunization with naked plasmid DNAs in which various HCV gene cassettes have been placed under the control of a cytomegalovirus (CMV) promoter. The HCV nucleocapsid (C) gene has received the most attention so far since it is the most conserved gene between different HCV isolates (BUKH et al. 1994), C-specific $CD4^+$ T helper responses have been associated with amelioration of disease (BOTARELLI et al. 1993; LECHMANN et al. 1996), and in addition, MHC class I-restricted $CD8^+$ CTL epitopes are known to exist within this protein (WONG et al. 1998; NELSON et al. 1997). Various investigators have shown that i.m. inoculation of naked DNA encoding HCV C can generate humoral immune responses, as well as lymphoproliferative responses and CTL activity (LAGGING et al. 1995; TOKUSHIGE et al. 1996; INCHAUSPÉ et al. 1997). Interestingly, co-administration of plasmid DNAs encoding cytokines such as interleukin (IL)2 or granulocyte/macrophage colony-stimulating factor (GM-CSF) have been shown to augment all of these immune responses to C (GEISSLER et al. 1997).

Naked plasmid DNA expressing HCV gpE2 has also been used successfully to induce anti-gpE2 antibodies in immunized mice. While i.m. injection of DNA was successful at inducing seroconversion, substantially higher seroconversion rates and antibody titers were obtained when the DNA was administered by gene-gun intraepidermally (NAKANO et al. 1997). Interestingly, while various subdomains of gpE2 were able to generate anti-gpE2 antibody responses, only DNA encoding a COOH-terminal truncated form of gpE2 (terminating at amino acid 674) was capable of generating antibodies that could bind the gpE1/gpE2 heterodimer (NAKANO et al. 1997).

Gene therapy programs involving gene delivery via various defective viral vectors also offer very attractive possibilities for HCV vaccination. In addition to the same benefits existing for naked DNA immunization, an appropriate viral vector may result in wider and more efficient expression and better presentation to the immune system (especially if antigen presenting cells can be targeted). The use of a defective RNA viral vector (not replicating through a DNA intermediate) would also completely remove safety concerns connected with potential integration of DNA vectors into the host genome (DRIVER et al. 1998).

Immunization of BALB/c mice with a replication-defective recombinant adenovirus expressing the HCV C and gpE1 genes has been shown to induce CTL responses to both C and gpE1 epitopes. Importantly, CTL activity was detectable for more than 100 days, suggesting that this or a similar vaccination method may result in enduring immunity to HCV (BRUNA-ROMERO et al. 1997).

It is clear that there is much potential in the use of naked or vectored HCV nucleic acids in vaccine development. Future challenges in this area include reproducing in primates the broad immune responses so far observed in mice and proving efficacy in the chimpanzee challenge model. The delivery of the HCV nucleic acid needs to be optimized since most mouse protocols involve the injection of several hundred micrograms of DNA in order to get generally weak immune responses. In particular, the targeting of naked or vectored HCV nucleic acids to antigen presenting cells (APCs) represents an important area of research (BOYLE et al. 1998). The development of an HCV nucleic acid vaccine may also be associated with some specific safety concerns. Firstly, part of the NS3 gene has been shown to transform mouse 3T3 cells and to be tumorigenic in nude mice (SAKAMURO et al. 1995). When co-transfected with *ras*, HCV C transforms primary rat embryo fibroblasts (RAY et al. 1996a) and HCV C has also been reported to transcriptionally regulate various cellular and viral genes (RAY et al. 1995), to repress the p53 promoter (RAY et al. 1997), to induce steatosis in transgenic mice (MORIYA et al. 1997), and to bind to a member of the tumor necrosis factor (TNF) receptor family (with unknown biological effects; MATSUMOTO et al. 1997; CHEN et al. 1997). The HCV C protein has also been suggested to affect apoptosis (RAY et al. 1996b; RUGGIERI et al. 1997). These observations need to be considered carefully in the development of an HCV genetic vaccine. Finally, any HCV vaccine, especially one capable of inducing a CTL response, would have to be carefully evaluated for its effects when administered to pre-existing HCV carriers, since there will be the potential for inflammatory responses within pre-infected livers. However, boosting HCV CTL's in patients may also result in improved responses to interferon therapy (NELSON et al. 1998).

5 Concluding Remarks

Complete protection of chimpanzees from experimental challenge following vaccination with recombinant gpE1/gpE2 encourages further investigations of this vaccine strategy. It will be necessary in the future to demonstrate efficacy at least against viruses of the same subtype as the originating vaccine strain. Protection against other HCV subtypes and types may be obtained either by a demonstrable cross-protective activity or by the inclusion of vaccine antigens from additional genotypes. Enduring immunity also needs to be demonstrated in vaccinees. At the T cell levels, it seems likely that HCV-specific $CD4^+$ and $CD8^+$ T cell responses are important for resolution of acute infection and prevention of chronicity. Since HCV-associated disease is mainly a sequela of chronic infection, an HCV vaccine would still be effective if it only stimulates recovery from acute infection. Given the known heterogeneity of HCV, it will obviously be important to vaccinate and prime $CD4^+$ and $CD8^+$ T cell responses to many conserved viral epitopes in association with multiple MHC alleles representative of the general population. It is then to be hoped that the vaccine will be effective at priming an earlier, stronger and broader immune response in newly infected vaccinees than would occur in unvaccinated individuals, leading to viral clearance in most vaccinees. It is also to be hoped that in chronically infected HCV patients, boosting and broadening the humoral and cellular immune responses may ameliorate the course of infection and disease either alone or in combination with antiviral drugs. In conclusion, there are many questions that remain to be answered in this challenging area of HCV research and development.

References

Adams G, Kuntz S, Rabalais G, Bratcher D, Tamburro CH et al. (1997) Natural recovery from acute hepatitis C virus infection by agammaglobulinemic twin children. Pediatr Infect Dis J 16:533–534

Allander T, Beyene A, Jacobson SH, Grillner L, Persson MAA (1997) Patients infected with the same hepatitis C virus strain display different kinetics of the isolate-specific antibody response. J Infect Dis 175:26–31

Alter HJ (1989) Chronic consequences of non-A, non-B hepatitis. In: Seef LB, Lewis JH (eds) Current prospective in hepatology. Plenum, New York, pp 83–97

Alter MJ (1993) The detection, transmission, and outcome of hepatitis C virus infection. Infect Agents Dis 2:155–166

Battegay M, Fikes J, Di Bisceglie AM, Wentworth PA, Sette A et al. (1995) Patients with chronic hepatitis C have circulating cytotoxic T cells which recognize hepatitis C virus-encoded peptides binding to HLA-A2.1 molecules. J Virol 69:2462–2470

Baumert TF, Ito S, Wong DT, Liang TJ (1998) Hepatitis C virus structural proteins assemble into virus-like particles in insect cells. J Virol 72:3827–3836

Bjoro K, Froland SS, Yun Z, Samdal HH, Haaland T (1994) Hepatitis C infection in patients with primary hypogammaglobulinemia after treatment with contaminated immune globulin. N Engl J Med 331:1607–1611

Botarelli P, Brunetto MR, Minutello MA et al. (1993) T-lymphocyte response to hepatitis C virus in different clinical courses of infection. Gastroenterology 104:580–587

Boyle JS, Brady JL, Lew AM (1998) Enhanced responses to a DNA vaccine encoding a fusion antigen that is directed to sites of immune induction. Nature 392:408–411

Bruna-Romero O, Lasarte JJ, Wilkinson G, Grace K, Clarke B et al. (1997) Induction of cytotoxic T-cell response against hepatitis C virus structural antigens using a defective recombinant adenovirus. Hepatology 25:470–477

Bukh J, Emerson SU, Purcell RH (1997) Genetic heterogeneity of hepatitis C virus and related viruses. In: Rizzetto M, Purcell RH, Gerin JL, Verme G (eds) Viral hepatitis and liver disease. Edizioni Minerva Medica, Torino, pp 167–175

Bukh J, Purcell RH, Miller RH (1993) At least 12 genotypes of hepatitis C virus predicted by sequence analysis of the putative E1 gene of isolates collected worldwide. Proc Natl Acad Sci USA 90:8234–8238

Bukh J, Purcell RH, Miller RH (1994) Sequence analysis of the core gene of 14 hepatitis C virus genotypes. Proc Natl Acad Sci USA 91:8239–8243

Cerny A, McHutchison JG, Pasquinelli C, Brown ME, Brothers MA et al (1995) Cytotoxic T lymphocyte response to hepatitis C virus-derived peptides containing the HLA A2.1 binding motif. J Clin Invest 95:521–530

Chen C-M, You L-R, Hwang L-H, Wu Lee Y-H (1997) Direct interaction of hepatitis C virus core protein with the cellular lymphotoxin-β receptor modulates the signal pathway of the lymphotoxin-β receptor. J Virol 71:9417–9426

Chien DY, Choo Q-L, Ralston R, Spaete R et al. (1993) Persistence of HCV despite antibodies to both putative envelope glycoproteins. Lancet 342:933

Choo Q-L, Kuo G, Ralston R et al (1994) Vaccination of chimpanzees against infection by the hepatitis C virus. Proc Natl Acad Sci USA 91:1294–1298

Choo Q-L, Richman KH, Han JH et al. (1991) Genetic organization and diversity of the hepatitis C virus. Proc Natl Acad Sci USA 88:2451–2455

Christie JML, Healey CJ, Watson J, Wong VS, et al. (1997) Clinical outcome of hypogammaglobulinaemic patients following outbreak of acute hepatitis C: 2 year follow-up. Clin Exp Immunol 110:4–8

Conry-Cantilena C, Vanraden M, Gibble J, Melpoler J, Shakil AO et al. (1996) Routes of infection, viremia, and liver disease in blood donors found to have hepatitis C virus infection. N Engl J Med 334:1691–1696

Cooper S, Erickson AL, Adams EJ, Kansspan J, Weiner AJ et al. (1999) Analysis of a successful immune response against hepatitis C virus. Immunity 10:439–449

Deleersnyder V, Pillez A, Wychowski C, Blight K, Xu J et al. (1997) Formation of native hepatitis C virus glycoprotein complexes. J Virol 71:697–704

Diepolder HM, Gerlach J-T, Zachoval R, Hoffman RM, Jung M-C et al. (1997) Immunodominant CD4[+] T-cell epitope within nonstructural protein 3 in acute hepatitis C virus infection. J Virol 71:6011–6019

Diepolder HM, Zachoval R, Hoffman RM, Wierenga EA, Santantonio T et al. (1995) Possible mechanism involving T-lymphocyte response to non-structural protein 3 in viral clearance in acute hepatitis C virus infection. Lancet 346:1006–1007

Donelly JJ, Friedman A, Martinez D, Montgomery DL, Shiver JW et al. (1995) Preclinical efficacy of a prototype DNA vaccine: enhanced protection against antigenic drift in influenza virus. Nat Med 1:583–587

Driver DA, Polo JM, Belli BA, Banks TA, Mangala H, Dubensky TW Jr (1998) Plasmid DNA-based alphavirus expression vectors for nucleic acid immunization. Mol Ther 1:510–520

Dubuisson J, Hsu HH, Cheung RC, Greenberg HB, Russell DG, Rice CM (1994) Formation and intracellular localization of hepatitis C virus envelope glycoprotein complexes expressed by recombinant vaccinia and Sindbis viruses. J Virol 68:6147–6160

Farci P, Alter HJ, Govindarajan S et al. (1992) Lack of protective immunity against reinfection with hepatitis C virus. Science 258:135–140

Farci P, Shimoda A, Wong D, Cabezon T, De Gioannis D et al. (1996) Prevention of hepatitis C virus infection in chimpanzees by hyperimmune serum against the hypervariable region 1 of the envelope 2 protein. Proc Natl Acad Sci USA 93:15394–15399

Féray C, Gigai M, Samuel D, Ducot B, Maissonneuve P et al. (1998) Incidence of hepatitis C in patients receiving different preparations of hepatitis B immunoglobulins after liver transplantation. Ann Intern Med 128:810–816

Geissler M, Gesien A, Tokushige K, Wands JR (1997) Enhancement of cellular and humoral immune responses to hepatitis C virus core protein using DNA-based vaccines augmented with cytokine-expressing plasmids. J Immunol 158:1231–1237

Hijikata M, Kato N, Ootsuyama Y, Nakagawa M, Shimotohno K (1991) Gene mapping of the putative structural region of the hepatitis C virus genome by in vitro processing analysis. Proc Natl Acad Sci USA 88:5547–5551

Houghton M, Choo Q-L, Chien D, Kuo G, Weiner A et al. (1997) Development of an HCV vaccine. In: Rizzetto M, Purcell RH, Gerin JL, Verme G (eds) Viral hepatitis and liver disease. Edizioni Minerva Medica, Torino, pp 656–659

Inchauspé G, Major ME, Nakano I, Vitvitski L, Trépo C (1997) DNA vaccination for the induction of immune responses against hepatitis C virus proteins. Vaccine 15:853–856

Ishii K, Rosa D, Watanabe Y, Katayama T, Harada H et al. (1998) High titers of envelope neutralizing antibodies correlate with natural resolution of chronic hepatitis C. Hepatology (in press)

Kato N, Ootsuyama Y, Sekiya H et al. (1994) Genetic drift in hypervariable region 1 of the viral genome in persistent hepatitis C virus infection. J Virol 68:4776–4784

Kikughi K, Tateda A (1980) A trial for prevention of serum hepatitis with intravenous human gamma-globulin (venoglobulin). J Jpn Soc Blood Transfus 24:2–8

Kita H, Moriyama T, Kaneko T et al. (1993) HLA B44-restricted cytotoxic T lymphocytes recognizing an epitope on hepatitis C virus nucleocapsid protein. Hepatology 18:1039–1044

Knodell RG, Conrad ME, Ginsberg AL, Bell CJ (1976) Efficacy of prophylactic gamma-globulin in preventing non-A, non-B posttransfusion hepatitis. Lancet 1:557–561

Knodell RG, Conrad NE, Ishak KG (1977) Development of chronic liver disease after acute non-A, non-B post-transfusion hepatitis: role of gamma globulin prophylaxis in its prevention. Gastroenterology 72:902–909

Kobayashi M, Tanaka E, Matsumoto A, Ichijo T, Kiyosawa K (1997) Antibody response to E2/NS1 hepatitis C virus protein in patients with acute hepatitis C. J Gastroenterol Hepatol 12:73–76

Krawczynski K, Alter MJ, Tankersley DL et al. (1993) Studies on protective efficacy of hepatitis C immunoglobulins (HCIG) in experimental hepatitis C virus infection. Hepatology 18:110

Lagging LM, Meyer K, Hoft D, Houghton M, Belshe RB, Ray R (1995) Immune responses to plasmid DNA encoding the hepatitis C virus core protein. J Virol 69:5859–5863

Lai ME, Mazzoleni AP, Argiolu F, Virgilis SD, Balesterieri A et al. (1994) Hepatitis C virus in multiple episodes of acute hepatitis in polytransfused thalassaemic children. Lancet 343:388–390

Lechmann M, Ihlenfeldt HG, Braunschweiger I, Giers G, Jung G et al. (1996) T- and B-cell responses to different hepatitis C virus antigens in patients with chronic hepatitis C infection and in healthy anti-hepatitis C virus-positive blood donors without viremia. Hepatology 24:790–795

Lechner S, Rispeter K, Meisel H, Kraas W, Jung G et al. (1998) Antibodies directed to envelope proteins of hepatitis C virus outside of hypervariable region 1. Virology 243:313–321

Manickan E, Rouse RJ, Yu Z, Wire WS, Rouse BT (1995) Genetic immunization against herpes simplex virus: Protection is mediated by $CD4^+$ T lymphocytes. J Immunol 155:259–265

Martell M, Esteban JI, Quer J et al. (1992) Hepatitis C virus (HCV) circulates as a population of different but closely related genomes: quasispecies nature of HCV genome distribution. J Virol 66:3225–3229

Martins LP, Lau LL, Asano MS, Ahmed R (1995) DNA vaccination against persistent viral infection. J Virol 69:2574–2582

Matsumoto M, Hsieh T-Y, Zhu N, vanArsdale T, Hwang AB et al. (1997) Hepatitis C virus interacts with the cytoplasmic tail of lymphotoxin-β receptor. J Virol 71:1301–1309

Michalak J-P, Wychowski C, Choukhi A, Meunier J-C, Ung S et al. (1997) Characterization of truncated forms of hepatitis C virus glycoproteins. J Gen Virol 78:2299–2306

Minutello MA, Pileri P, Unutmaz et al. (1993) Compartmentalization of T lymphocytes to the site of disease: intrahepatic $CD4^+$ T cells specific for the protein NS4 of hepatitis C virus in patients with chronic hepatitis C. J Exp Med 178:17–25

Missale G, Bertoni R, Lamonaca V, Valli A, Massari M et al. (1996) Different clinical behaviors of acute hepatitis C virus infection are associated with different vigor of the anti-viral cell-mediated immune response. J Clin Invest 98:706–714

Moriya K, Yotsuyanagi H, Shintani Y, Fujie H, Ishibashi K et al. (1997) Hepatitis C virus core protein induces hepatic steatosis in transgenic mice. J Gen Virol 78:1527–1531

Nakano I, Maertens G, Major ME, Vitvitski L, Dubuisson J et al. (1997) Immunization with plasmid DNA encoding hepatitis C virus envelope E2 antigenic domains induces antibodies whose immune reactivity is linked to the injection mode. J Virol 71:7101–7109

Nelson DR, Marousis CG, Davis GL, Rice CM, Wong J et al. (1997) The role of hepatitis C virus-specific cytotoxic T lymphocytes in chronic hepatitis C. J Immunol 158:1473–1481

Nelson DR, Marousis CG, Ohno T, Davis GL, Lau JYN (1998) Intrahepatic hepatitis C virus-specific cytotoxic T lymphocyte activity and response to interferon alfa therapy in chronic hepatitis C. Hepatology 28:225–230

Ogata N, Alter HJ, Miller RH, Purcell RH (1991) Nucleotide sequence and mutation rate of the H strain of hepatitis C virus. Proc Natl Acad Sci USA 88:3392–3396

Piazza M, Sagliocca L, Tosone G, Guadagnino V, Stazi MA et al. (1997) Sexual transmission of the hepatitis C virus and efficacy of prophylaxis with intramuscular immune serum globulin. Arch Intern Med 157:1537–1544

Prince AM, Brotman B, Huima T, Pascual D, Jaffery M et al. (1992) Immunity in hepatitis C infection. J Infect Dis 165:438–443

Ralston R, Thudium K, Berger K et al. (1993) Characterization of hepatitis C virus envelope glycoprotein complexes expressed by recombinant vaccinia viruses. J Virol 67:6753–6761

Ray RB, Lagging LM, Meyer K, Steele R, Ray R (1995) Transcriptional regulation of cellular and viral promoters by the hepatitis C virus core protein. Virus Res 37:209–220

Ray RB, Lagging LM, Meyer K, Ray R (1996a) Hepatitis C virus core protein cooperates with ras and transform primary rat embryo fibroblasts to tumorigenic phenotype. J Virol 70:4438–4443

Ray RB, Meyer K, Ray R (1996b) Suppression of apoptotic cell death by hepatitis C virus core protein. Virology 226:176–182

Ray RB, Steele R, Meyer K, Ray R (1997) Transcriptional repression of p53 promoter by hepatitis C virus core protein. J Biol Chem 272:10983–10986

Rehermann B, Chang K-M, McHutchison J, Kokka R, Houghton M et al. (1996a) Differential cytotoxic T-lymphocyte responsiveness to the hepatitis B and C viruses in chronically infected patients. J Virol 70:7092–7102

Rehermann B, Chang K-M, McHutchison JG, Kokka R, Houghton M, Chisari FV (1996b) Quantitative analysis of the peripheral blood cytotoxic T lymphocyte response in patients with chronic hepatitis C virus infection. J Clin Invest 98:1432–1440

Rosa D, Campagnoli S, Moretto C, Guenzi E, Cousens L et al. (1996) A quantitative test to estimate neutralising antibodies to the hepatitis C virus: cytofluorimetric assessment of envelope glycoprotein 2 binding to target cells. Proc Natl Acad Sci USA 93:1759–1763

Ruggieri A, Harada T, Matsuura Y, Miyamura T (1997) Sensitization to fas-mediated apoptosis by hepatitis C virus core protein. Virology 229:68–76

Sakamuro D, Furukawa T, Takegami T (1995) Hepatitis C virus nonstructural protein NS3 transforms NIH3T3 cells. J Virol 69:3893–3896

Sanchez-Quijano A, Lisen E, Diaz-Torres MA, Rivera F, Pineda JA et al. (1988) Prevention of post-transfusion non-A, non-B hepatitis by non-specific immunoglobulin in heart surgery patients. Lancet 1:1245–1249

Shimizu YK, Hijikata M, Iwamoto A, Alter HJ, Purcell RH et al. (1994) Neutralizing antibodies against hepatitis C virus and the emergence of neutralization escape mutant viruses. J Virol 68:1494–1500

Simmonds P, Mellor J, Nuchprayoon C, Tanprasert S, Smith DB (1997) Molecular epidemiology and classification of variants of hepatitis C virus found in South East Asia. In: Rizzetto M, Purcell RH, Gerin JL, Verme G (eds) Viral hepatitis and liver disease. Edizioni Minerva Medica, Torino, pp 187–194

Spaete RR, Alexander D, Rugroden ME et al. (1992) Characterization of the hepatitis C virus E2/NS1 gene product expressed in mammalian cells. Virology 188:819–830

Tang D, DeVit M, Johnston SA (1992) Genetic immunization is a simple method for eliciting an immune response. Nature 356:152–154

Thomssen R, Bonk S, Propfe C, Heermann K-H, Köchel HG et al. (1992) Association of hepatitis C virus in human sera with β-lipoprotein. Med Microbiol Immunol 181:293–300

Tokushige K, Wakita T, Pachuk C, Moradpour D, Weiner DB et al. (1996) Expression and immune response to hepatitis C virus core DNA-based vaccine constructs. Hepatology 24:14–20

Tsai S-L, Liaw Y-F, Chen M-H, Huang C-Y, Kuo GC (1997) Detection of type 2-like T-helper cells in hepatitis C virus infection: implications for hepatitis C virus chronicity. Hepatology 25:449–458

Ulmer JB, Donnelly JJ, Parker SE, Rhodes GH, Felgner PL et al. (1993) Heterologous protection against influenza by injection of DNA encoding a viral protein. Science 259:1745–1749

Wang B, Merva M, Dang K, Ugen KE, Boyer J et al. (1994) DNA inoculation induces protective in vivo immune responses against cellular challenge with HIV-1 antigen-expressing cells. AIDS Res Hum Retroviruses 4:S35

Weiner A, Erickson AL, Kansopon J, Crawford K, Muchmore E et al. (1995) Persistent hepatitis C virus infection in a chimpanzee is associated with emergence of a cytotoxic T lymphocyte escape variant. Proc Natl Acad Sci USA 92:2755

Weiner AJ, Geysen HM, Christopherson C et al. (1992) Evidence for immune selection of hepatitis C virus (HCV) putative envelope glycoprotein variants: potential role in chronic HCV infections. Proc Natl Acad Sci USA 89:3468-3472

Williams WV, Boyer JD, Merva M, Livolsi V, Wilson B et al. (1993) Genetic infection induces protective in vivo immune responses. DNA Cell Biol 12:675-683

Wolff JA, Malone RW, Williams P, Chong W, Acsadi G et al. (1990) Direct gene transfer into mouse muscle in vivo. Science 247:1465-1468

Wong DKH, Dudley DD, Afdhal NH, Dienstag J, Rice CM et al. (1998) Liver-derived CTL in hepatitis C virus infection: breadth and specificity of responses in a cohort of persons with chronic infection. J Immunol 160:1479-1488

Xiang ZQ, Spitalnik S, Tran M, Wunner WH, Cheng J, Ertl HC (1994) Vaccination with a plasmid vector carrying the rabies virus glycoprotein gene induces protective immunity against rabies virus. Virology 199:132-140

Yokoyama M, Zhang J, Whitton JL (1995) DNA immunization confers protection against lethal lymphocytic choriomeningitis virus infection. J Virol 69:2684-2688

The GB Viruses

J.N. Simons, S.M. Desai, and I.K. Mushahwar

1	The GB Agent	342
2	Discovery of the GB Viruses	344
2.1	Identification of GBV-A and GBV-B	344
2.2	Identification of GBV-C	346
2.3	Identification of Hepatitis G Virus	347
2.4	Identification of GBV-A Variants	348
3	Molecular Analyses	349
3.1	Sequence Characteristics	350
3.1.1	5′ NTRs	350
3.1.2	Structural Proteins	351
3.1.3	Nonstructural Proteins	353
3.1.4	3′ NTRs	354
3.2	Phylogenetic Relationships	355
3.2.1	GBVs' Relationship to the *Flaviviridae*	355
3.2.2	GBV-C Genotypes	355
3.2.3	GBV-A Variants	358
4	Biochemical Analyses	359
4.1	Translation Initiation	359
4.1.1	GBV-B	359
4.1.2	GBV-A and GBV-C	359
4.2	Proteases	360
4.2.1	GBV-B	360
4.2.2	GBV-C	361
5	Physical Analyses	361
6	Detection Methods	362
6.1	Viral RNA	363
6.2	Antiviral Antibodies	363
7	Route of Infection	364
8	Site(s) of Replication	365
9	Association with Disease	366
10	Immunity and Treatment	368
11	Future Direction	368
	References	369

Virus Discovery Group, Experimental Biology Research, Abbott Laboratories, Dept. 90D, L3, 1401 Sheridan Rd., North Chicago, IL 60064-6269, USA

1 The GB Agent

The lack of animal models of infection and in vitro systems for propagating hepatic viruses hindered much of the early work on the causative agents of viral hepatitis in humans. The use of volunteers had provided valuable information (BOGGS et al. 1970; RIGHTSEL et al. 1961). However, ethical considerations limited experimentation in humans. Suitable animal models were needed to better characterize these viruses.

The first successful nonhuman primate model of viral hepatitis was reported in the mid 1960s. DEINHARDT et al. (1967) reported that sera from two human sources, WW-55 and GB, induced hepatitis in inoculated tamarins, small South American primates of the *Sanguinus* genus. WW-55 was plasma from a 28-year-old male obtained from blood donated 1 day prior to developing hepatitis with jaundice. This sample induced hepatitis in approximately 50% of the human volunteers inoculated (RIGHTSEL et al. 1961). Six of nine WW-55-inoculated tamarins developed biochemical evidence of hepatitis between 35 and 61 days post-inoculation. Two subsequent passages of this material were made with ten of the 14 inoculated animals developing chemical and histological evidence of hepatitis. The second inoculum, GB, was from a 34-year-old surgeon who developed acute hepatitis (total serum bilirubin of 14.4mg/dl and icterus of 4 weeks duration). GB serum obtained on the third day of jaundice was inoculated into four tamarins. All four animals developed abnormal hepatic tests 16–40 days post-inoculation. Five additional passages of this "GB" agent were made with almost all of the animals inoculated showing evidence of hepatitis.

Subsequent to the original Deinhardt study, several inocula of human origin have induced hepatitis in tamarins, most notably MS-1 and the Berlin agent. MS-1, a hepatitis A virus (HAV) inoculum, originated from a pool of 22 serum specimens obtained from ten children with hepatitis (KRUGMAN et al. 1967). The Berlin agent originated from the sera of patients with acute infectious hepatitis that had been passaged several times in primary pig kidney cells. Material from the 11th and 13th passage in tissue culture induced hepatitis in inoculated tamarins (KOHLER and APODACA 1968). The antigenic relationship of these different inocula was addressed in cross-challenge experiments (HOLMES et al. 1973). As one would expect, prior inoculation with GB prevented a subsequent bout of hepatitis induced by GB, and prior inoculation with MS-1 prevented hepatitis induced by a subsequent inoculation of MS-1. However, these inocula did not confer cross-protection; GB-infected animals were still susceptible to MS-1-induced hepatitis, and MS-1 infected animals were not protected from challenge with GB. These data demonstrated that GB and MS-1 were antigenically distinct. In contrast, when a GB-infected animal was challenged with the Berlin agent, no hepatitis was noted. Conversely, prior infection with the Berlin agent prevented hepatitis induced by GB. These results suggest that the GB and Berlin agents are antigenically related. Similar results were obtained with WW-55. Thus, there appears to be two antigenically distinct groups of hepatitis agents in tamarins: MS-1 (which is an HAV source) and GB (which is related to the Berlin agent and WW-55).

The GB agent does not appear to be transmitted easily within tamarin colonies (PARKS et al. 1969). However, evidence of oral transmission similar to that noted for HAV has been reported. Specifically, tamarins that ingested infectious GB passage material developed hepatitis (DEINHARDT et al. 1975a). In addition, a laboratory technician (JW) developed hepatitis 28 days after accidentally swallowing infectious tamarin serum. Four of 12 tamarins inoculated with JW serum developed hepatitis that was passaged two additional times. However, JW worked with blood daily and was hepatitis B surface antigen (HBsAg)-positive. Thus, the etiologic agent of hepatitis in JW and the infected animals may not have been the GB agent (DEINHARDT et al. 1975a,b). Further complicating the issue was the finding that a fecal extract taken from a GB infected tamarin during the acute phase of hepatitis did not induce hepatitis upon intravenous inoculation into additional animals (KARAYIANNIS et al. 1989). Thus, although infectious by a per oral route, the GB agent may not be shed in the feces.

Several additional studies were undertaken to better characterize the nature of the GB agent. Filtration experiments demonstrated that 100% of the animals inoculated with a 50nm filtrate of the GB agent and 80% of the animals inoculated with a 25nm filtrate developed hepatitis (PARKS et al. 1969). In addition, the GB agent was thought to be an enveloped virus as treatment of the inoculum with 50% ether (4°C, 18h) reduced infectivity by 75% (PARKS et al. 1969). Consistent with being a small enveloped virus, the buoyant density of the GB agent in cesium chloride has been reported to be 1.21g/ml (DEINHARDT et al. 1975b; HOLMES et al. 1975).

Electron microscopy studies of the GB agent have been equivocal. ALMEIDA et al. (1976) found aggregates of 20–22nm particles in infectious plasma that had been treated with convalescent serum. The size was consistent with the filtration data, and the aggregation with convalescent serum suggested that these particles were the GB virus. However, these aggregates were also found in infectious plasma alone (ALMEIDA et al. 1976). Similarly, serum from animals infected with the Berlin agent also contains 22nm particles, and 22–24nm particles were detected transiently in the feces of a GB-infected animal. However, others have detected these particles in normal tamarins (APPLETON and DIENSTAG 1977). In contrast to the 22nm particles observed in the serum and feces of infected tamarins, DIENSTAG et al. (1976) detected 34–36nm particles at 1.4g/ml in cesium chloride gradients of acute phase liver homogenates. Surprisingly, these particles could be aggregated with both pre-inoculation and convalescent serum from a GB infected animal, but not with saline or convalescent serum of the surgeon from whom the GB agent was first isolated. However, due to the lack of specific aggregation with convalescent sera, neither the 22nm nor 34nm particles have been definitively associated with the GB agent.

Although the GB agent originated from the inoculation of tamarins with serum from a human suffering from hepatitis, some have questioned whether it is a human virus. PARKS and MELNICK (1969) have proposed that the GB agent is an indigenous (possibly latent) virus of tamarins which was activated upon inoculation of human serum and, once activated, easily passaged in tamarins. This was surmised from several lines of evidence. First, not all tamarins were found to be susceptible to the GB agent, suggesting that the resistant animals may have developed immunity to GB from prior exposure in the wild. Second, some control tamarins developed spontaneous hepatitis that could be passaged much like the GB agent. Third, passage material from a case of spontaneous tamarin hepatitis, WSP-5, was identical to GB passage material in cross-challenge experiments. Unless contamination is invoked to explain these data, the cross-challenge experiment implicates an indigenous tamarin virus as the etiologic agent in the GB passage material, rather than a human hepatitis virus. It should be noted that several other researchers have used tamarins, none of whom reported cases of spontaneous hepatitis (DEINHARDT et al. 1970). In addition, WW-55, which is related to GB in cross-challenge experiments causes hepatitis in human volunteers (RIGHTSEL et al. 1961), suggesting a human origin of the GB agent. Thus, the results of PARKS and MELNICK (1969) appear to be the exception rather than the rule. However, with the discoveries of HAV and hepatitis B virus (HBV) in the early 1970s and other candidates for non-A, non-B hepatitis, further study of the GB agent was not widely pursued into the 1980s.

2 Discovery of the GB Viruses

2.1 Identification of GBV-A and GBV-B

With the discovery of hepatitis C virus (HCV, CHOO et al. 1989) and hepatitis E virus (HEV, REYES et al. 1990) and the development of sensitive antibody and PCR tests for these agents, it became apparent that not all cases of non-A, non-B hepatitis are ascribed to these agents. For instance, ALTER (1994) reported no evidence of HCV infection in 12 of 98 cases from the NIH non-A, non-B hepatitis transfusion series. Similarly, a study of blood donors with elevated serum alanine transaminase (ALT) levels revealed 8% with unknown etiology (MARCELLIN et al. 1993), and 7% of post-transfusion hepatitis patients in France appeared to be cryptogenic (THIERS et al. 1993). Additional studies of fulminant hepatitis and hepatitis-associated aplastic anemia failed to implicate known agents of hepatitis (HIBBS et al. 1992; SALLIE et al. 1994). Together these studies, and others, have suggested the existence of a non-A, non-B, non-C, non-E (non-A-E) hepatitis agent. With the origin of GB's hepatitis unresolved, we decided to reexamine the GB agent with the goal of identifying the viral genome in infected animals. The specific diagnostic tools obtained from this work would allow us to assess the role of the GB agent in cases of non-A-E hepatitis.

The method employed to clone the GB agent was a subtractive PCR technique known as representational difference analysis (RDA). This technique, originally designed to clone the difference between two complex but related genomes (LISITSYN et al. 1993), was modified to selectively amplify unique nucleic acid sequences present in infectious tamarin plasma but absent from pre-inoculation plasma (SIMONS et al. 1995a).

Preinoculation and infectious plasma were obtained from a GB-infected tamarin, T-1053 (SCHLAUDER et al. 1995a). Because the genome of the GB agent could be single or double stranded DNA or RNA, the nucleic acids extracted from the T-1053 plasma were converted to double-stranded DNA by randomly primed reverse transcription and second strand synthesis. Double-stranded DNA from the infectious and preinoculation plasma was subjected to RDA. Products obtained following the second and third rounds of selective amplification were analyzed (SIMONS et al. 1995a). Ten of 11 unique RDA-derived clones were found to be exogenous from tamarin, human, yeast, and *E. coli* genomic DNA by PCR. These clones were absent from pre-inoculation plasma but were present as RNA molecules in the cloning source, T-1053, and additional GB-infected animals. Thus, these clones appeared to be derived from viral sequences present in the infectious tamarin plasma. Additional sequences contiguous with the RDA-derived clones were obtained from several PCR-based experiments. Surprisingly, analysis of the sequences demonstrated that acute phase T-1053 plasma contained two distinct RNA genomes. These genomes both possess limited sequence identity to each other and to members of the HCV group. These data demonstrated the presence of two distinct viruses in the GB-infected tamarin T-1053, namely GB virus A (GBV-A) and GB virus B (GBV-B). Furthermore, the existence of two viruses is supported by studies demonstrating that the two RNA genomes can be filtered, diluted and passaged separately in tamarins (SCHLAUDER et al. 1995a).

Animal passage studies indicate that GBV-B alone is sufficient to cause hepatitis in infected animals. In contrast, hepatitis does not occur in GBV-A-only infected tamarins (SCHLAUDER et al. 1995a). These results, together with finding GBV-B RNA but not GBV-A RNA by northern hybridization of liver from a dually infected tamarin (SIMONS et al. 1995a), strongly suggest that GBV-B causes hepatitis in infected tamarins. Because GBV-B, and not GBV-A, induces biochemical hepatitis, it is likely that much of the early GB work (which followed liver enzyme levels in the serum) may only relate to GBV-B.

A simple reverse transcription-polymerase chain reaction (RT-PCR) assay of the original GB serum (3 days post-presentation) inoculated into tamarins would answer the question of which virus caused hepatitis in the patient, GB. Unfortunately, it appears that this inoculum no longer exists. GB serum 3 and 8 weeks post-presentation have consistently tested negative for GBV-A and GBV-B in RT-PCR assays (SCHLAUDER et al. 1995a). The absence of RT-PCR signal in these sources may be due to the transient nature of viremia in GB and/or RNA degradation. Degradation appears likely since this GB serum had been stored at −20°C for almost 30 years. Thus, we were not able to link GBV-A or GBV-B directly to the original patient, GB. In addition, neither of these viruses has been found in many other human samples that have been tested.

An indirect link of GBV-A, but not GBV-B, to the original patient is found in serologic analysis of 8 week post-presentation GB serum. Specifically, two overlapping proteins from the NS5 region of GBV-A exhibited weak reactivity with GB serum in Western blot assays (PILOT-MATIAS et al. 1996a). In contrast, no immunoreactivity to GBV-B proteins was detected with this serum. However, the absence of pre-presentation serum from GB prevents us from determining whether the GBV-A seroreactivity detected in GB resulted from a seroconversion event or whether this seroreactivity reflects the presence of crossreactive antibodies induced by an antigen unrelated to GB's hepatitis.

2.2 Identification of GBV-C

The cloning of two complete virus genomes from the GB agent provided tools to investigate whether cases of non-A-E hepatitis are attributed to GBV-A and/or GBV-B. To determine seroprevalence to these viruses in various human populations, regions of GBV-A and GBV-B genomes were expressed as recombinant proteins in *E. coli*. Antigenic regions were identified in a Western blot format using convalescent tamarin plasma, convalescent plasma from the surgeon, GB, or various human plasma from individuals with or "at risk" for non-A-E hepatitis (PILOT-MATIAS et al. 1996a). These antigenic proteins were incorporated into indirect enzyme-linked immunosorbent assays (ELISAs). Nine different ELISAs, each utilizing a single recombinant GBV protein, were developed to detect antibodies to GBV-A nonstructural or GBV-B structural and nonstructural proteins.

Some sera from populations at risk for viral hepatitis were seroreactive to both GBV-A and GBV-B proteins. In addition, many of these dually reactive samples were positive for both IgG and IgM class antibodies. Because IgM class antibodies are indicative of recent infection, RT-PCRs were performed to detect GBV-A or GBV-B sequences in these sera. Despite using various primer sets from both viruses, we failed to identify any RNA-positive specimens. This may have been due to low viral titer, the clearance of the virus from the serum, or the lack of appropriate primer pairs. Alternatively, the serologic results noted in these samples may have resulted from cross-reactive antibodies induced by infection with a related virus. To account for this possibility, degenerate oligonucleotides were designed to amplify a segment of the putative helicase gene (NS3) from GBV-A, GBV-B or HCV-1 in hemi-nested RT-PCR (SIMONS et al. 1995b). Because these primers amplify a region from the GBV-A, GBV-B and hepatitis C viruses, it was reasoned that they might also be able to amplify a homologous region from a virus responsible for the seroreactivity in these GBV-A/GBV-B-immunoreactive humans.

When the IgM-positive samples were tested with the degenerate helicase primers, a product of the expected size was amplified from the serum of an individual from West Africa (SIMONS et al. 1995b). Sequence analysis of this product revealed limited identity at both the nucleotide (59.0%, 53.7% and 47.9%) and amino acid (64.2%, 57.3% and 50.4%) levels with GBV-A, HCV-1 and GBV-B, respectively (Fig. 1). Most notably, of the 42 amino acids conserved between GBV-A, GBV-B and HCV-1 in this region, 39 were also conserved in the product from the West African sample. Included in the 39 conserved residues are the residues conserved in all supergroup II RNA helicases (KOONIN and DOLJA 1993). These

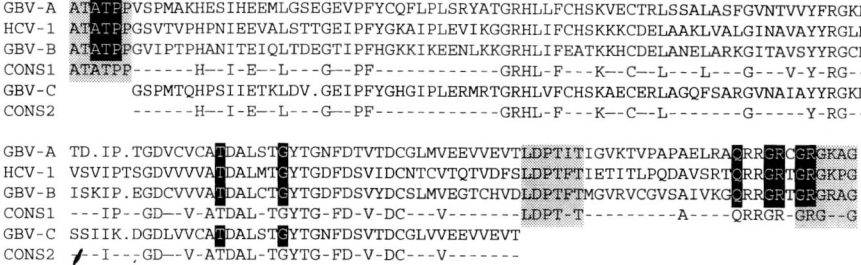

Fig. 1. Amino acid sequence alignment of the NS3 helicase domains. Residues conserved in all supergroup II RNA helicases (KOONIN and DOLJA 1993) are indicated (*black background*). Residues conserved between GBV-A, GBV-B and HCV-1 (CONS1) and the areas of high nucleotide sequence conservation where degenerate PCR primers were designed (*stippled*) are shown. Also displayed are the predicted translation product of the PCR product isolated from the West African individual (GBV-C) and the residues conserved between this sequence and GBV-A, GBV-B and HCV-1 (CONS2)

data suggested that the product obtained with the degenerate helicase primers was derived from a similar region of a GBV/HCV-like virus. Phylogenetic analysis of these alignments, together with additional members of the Flaviviridae, are consistent with this sequence being derived from a virus which is more closely related to GBV-A than to GBV-B, or any of the HCV genotypes (see Sect. 3.2.1). Because of its similarity to GBV-A, this virus has been named GB virus C (GBV-C).

Sequences both up- and downstream of the initial GBV-C clone were obtained using the methods previously employed to clone GBV-A and GBV-B (LEARY et al. 1996a). Analysis of the overlapping clones finds that GBV-C possesses a genome at least 9377 nucleotides (nts) in length. However, because the GBV-C genome has not been circularized at the RNA level, it is unknown whether the entire sequence of GBV-C has been determined.

2.3 Identification of Hepatitis G Virus

Subsequent to the initial discovery of GBV-C (SIMONS et al. 1995b), LINNEN et al. (1996) described the identification of an independent isolate of GBV-C that they termed hepatitis G virus (HGV). Pursuing the identification of a causative agent for cryptogenic post-transfusion and community-acquired hepatitis, molecular cloning was performed with plasma from a patient designated PNF2161. This individual was originally diagnosed with non-A, non-B hepatitis in the CDC Sentinel Counties Study of Viral Hepatitis (ALTER et al. 1992), based on consistently negative results with a first-generation immunoassay (Ortho HCV ELISA Test System). However, subsequent tests (both second-generation ELISA and a PCR assay for the 5′ nontranslated region of HCV) demonstrated that PNF2161 was chronically infected with HCV.

The molecular cloning of HGV employed λgt11 immunoscreening, similar to that used in the identification of HCV (CHOO et al. 1989). Immunoscreening the

PNF2161 library with PNF2161 plasma identified several HCV-specific clones as well as clones which contained unique sequences. Using a clone-specific PCR assay, one clone (470-20-1) was found to be exogenous to *E. coli*, *S. cerevisiae* and human genomes and absent from the plasma of health control subjects. In addition, analysis of serial dilutions of PNF2161 revealed the 470-2-1 was present at approximately 10^6 copies/ml as an RNA molecule.

From the initial 470-20-1 clone, an anchored PCR technique was used to extend this clone to a total of 9392 nt. When the HGV sequence was compared to GBV-A and GBV-B, these viruses were found to possess significant similarity. Moreover, comparison of HGV with the original 331 nt reported for GBV-C (SIMONS et al. 1995b) found that these sequences were nearly identical (85% and 100% at the nucleotide and amino acid levels, respectively) suggesting that these viruses are very closely related (LINNEN et al. 1996). Immediately prior to the publication of HGV, LEARY et al. (1996a) reported the near full-length sequence of GBV-C. Comparison of these two sequences revealed that GBV-C and HGV are indeed separate isolates of the same virus (ZUCKERMAN 1996).

Similar to the GBVs, HGV contains a single long open reading frame (ORF) encoding a polyprotein (see Sect. 3.1). Surprisingly, the amino acid sequence of the 470-20-1 immunoreactive clone is not encoded in this long ORF, but instead is found in the complementary (i.e. negative) strand. This fact, coupled with the immunoreactivity of other non-A-E hepatitis sera with the clone, lead the authors to suggest that the 119 amino acid negative-strand ORF which contains 470-20-1 may be expressed in infected individuals. However, this is unlikely for two reasons: first, the "ORF" does not contain an initiator methionine codon; second, the ORF is not conserved in additional viral isolates. Thus, it would appear that serendipitous antibodies unrelated to the HGV infection in PNF2161 identified the original 470-20-1 clone.

2.4 Identification of GBV-A Variants

Following the isolation of GBV-A and GBV-B, SCHLAUDER et al. (1995b) reported that several apparently healthy tamarins (*Saguinus labiatus*) appeared to be infected with a variant of GBV-A in the absence of any known exposure to the GB viruses. Using GBV-A-specific primers designed within the 5' nontranslated region (NTR), a gene product of 346 base pairs (bp) was detected in six of 17 uninoculated tamarins. These products were at least 94% identical to one another, but only 79% identical to the prototype GBV-A isolated from the tamarin, T-1053. The study of these GBV-A variants was further extended by LEARY et al. (1996b) who identified additional variants in *S. labiatus* (termed GBV-A_{lab}), as well as in *Callithrix* sp. (GBV-A_{mx}, where mx refers to marmoset cross), and *Aotus trivirgatus* (GBV-A_{tri}) with no history of exposure to an infectious agent. Within each primate species, the 5' NTRs were virtually identical to one another (95%), while these sequences were only 52%–79% identical between species and the prototype GBV-A. Thus, the

GBV-A variant sequences cluster on the basis of the primate species from which they were isolated.

Using primers designed within the 5' NTR, as well as the NS3 region of GBV-A and related viruses, BUKH and APGAR (1997) confirmed the presence of species-specific GBV-A variants in *S. labiatus, Callithrix jacchus* and *A. trivirgatus*. In addition, they reported the presence of GBV-A variants in other New World monkeys including *S. mystax, S. nigricollis,* and *S. oedipus*. Again, the sequences of these variants clustered based on the species from which they were isolated.

It is of interest that tamarins naturally infected with GBV-A_{lab} appear to be at least partially resistant to experimental infection with GBV-A. Of the four tamarins inoculated with H205 (a source of GBV-A and GBV-B), all became infected with GBV-B, while only one became dually infected with GBV-A and GBV-B (SCHLAUDER et al. 1995a). The three animals that appeared to be resistant to GBV-A infection were GBV-A_{lab}-viremic prior to inoculation. In contrast, a tamarin harboring GBV-A_{lab} became infected with both GBV-A and GBV-B after being experimentally inoculated with a high titer GBV-A/GBV-B inoculum (LEARY et al. 1996b). Others (BUKH et al. 1996) have noted similar interference. Thus, it may be that marginal protection to GBV-A infection is afforded by indigenous GBV-A variants, though animals challenged with high titers GBV-A are susceptible to infection.

The GBV-A variants identified in uninoculated New World primates suggest an origin for GBV-A identified in the GB agent. Specifically, three of five GBV-A variants isolated from *S. nigricollis* were at least 97.8% identical to the GBV-A prototype virus within the 5' NTR (BUKH and APGAR 1997). Because many of the early passages of the GB agent were in *S. nigricollis* (DEINHARDT et al. 1967), and because the GBV-A variants can persistently infect their hosts (BUKH and APGAR 1997), it is likely that the GBV-A was acquired during the passage of the GB agent in *S. nigricollis*. As such, the prototype GBV-A may actually be GBV-A_{nig}.

Sequences from GBV-A variants from a tamarin (GBV-A_{lab}), an owl monkey (GBV-A_{tri}), and a marmoset (GBV-A_{mx}) have been extended downstream of the 5' NTRs to a total of 9550, 9625 and 9586 nt, respectively (ERKER et al. 1998; LEARY et al. 1997). Based on comparative analysis to the GBV-A prototype, additional sequence that have yet to be deciphered may exist at the extreme 5' and 3'ends of these genomes.

3 Molecular Analyses

The GB viruses contain RNA genomes of approximately 9500 nucleotides in length (Fig. 2, Table 1). These genomes contain long 5' NTRs, single long ORFs that encode a large polyprotein, and long 3' NTRs. The NH_2-terminal one-third of the viral polyproteins contains the structural proteins followed by the non-structural proteins in the COOH-terminal two-thirds of the polyprotein. Based on the similarity of the GBVs and other members of the *Flaviviridae*, the RNA

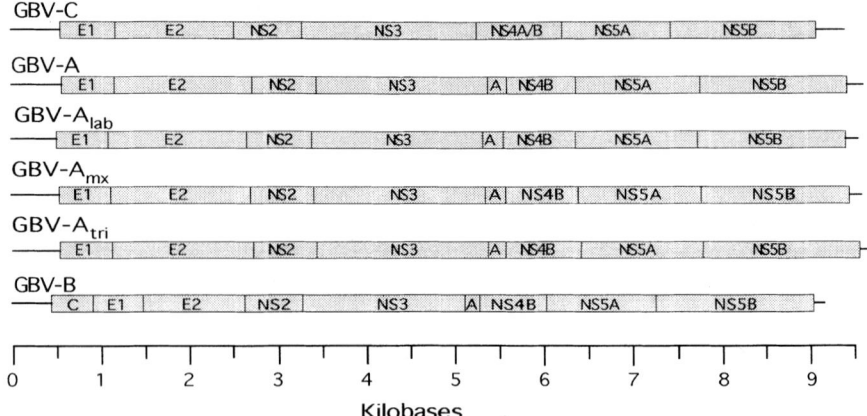

Fig. 2. Genomic maps of the GB viruses. Sequence alignments fail to predicted the cleavage site between GBV-C's NS4A and 4B gene products

Table 1. GB Virus Genome Characteristics

	GBV-A	GBV-A$_{lab}$	GBV-A$_{mx}$	GBV-A$_{tri}$	GBV-C	GBV-B	HCV-1
Genome size (nt)	9653	9550*	9586*	9625*	9377	9143	9401
5′ NTR (nt)	593	508*	539*	528*	551	445	341
3′ NTR (nt)	198	135*	137*	79*	315	103	24
Polyprotein (aa)	2954	2967	2970	3005	2842	2864	3011
Core (aa)	–	–	–	–	–	156	191
E1/E2 N-linked glycosylation	1/3	2/4	1/4	1/6	1/3	3/6	5/11

aa, amino acids; nt, nucleotides.
The asterisks denote that 5′ and 3′ extension experiments have not been performed on these isolates. Therefore, additional genomic sequences are likely to exist. Isolates displayed: GBV-A (U22303), GBV-A$_{lab}$ (U94421), GBV-A$_{mx}$ (AF023424), GBV-A$_{tri}$ (AF023425), GBV-C (U36380), GBV-B (U22304) and HCV-1 (M62321).

genome is assumed to be of positive polarity, however, only GBV-B has been demonstrated to be positive-stranded using strand-specific probes (A.S. Muerhoff, unpublished observation). In addition, the proteolytic processing of the viral polyproteins is thought to be mediated by both host-encoded and virus-encoded proteases.

3.1 Sequence Characteristics

3.1.1 5′ NTRs

The GBVs possess 5′ NTRs of between 445 and 593 nt. Based on sequence comparisons, these 5′ NTRs can be divided into two groups: the GBV-B/HCV group and the GBV-A/GBV-C group. Within each group, the 5′ NTRs vary by no more than 50% and possess local regions of high sequence identity. In contrast, no significant sequence identity is detected between the 5' NTRs of the GBV-B/HCV and GBV-A/GBV-C groups.

The GBV 5' NTR sequences are thought to fold into discrete structures (HONDA et al. 1996; PICKERING et al. 1997a; SIMONS et al. 1996a). The predicted GBV-B structure is similar to that found in HCV and the pestivirus 5' NTRs. This is most evident within the ~220 nt located immediately upstream of the long ORFs despite the fact that these sequences are ~50% identical. Similarly, the predicted structures of GBV-A and GBV-C are much the same while sharing only ~50% nucleotide identity (Fig. 3). These putative NTR structures appear to be involved in translation initiation (see Sect. 4.1).

3.1.2 Structural Proteins

The structural proteins of the GBVs can be divided into a GBV-B/HCV group and a GBV-A/GBV-C group similar to the 5' NTRs. GBV-B encodes a basic (pI 11.1) core protein of 156 amino acids followed by two putative envelope glycoproteins,

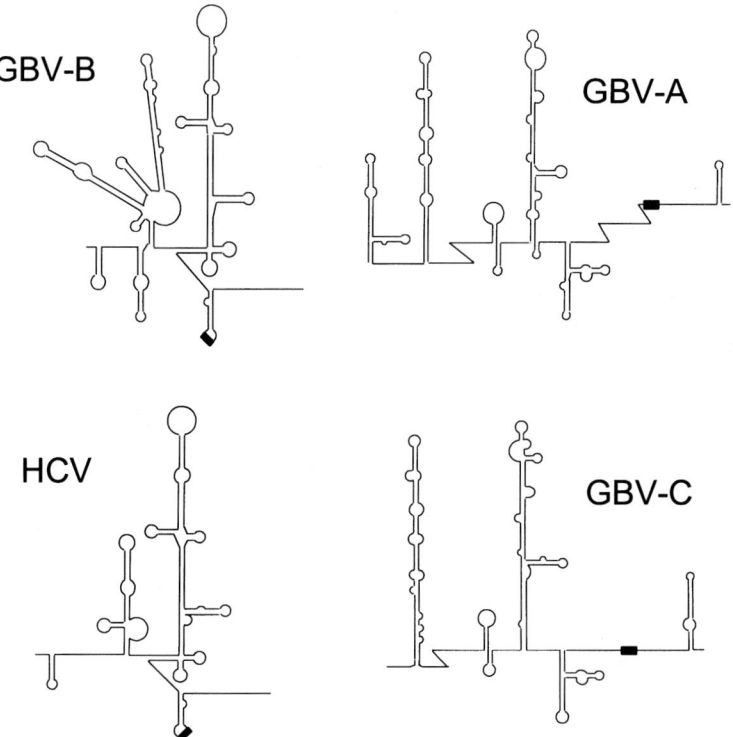

Fig. 3. Predicted 5' end structures. The secondary RNA folding predicted for GBV-B and HCV (HONDA et al. 1996), and GBV-A and GBV-C (SIMONS et al. 1996a) are displayed. The position of the initiating AUG codon (*black rectangle*) is shown for each structure

E1 and E2, containing three and six potential N-linked glycosylation sites, respectively. The analogous region in HCV-1 contains a core protein of 191 amino acids (pI 11.9) and glycoproteins E1 and E2 with five and 11 potential N-linked glycosylation sites, respectively. Similar to HCV, the host signal protease is thought to liberate core, E1 and E2 from the GB-B viral polyprotein. The basic core protein followed by envelope protein(s) seen in GBV-B is similar to the structural proteins found in other members of the *Flaviviridae* (MUERHOFF et al. 1995).

The GBV-A/GBV-C group of viruses encode E1-like and E2-like proteins based on sequence similarities with HCV and GBV-B. As with GBV-B, the existence of putative signal sequences for the insertion of the E1 and E2 proteins into the host cell membrane suggest that these proteins are liberated from the viral polyprotein by the host signal peptidase. However, unlike GBV-B and HCV, there does not appear to be a basic core region upstream of the E1 glycoprotein in the GBV-A/GBV-C group. It was originally hypothesized that GBV-A, which contains a short, neutral peptide upstream of E1, either lacks a core or that the high GC content in the first 650 nt obscured the identification of the ORF encoding the core (MUERHOFF et al. 1995). For GBV-C, different isolates contained ORFs of varying lengths upstream of E1 (MUERHOFF et al 1996). Although some of the longer predicted peptides upstream of E1 are basic, the longest (71 residues) was still much shorter that the core proteins of GBV-B or HCV. Again, it was suggested that GBV-C contained a defective or deleted core region, or that the true reading frame was being masked by the high GC content (LEARY et al. 1996a; LINNEN et al. 1996).

The possibility that the true reading frame containing the core was being masked by sequence compressions or deletions, or that the original cloning sources contained defective viruses with deleted core regions, was addressed by studying the 5′ sequences of several GBV-C (and GBV-A-like) isolates. Examination of numerous GBV-C sequences (MUERHOFF et al. 1996a, 1997; LINNEN et al. 1997; TAKAHASHI et al. 1997a) demonstrated different length ORFs upstream of the putative E1 sequence (from 18 to 135 amino acids). The different sized ORFs arise from numerous point deletions that occur upstream of the AUG (Met) codon positioned at the NH_2-terminal of the putative E1 signal sequence. In fact, the only conserved initiator codon present in all of the isolates reported to date is located immediately upstream of the E1 signal sequence. Similar to the various GBV-C isolates, the ORFs of some GBV-A-like viruses extend upstream of E1 for different lengths. However, all of the GBV-A-like viruses possess a conserved AUG initiator codon immediately upstream of the putative E1 signal sequence (ERKER et al. 1998; LEARY et al. 1997). In vitro translation experiments (see Sect. 4.1.2) suggest that this conserved AUG codon is the site of translation initiation in the GBV-A/GBV-C group. Thus, the absence of a core-like protein upstream of E1 appears to be a common feature of all GBV-A and GBV-C isolates examined to date. The lack of a core-like protein upstream of E1 distinguishes GBV-A and GBV-C from other members of the *Flaviviridae*. In fact, searches of all six potential reading frames from the various GBV-A and GBV-C genomes do not reveal a conserved ORF encoding a core-like protein. Thus, these viruses appear distinct from enveloped viruses in general because they do not appear to encode a basic protein to mediate the packaging of the viral nucleic acid into the virion envelope. Infectious particles without a core have been generated artificially using the vesicular stomatitis virus glycoprotein (ROLLS et al. 1994), and core-less but enveloped HCV negative strand RNA has been detected in human serum (SHINDO et al. 1994). Thus, it is possible that the GBV-A/GBV-C group of viruses are truly core-less enveloped viruses. However, it is possible that these viruses appropriate a cellular protein (or a core from a coinfecting virus) for viral assembly. Physical characterization of the virions should resolve this issue.

Another feature that differentiates the GBV-A/GBV-C group from GBV-B/HCV is the relative paucity of potential N-linked glycosylation sites in the envelope glycoproteins (Table 1). It is interesting to note that many of the glycosylation sites found in the GBV-A/GBV-C group are conserved in GBV-B and HCV. In addition, the many conserved Cys, Pro and Gly residues between all of the GBVs and HCV suggest that these envelope glycoproteins may be structurally very similar.

Sequences obtained from a large number of viral isolates have allowed the examination of the heterogeneity present in the structural proteins of GBV-C. Relative to the rest of the genome, the greatest sequence divergence is seen in the NH_2-terminal of E2 at ~10% (ERKER et al. 1996). Unlike HCV, the amino acid substitutions seen between isolates do not cluster into discernible hypervariable regions in the envelope glycoproteins of GBV-C (ERKER et al. 1996; LIM et al. 1997). While both GBV-C and HCV can persistently infect humans, the apparently random amino acid substitutions seen in GBV-C structural proteins and the lack of a consistently detectable serologic response in viremic individuals suggests that GBV-C does not employ antigenic variation as a mechanism for persistence.

3.1.3 Nonstructural Proteins

Based on sequence comparisons with HCV, the nonstructural proteins of the GBVs contain of two proteases, an RNA helicase, and an RNA-dependent RNA polymerase in addition to other proteins of unknown function. A putative zinc-dependent thiol protease appears to be present in the NS2/NS3 gene product based on the conserved His and Cys residues necessary for the autolytic cleavage of HCV NS2 from NS3 (LEARY et al. 1996a; MUERHOFF et al. 1995). Another putative protease, a chymotrypsin-like serine protease appears to be present in the NH_2-terminal one-third of NS3. The His, Asp and Ser residues required for catalytic activity of the homologous HCV protease are conserved in all of the GBVs (Fig. 4A). This putative protease is thought to be responsible for cleaving the remainder of the downstream viral polyprotein (see Sect. 4.2.1). The RNA helicase region of the GBVs contains residues conserved in all supergroup II RNA helicases from positive-strand RNA viruses (Fig. 1). Similarly, the RNA-dependent RNA polymerase region of the GBVs share residues held in common with all supergroup II RNA polymerases (Fig. 4B).

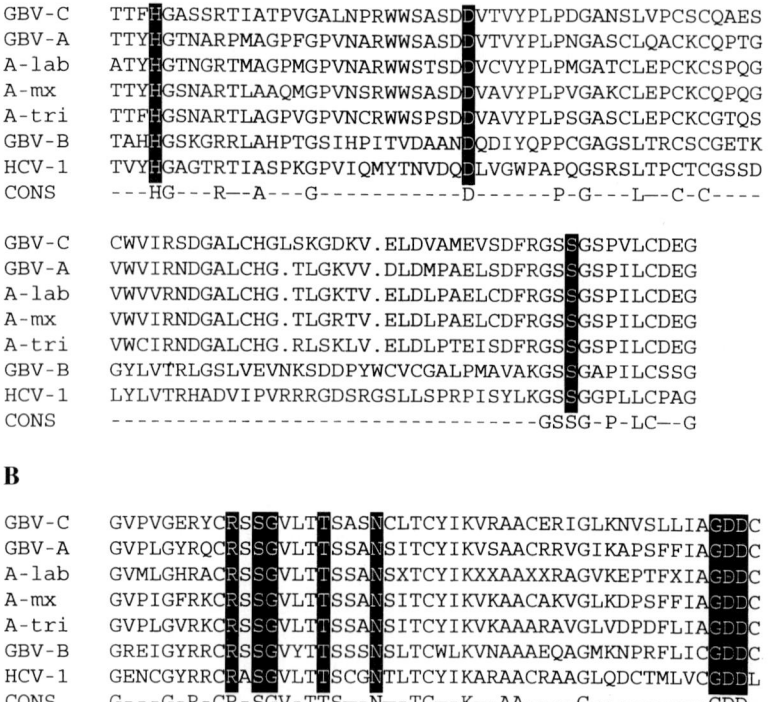

Fig. 4A,B. GBV nonstructural protein alignments. **A** NS3 protease domain. The predicted catalytic His-Asp-Ser residues (*black background*) are shown. **B** NS5B RNA-dependent RNA polymerase domain. The residues conserved in the supergroup II RNA replicases (KOONIN and DOLJA 1993) are displayed (*black background*)

3.1.4 3′ NTRs

The 3′ NTRs of the GBVs can be divided into groups similar to that seen with the 5′ NTR and structural genes. GBV-B 3′ NTR contains a short region of 30 bases followed by a stretch of polyU and an additional 50 residues downstream of the polyU. The presence of the additional sequence in GBV-B was surprising because sequence downstream of the polyU or polyA tract had not been described for HCV at that time. In fact, when we repeated the RNA circularization/RT-PCR that identified the downstream sequence in GBV-B on two HCV sources, we failed to identify sequence 3′ of the polyU tracts (MUERHOFF et al. 1995). Subsequent to the identification of the X tail in HCV (KOLYKHALOV et al. 1996; TANAKA et al. 1995, 1996), comparison between the HCV and GBV-B sequences reveals no similarity at the sequence or structure levels. The fact that the 3′ tail of GBV-B is approximately half the size of HCV may suggest that additional sequences may remain to be identified for GBV-B. In contrast to GBV-B and HCV, the GBV-A/GBV-C group of viruses do not possess a polyU or polyA sequence in the 3' NTR. Instead these viruses encode 198–315 nt sequences that are highly conserved between virus isolates within a group. Comparison between GBV-C isolates finds that the 3' NTR is the most conserved region of the virus genome. Several covariant changes suggest that this region is highly structured (ERKER et al. 1996).

3.2 Phylogenetic Relationships

Limited sequence identity between the GBVs and HCV suggests that these viruses are related. However, phylogenetic analysis of these sequences provide a more accurately delineation of these relationships. In addition, these analyses can be rendered schematically to give a visual picture of the relationship of the GBVs to other members of the *Flaviviridae* (flavivirus, pestivirus and HCV groups), as well as to each other.

3.2.1 GBVs' Relationship to the *Flaviviridae*

The relative evolutionary distances between the GB viruses and other members of the *Flaviviridae* are readily apparent upon inspection of an unrooted phylogenetic tree. The phylogenetic trees generated from sequence alignments of the RNA helicase regions (Fig. 5A) and RNA-dependent RNA polymerase regions (not shown) are similar (MUERHOFF et al. 1995 and unpublished data). The HCV genotypes are contained on a major branch as are the flavi-, pestiviruses. The GB viruses are contained on one or two major branches in the unrooted trees produced from the helicase and polymerase alignments. In both trees, GBV-A and GBV-C share a common ancestor distinct from GBV-B. The unrooted trees graphically demonstrate the significant degree of divergence of the GB viruses from the ancestor common to both the HCV lineage and the GB lineage. The divergence between GBV-A, GBV-B and GBV-C and other *Flaviviridae* members, including the hepatitis C group, demonstrates that the GB viruses cannot be considered genotypes of HCV. This analysis suggests that the GB agents may be classified into separate genera within the *Flaviviridae* or into subgenera in a genus including the hepatitis C viruses. However, the lack of a core-like protein in GBV-A and GBV-C may necessitate the placement of these viruses in their own family.

3.2.2 GBV-C Genotypes

GBV-C RNA-positive human plasma and sera have been identified from individuals worldwide. The existence of geographically diverse samples from a variety of different clinical sources provides material from which the diversity of GBV-C can be studied. It was anticipated that these studies may reveal the existence of phylogenetic groups (i.e. genotypes) of GBV-C and that different GBV-C groups may display specific biological properties and clinical manifestations similar to that found with distinct HCV types.

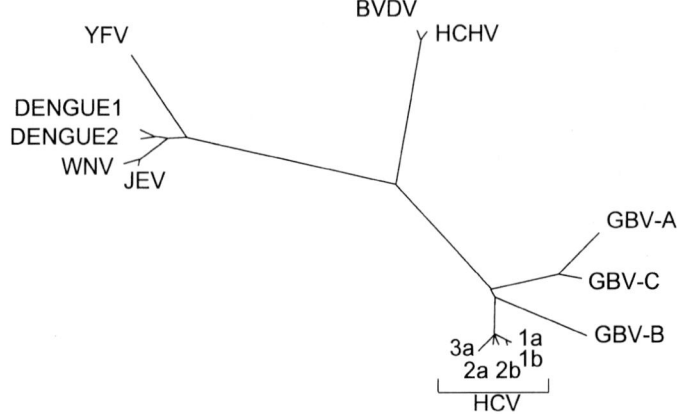

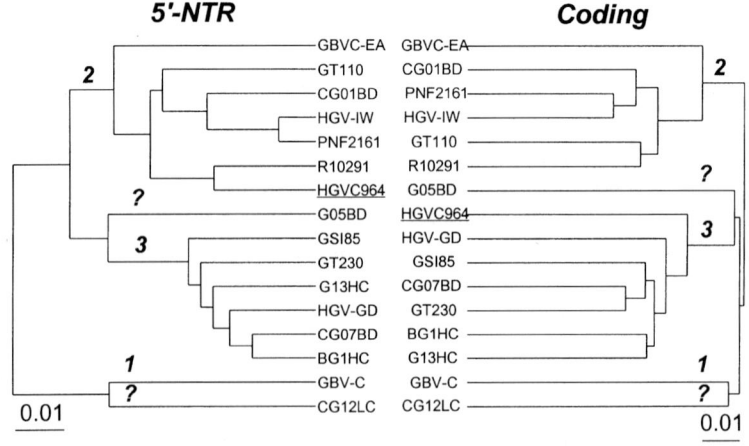

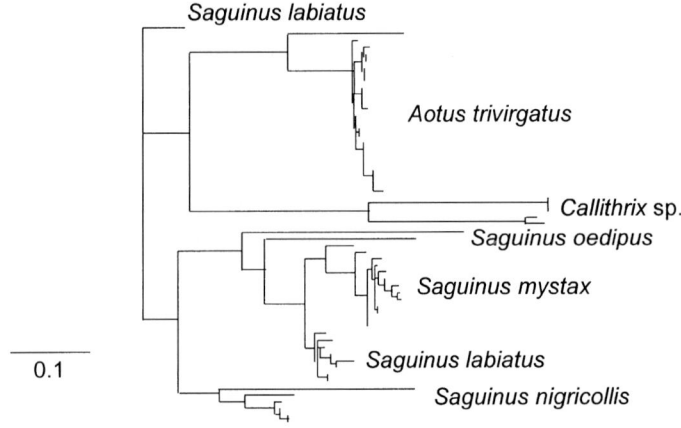

The first phylogenetic studies of GBV-C focused on short coding sequences in NS3 (~118 nt) or NS5b (~354 nt) where the initial PCR primers for the detection of this virus were located (LEARY et al. 1996c; LINNEN et al. 1996; SIMONS et al. 1995b). From these studies, no consistent phylogenetic relationship was noted between isolates from different parts of the world (BERG et al. 1996; KAO et al. 1996; SCHREIER et al. 1996; TSUDA et al. 1996; BROWN et al. 1997a; KINOSHITA et al. 1997; PICKERING et al. 1997b; VIAZOV et al. 1997a). This is in marked contrast to HCV in which the analysis of subgenomic fragments as small as 222 nt can distinguish between HCV types, subtypes and isolates (SIMMONDS et al. 1994, 1996; OHBA et al. 1995).

In contrast to subgenomic coding regions, the analysis of the 5' NTR of GBV-C demonstrates the existence of at least three major groups and two subgroups (MUERHOFF et al. 1996a; AN et al. 1997; FUKUSHI et al. 1996; GONZALEZ-PEREZ et al. 1997; SMITH et al. 1997). The evolutionary relationship between some of these isolates is presented graphically in Fig. 5B. Using the nomenclature of MUERHOFF et al. (1996a), group 1 consists of isolates found predominantly among individuals from West Africa and includes the original GBV-C isolate. Similarly, group 3 isolates have been found predominantly in individuals from Japan. Groups 2a and 2b isolates have been obtained primarily from individuals from North America, Europe and East Africa, and do not demonstrate a strict geographic distributions similar to those observed for group 1 and 3 isolates. In addition, isolates from group 2 are more closely related to each other than to either group 1 or group 3 isolates. The existence of groups 1, 2, 2a, 2b and 3 are supported by bootstrap values of 100%, 100%, 80%, 82% and 99%, respectively (MUERHOFF et al. 1997). These groupings have been confirmed by LINNEN et al. (1997), with Muerhoff's groups 1, 2 and 3 corresponding to Linnen's groups III, II and I, respectively. Analyses of the complete coding regions from isolates in each of these groups, as well as long sequences generated by joining separate coding regions from the same isolates, confirms the existence of phylogenetic groupings of GBV-C (MUERHOFF et al. 1997; TAKAHASHI et al. 1997b).

The inability to group GBV-C isolates using short coding regions indicates a very different pattern of variation from that reported for HCV, where consistent phylogenetic relationships are found for many subgenomic regions including E1

◄───

Fig. 5A–C. Phylogenetic trees. **A** Phylogenetic relationship of the NS3 RNA helicase domains of the GBVs and other members of the Flaviviridae. Flaviviruses: Japanese encephalitis virus (JEV), West Nile virus (WNV), yellow fever virus (YFV) and dengue virus types 1 and 2; pestiviruses: hog cholera virus (HCHV) and bovine viral diarrhea virus (BVDV); HCV group: genotypes 1a, 1b, 2a, 2b, and 3a. **B** Phylogenetic relationship of GBV-C isolates. The geographic origin and genotype designation (according to Muerhoff) of various GBV-C isolates. The *left side* displays the phylogenetic tree generated from an alignment of the of the 5' NTRs, while the *right side* displays the phylogenetic tree generated from an alignment of the complete open reading frames. Genotype designation are as determined previously by MUERHOFF et al. (1997). The *question marks* designate isolates that do not appear to be associated with any of the three major groups and thus may define new types or subtypes of GBV-C. **C** Phylogenetic relationship of the 5' NTRs of GBV-A variants. The New World monkey species from which the GBV-A variant was isolated is indicated. The prototype GBV-A isolated from the GB agent is the bottom branch of this tree

and NS5b (BUKH et al. 1995; OHBA et al. 1995; SIMMONDS et al. 1994). Part of the reason for this difference may be the high degree of conservation noted between GBV-C isolates at both the nucleotide and amino acid levels relative to HCV. Indeed, a detailed comparison of the distances between the GBV-C genomes at nonsynonymous and synonymous sites indicates that there is a strong selection against nonsynonymous substitutions throughout the genome. This is reflected in the high degree of amino acid sequence conservation among isolates from widely separated geographic regions. However, different subtypes of HCV are closer to saturation at synonymous sites when compared to GBV-C (and thus present a high level of background "noise" in the phylogenetic analysis), yet these HCV subtypes can be distinguished by the analysis of substitutions within a 222 nt fragment of NS5b or a 300 nt fragment of E1 (MUERHOFF et al. 1997). Thus, the inability to distinguish GBV-C groups using fragments of this size suggests that the GBV-C genome may be restricted at both synonymous and nonsynonymous positions.

The biological significance of the GBV-C groupings noted above is unclear. Groups 1, 2 and 3 differ by only 2.5%–4.6% at the amino acid level over their complete coding regions. This divergence is much lower than that found for individual genes between different types of papillomaviruses (10%–17%), subtypes of HIV (12%–36%) or serotypes of poliovirus (20%–30%), dengue virus (23%–38%) and vesicular stomatitis virus (30%–50%). In comparison with full length HCV polyproteins, the GBV-C divergence is in the range found between epidemiologically distinct isolates of the same HCV subtype (3.5%–8.1%) and much less than the divergence found between HCV subtypes (13%–19%) or HCV types (23%–29%). As noted above, different HCV types may display different clinical features and antigenicity profiles, although similar differences have not been associated between different HCV subtypes or isolates. Thus, comparison with HCV and other viruses suggests that the limited diversity between different GBV-C groups is unlikely to be associated with biological or clinical differences (SMITH et al. 1997).

3.2.3 GBV-A Variants

Similar to the GBV-C isolates described above, GBV-A variants isolated from New World monkeys can also be analyzed phylogenetically to determine the relationship of these viruses to one another. As shown in Fig. 5C, nucleotide sequences from the 5' NTR of the GBV-A variants (LEARY et al. 1996b; BUKH and APGAR 1997) were aligned and used to generate an unrooted phylogenetic tree. It is clear from this analysis that sequences isolated from the distinct species are dramatically different from one another while isolates obtained from animals within a species are virtually identical to one another. Thus, the GBV-A variants genetically cluster on the basis of the species from which they were isolated. The lone exception to this general rule is a single isolate from *S. labiatus* that cannot be explained at this time. On the basis of criteria used to distinguish HCV types from subtypes (BUKH et al. 1995), each of the species specific isolates would be considered a distinct type of GBV-A. The biological significance of these differences in GBV-A is not yet understood and will require further study.

4 Biochemical Analyses

4.1 Translation Initiation

The long 5' NTRs of the GBVs have been grouped by sequence comparison and predicted secondary structure into the GBV-B/HCV and the GBV-A/GBV-C groups (see Sect. 3.1.1). As with other viruses that contain long 5' NTRs such as picornaviruses and pestiviruses, the long 5' NTR of HCV contains an internal ribosome entry site (IRES) or ribosome landing pad which plays a role in translation initiation (RIJNBRAND et al. 1995; TSUKIYAMA-KOHARA et al. 1992; WANG et al. 1993). Specifically, the RNA comprising this *cis*-acting IRES forms a highly ordered structure that binds *trans*-acting cellular translation factors which, in turn, bind the 40S ribosome subunit at an internal site on the viral message, often hundreds of nucleotides downstream of the 5' end of the molecule, to initiate translation. Such translation initiation occurs independent from the 5' end of the message and can be functionally assessed by insertion of the sequence between two cistrons of a bicistronic message. If the intercistronic sequence contains an IRES, there is significant translation of the downstream cistron which is typically independent of the translational activity of the upstream cistron.

4.1.1 GBV-B

The grouping of the 5' NTR of GBV-B and HCV suggested that the GBV-B might contain an IRES element in the 5' NTR. This was confirmed using bicistronic vectors. In vitro transcription translation (IVTT) reaction mixtures programmed with bicistronic vectors containing the 5' NTR of GBV-B as the intercistronic sequence generated moderate amounts of the downstream cistron gene product. The amount of downstream product was between those found in IVTT reactions programmed with vectors containing the strong IRES elements from encephalomyocarditis virus and the weak IRES element from hepatitis A virus (J.N.S, unpublished data). These in vitro experiments confirmed the presence of a moderately strong IRES element in the 5' NTR of GBV-B.

4.1.2 GBV-A and GBV-C

The study of the 5' NTRs from GBV-A and GBV-C was complicated by absence of sequence similarity in this region with a previously studied virus, and the aforementioned fact that different potential initiation sites have been reported for different isolates (see Sect. 3.1.2). To define the site of translation initiation biochemically, the 5' ends of GBV-A and GBV-C through the Asn-Cys-Cys encoded motif of the envelope protein E1 were fused in-frame with the bacterial chloramphenicol acetyltransferase (CAT) gene. IVTT reactions programmed with these constructs generated CAT fusion proteins consistent with translation initiation occurring at the AUG codon immediately upstream of the putative E1 signal sequence (SIMONS et al. 1996a). Translation initiation from the AUG immediately upstream of the E1 signal sequence was confirmed by site directed mutagenesis of the GBV-A-CAT and GBV-C-CAT constructs and by Edman degradation of the GBV-CAT fusion proteins (SIMONS et al. 1996a). The site of translation initiation in the IVTT reactions is the same AUG codon that is conserved in all of the GBV-C and GBV-A variants reported to date. This appears to be true whether or not the ORF extends upstream of this site.

With the site of translation initiation defined, attempts were made to determine whether the 5' NTRs from GBV-A and GBV-C contained IRES elements. Suggestive evidence for the presence of IRES elements was obtained with the monocistronic GBV-CAT constructs. Specifically, removal of the GBV coding sequence from these constructs abrogated protein production in the IVTT reactions. The coding sequence eliminated in these constructs is predicted to contain a hairpin structure (downstream of the initiation site in Fig. 3). When the 5' ends of GBV-A or GBV-C were placed in the intercistronic space of a bicistronic constructs, low levels of the downstream cistron gene product were detected in IVTT reactions. The amount of protein generated with the GBV constructs was lower that that found in similar constructs containing the weak IRES element of HAV suggesting that GBV-A and GBV-C contain extremely weak IRES elements (SIMONS et al. 1996a).

The low activity of the GBV-A/GBV-C IRES observed in vitro might reflect the true activity of these elements in an infected host. Here, the limited amount of viral proteins produced from a weak IRES might be below the threshold of recognition by the host immune system, thus promoting virus persistence. Alternatively, the IVTT reactions might be missing a host cell factor that enhances translation, or the constructs used to program the IVTT reactions may require additional (yet to be identified?) GBV sequence. Further studies are required to distinguish between these different possibilities.

4.2 Proteases

4.2.1 GBV-B

The NS3 region of HCV contains a zinc-binding serine protease domain that has been shown to be responsible for processing downstream nonstructural proteins (GRAKOUI et al. 1993; TOMEI et al. 1993). Analysis of the three dimensional structure of HCV NS3 (KIM et al. 1996; LOVE et al. 1996) suggests that amino acid Phe-154 in the S_1 pocket defines the preference of the enzyme for a cysteine in the P1 substrate position. As stated above (see Sect. 3.1.3), the residues required for catalysis and zinc binding in the HCV protease are conserved in the homologous NS3 regions of the GBVs. In addition, GBV-B contains the homologous Phe-154, which is not present in GBV-A, GBV-C or the GBV-A variants. This sequence conservation suggests that the GBV-B protease might share substrate specificity with HCV NS3. To test this, SCARSELLI et al. (1997) expressed the GBV-B NS3 protease domain in E. coli and characterized the purified protein. The GBV-B protease cleaved HCV substrates NS4A/B, NS4B/5A and NS5A/B. The cleavage of the NS4B/5A substrate was surprising because the HCV protease only cleaves this substrate in the presence of the NS4A cofactor. The HCV NS4A cofactor had no effect on the activity of the GBV-B protease. The enzyme displayed similar activity on peptide substrates with respect to HCV and cleaved after cysteine residues as had been predicted (MUERHOFF et al. 1995). In addition, the classic serine protease inhibitors reduced the activity of the GBV-B protease. The overlapping substrate specificities between GBV-B and HCV suggest that GBV-B infection in tamarins may provide an alternative to chimpanzee models for evaluating HCV protease inhibitors in vivo (SCARSELLI et al. 1997).

4.2.2 GBV-C

Sequence similarity with HCV suggested that GBV-C would possess two proteases: one located in NS2 responsible for the cleavage of the NS2/NS3 junction, the other located in NS3 and responsible for the remaining downstream cleavages. To assess the activity of these two proteases, BELYAEV et al. (1998) expressed various lengths of the GBV-C nonstructural region in insect cells and examined the resultant products. Wild-type constructs that expressed the NS2-NS3-NS4A region of GBV-C produced free NS2, NS3 and NS4A. In addition, when NS4B/NS5A and NS5A/NS5B proteins were expressed in insect cells in *trans*, these substrates were cleaved as well. Similar to what is found for HCV NS3 protease (BARTENSCHLAGER et al. 1994; LIN et al. 1994), the NS4B/NS5A product required the presence of NS4A for processing, while the NS5A/NS5B product did not require the NS4A cofactor.

Several site-specific mutations were made in the NS2-NS3-NS4A construct to elucidate the roles of the individual GBV-C protease activities. Single amino acid substitutions to the homologous NS2 His or Cys residues required for activity of the HCV enzyme eliminate the *cis*-cleavage at the GBV-C NS2/NS3 junction without affecting cleavage at the NS5A/NS5B site. In contrast, a single amino acid substitution at the GBV-C NS3 Ser homologous to the active site residue in HCV NS3 abolished cleavage at the NS3/NS4A and the NS5A/NS5B junctions but did not affect processing at the NS2/NS3 junctions (BELYAEV et al. 1998). Thus, the GBV-C NS2 protease mediates cleavage at the NS2/NS3 site, while the NS3 serine protease mediates cleavage at the downstream sites.

5 Physical Analyses

Biophysical characterization of the GBVs has been hampered by the lack of adequate cell culture or animal models to generate large amounts of highly purified virions. At present, the best source of native virus is plasma or serum obtained for GBV-C-viremic humans. With viral titers being $< 10^9$/ml, the virions must be tracked by RT-PCR of the viral genomic RNA. Immunoprecipitation experiments suggest that GBV-C is associated with low density lipoproteins in plasma but not human antibodies because anti-apolipoprotein A1 and anti-apolipoprotein B antibodies shift GBV-C PCR signal to the precipitate while anti-human IgG antibody does not (HIJIKATA and MISHIRO 1996; SATO et al. 1996). Lectin columns bind GBV-C consistent with the presence of envelope glycoproteins (SATO et al. 1996). Filtration studies suggest that the virion is between 50 and 100nm in diameter based on the PCR titer being unchanged after passing through a 100nm filter and being reduced 10- to 1000-fold after passing through a 50nm filter (MELVIN et al. 1998). These findings are consistent with infectivity data which demonstrate that GBV-A and GBV-B can pass through a 100nm filter (SCHLAUDER et al. 1995a).

Physical evidence for a virion core was sought by density gradient ultracentrifugation. Previous studies of HCV and other flaviviruses demonstrated that the lipid envelopes of these viruses can be removed by detergent treatment to reveal the nucleocapsid component of these virions. For HCV, the free nucleocapsid is detected by the PCR signal shifting to an increased buoyant density after detergent treatment. Untreated GBV-C forms a broad peak around 1.08–1.10g/ml in sucrose density gradients. However, unlike HCV, treatment of GBV-C with numerous detergents using various amounts and incubation times appear to either completely abrogate PCR detection or fail to affect the buoyant density of the virus (SATO et al. 1996; MELVIN et al. 1998). The presence of an RNase inhibitor during the detergent treatment and subsequent centrifugation restores PCR detection. However, the density of the resulting product is indistinguishable from free RNA in sucrose gradients. Using isopycnic cesium chloride gradients, MELVIN et al. (1998) demonstrated that the detergent-treated GBV-C was >1.5g/ml. This value is more consistent with a free RNA species (>1.9g/ml) than with nucleocapsids (1.21–1.32g/ml). In addition, the detergent-treated virus could not pass through a 200nm filter, indicating that the GBV-C RNA was present in an extended or aggregated form (MELVIN et al. 1998). These findings do not support the existence of a nucleocapsid structure in the GBV-C virion.

6 Detection Methods

Detection methods of GBVs have focused on the human virus GBV-C. In contrast to what is found for HCV, the identification of serologic markers indicative of ongoing GBV-C infection has not been successful. Thus, the identification of viremic samples has relied on the detection of viral RNA. Several nucleic acid-based assays have been developed for this purpose. Although labor-intensive, many of these assays are sensitive and specific. Initial steps have been made to format these assays in a more user-friendly manner. In addition, an antibody marker of viral clearance has been identified.

6.1 Viral RNA

Detection techniques for the identification of GBV-C viremic samples have evolved over time. The first generation RT-PCR assays utilizes oligonucleotide primers to amplify the NS3 and NS5 domains of the genome (LEARY et al. 1996c; LINNEN et al. 1996). However, because the targets of these assays are coding regions, silent mutations noted in the various GBV-C isolates may result in a number of false-negative samples. The second generation RT-PCR assays target highly conserved regions within the 5' NTR of GBV-C (MUERHOFF et al. 1996b; BHARDWAJ et al. 1997; CANTALOUBE et al. 1997; FORNS et al. 1997; KAO et al. 1997a; ZHANG et al. 1997). These assays are more sensitive and specific than the first generation assays. However, both the first and second generation assays rely on Southern hybridization for the detection of specific PCR products. The inclusion of Southern hybridization for the detection of GBV-C-specific PCR products results in assays that are tedious, labor-intensive and time consuming.

Third generation RT-PCR assays have been developed which permit a higher level of through-put in conjunction with ease of use (LEARY et al. 1996d; SCHLUETER et al. 1996). In these assays, GBV-C-specific PCR products are captured by oligomer hybridization for the detection of sequences within the 5'NTR. The assay described by SCHLUETER et al. (1996) utilizes a two step RT-PCR assay in which digoxigenin-labeled dUTP is incorporated during the PCR amplification. A biotinylated capture oligomer hybridizes with any GBV-C-specific products that may be present. The GBV-C/oligomer complex is captured on a streptavidin-coated surface. After extensive washing, captured PCR product is detected immunologically using an anti-digoxigenin–peroxidase conjugate (SCHLUETER et al. 1996). In contrast, LEARY et al. (1996d) describe a single tube assay that only requires the addition of serum-derived nucleic acid to pre-aliquoted reaction vials. Reverse transcription, PCR amplification and oligomer hybridization occur in the same tube containing recombinant *Thermus thermophilus* (rTth) polymerase, adamantane-labeled sense and anti-sense oligonucleotide primers for the 5'NTR of GBV-C, and carbazole-labeled capture oligomer. Detection of GBV-C product employs the automated LCx detection system (Abbott Laboratories, Abbott Park, IL) utilizing a microparticle enzyme immunoassay. These third generation assays provide data within ~5h, as opposed to the 2 days minimum required for the first- or second-generation RT-PCRs.

6.2 Antiviral Antibodies

Five different prokaryotically expressed recombinant GBV-C proteins were employed for serologic studies utilizing ELISA and Western blot (DAWSON et al. 1996; PILOT-MATIAS et al. 1996a). Although some regions of GBV-C have been identified as immunogenic, efforts to develop a screening immunoassay for the detection of GBV-C antibodies have been unsuccessful because only ~25% of GBV-C infected individuals develop antibodies to these prokaryotically-expressed proteins. Most of the seropositive individuals produce antibodies against only a single antigen, and no single antigen has been identified thus far that is consistently recognized by individuals exposed to GBV-C (DAWSON et al. 1996).

The phylogenetic relationship between GBV-C and HCV suggested that useful markers of GBV-C exposure might be found in regions homologous to those identified for HCV. Several groups have demonstrated the utility of glycosylated HCV E2 as a target of virus-specific antibodies (LESNIEWSKI et al. 1995; LOK et al. 1993; ZAAIJER et al. 1994). With this in mind, PILOT-MATIAS et al. (1996b) expressed COOH-terminal truncated GBV-C E2 in mammalian cells. This protein was efficiently secreted from the mammalian cells in a glycosylated form. A sensitive and specific radioimmunoprecipitation assay (RIPA) was developed with the secreted GBV-C E2 to detect antibodies to this protein. In a transfusion transmission study of GBV-C, the E2 RIPA identified six seroconversion events out of seven GBV-C-transfused individuals. Interestingly, the presence of an anti-E2 response corresponded to the loss of detectable viral RNA from the plasma of these patients. The one case of GBV-C infection without seroconversion was from an individual who remained RNA-positive for the duration of the study (PILOT-MATIAS et al. 1996b). Thus, the presence of anti-E2 appears to be a marker of past GBV-C exposure.

To facilitate a higher level of through-put, GBV-C E2 protein has been expressed in Chinese hamster ovary cells, purified and used in solid-phase ELISAs (DILLE et al. 1997; LOU et al. 1997; SUROWY et al. 1997; TACKE et al. 1997a,b). These GBV-C E2 ELISAs perform similarly to the RIPA described above. Antibodies to GBV-C E2 appear to be long-lived (greater than 3 years) with a fairly constant titer (ranging in reciprocal endpoint dilution from 336 to 21,504) and appear to be associated with recovery (clearance) of GBV-C infection (GUTIERREZ et al. 1997). Analysis of serial specimens with assays for both anti-E2 and viral RNA suggest that a GBV-C infection follows one of two paths – acute infection followed by recovery (appearance of GBV-C E2 antibody), or acute infection progressing to chronicity (persistence of GBV-C RNA). Thus, total incidence of GBV-C exposure should take into account both the number of PCR positive samples (i.e. viremic) and anti-E2 positive samples (i.e. previously infected but cleared) in a given population. These data suggest that the exposure to GBV-C is much higher than determined by RT-PCR studies alone and that testing for GBV-C E2 antibodies should greatly extend the ability of RT-PCR to define the epidemiology and clinical significance of GBV-C.

7 Route of Infection

Epidemiological studies suggest that the predominant route of transmission for GBV-C is parenteral. Indeed, several documented cases of transmission through contaminated blood and blood products have been published (JARVIS et al. 1996;

ROTH et al. 1997; SHIMIZU et al. 1997). However, since recent reports have shown that both human saliva (CHEN et al. 1997) and semen (PERSICO et al. 1996) of GBV-C infected individuals are positive for GBV-C RNA, the possibility exists that both horizontal and vertical transmissions may play a key role in the spread of GBV-C infections. In fact, vertical transmission cases of GBV-C from infected mothers to newborns (FEUCHT et al. 1996; FISCHLER et al. 1997; VIAZOV et al. 1997b) as well as intraspousal transmissions have been documented (KAO et al. 1997b; SARRAZIN et al. 1997).

8 Site(s) of Replication

One of the key questions that remains to be answered for GBVs is their sites of replication. The first evidence for possible sites of replication came from the original GBV-A and GBV-B discovery (SIMONS et al. 1995a). Equivalent end-point dilution PCR titers were obtained for both GBV-A and GBV-B in plasma used for cloning (from tamarin T-1053). However, only GBV-B was detected by northern blot hybridization of T-1053 liver RNA. In fact, northern hybridization with strand-specific probes detected GBV-B negative-strand RNA, a presumed replicative intermediate. In contrast, several different GBV-A-specific probes failed to hybridize to T-1053 liver RNA. Thus, although equivalent levels of GBV-A and GBV-B were present in the plasma of T-1053, these data suggest that GBV-B replicated in T-1053 liver and GBV-A replicated elsewhere. It is possible that GBV-A replicated in focal sites in T-1053 liver apart from the region that was tested. However, a more extensive search of tissues for GBV-A RNA needs to be pursued to determine conclusively its site of replication.

Locating GBV-C's site of replication has been an area of intense interest. Unfortunately, only a few tissues have been examined with equivocal results. Using PCR, SHENG et al. (1997) found that washed peripheral blood mononuclear cells (PBMCs) were GBV-C positive by RT-PCR. Similarly, washed hepatocytes taken from a GBV-C positive individual have been reported to be RT-PCR positive (LOPEZ-ALCOROCHO et al. 1997). However, because PBMCs and hepatocytes are bathed in serum, it is possible that the positive results noted in these studies were due to virus bound to cells and not active replication. Several tissues from one GBV-C/HIV-coinfected patient have been examined. RNA prepared from liver, spleen, brain, lung, muscle, skin, and bone marrow was analyzed for the presence of glyceraldehyde-3-phosphate dehydrogenase (GAP-DH) mRNA and GBV-C RNA by RT-TAQMAN (Perkin Elmer-Applied Biosystem) PCR. The only GBV-C-positive tissue in this patient was liver with greater than 8×10^5 copies of GBV-C RNA per 2×10^4 copies of GAPDH mRNA. In this same patient, GBV-C RNA was localized to individual hepatocytes by in situ RT-PCR (MUSHAHWAR et al. 1998). Similar analysis of liver tissue from two additional patients (SEIPP et al. 1999) corroborate these findings.

GBV-C is presumed to be a positive-strand virus. As such, negative-strand RNA would be a replicative intermediate. With this in mind, negative-strand-specific RT-PCR assays have been used to examine the presence of replicative strand in these tissues. The results obtained with negative-strand-specific RT-PCR have not been consistent. MADEJON et al. (1997) and SAITO et al. (1997) detect negative strand GBV-C RNA in the PBMCs from one of 13 patients tested, suggesting that the PCR signal noted in this population may be due to cell-bound virus rather than replication occurring in these cells. These same two studies find GBV-C negative-strand RNA in liver biopsies from 12 of the 13 infected patients. In contrast, two studies by LASKUS et al. (1997a,b) did not find GBV-C negative-strand RNA in the liver of seven GBV-C-only infected patients or nine GBV-C/HCV co-infected patients.

It should be noted that strand-specific detection of RNA is fraught with problems such as false priming of the incorrect strand or self-priming related to RNA structure (LANFORD et al. 1994). All of the negative-strand studies utilized methods to reduce false priming and self-priming events such as chemical modification of 3′ ends (MADEJON et al. 1997; SAITO et al. 1997) or high temperature cDNA synthesis (LASKUS et al. 1997a,b) and control reactions were performed. However, only LASKUS et al. (1997a,b) qualified their reactions using in vitro-derived templates and provided end-point titration data. Clearly, more tissues need to be examined with assays of known sensitivity and specificity before GBV-C's tissue tropism can be demonstrated.

9 Association with Disease

Current epidemiologic data (reviewed in SIMONS et al. 1996b) show the presence of GBV-C RNA in sera obtained from a variety of sources. These include sera from hemophiliacs, thalassemic patients, intravenous drug abusers, multiply transfused individuals, transfusion associated hepatitis cases, volunteer blood donors with both normal and elevated serum transaminase levels, chronic hepatitis B (HBV) and HCV carriers, acute and chronic non-A-E hepatitis patients, kidney, liver and bone marrow transplant recipients and donors, fulminant hepatitis cases, and patients on maintenance hemodialysis. Analyses of serum specimens collected from healthy volunteer blood donors from different global areas confirmed the presence of GBV-C RNA in a remarkable 1%–4% of the specimens (DAWSON et al. 1996; FIORDALISI et al. 1996; MASUKO et al. 1996; MOAVEN et al. 1996; STARK et al. 1996; WU et al. 1997).

Most GBV-C infections are mild, transient and self-limiting, with slight or no elevation of serum ALT levels. A majority of these subclinical cases resolve after loss of serum GBV-C RNA with a concomitant appearance of antibody to envelop E2 of GBV-C (anti-GBV-C E2, or GBV-C E2 antibody) as reported by several investigators (DILLE et al. 1997; GUTIERREZ et al. 1997; LEFRERE et al. 1997; TACKE et al. 1997b). These types of GBV-C infections are hardly noticeable and very difficult to evaluate when studied in multitransfused patients and/or patients with a more serious and superimposed HBV or HCV infection.

GBV-C is capable of inducing persistent infection in about 15%–25% of GBV-C infected individuals. MASUKO et al. (1996) followed retrospectively eight hemodialysis patients with GBV-infection for 7–16 years. In two patients, the virus was present at the start of hemodialysis. One had a history of transfusion and GBV-C RNA persisted over a period of 16 years; the other cleared GBV-C RNA after 10. In five patients, GBV-C RNA was first detected 3–20 weeks after blood transfusion and persisted for up to 13 years. Elucidating the viral mechanisms that lead to the establishment and maintenance of the persistent state is crucial for our understanding of the pathogenesis of GBV-C (KUHN 1996).

The role of GBV-C in the etiology of fulminant hepatitis is not yet fully established. Further convincing studies are needed to confirm a definite link between fulminant hepatitis and GBV-C infection. One Japanese study (YOSHIBA et al. 1995) documented the presence of GBV-C RNA in three of six (50%) of fulminant hepatitis patients without evidence of infection with known hepatitis viruses. Since that report, questions were raised concerning the association of GBV-C with acute liver failure (ALTER 1996). Specifically, of whether GBV-C was an "innocent bystander" transmitted through transfusions given to the 3 patients prior to the onset of fulminant hepatitis. Additional studies (YOSHIBA et al. 1996), however, showed that only few of the 63 fulminant hepatitis patients studied so far had received therapeutic transfusion prior to the onset of fulminant hepatitis, but definitely not all. In a similar study also carried out in Japan (TAMEDA et al. 1997), GBV-C RNA was detected in three (20%) of 15 patients with HBV infection and three (12%) of 25 patients without markers of hepatitis A-E infection. Overall, GBV-C RNA was detected in six of 44 (14%) patients with fulminant hepatitis at a frequency significantly higher ($p < 0.001$) than that in three of 326 (0.9%) of blood donors matched for age with the patients. Of the six patients with GBV-C RNA, only three (50%) had a history of transfusion and all of these were coinfected with HBV. These results, according to the authors, indicate a role of GBV-C in inducing fulminant hepatitis either by itself or in concert with other hepatitis viruses.

A unique study (FIORDALISI et al. 1996) showing histological features in liver samples from patients infected with GBV-C alone has been reported. GBV-C was implicated in a significant number of acute and chronic cases of non-A-E hepatitis. Among the six chronic hepatitis patients positive for GBV-C RNA, the histology of the liver samples revealed chronic active hepatitis in one patient and chronic persistent hepatitis in five others. All chronic patients had elevated ALT levels between 89 and 478U/l. In contrast, among the 11 acute hepatitis cases positive for GBV-C RNA, the ALT levels varied between 615 and 2477U/l.

COLOMBATTO et al. (1996) studied GBV-C in 67 patients with liver disease without any markers for hepatitis A-E. They report that the spectrum of liver disease associated with GBV-C infection in these patients is wide, with a large variety of serious histologic liver lesions (steatosis, fibrosis and cirrhosis). Of interest, nonspecific inflammatory bile duct lesions were found in 50% of patients with only GBV-C infection. They also suggest that GBV-C infection is present significantly more often with elevated cholestatic enzymes, namely, γ-glutamyl transpeptidase and alkaline phosphatase.

Again, the above-mentioned studies point out to the importance of studying individuals positive for GBV-C RNA without concomitant HBV or HCV coinfection.

So far, several reports have shown the lack of association or involvement of GBV-C infection with many other known diseases. These include autoimmune liver disease (HERINGLAKE et al. 1996), aplastic anemia (BROWN et al. 1997b; MORIYAMA et al. 1997), hepatocellular carcinoma (KANDA et al. 1997; KUBO et al. 1997; TAGGER et al. 1997), non-Hodgkin's lymphoma (ZIGNEGO et al. 1997), porphyria cutanea tarda (FARGION et al. 1997), and oral cancer and oral lichen planus (NAGAO et al. 1997).

10 Immunity and Treatment

Of importance, is the recent evidence for the protective immunity of anti-GBV-C E2 among kidney recipients (MUSHAHWAR et al. 1998). The data show that anti-GBV-C E2 positive kidney recipients failed to acquire GBV-C infection after transplantation from GBV-C RNA-positive donors, while anti-GBV-C E2 negative kidney recipients became viremic after transplantation from GBV-C RNA-positive donors. Thus, the presence of anti-GBV-C E2 appears to be protective. This conclusion is consistent with the paucity of GBV-C RNA-positive, anti-E2-positive samples that have been identified. Because high titers of GBV-C E2 antibodies appear to be protective in humans, the possibility of using human or animal immune serum globulin for prophylactic purposes is intriguing. Other possibilities such as use of recombinant GBV-C E2 glycoprotein or synthetic subunit vaccines should be explored if in the future GBV-C is proven to be hazardous to the public.

Several studies (GOESER et al. 1997; NAGAYAMA et al. 1997) have shown that GBV-C is sensitive to interferon (IFN) while the patient is on treatment. However, most cases relapse upon withdrawal of therapy. Future studies should concentrate on many key variables for the establishment of an effective IFN therapy of GBV-C infected individuals. These variables should include such parameters as the duration and level of IFN treatment, viral load, and different GBV-C genotypes.

11 Future Direction

Much has been accomplished since the original discovery of the GB viruses in 1995. However, a better understanding of these viruses will require additional studies.

Foremost among these should be the resolution of the site(s) of replication for GBV-C. Tissue availability has limited the initial studies described above to circulating lymphocytes and liver biopsy material. Clearly, additional tissues need to be examined. Based on similar molecular characteristics, it might be assumed that GBV-C and the GBV-A group replicates in similar tissues. Thus, an extensive analysis of tissues from GBV-A-infected monkeys may identify the site(s) of replication for this virus and also suggest human tissues to be examined for GBV-C tropism.

Finally, the association of GBV-C (and GBV-A) with disease needs to be examined more thoroughly in the absence of co-infecting viruses. There are cases of severe disease in individuals solely infected with GBV-C in addition to the many clinically normal GBV-C carriers. Here, GBV-C may be similar to HCV in that there is a disease spectrum associated with infection ranging from subclinical infections (a majority of the GBV-C cases) to fulminant liver failure (very rare for GBV-C). Again, knowing the site(s) of replication will allow one to examine histopathology in infected patients. This, in turn, may implicate GBV-C as the etiologic agent in additional diseases of unknown origin.

Acknowledgements. The authors acknowledge the help of Drs. A. Scott Muerhoff and Thomas P. Leary for providing unpublished phylogenetic trees of the GBV-C genotypes and the GBV-A variants, respectively.

References

Almeida JD, Deinhardt F, Holmes AW, Peterson DA, Wolfe L, Zuckerman AJ (1976) Morphology of the GB hepatitis agent. Nature 261:608–609
Alter HJ (1994) Transfusion transmitted hepatitis C and non-A, non-B, non-C. Vox Sang 67:19–24
Alter HJ (1996) The cloning and clinical implications of HGV and HGBV-C. N Engl J Med 334: 1536–1537
Alter MJ, Margolis HS, Krawczynski K, Judson FN, Mares A, Alexander WJ, Hu PY, Miller JK, Gerber MA, Sampliner RE, Meeks EL, Beach MJ (1992) The natural history of community-acquired hepatitis C in the United States. N Engl J Med 327:1899–1905
An P, Wei L, Wu X, Yuhki N, O'Brien SJ, Winkler C (1997) Evolutionary analysis of the 5'-terminal region of hepatitis G virus isolated from different regions in China. J Gen Virol 78:2477–2482
Appleton H, Dienstag JL (1977) Virus particles in marmoset hepatitis. Nature 267:729–730
Bartenschlager R, Ahlborn-Laake L, Mous J, Jacobsen H (1994) Kinetic and structural analyses of hepatitis C virus polyprotein processing. J Virol 68:5045–5055
Belyaev AS, Chong S, Novikov A, Kongpachith A, Masiarz FR, Lim M, Kim JP (1998) Hepatitis G virus encodes protease activities which can effect processing of the virus putative nonstructural proteins. J Virol 72:868–872
Berg T, Dirla U, Maumann U, Heuft H-G, Kuther S, Lobeck H, Schreier E, Hopf U (1996) Responsiveness to interferon alpha treatment in patients with chronic hepatitis C coinfected with hepatitis G virus. J Hepatol 25:763–768
Bhardwaj B, Qian K, Detmer J, Mizokami M, Kolberg JA, Urdea MS, Schlauder G, Linnen JM, Kim JP, Davis GL, Lau JY (1997) Detection of GB virus-C/hepatitis G virus RNA in serum by reverse transcription polymerase chain reaction. J Med Virol 52:92–96
Boggs JD, Melnick JL, Conrad ME, Felsher BF (1970) Viral hepatitis: Clinical and tissue culture studies. JAMA 214:1041–1046
Brown KE, Wong S, Buu M, Binh TV, Be TV, Young NS (1997a) High prevalence of GB virus C/ hepatitis G virus in healthy persons in Ho Chi Minh City, Vietnam. J Infect Dis 175:450–453

Brown KE, Wong S, Young NS (1997b) Prevalence of GBV-C/HGV, a novel 'hepatitis' virus, in patients with aplastic anaemia. Br J Haematol 97:492–496

Bukh J, Miller RH, Purcell RH (1995) Genetic heterogeneity of hepatitis C virus: Quasispecies and genotypes. Semin Liver Dis 5:41–63

Bukh J, Apgar CL, Purcell RH (1996) Natural history of GBV-A and GBV-B in animal models: discovery of indigenous *Flaviviridae*-like viruses in several species of New World monkeys. In: Rizzetto M, Purcell RH, Gerin JL, Verme G (eds.) Viral hepatitis and liver disease: proceedings of the 9th triennial international symposium on viral hepatitis and liver disease. Edizioni Minerva Medica, Rome, pp 392–395

Bukh J, Apgar CL (1997) Five new or recently discovered (GBV-A) virus species are indigenous to New World monkeys and may constitute a separate genus of the Flaviviridae. Virology 229:429–436

Cantaloube JF, Charrel RN, Attoui H, Biagini P, De Micco P, De Lamballarie X (1997) Evaluation of four PCR systems amplifying different genomic regions for molecular diagnosis of GB virus C infections. J Virol Methods 64:131–135

Chen M, Sonnerborg A, Johansson B, Sallberg M (1997) Detection of hepatitis G virus (GB virus C) RNA in human saliva. J Clin Microbiol 35:973–975

Choo Q-L, Kuo G, Weiner AJ, Overby LR, Bradley DW, Houghton M (1989) Isolation of a cDNA clone derived from a blood-borne non-A, non-B viral hepatitis genome. Science 244:359–362

Colombatto P, Randone A, Civitico C, Gorin JM, Dolci L, Medaina N, Oliveri F, Verme G, Marchiaro G, Pagni R, Karayiannis D, Thomas HC, Hess G, Bonino F, Brunetto MR (1996) Hepatitis G virus RNA in the serum of patients with elevated gamma glutamyl transpeptidase and alkaline phosphatase: a specific liver disease. J Viral Hepatol 3:301–306

Dawson GJ, Schlauder GG, Pilot-Matias TJ, Thiele D, Leary TP, Murphy P, Rosenblatt JE, Simons JN, Martinson FEA, Gutierrez RA, Lentino JR, Pachucki C, Muerhoff AS, Widell A, Tegtmeier G, Desai S, Mushahwar IK (1996) Prevalence studies of GB virus-C using reverse-transcriptase-polymerase chain reaction. J Med Virol 50:97–103

Deinhardt F, Holmes AW, Capps RB, Popper H (1967) Studies on the transmission of disease of human viral hepatitis to marmoset monkeys. I. Transmission of disease, serial passage and description of liver lesions. J Exp Med 125:673–687

Deinhardt F, Holmes AW, Wolfe LG, Melnick JL, Parks WP (1970) Hepatitis in marmosets. J Infect Dis 121:351–354

Deinhardt F, Peterson D, Cross G, Wolfe L, Holmes AW (1975a) Hepatitis in marmosets. Am J Med Sci 270:73–80

Deinhardt F, Wolfe L, Peterson D, Cross GF, Holmes AW (1975b) The mythology of various hepatitis A virus isolates. Dev Biol Stand 30:390–404

Deinstag JL, Wagner JA, Purcell RH, London WT, Lorenz DE (1976) Virus-like particles and GB agent hepatitis. Nature 264:260–261

Dille BJ, Surowy TK, Gutierrez RA, Coleman PF, Knigge MF, Carrick RJ, Aach RD, Hollinger FB, Stevens CE, Barbosa LH, Nemo GJ, Mosley JW, Dawson GJ, Mushahwar IK (1997) An ELISA for detection of antibodies to the E2 protein of GB virus C. J Infect Dis 175:458–461

Erker JC, Simons JN, Muerhoff AS, Leary TP, Chalmers ML, Desai SM, Mushahwar IK (1996) Molecular cloning and characterization of a GB virus C isolate from a patient with non-A-E hepatitis. J Gen Virol 77:2713–2720

Erker JC, Desai SM, Leary TP, Chalmers ML, Montes CC, Mushahwar IK (1998) Genomic analysis of two GB virus A variants isolated from captive monkeys. J Gen Virol 79:41–45

Fargion S, Sampietro M, Fracanzani AL, Molteni V, Ticozz M, Mattioli A, Valsecchi C, Fiorelli G (1997) Hepatitis G virus in patients with porphyria cutanea tarda (PCT). J Hepatol 26 (Suppl 1):207

Feucht H-H, Zollner B, Polywka S, Laufs R (1996) Vertical transmission of hepatitis G. Lancet 347: 615–616

Fiordalisi G, Zanella I, Mantero G, Bettinardi A, Stellini R, Paraninfo G, Cadeo G, Primi D (1996) High prevalence of GB virus C infection in a group of Italian patients with hepatitis of unknown etiology. J Infect Dis 174:181–183

Fischler B, Lara C, Chen M, Sonnerberg A, Nemeth A, Sallberg M (1997) Genetic evidence for mother-to-infant transmission of hepatitis G virus. J Infect Dis 176:281–285

Forns X, Tan D, Alter HJ, Purcell RH, Bukh J (1997) Evaluation of commercially available and in-house reverse transcription-PCR assays for detection of hepatitis G virus or GB virus C. J Clin Microbiol 35:2698–2702

Fukushi S, Kurihara C, Ishiyama N, Okamura H, Hoshino FB, Oya A, Katayama K (1996) Nucleotide sequence of the 5′ noncoding region of hepatitis G virus isolated from Japanese patients: comparison with reported isolates. Biochem Biophys Res Commun 226:314–318

Goeser T, Seipp S, Wahl R, Muller HM, Stremmel W, Theilmann L (1997) Clinical presentation of GB-C virus infection in drug abusers with chronic hepatitis C. J Hepatol 26:498–502

Gonzalez-Perez MA, Norder H, Bergstrom A, Lopez E, Visona KA, Magnius LO (1997) High prevalence of GB virus C strains genetically related to strains with Asian origins in Nicaraguan hemophiliacs. J Med Virol 52:149–155

Grakoui A, McCourt DW, Wychowski C, Feinstone SM, Rice CM (1993) Characterization of the hepatitis C virus-encoded serine proteinase: determination of proteinase-dependent polyprotein cleavage sites. J Virol 67:2832–2843

Gutierrez RA, Dawson GJ, Knigge MF, Melvin SL, Heynen CA, Kyrk CR, Young CE, Carrick RJ, Schlauder GG, Surowy TK, Dille BJ, Coleman PF, Thiele DL, Lentino JR, Pachucki C, Mushahwar IK (1997) Seroprevalence of GB virus C and persistence of RNA and antibody. J Med Virol 53:167–173

Heringlake S, Tillmann HL, Cordes-Temme P, Trautwien C, Hunsmann G, Manns MP (1996) GBV-C/HGV is not the major cause of autoimmune hepatitis. J Hepatol 25:980–984

Hibbs JR, Frickhofen N, Rosenfeld SJ, Feinstone SM, Kojima S, Bacigalupo A, Locasciulli A, Andreas, Tzakis G, Alter HJ, Young NS (1992) Aplastic anemia and viral hepatitis: non-A, non-B, non-C? JAMA 267:2051–2054

Hijikata M, Mishiro S (1996) Circulating immune complexes that contain HCV but not GBV-C in co-infected hosts. Int Hepatol Comm 5:339–344

Holmes AW, Deinhardt F, Wolfe L, Froesner G, Peterson D, Casto B, Conrad ME (1973) Specific neutralization of human hepatitis type A in marmoset monkeys. Nature 243:419–420

Holmes AW, Cross GF, Peterson DA, Deinhardt F (1975) Differentiation of HAAg and HBAg: Identification of the hepatitis A virus. In: Greenwalt TJ, Jamieson GA (eds) Transmissible disease and blood transfusion, Grune and Stratton, New York, pp 33–42

Honda M, Brown EA, Lemon SM (1996) Stability of a stem-loop involving the initiator AUG controls the efficiency of internal initiation of translation on hepatitis C virus RNA. RNA 2:955–968

Jarvis LM, Davidson F, Hanley JP, Yap PL, Ludlam CA, Simmonds P (1996) Infection with hepatitis G virus among recipients of plasma products. Lancet 348:1352–1355

Kanda T, Yokosuka O, Imazeki F, Tagawa M, Ehata T, Saisho H, Omata M (1997) GB virus-C RNA in Japanese patients with hepatocellular carcinoma and cirrhosis. J Hepatol 27:464–469

Kao J-H, Chen P-J, Hsiang S-C, Chen W, Chen D-S (1996) Phylogenetic analysis of GB virus C: comparison of isolates from Africa, North America, and Taiwan. J Infect Dis 174:410–413

Kao JH, Chen PJ, Chen W, Hsiang SC, Lai MY, Chen DS (1997a) Amplification of GB virus-C/hepatitis G virus RNA with primers from different regions of the viral genome. J Med Virol 51:284–289

Kao JH, Liu CJ, Chen PJ, Chen W, Hsiang SC, Lai MY, Chen DS (1997b) Interspousal transmission of GB virus-C/hepatitis G virus: a comparison with hepatitis C virus. J Med Virol 53:348–353

Karayiannis P, Petrovic LM, Fry M, Moore D, Enticott M, McGarvey MJ, Scheuer PJ, Thomas HC (1989) Studies of GB hepatitis agent in tamarins. Hepatology 9:186–192

Kim JL, Morgenstern KA, Lin C, Fox T, Dwyer MD, Landro JA, Chambers SP, Markland W, Lepre CA, O'Malley ET, Harbeson SL, Rice CM, Murcko MA, Caron PR, Thomson JA (1996) Crystal structure of the hepatitis C virus NS3 protease domain complexed with a synthetic NS4A cofactor peptide. Cell 87:343–355

Kinoshita T, Miyake K, Nakao H, Tanake T, Tsuda F, Okamoto H, Miyakawa Y, Mayumi M (1997) Molecular investigation of GB virus C infection in hemophiliacs in Japan. J Infect Dis 175:454–457

Kohler H, Apodaca J (1968) Hepatitis bei Affen nach Verimpfung von Adenoviren aus Menschen mit Virushepatitis: I. Bestimmung der Enzymaktivitäten im Verlauf der Erkrankung. Zentralbl Bakteriol Orig 206:1–19

Kolykhalov AA, Feinstone SM, Rice CM (1996) Identification of a highly conserved sequence element at the 3' terminus of hepatitis C virus genome RNA. J Virol 70:3363–3371

Koonin EV, Dolja VV (1993) Evolution and taxonomy of positive-strand RNA viruses: implications of comparative analysis of amino acid sequences. Crit Rev Biochem Mol Biol 28:375–430

Krugman S, Giles JP, Hammond J (1967) Infectious hepatitis: Evidence for two distinctive clinical, epidemiological and immunological types of infection. JAMA 200:365–373

Kubo S, Nishiguchi S, Kuroki T, Hirohashi K, Tanaka H, Tsukamoto T, Shuto T, Kinoshita H (1997) Poor association of GBV-C viremia with hepatocellular carcinoma. J Hepatol 27:91–95

Kuhn J (1996) Introduction to diagnosis of persistent viral infections. Intervirology 39:139

Lanford RE, Sureau C, Jacob JR, White R, Fuerst TR (1994) Demonstration of in vitro infection of chimpanzee hepatocytes with hepatitis C virus using strand-specific RT/PCR. Virology 202:606–614

Laskus T, Radkowski M, Wang L-F, Vargas H, Rakela J (1997a) Lack of evidence for hepatitis G virus replication in the livers of patients coinfected with hepatitis C and G viruses. J Virol 71:7804–7806

Laskus T, Wang L-F, Radkowski M, Jang SJ, Vargas H, Dodson F, Fung J, Rakela J (1997b) Hepatitis G virus infection in American patients with cryptogenic cirrhosis: no evidence for liver replication. J Infect Dis 176:1491–1495

Leary TP, Muerhoff AS, Simons JN, Pilot-Matias TJ, Erker JC, Chalmers ML, Schlauder GG, Dawson GJ, Desai SM, Mushahwar IK (1996a) Sequence and genomic organization of GBV-C: a novel member of the Flaviviridae associated with human non-A-E hepatitis. J Med Virol 48:60–67

Leary TP, Desai SM, Yamaguchi J, Chalmers ML, Schlauder GG, Dawson GJ, Mushahwar IK (1996b) Species-specific variants of GB virus A in captive monkeys. J Virol 70:9028–9030

Leary TP, Muerhoff AS, Simons JN, Pilot-Matias TJ, Erker JC, Chalmers ML, Schlauder GG, Dawson GJ, Desai SM, Mushahwar IK (1996c) Consensus oligonucleotide primers for the efficient detection of GB Virus C. J Virol Methods 56:119–121

Leary TP, Desai SM, Erker JC, Mushahwar IK (1997) The sequence and genomic organization of a GB virus A variant isolated from captive tamarins. J Gen Virol 78:2307–2313

Leary TP, Muerhoff AS, Schlauder GG, Marshall RL, Pilot-Matias TJ, Simons JN, Dawson GJ, Desai SM, Mushahwar IK (1996d) Molecular diagnostic assays for the detection of GB virus C infection. In: Rizzetto M, Purcell RH, Gerin JL, Verme G (eds.) Viral hepatitis and liver disease: proceedings of the 9th triennial international symposium on viral hepatitis and liver disease. Edizioni Minerva Medica, Rome, pp 353–357

Lefrere JJ, Loiseau P, Maury J, Lasserre J, Mariotti M, Ravera N, Lerable J, Lefevre G, Morand-Joubert L, Girot R (1997) Natural history of GBV-C/hepatitis G virus infection through the follow-up of GBV-C/hepatitis G virus-infected blood donors and recipients studied by RNA polymerase chain reaction and anti-E2 serology. Blood 90:3776–3780

Lesniewski R, Okasinski G, Carrick R, VanSant C, Desai S, Johnson R, Scheffel J, Moore B, Mushahwar I (1995) Antibody to hepatitis C virus second envelope (HCV-E2) glycoprotein: a new marker of HCV infection closely associated with viremia. J Med Virol 45:415–422

Lim MY, Fry K, Yun A, Chong S, Linnen J, Fung K, Kim JP (1997) Sequence variation and phylogenetic analysis of envelope glycoprotein of hepatitis G virus. J Gen Virol 78: 2771–2777

Lin C, Pragai BM, Grakoui A, Xu J, Rice CM (1994) The hepatitis C virus NS3 serine proteinase: trans-cleavage requirements and processing kinetics. J Virol 68:8147–8157

Linnen J, Wages J, Zhang-Keck Z-Y, Fry KE, Krawczynski KZ, Alter H, Koonin E, Gallagher M, Alter M, Hadziyannis S, Karayiannis P, Fung K, Nakatsuji Y, Shih JW-K, Young L, Piatak M, Hoover C, Fernandez J, Chen S, Zou J-C, Morris T, Hyams KC, Ismay S, Lifson JD, Hess G, Foung SKH, Thomas H, Bradley D, Margolis H, Kim JP (1996) Molecular cloning and disease association of hepatitis G virus: a transfusion-transmissible agent. Science 271:505–508

Linnen JM, Fung K, Fry KE, Mizokami M, Ohba K, Wages JM, Zhang-Keck Z-Y, Song K, Kim JP (1997) Sequence variation and phylogenetic analysis of the 5' terminus of hepatitis G virus. J Viral Hepatol 4:293–302

Lisitsyn N, Lisitsyn N, Wigler M (1993) Cloning the differences between two complex genomes. Science 259:946–951

Lok ASF, Chien D, Choo Q-L, Chan T-M, Chiu EKW, Cheng IKP, Houghton M, Kuo G (1993) Antibody response to core, envelope and nonstructural hepatitis C virus antigens: comparison of immunocompetent and immunosuppressed patients. Hepatology 18:497–502

Lopez-Alcorocho JM, Millan A, Garcia-Trevijano ER, Bartolome J, Ruiz-Moreno M, Otero M, Carreno V (1997) Detection of hepatitis GB virus type C RNA in serum and liver from children with chronic viral hepatitis B and C. Hepatology 25:1258–1260

Lou S, Qiu X, Tegtmeier G, Leitza S, Brackett J, Cousineau K, Varma A, Seballos H, Kundu S, Kuemmerle S, Hunt JC (1997) Immunoassays to study prevalence of antibody against GB virus C in blood donors. J Virol Methods 68:45–55

Love RA, Parge HE, Wickersham JA, Hostomsky Z, Habuka N, Moomaw EW, Adachi T, Hostomska Z (1996) The crystal structure of hepatitis C virus NS3 proteinase reveals a trypsin-like fold and a structural zinc binding site. Cell 87:331–342

Madejon A, Fogeda M, Bartolome J, Pardo M, Gonzalez C, Cotant T, Carreno V (1997) GB virus C RNA in serum, liver, and peripheral blood mononuclear cell from patients with chronic hepatitis B, C, and D. Gastroenterology 113:573–578

Marcellin P, Martinot-Peignoux M, Gabriel F, Branger M, Degott C, Elias A, Xu LZ, Larzul D, Erlinger S, Benhamou JP (1993) Chronic non-B, non-C hepatitis among blood donors assessed with HCV third generation tests and polymerase chain reaction. J Hepatol 19:167–170

Masuko K, Mitsui T, Iwano K, Yamazaki C, Okuda K, Meguro T, Murayama N, Inoue T, Tsuda F, Okamoto H, Miyakawa Y, Mayumi M (1996) Infection with hepatitis GB virus C in patients on maintenance hemodialysis. N Engl J Med 334:1485–1490

Melvin SL, Dawson GJ, Carrick RJ, Schlauder GG, Heynen CA, Mushahwar IK (1998) Biophysical characterization of GB virus C from human plasma. J Virol Methods 71:147–157

Moaven LD, Hyland CA, Young IF, Bowden DS, McCaw R, Mison L, Locarnini SA (1996) Prevalence of hepatitis G virus in Queensland blood donors. Med J Aust 165:369–371

Moriyama K, Okamura T, Nakano S (1997) Hepatitis GB virus C genome in the serum of aplastic anaemia patients receiving frequent blood transfusions. Br J Haematol 96:864–867

Muerhoff AS, Leary TP, Simons JN, Pilot-Matias TJ, Dawson GJ, Erker JC, Chalmers ML, Schlauder GG, Desai SM, Mushahwar IK (1995) Genomic organization of GBV-A and GBV-B: two new members of the *Flaviviridae* associated with GB-agent hepatitis. J Virol 69:5621–5630

Muerhoff AS, Simons JN, Leary TP, Erker JC, Chalmers ML, Pilot-Matias TJ, Dawson GJ, Desai SM, Mushahwar IK (1996a) Sequence heterogeneity within the 5′-terminal region of the hepatitis GB virus C genome and evidence for genotypes. J Hepatol 25:379–384

Muerhoff AS, Simons JN, Erker JC, Desai SM, Mushahwar IK (1996b) Conserved nucleotide sequences within the GB Virus C 5′ untranslated region: Design of PCR primers for detection of viral RNA. J Virol Methods 62:55–62

Muerhoff AS, Smith DB, Leary TP, Erker JC, Desai SM, Mushahwar IK (1997) Identification of GB virus C variants by phylogenetic analysis of 5′-untranslated and coding region sequences. J Virol 71:6501–6508

Mushahwar IK, Erker JC, Muerhoff AS, Desai SM (1998) Tissue tropism of GBV-C and HCV in immunocompromised patients and protective immunity of antibodies to GB virus C second envelope (GBV-C E2) glycoprotein. Hepatol Clin 6:23–27 (suppl. 1)

Nagao Y, Sata M, Noguchi S, Suzuki H, Mizokami M, Kameyama T, Tanikawa K (1997) GB virus infection in patients with oral cancer and oral lichen planus. J Oral Pathol Med 26:138–141

Nagayama R, Miyake K, Okamoto H (1997) Effect of interferon on GB virus C and hepatitis C virus in hepatitis patients with the co-infection. J Med Virol 52:156–160

Ohba K, Mizokami M, Ohno T, Suzuki K, Orito E, Ina Y, Lau JYN, Gojobori T (1995) Classification of hepatitis C virus into major types and subtypes based on molecular evolutionary analysis. Virus Res 36:201–214

Parks WP, Melnick JL (1969) Attempted isolation of hepatitis viruses in marmosets. J Infect Dis 120:539–547

Parks WP, Melnick JL, Voss WR, Singer DB, Rosenberg HS, Alcott J, Casazza AM (1969) Characterization of marmoset hepatitis virus. J Infect Dis 120:548–559

Persico T, Thiers V, Tuveri R, Di Fine M, Semprini AE, Brechot C (1996) Detection if hepatitis G/GB-C viral RNA but not HCV RNA in the different semen fractions of infected patients. Hepatology 24:226A

Pickering JM, Thomas HC, Karayiannis P (1997a) Predicted secondary structure of the hepatitis G virus and GB virus-A 5′ untranslated regions consistent with an internal ribosome entry site. J Viral Hepatol 4:175–184

Pickering JM, Thomas HC, Karayiannis P (1997b) Genetic diversity between hepatitis G virus isolates: analysis of nucleotide variation in the NS-3 and putative 'core' peptide genes. J Gen Virol 78:53–60

Pilot-Matias TJ, Muerhoff AS, Simons JN, Leary TP, Buijk SL, Chalmers ML, Erker JC, Dawson GJ, Desai SM, Mushahwar IK (1996a) Identification of antigenic regions in the GB hepatitis viruses GBV-A, GBV-B, and GBV-C. J Med Virol 48:329–338

Pilot-Matias TJ, Carrick RJ, Coleman PF, Leary TP, Surowy TK, Simons JN, Muerhoff AS, Buijk SL, Chalmers ML, Dawson GJ, Desai SM, Mushahwar IK (1996b) Expression of the GB virus C E2 glycoprotein using the Semliki forest virus vector system and its utility as a serologic marker. Virology 225:282–292

Reyes GR, Purdy MA, Kim JP, Luk K-C, Young LM, Fry KE, Bradley DW (1990) Isolation of a cDNA from the virus responsible for enterically transmitted non-A, non-B hepatitis. Science 247:1335–1339

Rightsel WA, Keltsch RA, Taylor AR, Boggs JD, McLean IW, Boggs JD, Capps RB, Weiss CF, McLean IW (1961) Status report on tissue-culture cultivated hepatitis virus I. Virology laboratory studies. II. Clinical trials. JAMA 177:671–682

Rijnbrand R, Bredenbeek P, van der Straaten T, Whetter L, Inchauspé G, Lemon S, Spaan W (1995) Almost the entire 5′ non-translated region of hepatitis C virus is required for cap-independent translation. FEBS Lett 365:115–119

Rolls MM, Webster P, Balba NH, Rose JK (1994) Novel infectious particles generated by expression of the vesicular stomatitis virus glycoprotein from a self-replicating RNA. Cell 79:497–506

Roth WK, Waschk D, Marx S, Tschauder S, Zeuzem S, Bialleck H, Weber H, Seifried E (1997) Prevalence of hepatitis G virus and its strain variant, the GB agent, in blood donations and their transmission to recipients. Transfusion 37:651–656

Saito S, Tanaka K, Kondo M, Morita K, Kitamura T, Kiba T, Numata K, Sekihara H (1997) Plus- and minus-stranded hepatitis G virus RNA in liver tissue and in peripheral blood mononuclear cells. Biochem Biophys Res Comm 237:288–291

Sallie R, Silva AE, Purdy M, Smith H, McCaustland K, Tibbs C, Portmann B, Eddleston A, Bradley D, Williams R (1994) Hepatitis C and E in non-A non-B fulminant hepatic failure: a polymerase chain reaction and serological study. J Hepatol 20:580–588

Sarrazin C, Roth WK, Zeuzem S (1997) Heterosexual transmission of GB virus-C/hepatitis G virus infection. Eur J Gastroenterol Hepatol 9:1117–1120

Sato K, Tanaka T, Okamoto H, Miyakawa Y, Mayumi M (1996) Association of circulating hepatitis G virus with lipoproteins for a lack of binding with antibodies. Biochem Biophys Res Comm 229:719–725

Scarselli E, Urbani A, Sbardellati A, Tomei L, De Francesco R, Traboni C (1997) GB virus B and hepatitis C virus NS3 serine proteases share substrate specificity. J Virol 71:4985–4989

Schlauder GG, Dawson GJ, Simons JN, Pilot-Matias TJ, Gutierrez RA, Heynen CA, Knigge MF, Kurpiewski GS, Buijk SL, Leary TP, Muerhoff AS, Desai SM, Mushahwar IK (1995a) Molecular and serologic analysis in the transmission of the GB hepatitis agents. J Med Virol 46:81–90

Schlauder GG, Pilot-Matias TJ, Gabriel GS, Simons JN, Muerhoff AS, Dawson GJ, Mushahwar IK (1995b) Origin of GB-hepatitis viruses. Lancet 346:447–448

Schlueter V, Schmolke S, Stark K, Hess G, Ofenloch-Haehnle B, Engel AM (1996) Reverse transcription-PCR detection of hepatitis G virus. J Clin Microbiol 34:2660–2664

Schreier E, Hohne M, Kunkel U, Berg T, Hopf U (1996) Hepatitis GBV-C sequences in patients infected with HCV contaminated anti-D immunoglobulin and among I.V. drug users in Germany. J Hepatol 25:385–389

Seipp S, Scheidel M, Hofmann WJ, Töx U, Theilmann L, Goeser T, Kallinowski B (1999) Hepatotropism of GB virus C (GBV-C): GBV-C replication in human hepatocytes and cells of human hepatoma cell lines. J Hepatol 30:570–579

Sheng L, Soumillion A, Peerlinck K, Verslype C, Lin L, van Plet J, Hess G, Vermylen J, Yap SH (1997) Hepatitis G viral RNA in serum and in peripheral blood mononuclear cells and its relation to HCV-RNA in patients with clotting disorders. Thromb Haemost 77:868–872

Shimizu M, Osada K, Okamoto H (1997) Transmission of GB virus C by blood transfusions during heart surgery. Vox Sang 72:76–78

Shindo M, Di Bisceglie AM, Akatsuka T, Fong TL, Scaglione L, Donets M, Hoofnagle JH, Feinstone SM (1994) The physical state of the negative strand of hepatitis C virus RNA in serum of patients with chronic hepatitis C. Proc Natl Acad Sci USA 91:8719–8723

Simmonds P, Mellor J, Sakuldamrongpanich T, Nuchaprayoon C, Tanprasert S, Holmes EC, Smith DB (1996) Evolutionary analysis of variants of hepatitis C virus found in South-East Asia: comparison with classifications based upon sequence similarity. J Gen Virol 77:3013–3024

Simmonds P, Smith DB, McOmish F, Yap PL, Kolberg J, Urdea MS, Holmes EC (1994) Identification of genotypes of hepatitis C virus by sequence comparisons in the core, E1 and NS-5 regions. J Gen Virol 75:1053–1061

Simons JN, Pilot-Matias TJ, Leary TP, Dawson GJ, Desai SM, Schlauder GG, Muerhoff AS, Erker JC, Buijk SL, Chalmers ML, VanSant CL, Mushahwar IK (1995a) Identification of two flavivirus-like genomes in the GB hepatitis agent. Proc Natl Acad Sci USA 92:3401–3405

Simons JN, Leary TP, Dawson GJ, Pilot-Matias TJ, Muerhoff AS, Schlauder GG, Desai SM, Mushahwar IK (1995b) Isolation of novel virus-like sequences associated with human hepatitis. Nat Med 1:564–569

Simons JN, Desai SM, Schultz DE, Lemon SM, Mushahwar IK (1996a) Translation initiation in GB viruses A and C: evidence for internal ribosome entry and implications on genome organization. J Virol 70:6126–6135

Simons JN, Desai SM, Mushahwar IK (1996b) The GB viruses: Isolation, characterization, diagnosis and epidemiology. Viral Hepat Rev 2:229–246

Smith DB, Cuceanu N, Davidson F, Jarvis LM, Mokili JLK, Hamid S, Ludlam CA, Simmonds P (1997) Discrimination of hepatitis G virus/GBV-C geographical variants by analysis of the 5′ non-coding region. J Gen Virol 78:1533–1542

Stark K, Bienzle U, Hess G, Engel AM, Hegenscheid B, Schluter V (1996) Detection of hepatitis G virus genome among injecting drug users, homosexual and bisexual men, and blood donors. J Infect Dis 174:1320–1323

Surowy TK, Leary TP, Carrick RJ, Knigge MF, Pilot-Matias TJ, Heynen C, Gutierrez RA, Desai SM, Dawson GJ, Mushahwar IK (1997) GB virus C E2 glycoprotein: expression in CHO cells, purification and characterization. J Gen Virol 78:1851–1859

Tacke M, Kiyosawa K, Stark K, Schlueter V, Ofenloch-Haehnle B, Hess G, Engel A (1997a) Detection of antibodies to a putative hepatitis G virus envelope protein. Lancet 349:318–320

Tacke M, Schmolke S, Schlueter V, Sauleda S, Esteban JI, Tanaka E, Kiyosawa K, Alter HJ, Schmitt U, Hess G, Ofenloch-Haehnle B, Engel AM (1997b) Humoral immune response to the E2 protein of hepatitis G virus is associated with long-term recovery from infection and reveals a high frequency of hepatitis G virus exposure among healthy blood donors. Hepatology 26:1626–1633

Tagger A, Donato F, Ribero ML, Chiesa R, Tomasoni V, Portera G, Gelatti U, Albertini A, Fasola M, Nardi G (1997) A case-control study on GB virus C/hepatitis G virus infection and hepatocellular carcinoma. Brescia HCC Study. Hepatology 26:1653–1657

Takahashi K, Hijikata M, Aoyama K, Hoshino H, Hino K, Mishiro S (1997a) Characterization of GBV-C/HGV viral genome: comparison among different isolates for a ~2Kb-sequence that covers entire E1 and most of 5′UTR and E2. Int Hep Commun 6:253–263

Takahashi K, Hijikata M, Hino K, Mishiro S (1997b) Entire polyprotein-ORF sequences of Japanese GBV-C/HGV isolates: implication for new genotypes. Hepatol Res 8:139–148

Tameda Y, Kosaka Y, Tagawa S, Takase K, Sawada N, Nakao H, Tsuda F, Tanaka T, Okamoto H, Miyakawa Y, Mayumi M (1997) Infection with GB virus C (GBV-C) in patients with fulminant hepatitis. J Hepatol 25:842–847

Tanaka T, Kato N, Cho M-J, Shimotohno K (1995) A novel sequence found at the 3′ terminus of hepatitis C virus genome. Biochem Biophys Res Commun 215:774–779

Tanaka T, Kato N, Cho M-J, Sugiyama K, Shimotohno K (1996) Structure of the 3′ terminus of the hepatitis C virus genome. J Virol 70:3307–3312

Thiers V, Lunel F, Valla D, Azar N, Fretz C, Frangeul L, Huraux J-M, Opolon P, Brechot C (1993) Post-transfusional anti-HCV-negative non-A non-B hepatitis (II) serological and polymerase chain reaction analysis for hepatitis C and hepatitis B viruses. J Hepatol 18:34–39

Tomei L, Failla C, Santolini E, De Francesco R, La Monica N (1993) NS3 is a serine protease required for processing of hepatitis C polyprotein. J Virol 67:4017–4026

Tsuda F, Hadiwandowo S, Sawada N, Fukuda M, Tanaka T, Okamoto H, Miyakawa Y, Mayumi M (1996) Infection with GB virus C (GBV-C) in patients with chronic liver disease or on maintenance hemodialysis in Indonesia. J Med Virol 49:248–252

Tsukiyama-Kohara K, Iizuka N, Kohara M, Nomoto A (1992) Internal ribosome entry site within hepatitis C virus RNA. J Virol 66:1476–1483

Viazov S, Riffelmann M, Khoudyakov Y, Fields H, Varenholz C, Roggendorf M (1997a) Genetic heterogeneity of hepatitis G virus isolates from different parts of the world. J Gen Virol 78:577–581

Viazov S, Riffelmann M, Sarr S, Ballauff A, Meisel H, Roggendorf M (1997b) Transmission of GBV-C/HGV from drug-addicted mothers to their babies. J Hepatol 27:85–90

Wang C, Sarnow P, Siddiqui A (1993) Translation of human hepatitis C virus RNA in cultured cells is mediated by an internal ribosome binding mechanism. J Virol 67:3338–3344

Wu RR, Mizokami M, Cao K, Nakano T, Ge XM, Wang SS, Orito E, Ohba K, Mukaide M, Hikiji K, Lau JY, Iino S (1997) GB virus C/hepatitis G virus infection in southern China. J Infect Dis 175:168–171

Yoshiba M, Okamoto H, Mishiro S (1995) Detection of the GBV-C hepatitis virus genome in serum from patients with fulminant hepatitis of unknown aetiology. Lancet 346:1131–1132

Yoshiba M, Inoue K, Sekiyama K (1996) Hepatitis GB virus C. N Engl J Med 335:1392–1393

Zaaijer HL, Vallari DS, Cunningham M, Lesniewski R, Reesnick HW, van der Poel CL, Lelie PN (1994) E2 and NS5: New antigens for detection of hepatitis C virus antibodies. J Med Virol 44:395–397

Zhang XH, Shinzawa H, Shao L, Ishibashi M, Saito K, Ohno S, Yamada N, Misawa H, Togashi H, Takahashi T (1997) Detection of hepatitis G virus RNA in patients with hepatitis B, hepatitis C, and non-A-E hepatitis by RT-PCR using multiple primer sets. J Med Virol 52:385–390

Zignego AL, Giannini C, Gentilini P, Bellesi G, Hadziyannis S, Ferri C (1997) Could HGV infection be implicated in lymphomagenesis? Br J Haematol 98:778–779

Zuckerman AJ (1996) Alphabet of hepatitis viruses. Lancet 347:558–559

Subject Index

A
acupuncture   30
acute infection   44
adenylate kinase   180, 182, 185, 189
alanine aminotransferase (ALT)   286–289
anal-receptive intercourse   32
anti-apolipoprotein
– A1   360
– B   360
antibodies   286, 294
anti-D Ig   27
anti-gpE2 antibodies   330
antigenic variation   351
anti-hemophilic factor, FVIII   2
apoptosis   68, 122, 124, 302
– core protein   122
arginine-rich   176, 180, 185, 189
ascitic fluid   26
ATPase activity   173, 175, 176, 178–180, 187
autoimmune liver disease   366

B
BiP   138
blood
– fluids   26
– transfusions   26
border disease virus   87, 92
bovine viral diarrhea virus   87, 95
buoyant density   15, 341, 360

C
c-fos   120, 121
c-myc   120–122
calnexin   140
calreticulin   138
cancer, oral   366
capsid/core protein   61
CAT (chloramphenicol acetyl transferase)   94, 96, 99
CD4   312
– responses   292
CD4+ T helper responses   300, 329
CD8   312

CD8+ CTL response   399, 302, 330
CD81   64
cell tropism   272
cerebrospinal fluid   26
chimpanzee   2
– model   279
chloramphenicol acetyl transferase (CAT)   94, 96, 99
chloroform-sensitive   13
chronic infection   45
chymotrypsin catalytic triad   153
chymotrypsin-like fold   154
cirrhosis   47, 365
– incidence   47
classical swine fewer virus (*see* Hog Cholera Virus)
cleavage kinetics   151
3′ cloverleaf (3′-X tail)   249
coding region   86, 94, 97
colonoscopic procedure   29
commercial sex workers   31
conformation-sensitive MAbs   137
conjunctiva   26
copy-back priming mechanism   246
core protein   93–98, 108, 109, 118–133
– c-fos   120, 121
– c-myc   120–122
– hepatitis B virus   120
– HepG2 cells   123, 124
– heterogeneous nuclear ribonucleoprotein K (hnRNP K)   121
– HIV   120
– p21 (WAF1)   120, 122
– phosphorylation   120, 121
– RNA helicase   121
core-like protein, absence of   348, 350
crystal structure   66
CTL (*see* cytotoxic T lymphocytes)
cystein protease   61
cytokine   301
cytotoxic T lymphocytes (CTL)   280, 293, 294, 300
– epitopes   306

C

cytotoxic T lymphocytes (CTL)
- escape variants 293, 313
- response 305, 316

D

Daudi cells 265
DEAD box element 179, 182
degenerate oligonucleotides 344
dimer-sized RNA products 245
directionality 177
disulfide-linked aggregates 63
divalent cation 175, 177
DNA immunization 334
double-stranded RNA-dependent kinase 70
drug use 28
dsRNA-activated protein kinase 204
dsRNA-binding motifs 205

E

E1 glycoproteins 63
E2 glycoproteins 63, 280
E2-NS2 136
E2-p7 136
eIF2 102, 104, 109, 110
eIF2α 205, 213
- phosphorylation 206
eIF3 87, 99, 101, 102, 104, 109, 110
EMCV 86, 102, 103
envelope glycoprotein 280, 281, 293
- antibodies 287–289, 293
epidemiology 33–36
ER
- chaperones 142
- retention signal 64, 145
escape
- mutants 293–295
- mutations 295
essential mixed cryoglobulinemia 46

F

FADD 124
Fas 122, 124, 125
Fas-ligand 310
fibrosis 47, 365
filtration 341, 360
Flaviviridae 56, 149
Flavivirus 56
fulminant hepatitis 364, 365

G

Gammagard 27
GB 340
- agents 56, 340, 341
- viruses
- - bile duct lesions 365
- - GBV- 5′ NTRs 349
- - nonstructural proteins 352
- - phylogenetic trees 354
- - virus A (GBV-A) 90, 93, 342
- - - GBV-A/GBV-C group 349
- - - variants 346, 356
- - virus B (GBV-B) 8, 90, 92, 93, 98–100, 105, 109, 110, 342
- - - GBV-B/HCV group 349
- - - protease 358
- - virus C (GBV-C) 90, 345
- - - antibodies 361
- - - C-specific PCR 361
- - - E2 protein 362
- - - epidemiology 364
- - - infections 364, 365
- - - protease 359
- - - proteins 361
- - - RNA-replication sites 363
GDD motif 231
genome organization 57
glycoprotein folding 141
glycosylation sites, N-linked 348, 351
gpE1 331
gpE2 331
- HVR 330

H

H strain of HCV 281
HCV (*see* hepatitis C virus)
health care providers 30
helicase 67, 351
- activity 172, 173, 175, 177–180
- assays 190
- inhibitors 191
- motifs 172
hemophilia 48
hemophiliacs, transfused 27
Hepacivirus 56
hepatitis 364, 365
- fulminant
- A 279
- - RNA 286
- - virus 285
- B 279
- - virus 120
- C 279
- - virus (HCV) 25, 61, 149, 299
- - - antibody positive donors 27
- - - gene cassettes 333
- - - genotypes 49
- - - H strain 281
- - - infection by risk groups 44
- - - persistence 312
- - - polyprotein processing 59–61
- - - replication 262

- - - specific antibodies   286, 330
- D   279
- E   279
- - virus   6
- G   279, 345
- - virus (HGV)   345
hepatocellular carcinoma   47, 62, 366
HepG2 cells   123, 124
heterogeneous nuclear ribonucleoprotein (hnRNP)
- I (*see* polypyrimidine tract binding protein)
- K   121
- L   102–104
HGV (*see* hepatitis G virus)
high-titered plasma   16
histological analysis   290
HLA class II molecules   300
hnRNP (*see* heterogeneous nuclear ribonucleoprotein)
Hog cholera virus (HoCV)   87, 90, 95, 96, 99, 104
human immunodeficiency virus (HIV)   28, 120
hypervariable region 1 (HVR1)   280
- antibodies   280, 281, 295
- immune escape   280, 291, 295
- sequence   292
- - variability   280, 281, 291–293, 295
- species   270

I
ICAM-1   310
IL-2   301
IL-4   301
IL-10   310
illicit drug use   28
immune response   299
immunodominant HCV sequences   304
immunoglobuline   27
immunoscreening   345
immunotherapy   315
individuals at risk   28
infection
- acute   44
- chronic   45
interferon   200, 201, 366
- α2b   50
- sensitivity determining region   198
internal ribosome entry site (IRES)   56, 85, 357, 358
ISDR   71, 202, 203, 213, 214

J
JNK   124

L
La   101–103, 110
lichen planus, oral   366

liver
- biopsies   287, 290, 291
- failure   305
- histology   286, 287, 290, 291
low density lipoproteins   360
LT$\alpha\beta$ 122, 123, 125, 126, 128, 129
LTβR   122, 123, 128, 129
luciferase   94, 96, 97
lymphoproliferative disorders   272
lymphotoxin-$\alpha\beta$ (LT$\alpha\beta$)   122, 123, 125, 126, 128, 129
lymphotoxin-β receptor (LTβR)   62, 122, 123, 128, 129

M
macrophages   301
mechanism of persistence   295
metal binding site   165
MHC
- class I restricted HCV epitopes   307
- class II   301
minus strand synthesis   227
miscellaneous percutaneous transmission   30
MOLT-4Ma cells   262
monomer   188
pYxxxAUG motif   98
MT-2 cells   263
mutation rate   293, 295

N
naked DNA   333
natural history   47, 48
needle stick   30
negative strand RNA   309
NFκB   123–125, 128
nitric oxide   301
- synthase   309
NK cells   301
non-*Hodgkin's* lymphoma   366
3' nontranslated region   58, 283
5' nontranslated region   57
nosocomial transmission   29
NS2 protein   64
NS2-3 protease   61
NS3   56
- helicase   68
- - domain   181
- - salt inhibition   178
- protein   66
- serine protease   61
NS3-NS4A
- proteinase-substrate interactions   156
- site   150
NS3-NS5B site   150
NS3/4A   172

NS4A 60, 172, 175, 178
- derived peptide 154
- protein 69
NS4B protein 69
NS5A 201, 202, 205, 212
- protein 69
NS5A-PKR interaction 212
NS5B
- hydrophobic COOH-terminal domain 239
- polymerase sequences 232
- protein 72
NTPase activity 176
nuclear factor kappa B (NFκB) 123–125, 128
nucleoside triphosphate 56
nucleotide sequence 26

O

oligonucleotide binding site 182, 187
oncogenesis, core protein 121
oral cancer and oral lichen planus 366
organ transplants 28

P

P1 substitutions 163
p21 (WAF1) 120, 122
p25/S9 102
p53 68
P58IPK 209, 210
p120/eIF3 102
palm subdomain 230
PcrA DNA helicase 181, 184, 185, 187
perinatal transmission 31
peripheral blood mononuclear cells 106, 107, 305
pestivirus 56, 87, 90, 92–95, 98–101, 105, 109, 110
pH 175
PH5CH cells 266
phosphate binding loop 178–180, 182, 184, 185
phosphorylation 120, 121
physicochemical properties 13
picornavirus 86, 90, 97, 98, 103, 107, 110
PKR 204, 212, 214
- degradation 210
- dimerization 209
poliovirus 86, 97, 98, 103
polyprotein processing 59
polyproteins 56
polypyrimidine tract binding protein (PTB) 58, 101–105, 110
porphyria cutanea tarda 46, 366
post-transfusion non-A, non-B hepatitis 1
48S preinitiation complex 105
prevalence 44
protease 172, 173, 175, 176, 351, 358, 359
protein-RNA interactions 249

3AB and 3CD^{pro} proteins 249
pseudoknot 93, 96, 101, 104
pyrimidine 98, 99, 101, 102, 105

R

radioimmunoprecipitation assay (RIPA) 362
recombinant NS5B polymerase 226, 244
reinfection 281
Rep DNA helicase 181, 184, 188
replication 172, 178
representational difference analysis 342
Rho-Gam 47
ribavirin 50
ribosomal protein S9 104
40S ribosome subunit 101
ribosome landing pad 357
ribosome scanning 85, 86, 94, 99, 100, 109, 110
- process 85, 86
risk groups 44
RNA
- binding 62
- - domain 176, 188
- copy number 286–289, 291, 294
- genomes 56
- helicase 121, 150
- inoculation 287
- levels 287
- structure 86, 87, 90, 93, 97, 98, 100
- titers 287
RNA-dependent
- RNA polymerase 72, 226
- RNA-helicase 351
RNA-stimulated ATPase 150

S

saliva 26, 31
secretory alkaline phosphatase 97
semen 31
seminal fluid 26
sequence evolution 291
serine
- protease 56
- proteinase domain 150
serum
- HCV RNA levels 287
- RNA 294
sexual transmission 31
SF1 179, 181, 183, 187
SF2 173, 179, 181, 187
signal peptidase 60
steatosis 62, 365
structure, 3D^{pol} 226, 230
substrate recognition 161
superfamily 173, 178, 185
synthetic peptide substrates 163

T

T cell receptor   301
T helper cells (Th)   300
- Th1
- - cytokines   300
- - phenotype   305
- Th2   301
- - cytokines   301
tagged PCR   309
TAP   302
TAP 1   302
tatooing   30
tears   26
terminal nucleotidyl transferase   244
tetraspanning   64
thalassemics, transfused   27
therapy   50
TNFR (*see* tumor necrosis factor receptor)
trans cleavage sites   162
transformation   68, 206, 215
transfused
- hemophiliacs   27
- thalassemics   27
transgenic mice   62
- core protein   120–122, 124
transmission   26, 265
- cofactors   32
- miscellaneous percutaneous   30
- nosocomial   29
- perinatal   31
- routes   27
- sexual   31
transplants   28

tumor necrosis factor receptor (TNFR)   62, 122–125, 129, 130
type I integral membrane proteins   63

U

ultrastructural alterations   11
unwinding activity   180, 187

V

vaccine   333
- studies   331
Vβ5.1-postive T cells   311
vertical transmission   363
viral
- interference   12
- like particles   332
- load   32
- persistence   328
- replication complexes   250
- tropism   106
virus transmission (*see also* transmission)   265

W

WAF1 (p21)   120, 122
WW-55   340

X

X-ray structures   154
X tail   352

Z

zinc   65, 165

Current Topics in Microbiology and Immunology

Volumes published since 1989 (and still available)

Vol. 201: **Kosco-Vilbois, Marie H. (Ed.):** An Antigen Depository of the Immune System: Follicular Dendritic Cells. 1995. 39 figs. IX, 209 pp. ISBN 3-540-59013-7

Vol. 202: **Oldstone, Michael B. A.; Vitković, Ljubiša (Eds.):** HIV and Dementia. 1995. 40 figs. XIII, 279 pp. ISBN 3-540-59117-6

Vol. 203: **Sarnow, Peter (Ed.):** Cap-Independent Translation. 1995. 31 figs. XI, 183 pp. ISBN 3-540-59121-4

Vol. 204: **Saedler, Heinz; Gierl, Alfons (Eds.):** Transposable Elements. 1995. 42 figs. IX, 234 pp. ISBN 3-540-59342-X

Vol. 205: **Littman, Dan R. (Ed.):** The CD4 Molecule. 1995. 29 figs. XIII, 182 pp. ISBN 3-540-59344-6

Vol. 206: **Chisari, Francis V.; Oldstone, Michael B. A. (Eds.):** Transgenic Models of Human Viral and Immunological Disease. 1995. 53 figs. XI, 345 pp. ISBN 3-540-59341-1

Vol. 207: **Prusiner, Stanley B. (Ed.):** Prions Prions Prions. 1995. 42 figs. VII, 163 pp. ISBN 3-540-59343-8

Vol. 208: **Farnham, Peggy J. (Ed.):** Transcriptional Control of Cell Growth. 1995. 17 figs. IX, 141 pp. ISBN 3-540-60113-9

Vol. 209: **Miller, Virginia L. (Ed.):** Bacterial Invasiveness. 1996. 16 figs. IX, 115 pp. ISBN 3-540-60065-5

Vol. 210: **Potter, Michael; Rose, Noel R. (Eds.):** Immunology of Silicones. 1996. 136 figs. XX, 430 pp. ISBN 3-540-60272-0

Vol. 211: **Wolff, Linda; Perkins, Archibald S. (Eds.):** Molecular Aspects of Myeloid Stem Cell Development. 1996. 98 figs. XIV, 298 pp. ISBN 3-540-60414-6

Vol. 212: **Vainio, Olli; Imhof, Beat A. (Eds.):** Immunology and Developmental Biology of the Chicken. 1996. 43 figs. IX, 281 pp. ISBN 3-540-60585-1

Vol. 213/I: **Günthert, Ursula; Birchmeier, Walter (Eds.):** Attempts to Understand Metastasis Formation I. 1996. 35 figs. XV, 293 pp. ISBN 3-540-60680-7

Vol. 213/II: **Günthert, Ursula; Birchmeier, Walter (Eds.):** Attempts to Understand Metastasis Formation II. 1996. 33 figs. XV, 288 pp. ISBN 3-540-60681-5

Vol. 213/III: **Günthert, Ursula; Schlag, Peter M.; Birchmeier, Walter (Eds.):** Attempts to Understand Metastasis Formation III. 1996. 14 figs. XV, 262 pp. ISBN 3-540-60682-3

Vol. 214: **Kräusslich, Hans-Georg (Ed.):** Morphogenesis and Maturation of Retroviruses. 1996. 34 figs. XI, 344 pp. ISBN 3-540-60928-8

Vol. 215: **Shinnick, Thomas M. (Ed.):** Tuberculosis. 1996. 46 figs. XI, 307 pp. ISBN 3-540-60985-7

Vol. 216: **Rietschel, Ernst Th.; Wagner, Hermann (Eds.):** Pathology of Septic Shock. 1996. 34 figs. X, 321 pp. ISBN 3-540-61026-X

Vol. 217: **Jessberger, Rolf; Lieber, Michael R. (Eds.):** Molecular Analysis of DNA Rearrangements in the Immune System. 1996. 43 figs. IX, 224 pp. ISBN 3-540-61037-5

Vol. 218: **Berns, Kenneth I.; Giraud, Catherine (Eds.):** Adeno-Associated Virus (AAV) Vectors in Gene Therapy. 1996. 38 figs. IX, 173 pp. ISBN 3-540-61076-6

Vol. 219: **Gross, Uwe (Ed.):** Toxoplasma gondii. 1996. 31 figs. XI, 274 pp. ISBN 3-540-61300-5

Vol. 220: **Rauscher, Frank J. III; Vogt, Peter K. (Eds.):** Chromosomal Translocations and Oncogenic Transcription Factors. 1997. 28 figs. XI, 166 pp. ISBN 3-540-61402-8

Vol. 221: **Kastan, Michael B. (Ed.):** Genetic Instability and Tumorigenesis. 1997. 12 figs. VII, 180 pp. ISBN 3-540-61518-0

Vol. 222: **Olding, Lars B. (Ed.):** Reproductive Immunology. 1997. 17 figs. XII, 219 pp.
ISBN 3-540-61888-0

Vol. 223: **Tracy, S.; Chapman, N. M.; Mahy, B. W. J. (Eds.):** The Coxsackie B Viruses. 1997. 37 figs. VIII, 336 pp.
ISBN 3-540-62390-6

Vol. 224: **Potter, Michael; Melchers, Fritz (Eds.):** C-Myc in B-Cell Neoplasia. 1997. 94 figs. XII, 291 pp.
ISBN 3-540-62892-4

Vol. 225: **Vogt, Peter K.; Mahan, Michael J. (Eds.):** Bacterial Infection: Close Encounters at the Host Pathogen Interface. 1998. 15 figs. IX, 169 pp.
ISBN 3-540-63260-3

Vol. 226: **Koprowski, Hilary; Weiner, David B. (Eds.):** DNA Vaccination/Genetic Vaccination. 1998. 31 figs. XVIII, 198 pp.
ISBN 3-540-63392-8

Vol. 227: **Vogt, Peter K.; Reed, Steven I. (Eds.):** Cyclin Dependent Kinase (CDK) Inhibitors. 1998. 15 figs. XII, 169 pp.
ISBN 3-540-63429-0

Vol. 228: **Pawson, Anthony I. (Ed.):** Protein Modules in Signal Transduction. 1998. 42 figs. IX, 368 pp. ISBN 3-540-63396-0

Vol. 229: **Kelsoe, Garnett; Flajnik, Martin (Eds.):** Somatic Diversification of Immune Responses. 1998. 38 figs. IX, 221 pp.
ISBN 3-540-63608-0

Vol. 230: **Kärre, Klas; Colonna, Marco (Eds.):** Specificity, Function, and Development of NK Cells. 1998. 22 figs. IX, 248 pp.
ISBN 3-540-63941-1

Vol. 231: **Holzmann, Bernhard; Wagner, Hermann (Eds.):** Leukocyte Integrins in the Immune System and Malignant Disease. 1998. 40 figs. XIII, 189 pp.
ISBN 3-540-63609-9

Vol. 232: **Whitton, J. Lindsay (Ed.):** Antigen Presentation. 1998. 11 figs. IX, 244 pp.
ISBN 3-540-63813-X

Vol. 233/I: **Tyler, Kenneth L.; Oldstone, Michael B. A. (Eds.):** Reoviruses I. 1998. 29 figs. XVIII, 223 pp.
ISBN 3-540-63946-2

Vol. 233/II: **Tyler, Kenneth L.; Oldstone, Michael B. A. (Eds.):** Reoviruses II. 1998. 45 figs. XVI, 187 pp.
ISBN 3-540-63947-0

Vol. 234: **Frankel, Arthur E. (Ed.):** Clinical Applications of Immunotoxins. 1999. 16 figs. IX, 122 pp.
ISBN 3-540-64097-5

Vol. 235: **Klenk, Hans-Dieter (Ed.):** Marburg and Ebola Viruses. 1999. 34 figs. XI, 225 pp. ISBN 3-540-64729-5

Vol. 236: **Kraehenbuhl, Jean-Pierre; Neutra, Marian R. (Eds.):** Defense of Mucosal Surfaces: Pathogenesis, Immunity and Vaccines. 1999. 30 figs. IX, 296 pp.
ISBN 3-540-64730-9

Vol. 237: **Claesson-Welsh, Lena (Ed.):** Vascular Growth Factors and Angiogenesis. 1999. 36 figs. X, 189 pp.
ISBN 3-540-64731-7

Vol. 238: **Coffman, Robert L.; Romagnani, Sergio (Eds.):** Redirection of Th1 and Th2 Responses. 1999. 6 figs. IX, 148 pp.
ISBN 3-540-65048-2

Vol. 239: **Vogt, Peter K.; Jackson, Andrew O. (Eds.):** Satellites and Defective Viral RNAs. 1999. 39 figs. XVI, 179 pp.
ISBN 3-540-65049-0

Vol. 240: **Hammond, John; McGarvey, Peter; Yusibov, Vidadi (Eds.):** Plant Biotechnology. 1999. 12 figs. XII, 196 pp.
ISBN 3-540-65104-7

Vol. 241: **Westblom, Tore U.; Czinn, Steven J.; Nedrud, John G. (Eds.):** Gastroduodenal Disease and Helicobacter pylori. 1999. 35 figs. XI, 313 pp.
ISBN 3-540-65084-9

Printing: Saladruck, Berlin
Binding: H. Stürtz AG, Würzburg